Methods in Molecular Biology

Series Editor
John M. Walker
School of Life and Medical Sciences
University of Hertfordshire
Hatfield, Hertfordshire, AL10 9AB, UK

For further volumes:
http://www.springer.com/series/7651

Microbial Lipid Production

Methods and Protocols

Edited by

Venkatesh Balan

Department of Engineering Technology, Biotechnology Program, College of Technology, University of Houston, Houston, TX, USA

Humana Press

Editor
Venkatesh Balan
Department of Engineering Technology
Biotechnology Program
College of Technology
University of Houston
Houston, TX, USA

ISSN 1064-3745 ISSN 1940-6029 (electronic)
Methods in Molecular Biology
ISBN 978-1-4939-9483-0 ISBN 978-1-4939-9484-7 (eBook)
https://doi.org/10.1007/978-1-4939-9484-7

This Humana imprint is published by the registered company Springer Science+Business Media, LLC, part of Springer Nature.
The registered company address is: 233 Spring Street, New York, NY 10013, U.S.A.

Preface

Lipids derived from conventional sources such as oil seeds include building blocks of cellular membranes. They are utilized to make substances used as energy storage, insulation, a method of cellular communication, and protection. With population growth and increasingly limited farmland, there is an ever-growing demand to develop alternate sources of lipids to meet our food and energy needs. Oleaginous microorganisms such as bacteria, algae, yeast, fungi, and thraustochytrids are a promising, largely untapped, resource for lipid production. Microbial lipids offer some unique advantages, such as high content of unique polyunsaturated fatty acids that are widely used for dietary purposes and of chemically modified lipids that confer desirable physical properties. Lipid content in microorganisms depends upon genetic constituents and environmental conditions such as pH, temperature, exposure to natural light, and nitrogen content in the media. There are several advantages of producing lipids in microorganisms such as lower cultivation costs, ability to produce a diverse range of lipids using genetic manipulation, and the possibility of producing lipids year-round with limited space and infrastructure without the need for agricultural land.

The 24 chapters of this book provide comprehensive routinely used methods for isolation and characterization of lipids and for their production using various oleaginous organisms. The methods presented consider sugars derived from different substrates including chemically pretreated agricultural residues, industrial residues containing lignin, food wastes, and industrial waste water in an approachable format. The authors have also provided detailed applications using oleaginous organisms to transform substrates into a variety of products including bio-crude, high-value fatty acids, biofuels such as biodiesel, neutral lipids, volatile fatty acids, and surfactants. Protocols are presented in a basic standard outline format conducive to adaptation to suit specific application needs. This book is aimed at the novice, and therefore each technique is complete and assumes no prior knowledge. Novel screening protocols to identify oleaginous organisms with exceptional lipid yields, genetic engineering approaches to increase microbial lipid content, methods to judge fuel quality from microbial lipids, and life cycle analysis are some of the unique topics that are covered in this book.

Much of the success of the series is due to the "Notes" section, which describes where common procedural problems are identified, and solutions are discussed along with alternative procedures. It is in this section where important, practical details are presented that are rarely included in other published works. This is how the authors have passed along their practical experience to help mentor readers. Most of the chapters are focused on specific laboratory methods, but the first chapter provides a comprehensive review about lipids derived from plants and various microorganisms, an overview of available analytical techniques to characterize lipids, and their applications in various processes.

It has been inspiring to see the natural diversity of lipid molecules with varying properties and functions synthesized by microorganisms using complex genetic and enzymatic machinery. Also, it is equally inspiring to see the numerous analytical techniques and protocols now available for lipid researchers to use to analyze structure and function and to pursue diverse applications. Though it was challenging to organize such a range of chapter topics to fit together in a common theme, all the contributing authors are commended for their expertise, ensuring each chapter was on topic, and for writing in a clear and direct style. I take this opportunity to thank both my wife and two daughters for sacrificing family time and motivating me to complete the book at times when my morale ebbed. I thank all my colleagues in the College of Technology and administrator at the University of Houston for their support and encouragement that made this book possible. Also, I thank Dr. Patricia Slinger and Dr. Bruce Dien from USDA ARS, Peoria, for providing the cover picture and also helping me to critically review several chapters in this book.

Houston, TX, USA ***Venkatesh Balan***

Cover Illustration Caption

Shown in the figure are cells of Saitoella coloradoensis strain NRRL YB-2330 containing numerous lipid granules. This strain was discovered to tolerate and produce abundant lipid when cultivated on ammonia fiber expansion (AFEX)-pretreated enzyme saccharified hydrolyzates of corn stover (Slininger et al. 2016; Dien et al. 2016). Saitoella, NRRL YB-2330, was isolated by L. J. Wickerham from insect frass collected in 1950 by staff of the U.S. Forest Service from an Engelmann spruce (Picea engelmannii) growing in the White River National Forest, Meeker, CO, USA. Taxonomically described by Kurtzman and Robnett (2012), NRRL YB-2330 is recognized as a member of the genus Saitoella and as a new species distinct from NRRLY-17804, type strain of S. complicata, the only other Saitoella species in collection.

- Kurtzman CP and Robnett CJ (2012) Saitoella coloradoensis sp. nov., a new species of the Ascomycota, subphylum Taphrinomycotina. Antonie van Leeuwenhoek 101:795–802.
- Slininger PJ, Dien BS, Kurtzman CP, Moser BR, Bakota EL, Thompson SR, O'Bryan PJ, Cotta MA, Balan V, Jin M (2016) Comparative lipid production by oleaginous yeasts in hydrolyzates of lignocellulosic biomass and process strategy for high titers. Biotechnology and Bbioengineering 113:1676–1690
- Dien BS, Slininger PJ, Kurtzman CP, Moser BR, O'Bryan PJ (2016a) Identification of superior lipid producing Lipomyces and Myxozyma yeasts. AIMS Environmental Science, 3:1–20.

Contents

Contributors

VENKATESH BALAN • *Department of Engineering Technology, Biotechnology Program, College of Technology, University of Houston, Houston, TX, USA*

G. FLORIN BARLA • *Faculty of Food Engineering, University of Suceava, Suceava, Romania; Tyton Biosciences, Danville, VA, USA*

TYLER J. BARZEE • *Department of Biological and Agricultural Engineering, University of California, Davis, Davis, CA, USA*

EBOIBI BLESSING • *Department of Chemical Engineering, Delta State University, Oleh, Nigeria*

KYRIA L. BOUNDY-MILLS • *Phaff Yeast Culture Collection, Department of Food Science and Technology, University of California, Davis, Davis, CA, USA*

TOM BRADLEY • *Narec Distributed Energy, CPTC, High Quay, Blyth, UK*

TOMAS CAJKA • *West Coast Metabolomics Center, University of California, Davis, Davis, CA, USA; Department of Metabolomics, Institute of Physiology CAS, Prague, Czech Republic*

SENTHIL CHINNASAMY • *Biotechnology Division, Aban Infrastructure Limited, Chennai, India*

JACOB D. CROWE • *Department of Chemical Engineering and Materials Science, Michigan State University, East Lansing, MI, USA*

FARHA DEEBA • *Department of Polymer and Process Engineering, IIT Roorkee, Saharanpur, India*

GOKSEL N. DEMIRER • *Central Michigan University, School of Engineering & Technology, Mount Pleasant, MI, USA*

BRUCE S. DIEN • *Bioenergy Research Unit, National Center for Agricultural Utilization Research, USDA-ARS, Peoria, IL, USA*

IRENE FAKANKUN • *Department of Biosystems Engineering, University of Manitoba, Winnipeg, MB, Canada*

OLIVER FIEHN • *West Coast Metabolomics Center, University of California, Davis, Davis, CA, USA*

ANNALIESE K. FRANZ • *Department of Chemistry, University of California, Davis, Davis, CA, USA; Agricultural and Environmental Chemistry Graduate Group, University of California, Davis, CA, USA*

ANTONIO L. GARAY • *Phaff Yeast Culture Collection, Department of Food Science and Technology, University of California, Davis, Davis, CA, USA; Pepsico, Plano, TX, USA*

NASEEM A. GAUR • *Yeast Biofuel Group, DBT-ICGEB Centre for Advanced Bioenergy Research, International Centre for Genetic Engineering and Biotechnology, New Delhi, India*

MANICKAM GAYATHRI • *Department of Microbiology, Centre of Excellence in Life Sciences, Bharathidasan University, Tiruchirappalli, TN, India*

EDUARDO J. GUDIÑA • *CEB–Centre of Biological Engineering, University of Minho, Braga, Portugal*

FATEMEH HEIDARI • *Department of Bioproducts and Biosystems Engineering, University of Minnesota, Saint Paul, MN, USA*

David B. Hodge • *Department of Chemical and Biological Engineering, Montana State University, Bozeman, MN, USA; Department of Civil, Environmental, and Natural Resources Engineering, Luleå University of Technology, Luleå, Sweden*
Bo Hu • *Department of Bioproducts and Biosystems Engineering, University of Minnesota, Saint Paul, MN, USA*
Singaram Jayanthi • *Government College of Technology, Coimbatore, India*
Umakanta Jena • *Department of Chemical and Materials Engineering, New Mexico State University, Las Cruces, NM, USA*
Mingjie Jin • *School of Environmental and Biological Engineering, Nanjing University of Science and Technology, Nanjing, China*
A. Daniel Jones • *Department of Biochemistry and Molecular Biology, Michigan State University, East Lansing, MI, USA; Department of Chemistry, Michigan State University, East Lansing, MI, USA*
Sandeep Kumar • *Department of Civil and Environmental Engineering, Old Dominion University, Norfolk, VA, USA*
Priya Kumari • *Yeast Biofuel Group, DBT-ICGEB Centre for Advanced Bioenergy Research, International Centre for Genetic Engineering and Biotechnology, New Delhi, India*
Cletus P. Kurtzman • *Mycotoxin Prevention and Applied Microbiology Research Unit, National Center for Agricultural Utilization Research, USDA-ARS, Peoria, IL, USA*
Rosemary K. Le • *Department of Chemical and Biomolecular Engineering, University of Tennessee, Knoxville, TN, USA*
David B. Levin • *Department of Biosystems Engineering, University of Manitoba, Winnipeg, MB, Canada*
Muyang Li • *Department of Agricultural and Biological Engineering, Michigan State University, East Lansing, MI, USA*
Wei Liao • *Anaerobic Digestion Research and Education Center (ADREC), Biosystems and Agricultural Engineering, Michigan State University, East Lansing, MI, USA*
Tongjun Liu • *Department of Bioengineering, Qilu University of Technology, Jinan, China*
Balakrishna Maddi • *Suganit Bio-Renewables LLC, Toledo, OH, USA*
Daniel Maga • *Fraunhofer Institute for Environmental, Safety and Energy Technology UMSICHT, Oberhausen, Germany*
Kristina M. Mahan • *Department of Chemical and Biomolecular Engineering, University of Tennessee, Knoxville, TN, USA*
Schonna R. Manning • *UTEX Culture Collection of Algae, Department of Molecular Biosciences, The University of Texas at Austin, Austin, TX, USA*
Maryam Mirzaei • *Department of Biosystems Engineering, University of Manitoba, Winnipeg, MB, Canada*
Sima Modiri • *Department of Energy and Environmental Biotechnology, National Institute of Genetic Engineering and Biotechnology (NIGEB), Tehran, Iran*
Gangatharan Muralitharan • *Department of Microbiology, Centre of Excellence in Life Sciences, Bharathidasan University, Tiruchirappalli, TN, India*
Yuvraj S. Negi • *Department of Polymer and Process Engineering, IIT Roorkee, Saharanpur, India*
David R. Nobles Jr. • *UTEX Culture Collection of Algae, Department of Molecular Biosciences, The University of Texas at Austin, Austin, TX, USA*

KAMBIZ AKBARI NOGHABI • *Department of Energy and Environmental Biotechnology, National Institute of Genetic Engineering and Biotechnology (NIGEB), Tehran, Iran*
VIKAS PRUTHI • *Department of Biotechnology, IIT Roorkee, Roorkee, India*
JOSHUA C. QUARTERMAN • *Bioenergy Research Unit, National Center for Agricultural Utilization Research, USDA-ARS, Peoria, IL, USA*
ARTHUR J. RAGAUSKAS • *Department of Chemical and Biomolecular Engineering, University of Tennessee, Knoxville, TN, USA; Biosciences Division, Oak Ridge National Laboratory, Oak Ridge, TN, USA; Department of Forestry, Wildlife and Fisheries, Center of Renewable Carbon, Institute of Agriculture, University of Tennessee, Knoxville, TN, USA*
LÍGIA R. RODRIGUES • *CEB–Centre of Biological Engineering, University of Minho, Braga, Portugal*
KAYLA M. RUDE • *Department of Chemistry, University of California, Davis, Davis, CA, USA; Agricultural and Environmental Chemistry Graduate Group, University of California, Davis, CA, USA*
SUMATHY SHUNMUGAM • *Department of Microbiology, Centre of Excellence in Life Sciences, Bharathidasan University, Tiruchirappalli, TN, India*
IRNAYULI R. SITEPU • *Phaff Yeast Culture Collection, Department of Food Science and Technology, University of California, Davis, Davis, CA, USA*
PATRICIA J. SLININGER • *Bioenergy Research Unit, National Center for Agricultural Utilization Research, USDA-ARS, Peoria, IL, USA*
ALEX D. SMITH • *Department of Chemical and Biological Engineering, University of Wisconsin, Madison, WI, USA*
XIAO-LING TANG • *Zhejiang University of Technology, Hangzhou, Zhejiang, China*
ARUN KUMAR THALLA • *Department of Civil Engineering, National Institute of Technology Karnataka, Mangalore, Karnataka, India*
STEPHANIE R. THOMPSON • *Bioenergy Research Unit, National Center for Agricultural Utilization Research, USDA-ARS, Peoria, IL, USA*
BRYAN UBANWA • *Department of Engineering Technology, Biotechnology Program, College of Technology, University of Houston, Houston, TX, USA*
SIBEL ULUDAG-DEMIRER • *Anaerobic Digestion Research and Education Center (ADREC), Biosystems and Agricultural Engineering, Michigan State University, East Lansing, MI, USA*
AGASTESWAR VADLAMANI • *Department of Chemical Engineering, The University of Toledo, Toledo, OH, USA*
SASIDHAR VARANASI • *Department of Chemical Engineering, The University of Toledo, Toledo, OH, USA*
SRIDHAR VIAMAJALA • *Department of Chemical Engineering, The University of Toledo, Toledo, OH, USA*
ZHIQIANG WEN • *School of Environmental and Biological Engineering, Nanjing University of Science and Technology, Nanjing, China*
DANIEL L. WILLIAMS • *Element Materials Technology, Plymouth, MI, USA*
ZHAOXIAN XU • *School of Environmental and Biological Engineering, Nanjing University of Science and Technology, Nanjing, China*
YA-PING XUE • *Zhejiang University of Technology, Hangzhou, Zhejiang, China*
YAN YANG • *Department of Bioproducts and Biosystems Engineering, University of Minnesota, Saint Paul, MN, USA*
CODY YOTHERS • *Department of Chemistry, University of California, Davis, Davis, CA, USA*

FARNAZ YUSUF • *Yeast Biofuel Group, DBT-ICGEB Centre for Advanced Bioenergy Research, International Centre for Genetic Engineering and Biotechnology, New Delhi, India*
HOSSEIN SHAHBANI ZAHIRI • *Department of Energy and Environmental Biotechnology, National Institute of Genetic Engineering and Biotechnology (NIGEB), Tehran, Iran*
RUI ZHAI • *School of Environmental and Biological Engineering, Nanjing University of Science and Technology, Nanjing, China*

Chapter 1

Microbial Lipid Alternatives to Plant Lipids

A. Daniel Jones, Kyria L. Boundy-Mills, G. Florin Barla, Sandeep Kumar, Bryan Ubanwa, and Venkatesh Balan

Abstract

Lipids are in high demand in food production, nutritional supplements, detergents, lubricants, and biofuels. Different oil seeds produced from plants are conventionally extracted to yield lipids. With increasing population and reduced availability of cultivable land, conventional methods of producing lipids alone will not satisfy increasing demand. Lipids produced using different microbial sources are considered as sustainable alternative to plant derived lipids. Various microorganisms belonging to the genera of algae, bacteria, yeast, fungi, or marine-derived microorganisms such as thraustochytrids possess the ability to accumulate lipids in their cells. A variety of microbial production technologies are being used to cultivate these organisms under specific conditions using agricultural residues as carbon source to be cost competitive with plant derived lipids. Microbial oils, also known as single cell oils, have many advantages when compared with plant derived lipids, such as shorter life cycle, less labor required, season and climate independence, no use of arable land and ease of scale-up. In this chapter we compare the lipids derived from plants and different microorganisms. We also highlight various analytical techniques that are being used to characterize the lipids produced in oleaginous organisms and their applications in various processes.

Key words Lipids, Oleaginous organisms, Analytical methods, Classification, Economics

1 Introduction

Lipids are defined as a group of molecules soluble in organic solvents but insoluble in water, and include fat-soluble vitamins (e.g., vitamin A, D, E, and K), monoglycerides, diglycerides, triglycerides, hydrocarbon-like substances (e.g., sterols, terpenes, waxes), and glycerophospholipids. The number of individual chemical forms of lipids that exist in nature far exceeds the tens of thousands, and as of this writing, the LIPID MAPS consortium (https://www.lipidmaps.org/data/structure/index.php) documents more than 21,000 curated lipids plus a comparable number of computationally generated lipid compounds. Lipids play essential roles in biological function including energy storage, signaling,

Venkatesh Balan (ed.), *Microbial Lipid Production: Methods and Protocols*, Methods in Molecular Biology, vol. 1995, https://doi.org/10.1007/978-1-4939-9484-7_1,

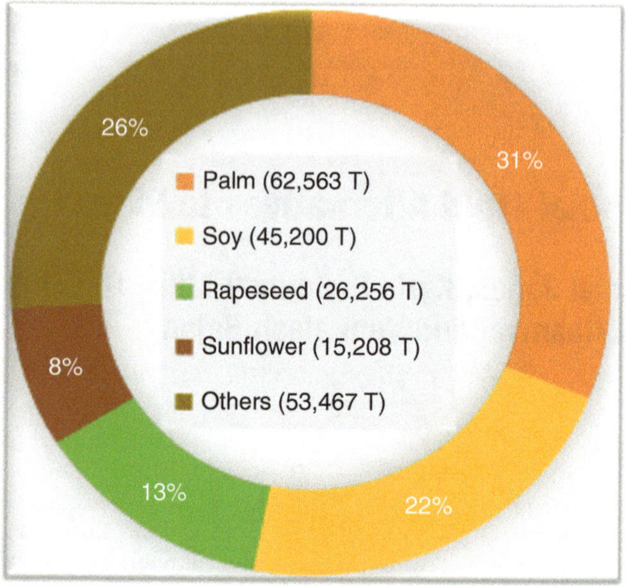

Fig. 1 World edible vegetable oil production in 2015. The amount of each source of oils is provided in bracket in Tons per year. Other oil that are produced in sizable amount include corn, coconut, olive, peanut, cotton, sesame, and mustard seeds or kernels

and as structural components of cell membranes. Most lipids are derived from oil seeds and about 350 crop species have been found suitable for lipid production [1]. Most common oilseeds used for extraction are presented in Fig. 1. Due to increased demand for lipids and limitations in producing lipids using oilseeds due to dwindling available cultivable land, alternate methods of producing lipids have been seriously considered since the 1980's [2]. Since then, microbe-derived lipids have been utilized in various applications including green fuel production, pharmaceuticals, cosmetics, food additives and biopolymers [3]. Many oleaginous microorganisms such as yeast, fungi, algae, bacteria and marine-derived thraustochytrids have the capability to produce an assortment of lipids (Fig. 2). Some of them include various forms of omega-3 and omega-6 fatty acids [4, 5], linoleic acid and its isomers [6]; arachidonic acid (ARA) and docosahexaenoic acid (DHA) [7]. In particular, dietary omega-3 lipids are known to have health benefits such as: improved cognition, anticarcinogenic, antiobesity, antidiabetic, antihypertensive, antiatherogenic, immunomodulatory, reduced arterial LDL-cholesterol and cardiovascular disease, anti-inflammatory, and osteosynthetic properties [8, 9]. In addition to food supplements, lipids are used in various cosmetic industries, automotive industry applications (e.g., lubricants), paint industry (e.g., coatings) and polymers, printing inks, leather processing,

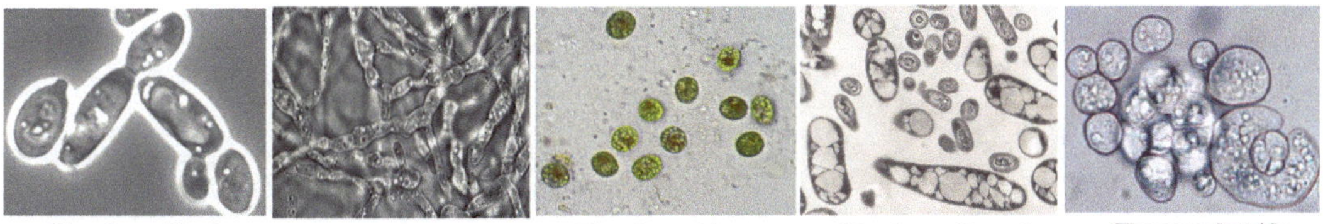

Fig. 2 Different types of oleaginous organisms that are currently being researched to produce high with lipids for commercial application

surfactants, solvents, hydraulic fluids, in pesticide/herbicide formulations, pharmacological and/or immunological agents, and biofuels. Lipids are used as is or altered by various chemical modifications, thermal treatments, catalytic hydrogenation, hydrolysis, or transesterification to produce biofuels.

2 Lipid Classification

Nature generates a wealth of lipids using a variety of biosynthetic pathways that cannot all be described in this discussion. When most of us think of lipids, we think of oils and fats that consist largely of triacylglycerols (TAGs) derived from the condensation of three fatty acids with glycerol. TAGs are neutral lipids normally found in lipid droplets rather than membranes. Membranes, on the other hand, rely on amphiphilic lipids that have polar and lipophilic structural units and form membrane bilayers that isolate cells from their surroundings and delineate organelles in eukaryotes. The structural diversity of some of the microbial lipid classes, illustrated in Fig. 3, derives from diversity in metabolic processes among microbes. Common membrane lipids include fatty acyls, glycerophospholipids, sterol lipids, ether lipids, prenol lipids, saccharolipids, glycerolipids, sphingolipids, and polyketides, which are abundant in plant chloroplasts and many prokaryotes. Examples are presented in Fig. 4. Membrane lipids in archaea often contain branched terpenoid ethers, rather than esters, such as GDGT-0. Terpenoid biosynthetic pathways are also important in the production of the hydrophobic tail of bacteriochlorophylls by photosynthetic bacteria, as well as fatty acid esters of the sterol ergosterol, abundant in yeast [10]. Triterpenoid lipids known as hopanoids (e.g., bacteriohopanetetrol) are common in many prokaryotes, whereas the terpenoid hydrocarbon botryococcene ($C_{34}H_{58}$) has drawn attention owing to its high abundance in the green microalga *Botryococcus braunii*. Rhamnolipids derived from glycosylation of medium-chain β-hydroxy fatty acid esters are biosurfactants produced by *Pseudomonas aeruginosa*. Many prokaryotes do not

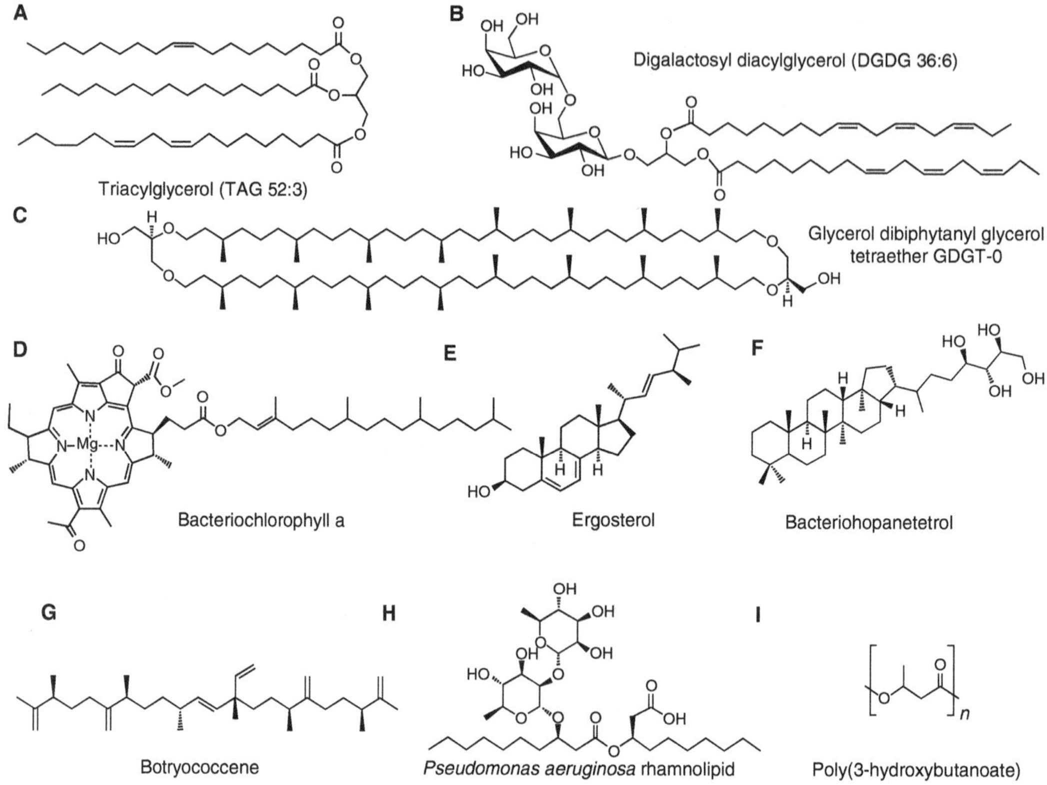

Fig. 3 Example structures illustrating the structural diversity of microbial lipids. (**a**) A triacylglycerol, (**b**) galactolipid, (**c**) prenyl ether lipid from archaea, (**d**) bacteriochlorophyll from photosynthetic prokaryotes, (**e**) ergosterol from yeast, (**f**) hopanoid from prokaryotes, (**g**) terpenoid hydrocarbon from the microalga *Botryococcus braunii*, (**h**) rhamnolipid from the bacterium *Pseudomonas aeruginosa*, and (**i**) polyhydroxyalkanoate from lipid droplets of many prokaryotes

produce significant amounts of TAGs, but instead accumulate organic solvent-soluble polyhydroxyalkanoates in lipid droplets that have generated interest as a source of biopolymers [11].

3 Plant Based Lipids

Vegetable oils extracted from seeds or fruit pulp are primarily mixtures of triglycerides which contain fatty acids in esterified forms. The four major sources of edible vegetable oil produced in the world include palm kernel, soybean, rapeseed, and sunflower seeds. Palm oil occupies a considerable market share (32%) and is mostly produced in Asian countries like Indonesia and Malaysia. The second largest source of vegetable oil is soybean (22%), with significant production in Brazil, Argentina, China and the USA. The third largest source of vegetable oil is rapeseeds (13%) produced largely in the European Union, China, India, Canada, and

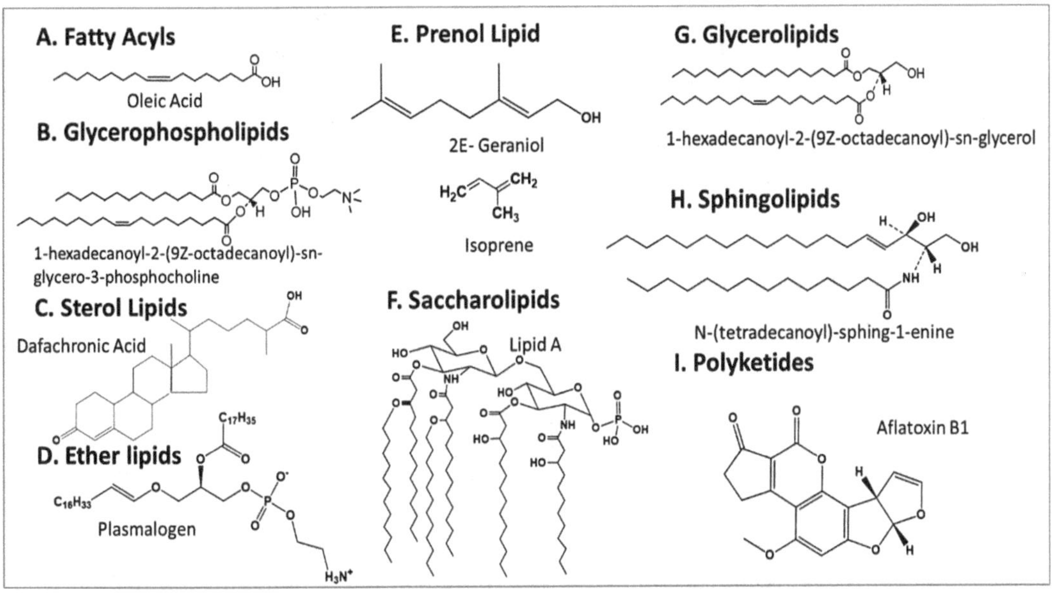

Fig. 4 Different types of lipids with specific example given in bracket. Here, (**a**) fatty acyls (oleic acid); (**b**) glycerophospholipids (1-hexadecanoyl-2-(9Z-octadecanoyl)-sn-glycero-3-phosphocholine); (**c**) sterol lipids (dafachronic acid); (**d**) ether lipids (plasmalogen); (**e**) prenol lipids (2E-geraniol, isoprene); (**f**) saccharolipids (lipid A); (**g**) glycerolipids (1-hexagonal-2-(9Z-octadecanoyl)-sn-glycerol); (**h**) sphingolipids (*N*-(tetradecanoyl)-sphing-1-enine); and (**i**) polyketides (aflatoxin B_1)

the USA [12]. Other vegetable oils that are produced in significant quantities around the world include oils derived from corn seeds, coconut kernel, olive pulp, peanut seeds, cotton seeds, sesame seeds, and mustard seeds. Minor quantities of edible vegetable oil obtained from almond, avocado, safflower, flax (Lind) seed, hemp seed, grape seed, and hazelnut.

Nonedible vegetable oils, known as second-generation feedstocks have become more attractive for biodiesel production and other industrial applications [13]. They include *Jatropha curcas*, *Calophyllum inophyllum*, *Sterculia feotida*, *Madhuca indica* (mahua), *Pongamia glabra* (koroch seed), *Linseed*, *Pongamia pinnata* (karanja), *Hevea brasiliensis* (Rubber seed), *Azadirachta indica* (neem), *Camelina sativa*, *Lesquerella fendleri*, *Nicotiana tabacum* (tobacco), *Deccan hemp*, *Ricinus communis L.* (castor), *Babassu*, *Simmondsia chinensis* (Jojoba), *Eruca sativa. L.*, *Cerbera odollam* (Sea mango), *Coriandrum sativum L* (Coriander), *Croton megalocarpus*, *Salmon oil*, *Pilu*, *Crambe*, *syringa*, *Schleichera triguga (kusum)*, *Stillingia*, *Shorea robusta* (sal), *Terminalia bellerica* Roxb., *Cuphea*, *Camellia*, *Champaca*, *Simarouba glauca*, *Garcinia indica*, rice bran, hingan (*Balanites*), *desert date*, *cardoon*, *Asclepias syriaca* (milkweed), *Guizotia abyssinica*, *Radish*, *mustard*, *Syagrus*, *Tung*, *Idesia polycarpa* var. *vestita*, *Alagae*, *Argemone mexicana L.* (*Mexican prickly poppy*), *Putranjiva roxburghii* (Lucky bean tree),

Sapindus mukorossi (Soapnut), *M. azedarach* (syringe), *Thevetia peruviana* (yellow oleander), *Copaiba*, Milk bush, Laurel, Cumaru, Andiroba, Piqui, *B. napus*, Tomato seed, *Zanthoxylum bungeanum*. Some of these seeds are poisonous due to the presence of aromatic lipids, toxic alkaloid ricin, and the highly potent toxalbumin ricin that interferes with protein and DNA synthesis [14].

4 Microbial Lipids and Their Sources

Most of the microbe's cell wall consists of lipids. However, only those microbes whose lipid content exceeds 20% are classified as 'oleaginous'. Prokaryotic microorganisms such as bacteria and cyanobacteria (blue-green algae) and eukaryotic microbial yeast and fungi are known to produce polyunsaturated fatty acid (PUFA) derivatives and phospholipids [15, 16]. The important difference between these two classes of organisms is that prokaryotes lack cellular compartments (organelles) and nuclear membranes. Both classes of organisms are capable of producing polyunsaturated omega-3 and omega-6 PUFAs that are reported to provide health benefits to humans (Table 1). These beneficial fatty acids are produced by microorganisms at industrially relevant conditions to meet global demand for these lipids. Details about some of these organisms are given below.

Overview of the composition of lipids present in different microbial sources is summarized in a chart [16]. It has long been recognized that cytosolic inclusions in bacteria contained lipids in organelles now termed lipid droplets (LDs) [17]. Additional investigations have suggested LDs are present throughout nature, appearing not only in higher plants and animals, but also in microorganisms including microalgae, fungi, and prokaryotes [18]. It is a common misconception that lipids are mainly fatty acids, but LDs contain relatively low levels of free fatty acids relative to ester forms such as TAGs. Some literature reports of fatty acid profiles specifically describe TAG fatty acids, while others reflect contributions of other fatty acid forms.

Perhaps the earliest observation of high lipid accumulation in prokaryotes was described in 1889, in which it was shown that 26–28% of the dry weight of the tubercle bacillus (*Mycobacterium tuberculosis*) was extracted in alcohol and ether [19]. A follow-up study in 1932 identified stearic (C18:0, nomenclature indicates the numbers of carbon atoms and double bonds) and palmitic (C18:0) acids released after hydrolysis, plus additional substances assigned as tuberculostearic (10-methyloctadecanoic) and phthioic (3,13,19-trimethyltricosanoic) acids [20]. Among bacteria, isolates from the genera *Mycobacterium*, *Rhodococcus*, and *Streptomyces* have also been noted for a high accumulation of TAGs [21].

Table 1
Composition of different fatty acids present in oil seeds/fruit kernels and in oleaginous organisms

Oil source	Relative % fatty acid composition							
	C14:0	C16:0	C16:1	C18:0	C18:1	C18:2	C18:3	Other
Oil seeds								
• Soybean		11		4	24	54	7	
• Rapeseed		4		2	62	22	10	
• Sunflower		7		5	19	68		
Tree fruit and Kernels								
• Palm		44		4	38	10	1	C4-C10(1%);C12:0(1%)
• Palm Kernel	16	8		3	15	2		C4-C10(1%);C12:0(1%)
• Coconut	18	9		3	6	2		C4-C10(4%);C12:0(48%)
• Olive		13	1	3	72	10	1	C20(1%)
Microorganisms								
A. Yeast								
• *Rhodosporidium toruloides*		18	3	3	66			C23:0(3%);C24:0(6%)
• *Rhodotorula glutinis*		37	1	3	47	8		
• *Lipomyces starkeyi*		34	6	5	51			
B. Fungi								
• *Mortierella isabellina*		29		3	55	3	3	
• *Mucor circinelloides*		22		4	38	10	15	
• *Pythium ultimum*		15		2	20	16	1	C4-C10 (7%); C20:1 (4%); C20:4 (15%); C20:5 (12%)
C. Bacteria								
• *Rhodococcus opacus*				3–19	6–74			
• *Acinetobacter calcoaceticus*		28	20	6	35			C12 (6%)
D. Microalgae								
• *Schizochytrium linacinum*	3–4	56–60		1–4				C22:5:2 (4-6%); C22:6 (29-35%)
• *Chorella* SP		7–19	10-9	1–4	8–9	1–14	16–19	C15 (5%); C16:2 (11%)
• *Chaetoceros muelleri*	18–40	5–40		0–25	0–4	0–5	0–5	C12 (6–20%); C16:2 (0–8%)
E. Thraustochytrids								
• *Schizochytrium* sp. *SR2I*	4	55		1				C22:5(6%),C22:6(30%)
• *Thraustochytrium aureum*	3	8			16	2	2	C20:4(3%), C22:6(52%)
• *Skeletonema costatum*	17	17	11		2	1		C18:4 (6%), C20:% (41%), C22:6 (7%)

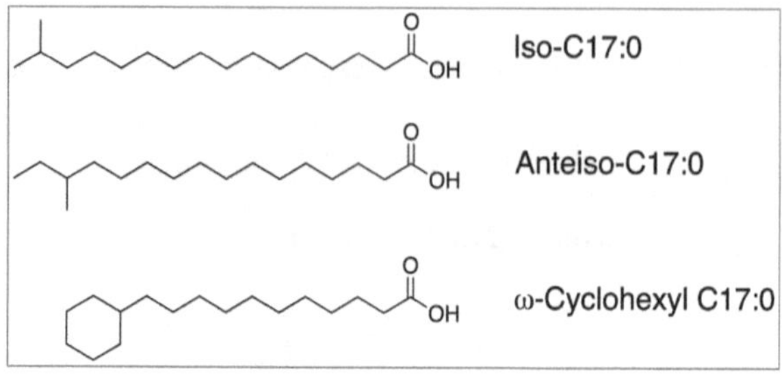

Fig. 5 Examples of branched iso-, anteiso-, and ω-alicyclic fatty acid components of prokaryotic lipids

Bacterial fatty acid composition differs from eukaryote fatty acid chemistry in that branched and cyclic fatty acids account for a substantial proportion of the fatty acid groups in membrane lipids [22]. Of particular industrial promise are the *iso*-branched fatty acids, which contain a methyl branch at the penultimate carbon, *anteiso*-branched fatty acids in which the methyl branch is two carbon atoms from the end of the chain, and ω-alicyclic fatty acids (Fig. 5). Complex lipids that incorporate branched fatty acids as esters are expected to influence membrane fluidity, and have lower melting points than analogous linear fatty acid esters.

Though TAGs are known in most eukaryotes, most prokaryotes are not known for accumulating TAGs except in the actinomycetes group [23]. When TAGs are of low abundance, fatty acid groups are important as components of complex membrane lipids including phospholipids. The genera *Mycobacterium*, *Rhodococcus*, and *Streptomyces* have been noted for TAG accumulation [21]. The current consensus is that most prokaryotes do not store neutral lipids in large quantities, but an accumulation of the specialized lipids polyhydroxyalkanoates (PHAs), which were first described in *Bacillus megaterium* in 1926 [24], is common.

4.1 Yeast Lipids

Some of the most-studied oleaginous yeast strains include *Lipomyces starkeyi* (ATCC 12659), *Apiotrichum curvatum* (ATCC 10567), *Cryptococcus albidus* (ATCC 56297), and *Rhodosporidium toruloides* (ATCC 204091) [25, 26]. Several carbon sources such as starch hydrolysate (glucose), biomass hydrolysate (glucose, xylose), glycerol, municipal organic waste have been used to produce oleaginous yeast under nutrient limiting conditions. Under nitrogen depriving conditions, protein and nucleic acid synthesis is shut down and carbon substrates are converted into storage lipids [27]. Depending on the type of carbon source available, different amounts of unsaturated lipid accumulation take place. For example in *R. glutinis*, the highest lipid accumulation of 53% was achieved using glycerol, and 25% achieved when xylose was used as carbon source [28]. Also, optimum accumulation of lipid occurred when the growth media carbon to nitrogen ratio is about 80:1. When the amount is greater than 80:1, lipid synthesis is altered, and greater production of citric acid occurs.

4.2 Lipids of Fungi

Popular oleaginous fungi include *Aspergillus terreus*, *Trichosporon fermentans*, *Claviceps purpurea*, *Tolyposporium* (genus of smut fungi), *Aspergillus oryzae* (*A-4*), *Mortierella alpina*, *Mortierella isabellina*, and *Mucor rouxii*. The fungi are explored since they produce special lipids rich in lipids containing docosahexaenoic acid (DHA), gamma-linolenic acid (GLA), eicosapentaenoic acid (EPA), and arachidonic acid (ARA) [29]. Mechanisms of lipid accumulation in fungi are similar to those of yeast. Carbon sources influence the production and composition of fatty acid esters in lipids due to differences in their metabolism. Levels of polyunsaturated fatty acids in fungal cells were correlated to the age of the mycelia [30]. Due to high saturated fatty acid content (up to 60%) in oleaginous fungi, fungal lipids are used to make cocoa butter substitutes [31].

4.3 Lipids of Eukaryotic Microalgae

Lipid accumulation is higher in microalgae, they grow faster than plants, with the potential to double biomass on the order of every 4–24 h and can be harvested daily. On average, the lipid content in algae could vary between 5% and 70% depending on species and growth conditions. The mechanism of lipid accumulation in algae is quite different than yeast and fungi. Microalgae use photoautotrophic mechanism (derives energy for food synthesis using light and is capable of using carbon dioxide as its principal source of carbon) [32]. Due to the dependence on sunlight intensity, dissolved oxygen concentration, pH, concentrations of nutrients such as nitrogen, phosphorous, silicon, and iron for producing lipids, seasonal variation and nutrient concentration in water significantly influence algae production. Some algae commonly used for lipid production include *Chlorella* spp., *Chlorella zofingiensis*, *Crypthecodinium cohnii*, *Chatoceros muelleri*, *Chlorella vulgaris*, *Arthrospira platensis*, and *Schizochytrium limacinum*. Some of the algae strains are reported to grow in both autotrophic and heterotrophic (light limiting) conditions. Heterotrophic cultivation on organic carbon sources has been used to overcome the issues of photonic energy delivery to cells in photoautotrophic growth. Algae grown under heterotrophic conditions use organic carbon sources for accumulating lipids [33]. Biomass productivity, cellular lipid content and overall lipid productivity are important parameters that influence the economics of lipid production in Algae. However, these requirements are not compatible to produce both a high growth rate of cells with high cell density. Lipid accumulation in algae is achieved by manipulating environmental stress (by limiting nitrogen, phosphorous, silicon, and iron) [34].

Eukaryotic microalgae share a significant amount of lipid biochemistry with plants, and form LBs that accumulate neutral lipids surrounded by a polar lipid surface [35]. Some of the most extensive characterization of LD lipids in microalgae has been performed for the genera *Chlamydomonas*, *Nannochloropsis*, *Chlorella*, and *Scenedesmus* [36]. Most microalgae produce small amounts of TAGs and large amounts of polar lipids under conditions where growth is optimal [37]. In contrast, *Nannochloropsis* sp. is known to accumulate significant yields of TAGs under regular or stress conditions [38]. Sixteen- and eighteen-carbon fatty acids are major lipid building blocks in these genera.

A recent paper performed a comprehensive examination of lipids in five microalgae genera [39]. Conversion of the total lipid extract to FAMEs allowed for comparisons of fatty acid composition in terms of both carbon chain lengths and degree of unsaturation. All five genera yielded abundant omega-3 FAMEs, though significant qualitative and quantitative differences in fatty acid content between genera were observed. FAME profiles highlighted a few key differences, most notably the high proportions of C20 and C22 PUFAs in *Nannochloropsis* and *Schizochytrium*.

4.4 Lipids of Prokaryotes (Bacteria)

Some of the promising gram-positive bacteria used to produce lipids include *Arthrobacter* sp. AK 19, *Gordonia* sp. DG, *Rhodococcus opacus* DSM 1069, and *R. opacus* PD630. Some promising gram-negative bacteria include *Serratia* sp. ISTD04 and *Synechococcus sp. HS01*. Cyanobacteria may represent another special case of prokaryotes, as storage of intracellular TAG has not yet been established in this group of microorganisms [40]. Some of the strains *Cyanobacterium aponinum* or *Synechococcus* sp. produce complex lipids that need to be hydrolyzed to release fatty acids. Most free fatty acids are toxic to nearly all organisms if their levels rise, so fatty acids are primarily stored as esters (e.g., triacylglycerols).

When compared to other oleaginous organisms, bacteria demonstrate high cell growth rates under simple cultivation conditions [41]. However, some bacteria accumulate oil only under special environmental conditions. Bacteria are cultivated in batches and are often affected by two variables: micronutrient and macronutrient limitations. It is interesting to note that in bacteria, the most abundant class of neutral lipids is polyhydroxyalkanoic acids. These lipids serve as intracellular carbon and energy storage compounds. Bacteria have been extensively researched and gene regulation mechanisms are understood to a greater extent. Due to this fact, it is easy to exploit modern metabolic engineering and genetic engineering tools to modify bacteria to improve oil accumulation. For economical production of lipids using oleaginous bacteria, one need inexpensive substrate, less external nutrient supplementation while growing, have high growth rate and produce high cell density biomass in short period of time with desired lipid concentration.

4.5 Lipids of Thraustochytrids

Thraustochytrids are large-celled but nonphotosynthetic marine oleaginous microorganisms. They are known to produce longer chain fatty acids [42]. Researchers are developing different isolation techniques for the cultivation of thraustochytrids. Novel approaches in the area of exploration for more diverse isolates having fast growth rates, metabolic engineering including gene cloning, and growing thraustochytrids on alternate low cost carbon sources are underway by varying pH, sea salt concentrations, and glucose [9, 43–45].

5 Comparisons of Plant Lipids and Microbial Lipids

Lipids derived from plants comprise esters of more than 300 different fatty acid structures (Https:plantfadb.org) with different chain lengths (primarily 14–18 carbon atoms) and varying numbers of double bonds or with functional groups such as hydroxyl, epoxy, ketone, acetylenic, and cyclopropyl [46]. Fatty acids are classified as (a) saturated, (b) monounsaturated, oleic acid, (c) polyunsaturated, (d) conjugated polyunsaturated; (e) hydroxylated, (f) epoxygenated, (g) acetylenic, (h) cyclopropane, and (i) cyclopentane (Fig. 6). The five major fatty acids most commonly found in oil seeds are: palmitic acid (C16:0), stearic acid (C18:0), oleic acid (C18:1ω9), linoleic acid (C18:2ω6), and α-linolenic acid (C18:3ω3). Commonly found storage lipids include wax esters, triacylglycerols, acetyl-triacylglycerols, and triacylglycerol estolides [47]. Both plant breeders and plant lipid biotechnologists have developed novel storage lipid biosynthetic pathways to increase yields of oil per unit area of land in crop species [48, 49]. Technologies like transcript profiling, monitoring of carbon and acyl fluxes during seed maturation, spatial mapping of lipid populations within seed tissues, and modular cloning of genes of interest are being pursued to understand lipid metabolism and engineer strategies to produce a higher amount of oil and in some cases designer single desirable fatty acids [50–53]. However, there are biological limitations in producing higher lipid amount in oilseeds crops [54].

When compared to lipids produced in plants, the oil quality and fatty acid profiles differ in microorganisms (Table 1). For example, microalgae produce polyunsaturated fatty acids (PUFAs), particularly the omega-3 with longer carbon chain lengths and greater degrees of unsaturation; while oleaginous yeast produce C20 fatty acids and omega-6 that has more commercial value (Fig. 7). Of different oleaginous organisms, yeast produce the maximum amount of lipids, while oleaginous fungi, algae, bacteria have lower lipid accumulation capacity [55]. Also, using cheap raw materials as carbon source will help to produce lipids in microorganisms that are cost competitive with plant-derived lipids. Different strains of organisms have been reported to produce lipids and

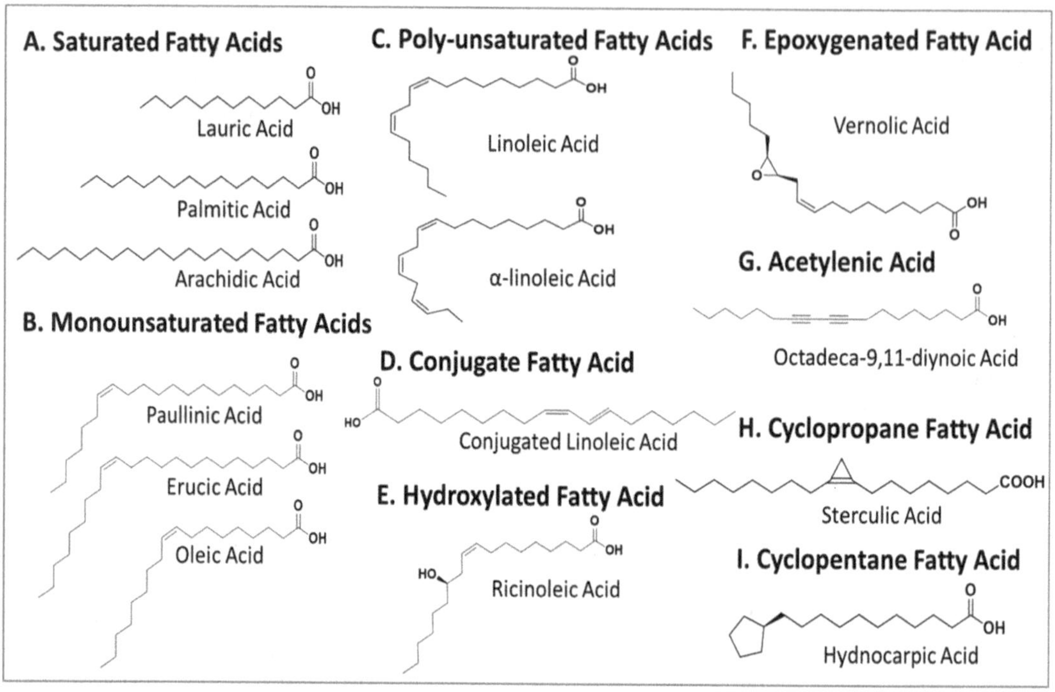

Fig. 6 Examples of structural diversity among fatty acids produced in oil seeds with examples given in bracket. Here, (**a**) saturated fatty acids (lauric acid, palmitic acid, arachidic acid); (**b**) monounsaturated fatty acids (paullinic acid, erucic acid, oleic acid); (**c**) polyunsaturated fatty acids (linoleic acid, α-linoleic acid); (**d**) conjugated fatty acid (conjugated linoleic acid); (**e**) hydroxylated fatty acid (Ricinoleic acid); (**f**) epoxygenated fatty acid (vernolic acid); (**g**) acetylenic acid (octadeca-9,11-diynooic acid); (**h**) cyclopropane fatty acid (sterculic acid); and (**i**) cyclopentane fatty acid (hydnocarpic acid)

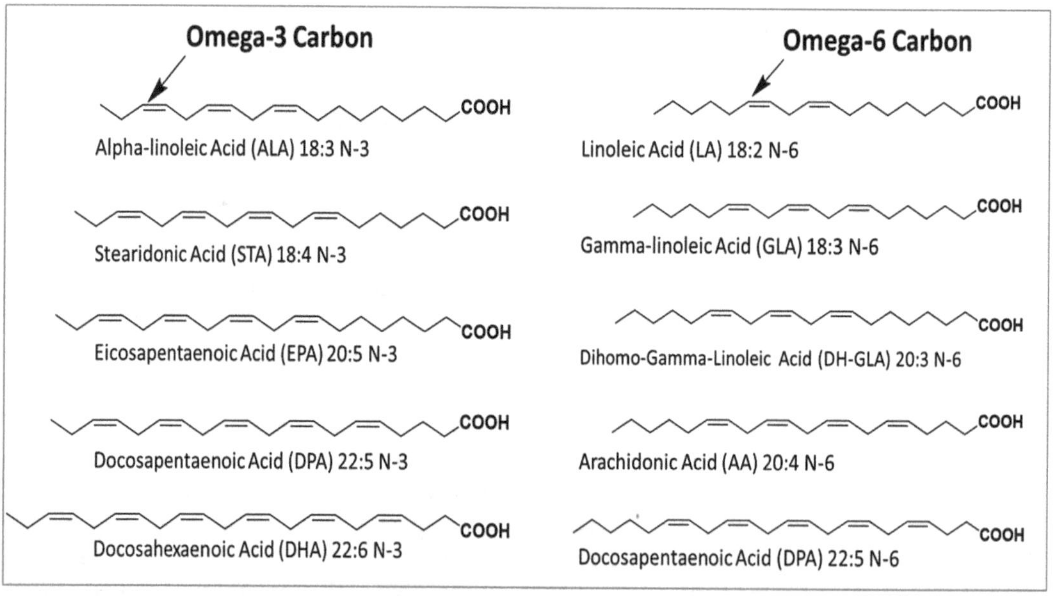

Fig. 7 Examples of structural diversity high value saturated fatty acids produced in oleaginous organisms. Here, omega-3 fatty acid (left) and omega-6 fatty acid (right)

choosing a specific strain and method of cultivation greatly influence the concentration of cells produced and lipid content in the cells [56]. Also, by manipulating metabolic pathways based on information derived from genomics, transcriptomics, metabolomics, and lipidomics techniques it has been demonstrated that lipid accumulation could be increased and designer lipids with high commercial value can be produced [57–59].

Producing lipid from oilseeds is dependent on the crop yield (kg/ha/year), while that in microorganisms is dependent on g/l/day (or kg/m^3/day or kg/m^3/year) (Table 2) [33].

Table 2
Lipid content, oil yield, and productivity in oil seeds/fruit kernels and in different oleaginous organisms

Feed-stock for producing lipids	Lipid content (% wt)	Oil yield (kg/ha/year)	Productivity (kg/m^3/year)		References
			Biomass	Lipid	
Oil seeds					
• Rape seed	45	591–664	NA	NA	[61]
• Soybean	20	450–506	NA	NA	[61]
• Palm	50	3004–5006	NA	NA	[61]
• Sunflower	45	517–664	NA	NA	[61]
Free fruit/kernels					
• Coconut	50	731–979	NA	NA	[61]
• Olive	15–35	101–292	NA	NA	[61]
• Palm	50	3005–5006	NA	NA	[61]
Yeast					
• *Rhodosporidium toruloides*	58–68	NA	3362	2120	[26]
• *Cryptococcus curvatus*	25–46	NA	1990	1154	[63]
• *Lipomyces starkeyi*	61–68	NA	636	410	[72]
Fungi					
• *Mortierella isabellina*	54–74	NA	NR	NR	[33]
• *Cunninghamella echinulata*	18–32	NA	NR	NR	[33]
Micro Algae					
• *Schizochytrium limacinum*, AT	15–60	NA	1065	404	[33]
• *Schizochytrium mangrovei*, HT	68	NA	732	498	[33]
• *Chaetoceros gracilis*, HT	50	NA	1044	525	[33]
Bacteria					
• *Arthrobacter* sp.	>40	NA	NR	NR	[33]
• *Acinetobacter calcoaceticus*	27–38	NA	NR	NR	[33]
• *Rhodococcus opacus*	24–26	NA	NR	NR	[33]
Thraustochytrids					
• *Schizochytrium* sp. *SR21*	33.3–38.6	NA	21.9–59.2	NR	[148]
• *T. aureum 34304*	41–75	NA	1.1–5.0	NR	[45]
• *T. roseum 28,210*	48.3–58.2	NA	6.1–17	NR	[44]

NA not applicable, *NR* not reported

Both autotrophic and heterotopic conditions (dependent on light) could be used to cultivate different oleaginous organisms. While the former gives high lipid content than the later due to variation in lighting requirements. Cultivating lipids using microbes in 10 m^3 volume will give yields similar to one hectare ($10^3 m^2$) area of oil palm, demonstrating the reduced space requirement for lipid production using microbes [60]. However, given the cost of production and processing the lipids using existing technology, production of plant lipids is economical when compared to microbes [61, 62]. Producing high-value lipids like omega-3, omega-6 used for food application and other fatty acids used in cosmetic applications will give microbial platforms an upper hand relative to plant-derived platforms [41, 56, 63–67]. Unusual fatty acids, such as ricinoleic acid and conjugated linoleic acid or polyunsaturated fatty acids and fatty acid-derived compounds such as γ-decalactone, hexanal, and dicarboxylic acids could be produced using yeast [67].

6 Advantages and Limitations of Producing Microbial Lipids

In terms of advantages, eukaryotic microorganisms produce a wide range of polyunsaturated fatty acids and widely accepted by public on ethical and moral grounds. The genetic machinery of these microorganisms is relatively simple to manipulate and synthesize a single desired fatty acid with high economic value. Techno economic analysis has helped to calculate the price for the microbial oil to be US$3000 per ton (excluding the cost of feedstock), with calculations based on previous work in New Zealand on the production of Cocoa Butter equivalent from lactose using *Candida curvata* which is not viable in the current market, where vegetable oil prices are sold around US$800–900 per ton [68, 69]. Producing high-value fatty acids (e.g., GLA, DHA, ARA) from microbes can fetch a market value between US$40,000 and 120,000 per ton which could be used for cosmetic application and for producing superfoods [64].

On the other hand, producing microbial lipids is very expensive when compared to producing them in plants due to low yield, low titer and low productivity of lipid, low tolerance to pretreatment degradation products [1, 70]. Three major costs associated with producing microbial lipids include (1) carbon and nutrient source, (2) operating costs of fermentation, and (3) post-processing costs associated with separating lipids from microorganisms. Carbon resources (e.g., glucose, xylose, glycerol, starch, cellulose hydrolysates) and nitrogen sources (e.g., peptone, yeast extract, or corn steep liquor) are used as feedstocks to produce microbial lipids using molds, yeast and bacteria [71]. The costs of producing the carbon and nitrogen resources influence the cost of producing microbial lipids. It will be relatively cheaper to produce lipids using cane sugar juice or molasses as less processing is involved. However, such strategy becomes expensive as one uses sugars derived from grains and lignocellulosic biomass due to costs associated with processing. Readily available feedstocks such as industrial food waste (e.g., bakery, food processing industry) and sugar-rich industrial wastewater (e.g., sago factory) could be used feedstocks for producing microbial lipids at reduced processing cost [72, 73]. Though the feedstock cost is insignificant, economies of scale will dictate the costs of lipid production. Though producing lipids using algae and Thraustochytrids using freshwater and industrial wastewater is considered as a viable alternative for mass producing lipids, the costs associated with separating and processing lipids from wet microorganisms presents a stumbling block to produce cost-competitive lipids [74, 75]. Some of the above mentioned challenges may be overcome by carrying our genetic modification and directed evolution of oleaginous microorganisms, process design and using advanced lipid extraction protocols.

7 Machineries Involved in Biosynthesis of Microbial Lipids

Understanding the cellular mechanisms responsible for lipid production in different organisms will aid in the selection of an appropriate organism suitable for a specific feedstock, production technology, product, season, and market. The mechanisms of synthesis of lipids have similarities and differences among the microbial taxa discussed in this book: algae, yeasts, bacteria and filamentous fungi. These differences are connected to genetic, structural and chemical properties of the organisms. For instance, plant cells have more organelles than fungal cells [76]. Furthermore, while plants have multiple genes encoding proteins with multiple enzymatic activities, fatty acid synthesis in some microalga pathways is encoded by single-copy genes [77].

Selection of appropriate organisms for production of a desired lipid starts with identification of organisms that synthesize desired products, under relevant conditions. Synthesis and storage of lipids is often triggered by environmental or nutritional conditions. For example, in oleaginous yeasts, cessation of replication and accumulation of storage lipids is triggered by depletion of a nutrient such as nitrogen when carbon is not limiting [78]. Optimal molar C-to-N ratios range from 30 to 80 depending on the yeast species and growth conditions [79]. In photosynthetic microalgae, light–dark cycles influence lipid accumulation: TAG synthesis and storage occurs during daylight, then the lipid stores are depleted to generate ATP during the night [80].

While oleaginous yeasts accumulate triacylglycerols (TAG) and/or sterol esters (SE) [81], some bacteria and archaea accumulate polyhydroxyalkanoates (PHA) and lower amounts of TAG and wax esters [23, 82]. Mechanistic comparisons are described in detail in the literature [81–86]. Some key points are summarized here.

The process of lipid accumulation in yeasts and other fungi involves four stages. Metabolic pathways involved are shown in Fig. 8. In the first stage, various strategies are used to generate pools of acetyl-CoA and NADPH, which are the building blocks of fatty acids and the reducing power for fatty acid synthesis, respectively [87, 88]. Two moles of NADPH are required for each mole of acetyl-CoA [89, 90]. In the case of oleaginous yeasts, many years of enzymatic and genetic studies led to the conclusion that two key enzymes in this stage, highlighted below, are responsible for enhanced lipid accumulation in oleaginous fungi.

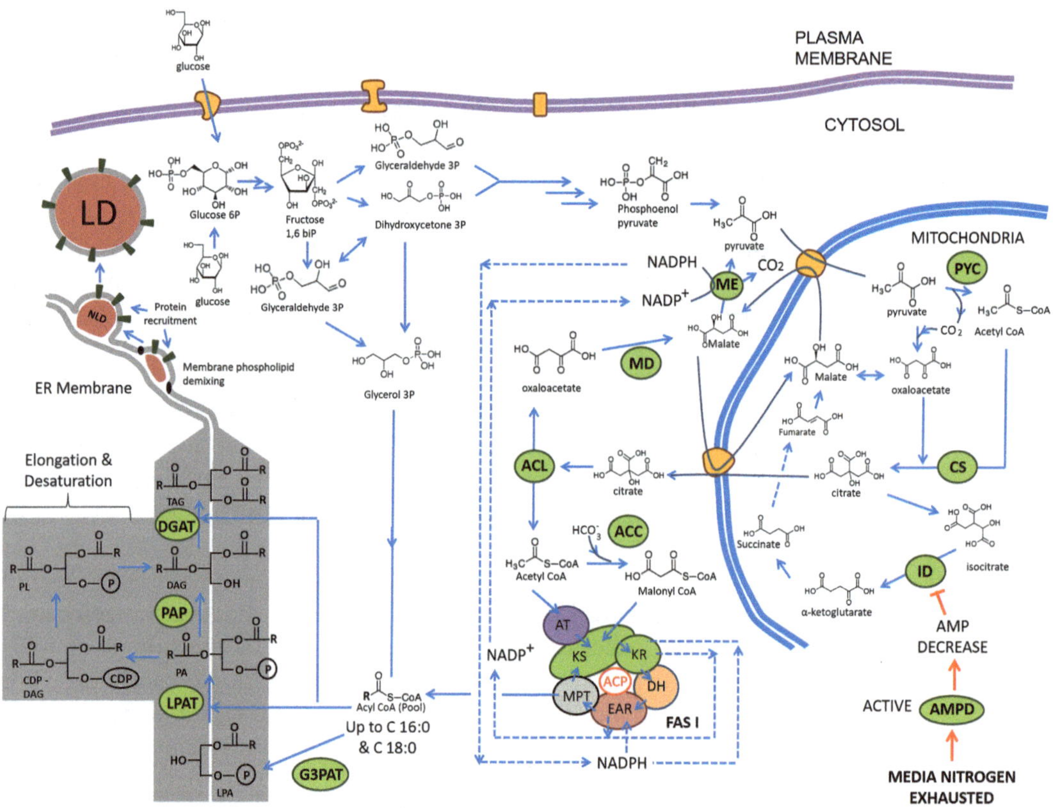

Fig. 8 Metabolic pathway of fatty acid synthesis, triacylglycerol (TAG) synthesis, and lipid droplet formation in yeasts. Abbreviations: *ME* malic enzyme (cytosolic), *ME 2* malic enzyme (endoplasmic reticulum membrane, location of desaturation), *ACL* ATP:citrate lyase, *ID* isocitrate dehydrogenase, *NLD* nascent lipid droplet, *LD* lipid droplet, *ER* endoplasmic reticulum, *LPA* lysophosphatidic acid, *PA* phosphatidic acid, *FAS I* fatty acid synthase I, *DAG* diacylglycerol, *TAG* triacylglycerol, *AMPD* AMP deaminase

- When nitrogen is depleted, AMP deaminase is activated, which sharply decreases cellular AMP levels [91].
- Reduced AMP levels decrease or stop the activity of isocitrate dehydrogenase (IDH) [88], resulting in decreased production of α-ketoglutarate and halting of the tricarboxylic acid cycle.
- Isocitrate accumulates inside the mitochondria.
- The enzyme aconitase transforms the isocitrate to citrate, which then accumulates in the mitochondria.
- The antiport protein citrate/malate translocase (CMT) transports citrate to the cytoplasm, where it is cleaved by ATP:citrate lyase (ACL) to form oxaloacetate and acetyl-CoA. The ACL enzyme is a key step because apparently it is exclusively found in oleaginous species [55, 87, 92].
- Oxaloacetate in the cytosol is reduced by malate dehydrogenase (MD) to malate, which is then converted to pyruvate by malic enzyme (ME). This reaction also generates NADPH later needed to convert acetyl-CoA to fatty acids [93].
- In a parallel reaction inside the mitochondrion, pyruvate is carboxylated to form oxaloacetate, then is reduced by MD to malate which is transported outside the mitochondrion and decarboxylated by malic enzyme (ME) to form malate. This reaction also generates NADPH. The ME enzyme is present in all fungi, but particularly abundant in oleaginous fungi and changes isoform upon nitrogen limitation [94].

In the second stage, fatty acid synthesis, acetyl-CoA is carboxylated in the cytosol or mitochondrion by acetyl-CoA carboxylase to form malonyl-CoA, then bound as a thiol ester to acyl carrier protein (ACP). Successive acetyl groups (as acetyl-CoA) are added by fatty acid synthase (FAS). FAS Type I, found in eukaryotic cytoplasm, is an integrated enzyme complex. The fungal form, Type IA, is an $\alpha_6\beta_6$ complex found in the cytoplasm, and the animal form, Type IB, is an α_2 dimer, also found in the cytoplasm. FAS Type II, found in prokaryotes and organelles of prokaryotic origin such as plastids and mitochondria, is a dissociated version: the units are independent [95, 96]. In yeasts, the steps that are repeated for fatty acid elongation performed by FAS I are as follows:

- Condensation of acetyl-CoA with malonyl-CoA by KS.
- Reduction by ketoacyl reductase (KR).
- Dehydration by dehydratase (DH).

The *FAS2* gene of oleaginous yeast *Rhodotorula toruloides* (formerly called *Rhodosporidium toruloides*) has two tandem acyl carrier protein domains, thought to improve the efficiency of the enzyme [92].

Depending on the type of organism, various strategies are used to release the full-length fatty acid from the enzyme. In yeasts, malonyl-palmitoyl transacylase (MPT) transfers the acyl chain from acyl ACP to acyl-CoA [88]. In algae, three mechanisms are used to release the acyl chain: hydrolysis by thioesterase to release a free fatty acid, transfer by acyltransferase in the chloroplast to either G3P or monoacylglycerol-3-phosphate, or release in the acyl-CoA form by acyl-coenzyme A synthetase (ACS). The location and activities of these enzymes determine the chain length and cellular location of lipids. Gram-positive and gram-negative bacteria utilize different mechanisms.

The third stage conversion to polar or neutral lipids depends on the type of organism and the cell's needs. The acyl chain can be coupled to a glycerol-3-phosphate backbone to form DAG or TAG, or can be converted to phosphatidylcholine, phosphatidylethanolamine or other types of phospholipids to form phospholipids or other polar lipids. Further elongation, desaturation, or hydroxylation can also occur. For example, in yeast, after release from FAS I, acyl-CoA is channeled to the ER membrane where transesterification of a G3P backbone occurs, called the Kennedy pathway. Further elongation and desaturation also occur in the ER membrane [95].

In the fourth stage, the product is directed to a cellular location. In oleaginous yeasts, TAG are stored in lipid bodies, which are assembled between the two membrane leaflets in the ER [97–99]. Similarities and differences are seen when comparing lipid synthesis and accumulation in yeast to other organisms. The chemical nature of the stored lipids varies greatly among organisms. Yeasts accumulate primarily TAG and/or sterol esters (SE) [81]. In contrast, most bacteria do not store significant quantities of TAG. Some exceptions are selected gram-positive actinobacteria such as *Dietzia*, *Mycobacterium*, *Nocardia*, *Rhodococcus* and *Streptomyces*, which readily accumulate TAG [100], and a few gram-negative species such as *Acinetobacter* accumulate small amounts of TAG as a storage lipid [23]. The primary storage lipids of gram-negative species are wax esters, while TAGs are membrane lipid precursors. TAG accumulation in archaea has not been reported [101]; instead these organisms synthesize lipids consisting of isoprenoid chains, having methyl branches [102]. Like bacteria, some archaea accumulate PHA under conditions of nutrient limitation where carbon is in excess. Similarities in genes indicate a horizontal transfer of genes from bacteria to archaea [103].

There are parallels in lipid accumulation mechanisms among diverse organisms. As discussed above for yeast TAG accumulation, four stages are observed for polyhydroxyalkanoate accumulation in prokaryotes [84]. In the first stage, a pool of cytoplasmic acetyl-CoA is generated. In the second stage, hydroxyalkanoate (HA) monomers are synthesized. In the third stage, the HA monomers are polymerized or copolymerized to form PHA chains. In the fourth stage, intracellular PHA lipid droplets are formed.

In eukaryotes, lipids accumulate in spherical structures called lipid droplets or lipid bodies, composed of the core neutral lipids (TAG and/or SE), surrounded by a monolayer consisting of polar lipids and inserted proteins that play functional, structural, and regulatory roles [97, 104]. Terminology differs among organisms: plants accumulate oleosomes or spherosomes, or plastoglobules if located inside a plastid. In humans, lipid droplets interact with other cellular organelles, and are related to the progression of metabolic diseases such as obesity, fatty liver, type 2 diabetes mellitus, and atherosclerosis; thus, lipid droplets are a dynamic functional organelle rather than a static storage unit [104].

In addition to these mechanistic differences, various oil-accumulating organisms utilize diverse feedstocks, exhibit a broad range of productivity, have different biosafety levels, and produce diverse coproducts or contaminants. These factors should be taken into consideration when selecting an organism for production of a desired lipid-based product.

8 Analytical Methods for Characterizing Lipids

Despite historical awareness of microbial lipids, their comprehensive characterization has been limited by technologies available for their characterization. Analytical chemists have developed several numerous methods and protocols to analyze individual or combinations of analytes from biological samples. Since biological samples contain multiple lipids, the expertise of chemists, analysts, statisticians and information technology personnel have come together and developed chromatographic, spectroscopic, and computational techniques to evaluate lipid composition [105, 106]. Sample preparation plays an essential role in dictating which lipids are preserved and detected. Biological samples are homogenized followed by using different extraction methods. The most common extraction approach employs nonpolar solvents or materials (single or mixed organic solvents, liquid-liquid extraction or solid phase extraction), often facilitated by energy-based (ultrasonic, microwave, pressurized assisted) acceleration of extraction. In some instances, crude total lipid extracts are analyzed without fractionation into lipid classes, while in other situations, lipid classes are separated using a specific separation method (e.g., chromatography) followed by subjecting separated lipids to various forms of analysis [107–110].

8.1 Chromatography and Separation Methods

Different types of chromatographic separations are widely used to separate lipids from other molecules for both qualitative and quantitative determination. Following chromatographic separation, a suitable detector such as a mass spectrometer is used to confirm the identity of molecules. Details about different separation methods are given below.

8.1.1 Thin-Layer Chromatography (TLC)

Even with the development of sophisticated modern instruments, much of the characterization of lipids relies heavily on thin-layer chromatography (TLC) to separate them according to lipid classes. This inexpensive method, usually based on silica as a stationary phase, is used in most laboratories for qualitative and quantitative analyses that combine variant forms within a class (e.g., different fatty acid groups within triacylglycerols) to give aggregate measures for each lipid class. Detection involves using appropriate standards, and visualization is usually performed by exploiting nonselective binding to a dye such as iodine vapor (brown spot), 2,7-dichlorofluorescein (yellow spot when illuminated using UV light), or rhodamine (pink spot) [111, 112]. Staining with 0.2% Amido Black 10B in 1 M NaCl is very sensitive, detecting lipids at nanogram quantities [113]. Non-destructive visualization of lipids offers the advantage that the spots may be collected, followed by elution of the lipids for more detailed structural or quantitative detection schemes that resolve individual components within each spot. Some destructive nonspecific staining methods such as using 50% sulfuric acid in methanol or water to char lipids to generate dark spots, or oxidative methods that employ either 5% potassium dichromate in 40% sulfuric acid or 3–6% cupric acetate in 8–10% phosphoric acid have been used [114].

Recent improvements in high-performance thin-layer chromatography (HPTLC) employ reduced particle sizes and slab thickness which increase the resolution of separation [115]. In sample cases the lipid samples are separated in two orthogonal directions using 2D TLC [114–116]. Resolution of lipid separations may be significantly increased by using electrochromatographic chamber and microfluidic paper-based devices [117]. Other new separation technique reported in the literature include using TLC followed by mass spectrometry (MS-TLC) [118] which has several advantages such as reduced oxidation, high resolution, and analysis of crude lipids without the necessity of converting them to fatty acid methyl esters.

8.1.2 Gas Chromatography

Gas chromatography is a widely used analytical method for volatile lipids, but not a widely used method for lipidomic analysis due to hydrolysis and derivatization requirement (e.g., fatty acid methyl esters) [119]. In addition to using mass spectrometer as detector after separating the lipids using GC (GC-MS), the flame ionization detector (FID) is commonly used for quantitative analysis of fatty acid content after catalyzed trans-esterification of triglycerides or other complex lipids [120, 121].

8.1.3 High-Performance Liquid Chromatography (HPLC)

This method has several advantages including quantitative analysis, high efficiency separation and selectivity. Particularly this technique is useful for analysis of low volatility lipids (e.g., polar complex lipids or TAGs). With considerable innovation in column technology and different separation media, the time taken to carry out lipid separations has considerably decreased particularly with automation. In addition, use of a mass spectrometer as detector in combination with liquid chromatography (LC-MS) has enhanced the range of lipids that are detected in a single analysis. Lipids often lack analytically useful optical absorbance properties, but may also be quantified after LC separation using refractive index detector (RID), evaporative light scattering detector (ELSD), or charged aerosol detector (CAD) [122]. Some of the commonly used chromatography separation include reverse phase, hydrophilic interaction, normal phase, and two-dimensional LC.

8.1.4 Supercritical Fluid Chromatography (SFC)

Using supercritical carbon dioxide as mobile phase is another popular method of analyzing lipids with wide range of polarities. Supercritical solvents offer more rapid analyte diffusion rates relative to liquid solvents, and this can be exploited to yield improved chromatographic separations. Two common SFC column types include open tubular column SFC (OTSFC) helpful to carry out high resolution separation with complex mixture of lipid samples and packed column SFC (PC-SFC) helpful for high analysis speed with large analyte capacity [123, 124].

8.1.5 Capillary Electrophoresis

This technique has several advantages when compared to HPLC such as nanoliter sample requirement with minimal organic solvent requirement for mobile phase. This technique can be integrated with MS [125]. Capillary zone electrophoresis (CZE) offers very high separation efficiency, requires that analytes be charged (e.g., not for neutral lipids), and has been used to analyze shorter fatty acids (C2–C14) [126]. Another interesting variation on this approach is capillary electrochromatography (CEC) which uses electrodrive to separate thermally labile lipids efficiently [127]. Separating linear saturated free fatty acids (C12–C31) containing mixture is possible using micellar electrokinetic chromatography (MEKC). Here high molecular weight fatty acids are efficiently solubilized using micellar systems when compared to cyclodextrins [128].

8.2 Spectroscopic Techniques for Lipid Analysis

Optical spectroscopy uses the interactions between radiated energy and matter to detect the functional groups in lipid molecules, and is usually nondestructive. Vibrational spectroscopy, a subset of optical spectroscopy, probes vibrational energies associated with molecular stretching and bending modes to yield spectra (fingerprints) characteristic of functional groups in lipids. A prominent example is infrared (IR) spectroscopy which measures absorption of infrared light and is sensitive to vibrations that change dipole moments. Raman spectroscopy complements IR, using nonelastic light scattering to detect vibrations that involve a change in polarizability. Another technique, nuclear magnetic resonance (NMR) spectroscopy, uses magnetic properties of atomic nuclei by exciting the nuclear spin in the presence of an external magnetic field and registers the electromagnetic radiation released during nuclear spin relaxation to probe nuclear environments, yielding information about molecular structure, dynamics, and reaction state. Both in vitro and in vivo samples may be analyzed noninvasively and nondestructively using NMR.

8.2.1 Raman and Infrared Spectroscopy

These two vibrational spectroscopy techniques are useful for detecting structural features of different diverse groups of lipids, that is, saturated/unsaturated fatty acids, triacylglycerols, cholesterol, cholesteryl esters, and phospholipids using vibrational spectra library [129]. In some cases, direct measurement of lipids within the cellular structure of interest, detailed structural information such as degree of unsaturation, length and branching of chain may be determined using both Raman and Infrared spectroscopy (IR) [130]. Raman spectra are usually acquired using 532 nm (visible light) and 1064 nm (near-infrared) light from lasers. In some cases Fourier Transformed (FT)-Raman spectra are used to get chemical information for the lipid samples in the area of tens of micrometers and the chemical information is averaged. Raman spectra have been documented for different classes of lipids in seven wavenumber ranges, that is, 3200–2700 (I), 1800–1600 (II), 1600–1400 (III), 1400–1200 (IV), 1200–1000 (V), 1000–800 (VI), and 800–400 cm^{-1} (VII) [131]. Raman spectra of the hydrocarbons present in lipids are usually found in the following regions: 1500–1400 cm^{-1} (scissoring and twisting vibrations of CH_2 groups), 1300–1250 cm^{-1}(scissoring and twisting vibrations of CH_3), 1200–1050 cm^{-1}(C–C stretching), and 3000–2800 cm^{-1} (C–H stretching). The Raman bands are due to group of frequencies which are considered as marker bands in lipids. For example, 1720–1750 cm^{-1} (the C=O vibration is marker for TAGs) and ~1650 cm^{-1} (the C=O vibration is marker for FAs). Cholesteryl esters were distinguished from cholesterol based on the presence of a band located at ~1740 cm^{-1} associated with the vibrations of the C=O group in the esters. In the case of phospholipids, a typical band at ~1090 cm^{-1} is observed (due to the P–O stretching vibrations). Lipid polymorphs could be identified using the fingerprint spectral region. On the other hand lipid structures differing in orientation of layers in the crystal show some dissimilarities in the low-wavenumber range. Some of the characteristic marker groups identified in lipids are as follows: (1) a band in the 1400–1430 cm^{-1} range observed for solid FAs (bending vibrations in the CH_2 groups), while such spectra is absent for other analyzed lipids; (2) a double band is found at ~1730 and 1743 cm^{-1} for solid triacylglycerols (C=O stretching vibrations); (3) a medium intensity band at ~700 cm^{-1} (for cholesterol) and low-intensity band at ~430 cm^{-1} are observed (for cholesteryl esters); (4) a band at ~720 cm^{-1} (for choline), 760 cm^{-1} (and ethanolamine moieties) and ~1085 cm^{-1} (phospholipids), while the intensity of the spectra is found to be lower for other lipids.

Raman spectroscopy is quite useful for calculating ratios of unsaturated to saturated fatty acid groups using the intensity of bands n(C=C)/n(CH_2) by using the ratio of intensity at 1655/1444 cm^{-1}. Other noticeable spectral markers are ~1260 cm^{-1}(=C–H deformation stretching) and ~3005 cm^{-1}(=C–H vibrations). The level of *Z* and *E* unsaturation in a wide range of fat-containing products can be estimated using Raman bands near 1655 and 1670 cm^{-1}. Both Raman and IR are capable of determining qualitative information in lipid analysis including: (1) cis/trans ratio, (2) degree of unsaturation (also known as mass unsaturation), (3) molar unsaturation (the ratio of C=C bonds per molecule), (4) the content of conjugated double bonds, (5) chain length, and (6) branching [132–134]. Both Raman and IR imagining help to identify triglyceride, cholesterol and its esters and their spatial distribution in cells [135].

8.2.2 Nuclear Magnetic Resonance (NMR) Spectroscopy

NMR spectroscopy is particularly useful for chemical structure determination because it provides information about chemical environments of individual atoms (from chemical shifts) and about how these atoms are connected to one another through spin–spin couplings. One-dimensional ^{1}H and ^{13}C NMR spectra and two-dimensional ^{1}H–^{1}H–COSY and ^{1}H–^{13}C–HSQC spectra were used to study the structure of naturally occurring lipids and their derivatives extracted from different microbes [136]. In some cases relative stereochemical configurations could be determined using this technique [137] as well as the structure of a novel fatty acid. Structures and dynamics of deuterated phospholipids in multilamellar vesicles were analyzed using solid state ^{31}P NMR and ^{2}H NMR spectroscopies [138]. Since NMR is noninvasive, it offers the advantage of not only determining composition (and differences between samples) but can also help monitor chemical changes of individual components inside microorganisms using ^{13}C metabolic enrichment and ^{13}C NMR spectroscopy. In one other study, ^{1}H magic angle spinning magnetic resonance spectroscopy (MRS) demonstrated alterations in lipids in cells and suggested that this can be a used to identify cell stress as well as a noninvasive, early indicator of cell death [139].

8.2.3 Mass Spectrometry and Hyphenated Techniques

A large proportion of lipids lack characteristic absorbance of ultraviolet or visible light wavelengths that might allow for their selective detection in the midst of complex extracts of biological material. Over the past few decades, mass spectrometry (MS) has emerged as the analytical method of choice for discovery of novel lipids owing to its sensitivity, selectivity of detection that comes from measuring molecular mass, and high information content about elemental formulas and chemical structures. The general approach to MS involves formation of ionized molecules in the gas phase followed by separation of ions by their mass-to-charge (m/z) ratios and measurements of the amounts of these ions. MS detection is readily coupled to gas chromatography (GC-MS), liquid chromatography (LC-MS), and both hyphenated techniques are widely used for lipid analysis. Sequential stages of mass analysis (MS-MS) may be performed on ionized lipids, with activation of ions of selected mass yielding fragment ions that are subsequently analyzed using second step of mass analysis. In this manner, structural information may be obtained for many different analytes.

Matrix-Assisted Laser Desorption/Ionization (MALDI) Mass Spectrometry

Mass spectrometric analyses of complex mixtures of lipids may be performed without prior fractionation using Matrix-Assisted Laser Desorption Ionization (MALDI), which employs a pulsed laser to evaporate and ionize compounds. The most common approach embeds sample molecules in a matrix material that absorbs the laser light. Upon irradiation, the sample surface undergoes rapid heating and explosive liftoff of compounds, some of which are intact ionized molecules. Owing to the pulsed nature of the ionization process, MALDI is frequently performed using mass analyzers that employ pulsed introduction of ions into the mass analyzer, such time-of-flight (TOF) or Fourier-transform mass spectrometers including electrostatic orbital trap and ion cyclotron resonance analyzers. An early demonstration of MALDI for lipid profiling involved depositing intact bacterial cells in 2,5-dihydroxybenzoic acid matrix, and yielded a series of ions attributed to bacterial phosphatidylethanolamine (PE) and phosphatidylglycerol (PG) lipids [140]. In a more recent paper, MALDI was used to profile TAGs in *Rhodococcus opacus* PD130 [141].

Gas Chromatography–Mass Spectrometry (GC-MS)

Given the widespread use of gas chromatography for profiling fatty acids after their conversion to FAMEs), GC-MS frequently plays more of a supportive qualitative role for confirmation of FAME assignments or for identification of novel or unexpected compounds then as a quantitative technique. The positions of double bonds in unsaturated fatty acid derivatives are often not apparent from their mass spectra. One successful approach for locating double bonds involves derivatizing the double bonds via reaction with iodine and dimethyl disulfide, which results in attachment of a methyl-thio (CH_3S–) group at each end of each double bond [142]. These derivatives undergo fragmentation upon electron ionization, cleaving the carbon-carbon bond(s) between methyl-thio groups and resulting in fragment ions whose masses indicate the original double bond position(s).

Liquid Chromatography–Mass Spectrometry

In the 1980s, gentle ionization methods for nonvolatile compounds including electrospray ionization (ESI) were paired with MS, allowing for wide adoption of liquid chromatography-mass spectrometry (LC-MS). ESI performs best when analyzing compounds that contain acidic or basic functional groups, though profiling of a wide range of neutral TAGs extracted from the microalgae *Chlamydomonas* and *Nannochloropsis* has been achieved with a reasonable quantitative agreement with GC-FID measurements of FAMEs [38]. A technique complementary to ESI, atmospheric pressure chemical ionization (APCI), ionizes molecules by first ionizing the LC solvent using a corona discharge, and the ionized solvent then transfers change, often via proton transfer, to the analyte. APCI offers suitable performance for many neutral lipids, whereas ESI is preferred for analysis of lipids with ionic head groups. Selective ionization for prenylquinones and carotenoids extracted from cyanobacteria and photosynthetic bacteria has been achieved using LC-MS and APCI in negative-ion mode, yielding true negatively charged molecular ions and minimal interference from more abundant polar lipids [143, 144]. Despite the remarkable range of lipids that can be detected using LC-MS, this approach remains underutilized in the investigation of microbial lipids, perhaps due to the vast complexity of lipids and the limited availability of lipid standards.

Shotgun Lipidomics Using Infusion and Tandem Mass Spectrometry

The beginning of the twenty-first century heralded the revolutionary idea that comprehensive analysis of the entire suite of lipids, known as the lipidome, was the most appropriate way to define lipid phenotypes [145]. Direct infusion of extracted lipids into the mass spectrometer circumvents the need to develop chromatographic separations. One successful approach has employed direct infusion and a variety of MS-MS scans to profile an assortment of different lipid classes [146]. Direct infusion with high mass resolution Fourier Transform Ion Cyclotron Resonance (FT-ICR) mass spectrometry was employed in a recent paper that profiled lipids from *Scenedesmus obliquus* and *Nannochloropsis salina* [147]. These analyses detected more than 1500 peaks in positive-ion mode and more than 900 in negative-ion mode, spanning at least 20 lipid classes. Mass accuracies were within less than 500 part-per-billion error in each polarity, and facilitated assignments of elemental formulas.

9 Conclusion

Though plant-derived lipids are conventionally used for various applications, expanding population and reduced cultivable land is driving a search for alternate resources to meet growing demands for lipids. Recently, several promising technologies have been developed to produce microbial lipids for high-value applications and research is underway to genetically alter microbes to produce desired lipids in high titers. In this introductory chapter, we have compared the lipids that are derived from plants and different microbes including bacteria, yeasts, molds, algae, and thraustochytrids [148]. The advantages of producing lipids from different microbial sources were discussed. Several chromatography methods for separating and analyzing lipids using spectroscopic methods ranging from NMR, IR, and Raman spectroscopies and mass spectrometry were discussed. Other chapters in this book provide more details about lipids derived using native and genetically engineered oleaginous organisms including characterization, separation techniques, and value addition. Following chapters provide detailed information about producing oleaginous organisms using lignocellulosic biomass following pretreatment and enzyme hydrolysis, wastewater, industrial waste, anaerobic digestion, using food waste and further processing to biocrude to produce biofuels and assessment of their quantity and quality using various analytical techniques. Life cycle analysis of producing microbial lipids and further processing to biofuels and further process the lipids to high-value products are also discussed in one of the chapters.

Acknowledgments

A.D.J. acknowledges support from the USDA National Institute of Food and Agriculture, Hatch project MICL02474. VB thanks the University of Houston and the State of Texas for startup funds.

References

1. Patel A, Shindhu DK, Arora N, Singh RP, Pruthi V, Pruthi PA (2015) Biodiesel production from non-edible lignocellulosic biomass of *Cassia fistula* L. fruit pulp using oleaginous yeast *Rhodosporidium kratochvilovae* HIMPA1. Bioresour Technol 197:91–98
2. Colin R, James PW (2002) The biochemistry and molecular biology of lipid accumulation in oleaginous microorganisms. Elsevier Science, New York
3. Bharathiraja B, Sridharan S, Sowmya V, Yuvaraj D, Praveenkumar R (2017) Microbial oil—a plausible alternate resource for food and fuel. Bioresour Technol 233:423–432
4. Christophe G, Kumar V, Nouaille R et al (2012) Recent developments in microbial oils production: a possible alternative to vegetable oils for biodiesel without competition with human food? Braz Arch Biol and Technol 55(1):29–46
5. Sakuradani E, Ando E, Ogawa A et al (2009) Improved production of various polyunsaturated fatty acids through filamentous fungus *Mortierella alpina* breeding. Appl Microbiol Biotechnol 84(1):1–10
6. Gurovic V, Soledad M, Gentili AR, Oliviera NL, SusanaRodríguez M (2014) Lactic acid bacteria isolated from fish gut produce conjugated linoleic acid without the addition of exogenous substrate. Process Biochem 49:1071–1077
7. Beligon V, Christophe G, Fontanille P et al (2016) Microbial lipids as potential source to food supplements. Curr Opin Food Sci 7:35–42
8. Gorissen L, De Vuyst L, Raes K et al (2012) Conjugated linoleic and linolenic acid production kinetics by bifidobacteria differ among strains. Int J Food Microbiol 155:234–240

9. Ward OP, Singh A (2005) Omega-3/6 fatty acids: alternative sources of production. Process Biochem 40:3627–3652
10. Zielińska A, Nowak I (2014) Fatty acids in vegetable oils and their importance in cosmetic industry. Chemix 68:103–110
11. Koller M, Maršálekc L, Dias MMDS, Braunegg G (2017) Producing microbial polyhydroxyalkanoate (PHA) biopolyesters in a sustainable manner. New Biotechnol 37:24–38
12. USDA, Foreign Agriculture Services (2017) Global oilseed demand growth forecast to outpace production.
13. Atabani AE, Silitonga AS, Ong HC, Mahlia TMI, Masjuki HH, Badruddin IA, Fayaz H (2013) Non-edible vegetable oils: a critical evaluation of oil extraction, fatty acid compositions, biodiesel production, characteristics, engine performance and emissions production. Renew Sust Energ Rev 18:211–245
14. Küçükugurluoglu Y, Karasalihoglu S, Vatansever U, Biner B, Acunaş B, Pala O (2005) Castor oil plant seed poisoning. Case Rep Clin Pract Rev 6:55–57
15. Carman GM, Han G-S (2009) Regulation of phospholipid synthesis in yeast. J Lipid Res 50 (Supplement):S69–S73
16. Thevenieau F, Nicaud J-M (2013) Microorganisms as sources of oils. Oilseeds Fats Crops Lipids 20(6):D603
17. Hanstein J (1880) Ueber die Gestaltungsvorgange in den Zellkerne bei der Theilung der Zellen. Botanische Abhandlungen aus dem Gebiet der Morphologie und Physiologie. Adolph Marcus, Bonn, p 4
18. Murphy DJ (2001) The biogenesis and functions of lipid bodies in animals, plants and microorganisms. Prog Lipid Res 40 (5):325–438
19. Hammerschlag A (1889) Bacteriologisch-Chemische Untersuchungen der Tuberkelbacillen. Monatsh Chem 10(1):9–18
20. Anderson RJ (1932) The chemistry of the lipoids of tubercle bacilli. Physiol Rev 12 (2):166–189
21. Alvarez HM, Pucci OH, Steinbüchel A (1997) Lipid storage compounds in marine bacteria. Appl Microbiol Biotechnol 47 (2):132–139
22. Kaneda T (1991) Iso-fatty and anteiso-fatty acids in bacteria—biosynthesis, function, and taxonomic significance. Microbiol Rev 55 (2):288–302
23. Alvarez H, Steinbüchel A (2002) Triacylglycerols in prokaryotic microorganisms. Appl Microbiol Biotechnol 60(4):367–376
24. Lemoigne M (1926) Produit de déshydratation et de polymérisation de l'acide ß-oxybutyrique. Bull Soc Chim Biol 8:770–782
25. Akhtar P, Gray JI, Asghar A (1998) Synthesis of lipids by certain yeast strains grown on whey permeate. J Food Lipids 5:283–297
26. Wu S, Hu C, Jin C, Zhao X, Zhao ZB (2010) Phosphate limitation mediated lipid production by *Rhodosporidium toruloides*. Bioresour Technol 101:6124–6129
27. Granger LM, Perlot P, Goma G, Pareilleux A (1993) Effect of various nutrient limitations on fatty acid production by *Rhodotorula glutinis*. Appl Microbiol Biotechnol 38:784–789
28. Easterling ER, French WT, Hernandez R, Licha M (2009) The effect of glycerol as a sole and secondary substrate on the growth and fatty acid composition of *Rhodotorula glutinis*. Bioresour Technol 100:356–361
29. Eroshin VK, Satroutdinov AD, Dedyukhina EG, Chistyakova TI (2000) Arachidonic acid production by Mortierella alpine with growth-coupled lipid synthesis. Proc Chem 35:1171–1175
30. Fakas S, Makri A, Mavromati M, Tselepi M, Aggelis G (2009) Fatty acid composition in lipid fractions lengthwise the mycelium of *Mortierella isabellina* and lipid production by solid state fermentation. Bioresour Technol 100:118–6120
31. Dyal SD, Narine SS (2005) Implications for the use of Mortierella fungi in the industrial production of essential fatty acids. Food Res Int 38:445–467
32. Cheirsilp B, Suwannarat W, Niyomdecha R (2011) Mixed culture of oleaginous yeast *Rhodotorula glutinis* and microalgae *Chlorella vulgaris* for lipid production from industrial wastes and its use as biodiesel feedstock. New Biotechnol 28:362–368
33. Sawangkeaw R, Ngamprasertsith S (2013) A review of lipid-based biomasses as feedstocks for biofuels production. Renew Sust Energy Rev 25:97–108
34. Li Q, Du W, Liu D (2008) Perspectives of microbial oils for biodiesel production. Appl Microbiol Biotechnol 80:749–756
35. Goold H, Beisson F, Pentier G, Li-Beisson Y (2015) Microalgal lipid droplets: composition, diversity, biogenesis and functions. Plant Cell Rep 34(4):545–555
36. de Jaeger L, Verbeek RE, Draaisma RB, Martens DE, Springer J, Eggink G, Wijffels RH (2014) Superior triacylglycerol (TAG) accumulation in starchless mutants of

Scenedesmus obliquus: (I) mutant generation and characterization. Biotechnol Biofuels 7:69

37. Du Z-Y, Benning C (2016) Triacylglycerol accumulation in photosynthetic cells in plants and algae. In: Nakamura Y, LiBeisson Y (eds) Lipids in plant and algae development, vol 86. Springer International Publishing, Cham, pp 179–205
38. Liu B, A. Vieler A, Li C, Jones AD, Benning C (2013) Triacylglycerol profiling of microalgae *Chlamydomonas reinhardtii* and *Nannochloropsis oceanica*. Bioresour Technol 146: 310-316
39. Yao L, Gerde JA, Lee SL, Wang T, Harrata KA (2015) Microalgae lipid characterization. J Agric Food Chem 63(6):1773–1787
40. Shruthi P, Rajeshwari T, Mrunalini BR, Girish V, Girisha ST (2014) Evaluation of oleaginous bacteria for potential biofuel. Int J Curr Microbiol Appl Sci 3(9):47–57
41. Zhang X, Agrawal A, San K-Y (2012) Improving fatty acid production in *Escherichia coli* through the overexpression of malonyl coA—acyl carrier protein transacylase. Biotechnol Prog 28:60–65
42. Gupta A, Barrow CJ, Puri M (2012) Omega-3 biotechnology: Thraustochytrids as a novel source of omega-3 oils. Biotechnol Adv 30:1733–1745
43. Ugalde G, Armenta RE, Kermanshahipour A, Sun Z, Berryman KT, Brooks MS (2018) Improvement of culture conditions for cell biomass and fatty acid production by marine thraustochytrid F24-2. J Appl Phycol 30:329–339
44. Singh A, Ward OP (1997) Microbial production of docosahexaenoic acid (DHA, C22:6). Adv Appl Microbiol 45:271–312
45. Bajpai PK, Bajpai P, Ward OP (1991) Optimization of production of docosahexaenoic acid (DHA) by Thraustochytrium aureum ATCC 34304. J Am Oil Chem Soc 68(7):509–514
46. Napier JA (2007) The production of unusual fatty acids in transgenic plants. Annu Rev Plant Biol 58:295–319
47. Baud S (2018) Seeds as oil factories. Plant Reprod 31:213–235
48. Barthole G, Lepiniec L, Rogowsky PM, Baud S (2012) Controlling lipid accumulation in cereal grains. Plant Sci 185–186:33–39
49. Marchive C, Nikovics K, To A, Lepiniec L, Baud S (2014) Transcriptional regulation of fatty acid production in higher plants: molecular bases and biotechnological outcomes. Eur J Lipid Sci Technol 116:1332–1343
50. Chen G-C, Su H-M, Lin Y-S, Tsou P-Y, Chyan J-H, Chao P-M (2016) A conjugated fatty acid present at high levels in bitter melon seed favorably affects lipid metabolism in hepatocytes by increasing NAD+/NADH ratio and activating PPARα, AMPK and SIRT1 signaling pathway. J Nutr Biochem 33:28–35
51. Bates PD (2016) Understanding the control of acyl flux through the lipid metabolic network of plant oil biosynthesis. Biochim Biophys Acta 1861:1214–1225
52. Nguyen HT, Park H, Koster KL, Cahoon RE, Nguyen HTM, Shanklin J, Clemente TE, Cahoon EB (2015) Redirection of metabolic flux for higher levels of omega-7 monounsaturated fatty acid accumulation in camelina seeds. Plant Biotechnol J 13:38–50
53. Ruiz-Lopez N, Haslam RP, Usher S, Napier JA, Sayanova O (2015) An alternative pathway for the effective production of the omega-3 long-chain polyunsaturates EPA and ETA in transgenic oilseeds. Plant Biotechnol J 13:1264–1275
54. Vanhercke T, Wood CC, Stymne S, Singh SP, Green AG (2013) Metabolic engineering of plant oils and waxes for use as industrial feedstocks. Plant Biotechnol J 11:197–210
55. Beopoulos A, Mrozova Z, Thevenieau F, Le Dall M-T, Hapala I, Papanikolaou S, Chardot T, Nicaud J-M (2008) Control of lipid accumulation in the yeast *Yarrowia lipolytica*. Appl Environ Microbiol 74 (24):7779–7789
56. Liang M-H, Jiang J-G (2013) Advancing oleaginous microorganisms to produce lipid via metabolic engineering technology. Prog Lipid Res 52:395–408
57. Dulermo T, Nicaud J-M (2011) Involvement of G3P shuttle and β-oxidation pathway into the control of TAG synthesis and lipid accumulation in Yarrowia lipolytica. Metab Eng 13:482–491
58. Groenewald M, Boekhout T, Neuvéglise C, Gaillardin C, van Dijck PW, Wyss M (2014) *Yarrowia lipolytica*: safety assessment of an oleaginous yeast with a great industrial potential. Crit Rev Microbiol 40:187–206
59. Tai M, Stephanopoulos G (2013) Engineering the push and pull of lipid biosynthesis in oleaginous yeast *Yarrowia lipolytica* for biofuel production. Metab Eng 15:1–9
60. Miao XL, Wu QY (2004) Bio-oil fuel production from microalgae after heterotrophic growth. Renew Energy 4:41–44

61. O'Brien RD (2008) Fats and oils: formulating and processing for applications, 3rd edn. CRC Press, Boca Raton
62. Papanikolaou S, Aggelis G (2011) Lipids of oleaginous yeasts. Part I: biochemistry of single cell oil production. Eur J Lipid Sci Technol 113:1031–1105
63. Hassan M, Blanc PJ, Pareilleux A, Goma G (1994) Selection of fatty acid auxotrophs from the oleaginous yeast *Cryptococcus curvatus* and production of cocoa-butter equivalents in batch mode. Biotechnol Lett 16:819–824
64. Jin M, Slininger PJ, Dien BS, Waghmode S, Moser BR, Orjuela A, Sousa LDC, Balan V (2015) Microbial lipid based lignocellulosic biorefinery: feasibility and challenges. Trends Biotechnol 33(1):43–54
65. Hussain MS, Rodriguez GM, Gao D, Spagnuolo M, Gambill L, Blenner M (2016) Recent advances in bioengineering of the oleaginous yeast *Yarrowia lipolytica*. AIMS Bioengineering 3(4):493–514
66. Carsanba E, Papanikolaou S, Erten H (2018) Production of oils and fats by oleaginous microorganisms with an emphasis given to the potential of the nonconventional yeast *Yarrowia lipolytica*. Crit Rev Biotechnol 38 (8):1230–1243
67. Ledesma-Amaro R, Nicaud J-M (2016) *Yarrowia lipolytica* as a biotechnological chassis to produce usual and unusual fatty acids. Prog Lipid Res 61:40–50
68. Ratledge C, Cohen Z (2008) Microbial and algal oils: do they have a future for biodiesel or as commodity oils? Lipid Technol 20:155–160
69. Koutinas AA, Chatzifragkou A, Kopsahelis N, Papanikolaou S, Kookos LK (2014) Design and technoeconomic evaluation of microbial oil production as a renewable resource for biodiesel and oleochemical production. Fuel 116:566–577
70. Lee JE, Vadlani PV, Min D (2017) Sustainable production of microbial lipids from lignocellulosic biomass using oleaginous yeast cultures. J Sus Bioenerg Sys 7:36–50
71. Bellou S, Triantaphyllidou I-E, Aggeli D, Elazzazy AM, Baeshen MN, Aggelis G (2016) Microbial oils as food additives: recent approaches for improving microbial oil production and its polyunsaturated fatty acid content. Cur Opin Biotechnol 37:24–35
72. Angerbauer C, Siebenhofer M, Mittelbach M, Guebitz GM (2008) Conversion of sewage sludge into lipids by *Lipomyces starkeyi* for biodiesel production. Bioresour Technol 99:3051–3056
73. Leiva-Candia DE, Pinzi S, Redel-Macias MD, Koutinas A, Webb C, Dorado MP (2014) The potential for agro-industrial waste utilization using oleaginous yeast for the production of biodiesel. Fuel 123:33–42
74. Chiu S-Y, Kao C-Y, Chen T-Y, Chang Y-B, Kou C-M, Lin C-S (2015) Cultivation of microalgal Chlorella for biomass and lipid production using wastewater as nutrient resource. Bioresour Technol 184:179–189
75. Dong T, Knoshaug EP, Pienkos PT, Laurens LML (2016) Lipid recovery from wet oleaginous microbial biomass for biofuel production: a critical review. Appl Energy 177:879–895
76. Ratledge C (1992) Microbial lipids: commercial realities or academic curiosities. In: Kyle D, Ratledge C (eds) Industrial applications of single cell oils. AOCS Press, Champaign, IL, pp 1–15
77. Liu B, Benning C (2013) Lipid metabolism in microalgae distinguishes itself. Curr Opin Biotechnol 24:300–309
78. Lundin H (1950) Fat synthesis by microorganisms and its possible applications in industry. J Inst Brew 56(1):17–28
79. Sitepu I, Garay L, Sestric R, Levin D, Block DE, German J, Boundy-Mills K (2014) Oleaginous yeasts for biodiesel: current and future trends in biology and production. J Biotechnol Adv 32(7):1336–1360
80. Radakovits R, Jinkerson RE, Darzins A, Posewitz MC (2010) Genetic engineering of algae for enhanced biofuel production. Eukaryot Cell 9(4):486–501
81. Athenstaedt K, Daum G (2006) The life cycle of neutral lipids: synthesis, storage and degradation. Cell Mol Life Sci 63:1355–1369
82. Wältermann M, Stöveken T, Steinbüchel A (2007) Key enzymes for biosynthesis of neutral lipid storage compounds in prokaryotes: properties, function and occurrence of wax ester synthases/acyl-CoA: diacylglycerol acyltransferases. Biochimie 89(2):230–242
83. Probst KV, Schulte LR, Durrett TP, Rezac ME, Vadlani PV (2016) Oleaginous yeast: a value-added platform for renewable oils. Crit Rev Biotechnol 36(5):942–955
84. Garay L, Boundy-Mills K, German J (2014) Accumulation of high value lipids in single cell microorganisms: a mechanistic approach and future perspectives. J Agric Food Chem 62 (13):2709–2727
85. Singh A, Nigam PS, Murphy JD (2011) Mechanism and challenges in

commercialization of algal biofuels. Bioresour Technol 102(1):26–34
86. Weete JD (2012) Lipid biochemistry of fungi and other organisms. Springer Science & Business Media, Berlin
87. Kosa M, Ragauskas AJ (2011) Lipids from heterotrophic microbes: advances in metabolism research. Trends Biotechnol 29 (2):53–61
88. Wynn JP, Hamid AA, Li Y, Ratledge C (2001) Biochemical events leading to the diversion of carbon into storage lipids in the oleaginous fungi *Mucor circinelloides* and *Mortierella alpina*. Microbiology 147(10):2857–2864
89. Ratledge C (2002) Regulation of lipid accumulation in oleaginous micro-organisms. Biochem Soc Trans 30:1047–1050
90. Ratledge C (2004) Fatty acid biosynthesis in microorganisms being used for single cell oil production. Biochimie 86(11):807–815
91. Ratledge C, Wynn J (2002) The biochemistry and molecular biology of lipid accumlation in oleaginous microorganisms. Adv Appl Microbiol 51:1–44
92. Zhu Z, Zhang S, Liu H, Shen H, Lin X, Yang F, Zhou YJ, Jin G, Ye M, Zou H (2012) A multi-omic map of the lipid-producing yeast *Rhodosporidium toruloides*. Nat Commun 3:1112
93. Zhang Y, Adams IP, Ratledge C (2007) Malic enzyme: the controlling activity for lipid production? Overexpression of malic enzyme in *Mucor circinelloides* leads to a 2.5-fold increase in lipid accumulation. Microbiology 153(7):2013–2025
94. Rodríguez-Frómeta RA, Gutiérrez A, Torres-Martínez S, Garre V (2013) Malic enzyme activity is not the only bottleneck for lipid accumulation in the oleaginous fungus Mucor circinelloides. Appl Mol Biotechnol 97(7):3063–3072
95. Schweizer E, Hofmann J (2004) Microbial type I fatty acid synthases (FAS): major players in a network of cellular FAS systems. Microbiol Mol Biol Rev 68(3):501–517
96. Lu Y-J, Zhang Y-M, Rock CO (2004) Product diversity and regulation of type II fatty acid synthases. Biochem Cell Biol 82 (1):145–155
97. Vieler A, Brubaker SB, Vick B, Benning C (2012) A lipid droplet protein of Nannochloropsis with functions partially analogous to plant oleosins. Plant Physiol 158 (4):1562–1569
98. Choudhary V, Jacquier N, Schneiter R (2011) The topology of the triacylglycerol synthesizing enzyme Lro1 indicates that neutral lipids can be produced within the luminal compartment of the endoplasmatic reticulum: Implications for the biogenesis of lipid droplets. Commun Integr Biol 4(6):781–784
99. Long AP, Manneschmidt AK, VerBrugge B, Dortch MR, Minkin SC, Prater KE, Biggerstaff JP, Dunlap JR, Dalhaimer P (2012) Lipid droplet de novo formation and fission are linked to the cell cycle in fission yeast. Traffic 13(5):705–714
100. Wältermann M, Hinz A, Robenek H, Troyer D, Reichelt R, Malkus U, Galla HJ, Kalscheuer R, Stöveken T, Von Landenberg P (2005) Mechanism of lipid-body formation in prokaryotes: how bacteria fatten up. Mol Microbial 55(3):750-763
101. Murphy DJ (2012) The dynamic roles of intracellular lipid droplets: from archaea to mammals. Protoplasma 249(3):541–585
102. Jacquemet A, Barbeau J, Lemiègre L, Benvegnu T (2009) Archaeal tetraether bipolar lipids: structures, functions and applications. Biochimie 91(6):711–717
103. Legat A, Gruber C, Zangger K, Wanner G, Stan-Lotter H (2010) Identification of polyhydroxyalkanoates in Halococcus and other haloarchaeal species. Appl Microbiol Biotechnol 87(3):1119–1127
104. Ding Y, Yang L, Zhang S, Wang Y, Du Y, Pu J, Peng G, Chen Y, Zhang H, Yu J (2012) Identification of the major functional proteins of prokaryotic lipid droplets. J Lipid Res 53(3):399–411
105. Checa A, Bedia C, Jaumot J (2015) Lipidomic data analysis: Tutorial, practical guidelines and applications. Anal Chim Acta 885:1–16
106. Jurowski K, Kochan K, Walczak J, Barańska M, Piekoszewski W, Buszewski B (2017) Analytical techniques in lipidomics: state of the art. Crit Rev Anal Chem 47:418–437
107. Fuchs B (2012) Analysis of phospolipids and glycolipids by thin-layer chromatography--matrix-assisted laser desorption and ionization mass spectrometry. J Chromatogr A 1259:62–73
108. Li Y, Wang J, Zhan L, Wleklinski M, Wang J, Xiong C, Liu H, Zhou Y, Nie Z (2016) The bridge between thin layer chromatography-mass spectrometry and high-performance liquid chromatography-mass spectrometry :The realization of liquid thin layer chromatography-mass spectrometry. J Chromatogr A 1460:181–189
109. Hsu F-F (2018) Mass spectrometry-based shotgun lipidomics – a critical review from

the technical point of view. Anal Bioanal Chem 410:6387–6409

110. Zheng X, Smith RD, Baker ES (2018) Recent advances in lipid separations and structural elucidation using mass spectrometry combined with ion mobility spectrometry, ion-molecule reactions and fragmentation approaches. Curr Opin Chem Biol 42:111–118

111. Rabiei Z, Bigdeli MR, Rasoulian B, Ghassempour A, Mirzajani F (2012) The Neuroprotection Effect of Pretreatment with Olive Leaf Extract on Brain Lipidomics in Rat Stroke Model. Phytomedicine 19:940–946

112. Dillon JT, Aponte JC, Tsai YJ, Huang Y (2012) Thin Layer Chromatography in the Separation of Unsaturated Organic Compounds Using Silver-Thiolate Chromatographic Material. J Chromatogr A 1251:240–243

113. Plekhanov AY (1999) Rapid staining of lipids on thin-layer chromatograms with amido black 10B and other water-soluble stains. Anal Biochem 271(2):186–187

114. Fuchs B, Suss R, Teuber K, Eibisch M, Schiller J (2011) Lipid Analysis by Thin-Layer chromatography - A Review of the Current State. J Chromatogr A 1218:2754–2774

115. Domınguez A, Jarne C, Cebolla VL, Galban J, Saviron M, Orduna J, Membrado L, Lapieza M-P, Romero E, Vicente IS, de Marcos S, Garriga R (2015) A Hyphenated Technique Based on High-Performance Thin Layer Chromatography. Chromatography 2:167–187

116. Tan G, Tian Y, Addy M, Cheng Y, Xie Q, Zhang B, Liu Y, Chen P, Ruan P (2017) Structural analysis of phosphatidylcholine using a thin layer chromatography-based method. Eur J Lipid Sci Technol 119:1600282. https://doi.org/10.1002/ejlt.201600282

117. Zarzycki PK, MM S˛c, Zarzycka MB, Bartoszuk MA, Wodarczyk E, Baran MJ (2011) Temperature-Controlled Micro-TLC: A Versatile Green Chemistry and Fast Analytical Tool for Separation and Preliminary Screening of Steroids Fraction from Biological and Environmental Samples. J Steroid Biochem Mol Biol 127:418–427

118. Kofeler HC, Fauland A, Rechberger GN, Trotzmuller M (2012) Mass Spectrometry Based Lipidomics: An Overview of Technological Platforms. Meta 2:19–38

119. Dodds ED, McCoy MR, Rea LD, Kennish JM (2005) Gas Chromatographic Quantification of Fatty Acid Methyl Esters: Flame Ionization Detection vs. Electron Impact Mass Spectrometry. Lipids 40:419–428

120. Roach JA, Yurawecz MP, Kramer JK, Mossoba MM, Eulitz K, Ku Y (2000) Gas Chromatography-High Resolution Selected-Ion Mass Spectrometric Identification of Trace 21:0 and 20:2 Fatty Acids Eluting with Conjugated Linoleic Acid Isomers. Lipids 35:797–802

121. Bamba T, Lee JW, Matsubara A, Fukusaki E (2012) Metabolic Profiling of Lipids by Supercritical Fluid Chromatography/Mass Spectrometry. J Chromatogr A 1250:212–219

122. Pati S, Nie B, Arnold RD, Cummings BS (2016) Extraction, Chromatographic and Mass Spectrometric Methods for Lipid Analysis. Biomed Chromatogr 30:695–709

123. Uchikata T, Matsubara A, Fukusaki E, Banba T (2012) High-Throughput Phospholipid Profiling System Based on Supercritical Fluid Extraction– Supercritical Fluid Chromatography/Mass Spectrometry for Dried Plasma Spot Analysis. J Chromatogr A 1250:69–75

124. Lee JW, Nishiumi S, Yoshida M, Fukusaki E, Bamba T (2013) Simultaneous Profiling of Polar Lipids by Supercritical Fluid Chromatography/Tandem Mass Spectrometry with Methylation. J Chromatogr A 1279:98–107

125. Gaspar A, Englmann M, Fekete A, Harir M, Schmitt-Kopplin P (2008) Trends in CE-MS 2005–2006. Electrophoresis 29:66–79

126. Otarola J, Lista AG, Band BF, Garrido M (2015) Capillary electrophoresis to determine entrapment efficiency of a nanostructured lipid carrier loaded with piroxicam. J Pharm Anal 5:70–73

127. Otieno AC, Mwongela SM (2008) Capillary Electrophoresis-Based Methods for the Determination of Lipids-A Review. Anal Chim Acta 624:163–174

128. Xiao Y, Mei J, He X, Cheng W (2006) Fractionation and High Performance Capillary Electrophoretic Analysis of Phospholipids. Chinese J Chromatogr 24:30–34

129. Wu H, Volponi JV, Oliver AE, Parikh AN, Simmons BA, Singh S (2011) In vivo lipidomics using single-cell Raman spectroscopy. Proc Natl Acad Sci U S A 108:3809

130. Samek O, Janas A, Pilat Z, Zemanek P, Nedbal L, Triska J, Kotas P, Trtilek M (2010) Raman Microspectroscopy of Individual Algal Cells: Sensing Unsaturation of Storage Lipids In Vivo. Sensors 10:8635–8651

131. Czamara K, Majzner K, Pacia Z, Kochan K, Kaczora A, Baranska M (2015) Raman

spectroscopy of lipids: a review. J Raman Spectrosc 46:4–20
132. Kucuk Baloglu F, Garip S, Heise S, Brockmann G, Severcan F (2015) FTIR Imaging of Structural Changes in Visceral and Subcutaneous Adiposity and Brown to White Adipocyte Trans-differentiation. Analyst 140:2205–2214
133. Kochan K, Maslak E, Krafft C, Kostogrys R, Chlopicki S, Baranska M (2015) Raman Spectroscopy Analysis of Lipid Droplets Content, Distribution and Saturation Level in Non-Alcoholic Fatty Liver Disease in Mice. J Biophotonics 8:597–609
134. Abramczyk H, Surmacki J, Kopec M, Olejnik AK, Lubecka-Pietruszewska K, Fabianowska-Majewska K (2015) The Role of Lipid Droplets and Adipocytes in Cancer. Raman Imaging of Cell Cultures: MCF10A, MCF7, and MDA-MB-231 Compared to Adipocytes in Cancerous Human Breast Tissue. Analyst 140:2224–2235
135. Wrobel TP, Mateuszuk L, Kostogrys RB, Chlopicki S, Baranska M (2013) Quantification of Plaque Area and Characterization of Plaque Biochemical Composition with Atherosclerosis Progression in ApoE/LDLR¡/¡ mice by FT-IR Imaging. Analyst 138:6645–6652
136. Quinn PJ, Rainteau D, Wolf C (2009) Lipidomics of the Red Cell in Diagnosis of Human Disorders. Methods Mol Biol 579:127–159
137. Tan HH, Makino A, Sudech K, Greimel P, Kobayashi T (2012) Spectroscopic Evidence for the Unusual Stereochemical Configuration of an Endosome-Specific Lipid. Angew Chem Int Ed Engl 51:533–535
138. Zhendre V, Grelard A, Garnier-Lhomme M, Buchoux S, Larijani B, Dufourc EJ (2011) Key Role of Polyphosphoinositides in Dynamics of Fusogenic Nuclear Membrane Vesicles. PLoS One 6:e23859
139. Mirbahai L, Wilson M, Shaw CS, McConville C, Malcomson RD, Kauppinen RA, Peet AC (2012) Lipid Biomarkers of Glioma Cell Growth Arrest and Cell Death Detected by ^{1}H Magic Angle Spinning MRS. NMR Biomed 25:1253–1262
140. Ishida Y, Madonna AJ, Rees JC, Meetani MA, Voorhees KJ (2002) Rapid analysis of intact phospholipids from whole bacterial cells by matrix-assisted laser desorption/ionization mass spectrometry combined with on-probe sample pretreatment. Rapid Commun Mass Spectrom 16(19):1877–1882
141. Waltermann M, Luftmann H, Baumeister D, Kalscheuer R, Steinbuchel A (2000) Rhodococcus opacus strain PD630 as a new source of high-value single-cell oil? Isolation and characterization of triacylglycerols and other storage lipids. Microbiology-Uk 146:1143–1149
142. Nichols PD, Guckert JB, White DC (1986) Determination of Monounsaturated Fatty-Acid Double-Bond Position and Geometry for Microbial Monocultures and Complex Consortia by Capillary GC-MS of Their Dimethyl Disulfide Adducts. J Microbiol Meth 5(1):49–55
143. Sakuragi Y, Zybailov B, Shen GZ, Jones AD, Chitnis PR, van der Est A, Bittl R, Zech S, Stehlik D, Golbeck JH, Bryant DA (2002) Insertional inactivation of the menG gene, encoding 2-phytyl-1,4-naphthoquinone methyltransferase of Synechocystis sp PCC 6803, results in the incorporation of 2-phytyl-1,4-naphthoquinone into the A(1) site and alteration of the equilibrium constant between A(1) and F-x in photosystem I. Biochemistry 41(1):394–405
144. Frigaard, NU, Maresca JA, Yunker CE, Jones AD, Bryant (2004). Genetic manipulation of carotenoid biosynthesis in the green sulfur bacterium *Chlorobium tepidum*. J Bacteriol 186(16):5210-5220
145. Han X, Gross RW (2003) Global analyses of cellular lipidomes directly from crude extracts of biological samples by ESI mass spectrometry: a bridge to lipidomics. J Lipid Res 44 (6):1071–1079
146. Welti R, Shah J, Li W, Li M, Chen J, Burke JJ, Fauconnier M-L, Chapman K, Chye ML, Wang X (2007) Plant lipidomics: Discerning biological function by profiling plant complex lipids using mass spectrometry. Front Biosci 12(1):2494–2506
147. Holguin FO, Schaub T (2013) Characterization of microalgal lipid feedstock by direct-infusion FT-ICR mass spectrometry. Algal Research-Biomass Biofuels and Bioproducts 2(1):43–50
148. Yokochi T, Honda D, Higashihara T, Nakahara T (1998) Optimisation of ocosahexaenoic acid production by *Schizochytrium limacinum* SR21. Appl Microbiol Biotechnol 49:72–76

Chapter 2

Laboratory Screening Protocol to Identify Novel Oleaginous Yeasts

Irnayuli R. Sitepu, Antonio L. Garay, Tomas Cajka, Oliver Fiehn, and Kyria L. Boundy-Mills

Abstract

Oleaginous microbes, which contain over 20% intracellular lipid, predominantly triacylglycerols (TG), by dry weight, have been discovered to have high oil content by many different protocols, ranging from simple staining to more complex chromatographic methods. In our laboratory, a methodical process was implemented to identify high oil yeasts, designed to minimize labor while optimizing success in identifying high oil strains among thousands of candidates. First, criteria were developed to select candidate yeast strains for analysis. These included observation of buoyancy of the yeast cell mass in 20% glycerol, and phylogenetic placement near known oleaginous species. A low-labor, semiquantitative Nile red staining protocol was implemented to screen numerous yeast cultures for high oil content in 96-well plates. Then, promising candidates were selected for more quantitative analysis. A more labor-intensive and quantitative gravimetric assay was implemented that gave consistent values for intracellular oil content for a broad range of yeast species. Finally, an LC-MS protocol was utilized to quantify and identify yeast triacylglycerols. This progressive approach was successful in identifying 30 new oleaginous yeast species, out of over 1000 species represented in the Phaff Yeast Culture Collection.

Key words Oleaginous microbes, Yeast, Triacylglycerol, Nile red, Gravimetric, Liquid chromatography–mass spectrometry, Lipidomics

1 Introduction

Oleaginous microbes are defined as those that accumulate at least 20% intracellular lipid by dry weight [1]. Some bacteria, fungi and algae are oleaginous. Of over 1600 species of yeasts (single-celled fungi), over 90 are oleaginous [2, 3], with the lipid content ranging from just over 20% up to 65%, depending on the species and growth conditions. These yeasts have been the subject of study for many decades both to understand the biology of lipid synthesis and storage, and to develop technologies to produce renewable, sustainable oleochemicals such as biodiesel [4–6].

Venkatesh Balan (ed.), *Microbial Lipid Production: Methods and Protocols*, Methods in Molecular Biology, vol. 1995, https://doi.org/10.1007/978-1-4939-9484-7_2,

Many studies have reported the results of screening yeasts or other microorganisms for ability to accumulate high intracellular oil, which is in the form of triacylglycerols (TG). Various strategies have been employed, with varying success rates. The two key elements of a screening strategy include selection of candidate strains, and selection of an appropriate assay or series of assays. Screening strategies that have employed a rational selection of candidate microbes had a higher rate of success than studies that used randomly selected microbes [5]. Studies that used randomly selected yeasts had a 3–10% hit rate. In contrast, more rational selection criteria resulted in much higher success rates. For example, Oguri et al. [7] discovered that relatives of known oleaginous species *Lipomyces starkeyi* are also high in oil. Gujjari et al. [8] discovered new oleaginous species by examining yeasts deposited in the ATCC patent repository for applications related to oil production.

Of the over 90 known oleaginous yeast species, 30 were discovered to be oleaginous in the Boundy-Mills laboratory since 2012, utilizing yeasts from the Phaff Yeast Culture Collection. Because the collection contains over 7500 yeast strains of over 1000 species, a stepwise strategy was implemented to determine which strains are oleaginous, starting with selection of a reasonable number of yeast strains to examine.

1. *Strains were selected for analysis based on numerous criteria.* These include buoyancy of the cell mass in 20% glycerol (detailed below), belonging to taxonomic clades containing known oleaginous species, or belonging to taxonomic clades not previously studied for oil content. The buoyancy observation was fortuitous. Cryopreservation of yeasts involves scraping a mass of cells off an agar plate and placing it in a tube of 20% glycerol. Usually, the cell mass sinks to the bottom of the tube. Rarely (about 5% of cases in our studies), the cell mass floats. We presumed that buoyancy may be due to high oil content, which turned out to be true: a large proportion of buoyant yeasts turned out to be high in oil.
2. *A simplified, high-throughput Nile red assay was developed*, to allow for relatively fast evaluation of lipid content. Below are detailed protocols for semiquantitative determination of intracellular lipids in microplate scale, and microscopic visualization of intracellular lipids in yeast.
3. *A gravimetric assay was developed* that resulted in consistent determinations of lipid mass as a percent of cell dry weight, applicable to a broad variety of yeast species including both ascomycetes and basidiomycetes.

4. *LC–MS chemical analysis of lipids has been adapted for yeast lipids.* Over the past decades, several approaches have been used for chemical analysis of lipids. Previous studies focused mainly on the determination of primary constituents of these compounds after hydrolysis by using gas chromatography with flame ionization detection (GC–FID) or more recently mass spectrometry (GC–MS). However, recent advances in liquid chromatography–mass spectrometry (LC–MS) instrumentation have permitted characterization of small molecules (including lipids) by analyzing the intact molecules without a need for hydrolysis and/or derivatization [9]. Furthermore, by utilizing tandem mass spectrometry (MS/MS), (1) the lipid class head group, (2) the lengths of carbon chains, and (3) the degree of unsaturation of fatty-acid components of these acyl groups are annotated with support of in silico MS/MS libraries [10].

2 Materials

2.1 Buoyancy Observation

1. Yeast grown on an agar plate of a high carbon, low nitrogen media such as potato dextrose agar (*see* **Note 1**).
2. 20% (w/v) glycerol, technical grade or better.
3. Small vials such as cryovials.
4. Inoculating loops or needles.

2.2 Nile Red Assays

2.2.1 Qualitative Microscopic Visualization of Intracellular Lipids in Yeast

Equipment

1. Calibrated pipettes 1–10 μL.
2. Amber glass vials (2 and 50 mL).
3. Aluminum foil.
4. Glass slides and cover slides.
5. Fluorescence microscope, equipped with a camera, a workstation and the appropriate software capable to visualize samples and acquire pictures in bright-field and fluorescence modes, such as a Carl Zeiss Axio Imager.A1 microscope for polarized light (Göttingen, Germany) with 50× dry and 100× oil immersion objective lenses, and a Texas Red fluorescence filter.
6. Fluorescence lamp, compatible with the microscope, capable of reaching 530/25 nm excitation and 590/35 nm emission, with a mercury lamp such as model LQ-HXP 120-Z (LEJ Leistungselektronik, Jena, Germany).
7. 10 μL calibrated pipette.

Reagents

1. Nile red (9-diethylamino-5-benzo[α]phenoxazinone) such as Acros Organics Cat. # 7385-67-3, NJ, USA.
2. Acetone.

3. 70% ethanol.
4. Sterile deionized water.
5. Immersion oil.

2.2.2 Quantitative Determination of Intracellular Lipids in Microplate Scale

Equipment

1. Biosafety cabinet.
2. Autoclave.
3. Rotary shaker incubator.
4. Roller drum.
5. Analytical balance.
6. Microplate reader with fluorescence capability such as Biotek Synergy 2 (Vermont, USA).
7. Aluminum foil.
8. Black flat bottom 96-well microplate such as Cat. # 3370, Costar Incorporated, New York, USA.
9. Viewseal™ optically transparent plate sealing film (Cat. # EK-46070, E&K Scientific, CA, USA).
10. 250 mL Erlenmeyer flasks.
11. 10, 200, and 1000 μL single-channel pipettors, 200 μL multichannel pipettor, and appropriate pipette tips.
12. Disposable reservoir.
13. Paper towels.
14. 15 mL conical tubes.
15. 5 mL amber vials and caps.
16. Biohazard bags.
17. Carbon rich, nitrogen poor growth medium such as potato dextrose broth (PDB).

Reagents (All Chemicals Should Be of Analytical Grade)

1. 100 mL autoclaved Medium A (glucose 30 g/L, yeast extract 1.5 g/L, NH_4Cl 0.5 g/L, KH_2PO_4 7.0 g/L, $Na_2HPO_4{\cdot}12H_2O$ 5.0 g/L, $MgSO_4{\cdot}7H_2O$ 1.5 g/L, $FeCl_3{\cdot}6H_2O$ 0.08 g/L, $ZnSO_4{\cdot}7H_2O$ 0.01 g/L, $CaCl_2{\cdot}2H_2O$ 0.1 g/L, $MnSO_4{\cdot}5H_2O$ 0.1 mg/L, $CuSO_4{\cdot}5H_2O$ 0.1 mg/L, $Co(NO_3)_2{\cdot}6H_2O$ 0.1 mg/L; pH 5.5 [11].
2. DMSO (dimethyl sulfoxide, Cat. # BP231-100, Fisher Chemicals, NJ, USA).
3. Nile red (9-diethylamino-5-benzo[α]phenoxazinone) such as Acros Organics Cat. # 7385-67-3, NJ, USA.
4. Acetone.
5. Deionized sterile water.

2.3 Gravimetric Assay

2.3.1 Equipment

1. Calibrated pipettors 10–200 μL and 100–1000 μL.
2. Polyethylene pipette tips.
3. 2 mL polyethylene screw-cap tubes.
4. 0.5 mm Zirconia Beads (Cat. # 11079105z, Biospec Products Inc., OK, USA).
5. 0.4 cm glass beads.
6. Bead beater MP Bio Fast Prep-24 homogenizer (MP Biomedicals, OH, USA).
7. Centrifuge.
8. 8 mL glass tubes (15 mL) with screw caps with Teflon septa.
9. 1.5 mL amber glass vials with screw caps.
10. Rotary vacuum concentrator (preferred). A nitrogen flushing system can be used instead.
11. High-precision analytical scale (ideally designed to weigh 0.01 mg).
12. Digital bottle-top glass dispenser for high precision solvent transfer.
13. Modified 20 mL syringe with a stainless-steel mesh fit to the bottom of the syringe to capture the beads.
14. 15 mL conical tube with visible scale.
15. 18 cm disposable Pasteur pipette and rubber bulb.
16. Ice and container.

2.3.2 Reagents (All Chemicals Should Be of Analytical Grade)

1. Solvents: Chloroform, methanol.
2. Deionized sterile water.
3. 0.9% sodium chloride (NaCl) in water.

2.4 LC-MS Assay

2.4.1 Equipment

1. Calibrated pipettors 10–200 μL and 100–1000 μL and appropriate tips.
2. Vortex.
3. Agilent 1290 Infinity LC system (Agilent Technologies, Santa Clara, CA, USA) with a pump, a column oven, and an autosampler.
4. Agilent 6550 iFunnel QTOFMS system (Agilent) with an electrospray (ESI) ion source operated in positive ion polarity (*see* **Note 2**).

2.4.2 Reagents

1. LC-MS-grade solvents: water, acetonitrile, isopropanol.
2. Mobile-phase modifiers: formic acid, ammonium formate.
3. Solvent for sample dilution: methanol.

2.4.3 Supplies

1. 1.5 mL Eppendorf tubes, uncolored.
2. Tips for organic solvents such as methanol.
3. Glass amber vial (2 mL volume).
4. Glass inserts for 2 mL standard opening vial (300 μL).
5. Screw caps for vials.
6. Acquity UPLC CSH C18 column (100 × 2.1 mm; 1.7 μm) (Waters, Milford, MA, USA).
7. Acquity UPLC CSH C18 VanGuard precolumn (5 × 2.1 mm; 1.7 μm) (Waters).

3 Methods

3.1 Buoyancy Observation

1. Inoculate yeasts on agar plates of a high carbon, low nitrogen medium such as potato dextrose agar (*see* **Note 1**).
2. Incubate aerobically at the appropriate temperature for that yeast species (often room temperature or 30 °C) for 5–10 days. Allow sufficient growth time: oleaginous yeasts accumulate oil in late exponential and stationary phase, after nitrogen is depleted from the medium.
3. Using a sterile inoculating loop or needle, aseptically scrape roughly 100 μL of yeast off the agar plate.
4. Gently place the yeast biomass in a tube of 20% glycerol. Do not disperse the cells. Observe whether the cell mass floats or sinks (Fig. 1). Buoyancy indicates the yeast may be of high oil content. False positives can result if the yeast colony is of a lacy texture, thus buoyant due to trapped air.

3.1.1 Qualitative Microscopic Visualization of Intracellular Lipids in Yeast

1. Use lab coat, nitrile glove and goggles at all times during this procedure. Prepare a Nile red stock solution by weighing 1 mg of powdered Nile red and diluting it to 20 mL with acetone in a 50-mL amber glass bottle (concentration: 50 μg/mL), vortex until all Nile red is dissolved. Wrap with foil and store in the refrigerator.
2. Prepare a 1:10 aliquot by transferring 100 μL of the stock solution described above into a 2 mL amber glass vial and adding 900 μL of acetone. (Concentration: 5 μg/mL.) Use the aliquot for visualization under the microscope. Store unused solution in the refrigerator (*see* **Note 3**).
3. Turn on microscope, microscope workstation and fluorescent lamp. Make sure instrument and workstation are connected.
4. Wipe glass slide with a Kimwipes moistened with 70% ethanol. Label slide on the edge.
5. Swirl culture gently to resuspend cells.

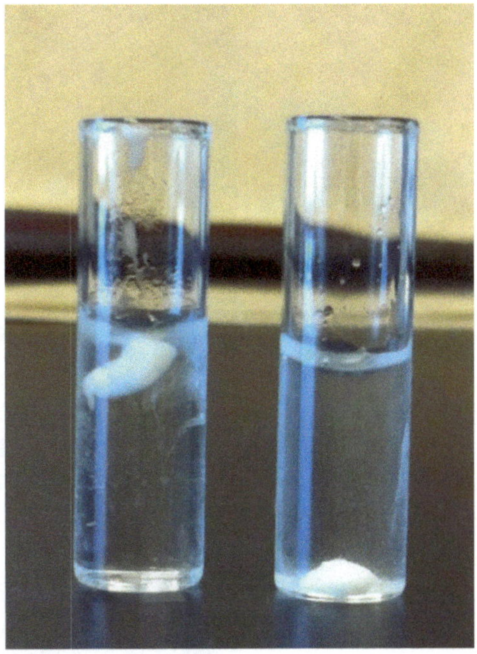

Fig. 1 Buoyant yeast cell mass of an oleaginous Lipomyces yeast strain (left) vs. nonbuoyant yeast cell mass of a nonoleaginous Kluyveromyces yeast strain (right) in 20% glycerol

6. Transfer 5 μL of culture into the center of the slide (*see* **Note 4**).
7. Add 10 μL of a fresh Nile red solution (*see* **Notes 5** and **6**).
8. Place slide cover on top of slide by having one side of the cover touch the edge of the liquid on the slide first, and then bringing it down slowly.
9. Visualize under the microscope first at 40× in bright-field mode. Adjust light, focus, filters and number of cells appropriately. Adjust software parameters and take a snapshot, with scale bar displayed.
10. Switch filter turret to the Texas Red fluorescent filter. Ensure the images have the same focus and location as in bright-field mode. Adjust if necessary.
11. Turn on fluorescent light and immediately take capture or three images (*see* **Note** 7).
12. Add scale bars to the pictures taken in both modes.
13. Remove the 40× objective and carefully add one drop of immersion oil on top of the slide cover, without moving the slide.
14. Visualize under the microscope at 100× in bright-field mode. Repeat **steps 6–9**. Figure 2 illustrates oleaginous yeast micrographs.
15. Discard all samples and supplies containing Nile red in biohazard waste.

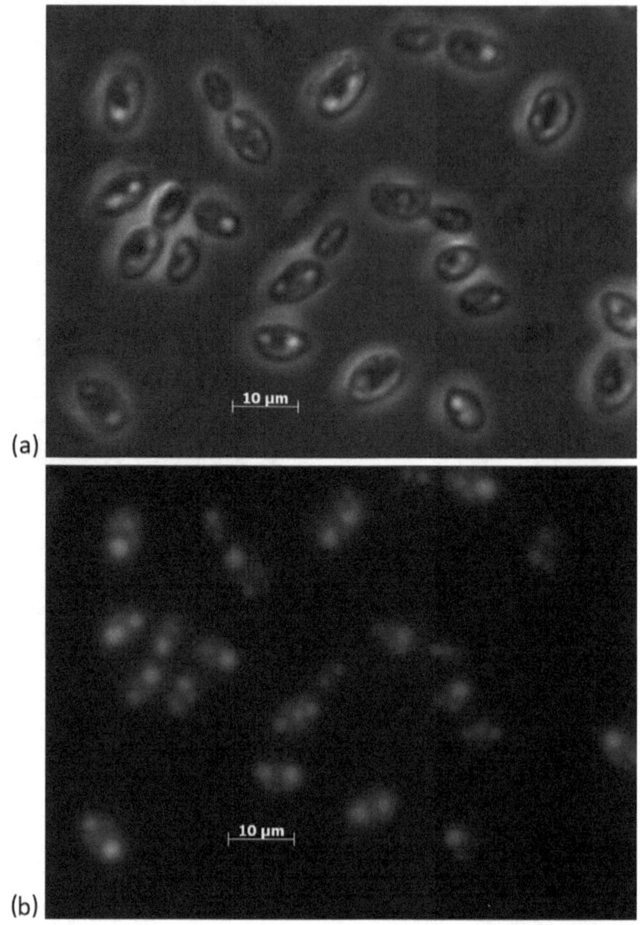

Fig. 2 Intracellular oil within cells of oleaginous yeast *Rhodotorula babjevae* UCDFST 04-877 cells stained with Nile Red, taken using a Carl Zeiss Axio Imager.A1 light microscope (Göttingen, Germany), under 100× magnification using immersion oil with bright field (**a**) and fluorescence (**b**)

3.1.2 Quantitative Determination of Intracellular Lipids in Microplate Scale

Preparing Yeast Culture Using Aseptic Technique [12]

1. Revive yeasts from −80 °C cryopreserved stocks by streaking on a PDA (potato dextrose agar) plate and incubate at room temperature for about 3–5 days (*see* **Note 8**).
2. Prepare seed culture by inoculating an 8 mL Potato Dextrose Broth (PDB) medium in an 18-mL test tube with one loopful fresh yeast colony and incubate it in the roller drum for 24 h.
3. Prepare experimental culture in 100 mL Medium A in a 250 mL Erlenmeyer flask capped with aluminum foil.
4. Inoculate 0.5 mL seed culture yeast to the flask and incubate the culture in a rotary shaker at 200 rpm at 30 °C for 5–7 days (*see* **Note 1**).
5. Harvest culture at stationary phase and adjust cell density using fresh Medium A to $OD_{600} = 1.0$ (*see* **Note 9**).

Nile Red Assay

1. Transfer 250 μL of this culture of $OD_{600} = 1.0$ to the 96-well black plate, each in quadruplicate wells.
2. Gently seal the plate with Viewseal™.
3. Read $OD_{600\ nm}$ using a 96-well microplate reader (Biotek Synergy 2, Vermont, USA).
4. Set microplate reader program to fluorescence at 530/25 nm excitation, at 590/35 nm emission and measure kinetically every 60 s for 20 min. The optic position is set from bottom to top (*see* **Note 10**).
5. Prepare DMSO–MedA (1:1, v/v) immediately before transferring and mix it by inverting gently. Open the seal and transfer 25 μL DMSO–MedA (1:1, v/v) to each well using multichannel pipettor and mix by pipetting up and down.
6. Prepare 0.05 mg/mL Nile red in acetone and dissolve it well by vortexing. Transfer 25 μL of solution to each well using a multichannel pipettor and mix by pipetting up and down (*see* **Note 11**).
7. Seal the plate loosely and perform kinetic fluorescence reading.
8. Plates, tips, seal, tissue paper, etc. that are in contact with the Nile red must be discarded as hazardous chemical waste.
9. For data analysis, select the highest fluorescence intensity (rfu value) regardless of the time when the Emax value is reached.

3.2 Gravimetric Assay

1. Weigh 20 mg of dry yeast cells ± 0.5 mg inside a labeled 2 mL screw-cap tube (*see* **Note 12**).
2. Attach the bottle top dispenser on the bottle of Folch reagent, set at 1.5 mL. Prime the dispenser, then add 1.5 mL of Folch reagent to each vial (*see* **Note 13**).
3. Add 0.5 mL of 0.5 mm zirconia beads followed by two 0.4 cm glass beads (*see* **Note 14**).
4. Cap the tubes and place them on ice in an ice bucket.
5. Homogenize the cells using a bead beater with five cycles of 30 s each, at 6.0 m/s (*see* **Note 15**), chilling on ice between each cycle.
6. Using the modified syringe, transfer the sample to a 15-mL conical tube, and wash the vial with 1.5 mL. Repeat three times to give a total volume of 6 mL of the entire Folch reagent.
7. Add 1.2 mL of 0.9% sodium chloride to the tube. Vortex gently for sufficient separation of phases.
8. Centrifuge the tubes at 4000 rpm (3220 × *g*) for 2 min. Three phases will be formed, the top one is the water, the middle is the cell pellet and the bottom is the chloroform layer containing the lipid.
9. Visually estimate and record the volume of the chloroform layer.

10. Carefully transfer as much of the chloroform phase as possible using a Pasteur pipette with a pipette filler into a new glass tube with screw cap and Teflon septum, without disturbing the pelleted cell debris layer in the middle phase (*see* **Note 16**).
11. Pipet 1.5 mL of the chloroform extract into a tared amber vial without screw cap.
12. Remove the solvent using a rotary vacuum concentrator.
13. Repeat **steps 10–12** until lipids from a total of 3 mL of chloroform phase are collected in the amber vial.
14. Stop solvent evaporation when the vial plus lipid weight reaches a constant weight. Record the weight and calculate the intracellular lipid fraction (expressed as a percent of the dry cell weight) as follows:

$$\%_{\mathrm{IL}} = \left[\frac{(m_{\mathrm{F}} - m_{\mathrm{t}})}{3} \times V_{\mathrm{C}}\right] \times 100$$

where

m_{F} = final mass of the vial containing the intracellular lipid residue.

m_{t} = mass of the tared vial (preweighed).

m_{s} = mass of the dry cells in grams.

$\%_{\mathrm{IL}}$ = percent of intracellular lipid.

V_{C} = total volume of the chloroform layer.

15. Store vials containing intracellular lipid at −80 °C until further use such as TG analysis (*see* Subheading 3.3).

3.3 LC–MS Assay

3.3.1 Sample Preparation

1. Prepare isolated lipids solution by adding methanol to the TG sample from Subheading 3.2, **step 15** to a concentration of 10 μg/mL solution in methanol.
2. Transfer an aliquot to a glass amber vial with a glass insert.
3. Cap the vial.
4. Perform LC-ESI(+)-QTOFMS analysis (*see* **Note 17**).

3.3.2 LC–MS Analysis

1. Prepare mobile phase (A): 60:40 (v/v) acetonitrile–water with 10 mM ammonium formate and 0.1% formic acid (*see* **Note 18**).
2. Prepare mobile phase (B): 90:10 (v/v) isopropanol–acetonitrile with 10 mM ammonium formate and 0.1% formic acid (*see* **Note 18**).
3. Needle wash solvent: isopropanol.
4. Tune the instrument according to manufacturer's recommendations.
5. Purge the LC system with mobile phases A and B, and needle wash solvent.

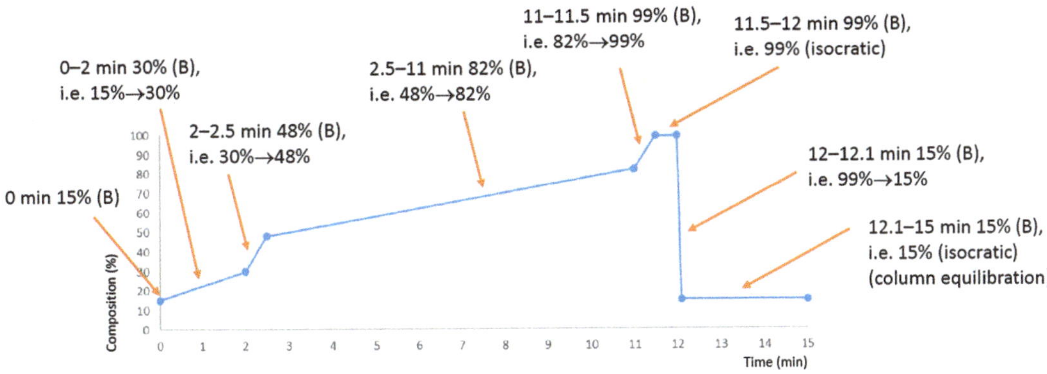

Fig. 3 Diagram of the LC gradient used for separation of lipids

6. Check the backpressure of the LC column (500–600 bar at 15% of mobile phase B).
7. Apply LC gradient for lipid separation (Fig. 3): 0 min 15% (B); 0–2 min 30% (B); 2–2.5 min 48% (B); 2.5–11 min 82% (B); 11–11.5 min 99% (B); 11.5–12 min 99% (B); 12–12.1 min 15% (B); 12.1–15 min 15% (B).
8. Set column flow and temperature: 0.6 mL/min; 65 °C.
9. Set injection volume: 3 μL.
10. Control sample temperature (autosampler): 4 °C.
11. ESI(+)-MS conditions for "discovery mode" (*see* **Note 19**): capillary voltage, +3.5 kV; nozzle voltage, +1 kV; gas temperature, 200 °C; drying gas (nitrogen), 14 L/min; nebulizer gas (nitrogen), 35 psi; sheath gas temperature, 350 °C; sheath gas flow (nitrogen), 11 L/min; MS^1 mass range, m/z 100–1700; MS/MS mass range, m/z 100–1700; MS^1, 10 spectra/s, 100 ms; MS/MS, 13 spectra/s, 77 ms; total cycle time, 0.5 s; collision energy, +20 eV; number of precursor ion per cycle, 4; mass range for selection of precursor ions, m/z 700–1100; isolation width, narrow (1.3 m/z); precursor threshold, 2000 counts; active exclusion, excluded after 3 spectra, release after 0.07 min.
12. ESI(+)-MS conditions for "routine mode" (*see* **Note 20**): capillary voltage, +3.5 kV; nozzle voltage, +1 kV; gas temperature, 200 °C; drying gas (nitrogen), 14 L/min; nebulizer gas (nitrogen), 35 psi; sheath gas temperature, 350 °C; sheath gas flow (nitrogen), 11 L/min; MS^1 mass range, m/z 100–1700; MS^1, 2 spectra/s, 500 ms; total cycle time, 0.5 s.
13. Use *Injector Cleaning* option in MassHunter Data Acquisition system to avoid carryover between sample injections: Time 1, 0.1 min (bypass); Time 2, 11.6 min (mainpass/bypass); Time 3, 13.0 min (mainpass/bypass) (*see* **Note 21**).

3.3.3 Data Processing and Reporting

1. Based on our previous analyses, the following TG species are typically detected in analyzed yeast lipid extracts (Table 1). TG lipids are detected as ammonium $[M+NH_4]^+$ and sodium $[M+Na]^+$ adducts. Use this information as starting point for further investigation.
2. Create targeted data processing method containing information on lipid name and adduct type, retention time (t_R), and m/z for each lipid species in Agilent MassHunter Quantitative Analysis software or in other software for quantitative analysis.
3. Correct retention times for particular lipids if needed and process the raw data set.
4. Export peak heights (h) for particular lipid species.
5. Sum intensities (peak heights) of $[M+NH_4]^+$ and $[M+Na]^+$ adducts for each lipid species (e.g., TG 54:1 (total) = TG 54:1 $[M+NH_4]^+$ + TG 54:1 $[M+Na]^+$).
6. Calculate a relative percentage (rel. %) for TG by using following equation for lipid i of sample j:

$$w_{ij,\%\mathrm{TG}} = \frac{h_{ij,\mathrm{TG}}}{\sum h_{j,\mathrm{TG}}} \times 100$$

7. Figure 4 shows examples of TG LC–MS lipid profiles.

Table 1
List of the most abundant TG detected in yeast lipid extracts (*see* Notes 22–24)

TG species	El. comp.	t_R (min)	m/z $[M+NH_4]^+$	m/z $[M+Na]^+$	TG acyl-chain annotation
TG 48:1	$C_{51}H_{96}O_6$	10.66	822.7551	827.7105	TG 14:0_16:0_18:1
TG 48:2	$C_{51}H_{94}O_6$	10.26	820.7394	825.6948	TG 14:0_16:0_18:2, TG 16:0_16:1_16:1
TG 48:3	$C_{51}H_{92}O_6$	9.91	818.7238	823.6792	TG 16:1_16:1_16:1
TG 50:1	$C_{53}H_{100}O_6$	11.06	850.7864	855.7418	TG 16:0_16:0_18:1
TG 50:2	$C_{53}H_{98}O_6$	10.69	848.7707	853.7261	TG 16:0_16:0_18:2
TG 50:3	$C_{53}H_{96}O_6$	10.30	846.7551	851.7105	TG 16:0_16:1_18:2, TG 16:1_16:1_18:1, TG 14:0_18:1_18:2
TG 50:4	$C_{53}H_{94}O_6$	9.91	844.7394	849.6948	TG 16:0_16:1_18:3, TG 14:0_18:2_18:2
TG 51:2	$C_{54}H_{100}O_6$	10.90	862.7864	867.7418	TG 15:0_18:1_18:1, TG 16:0_17:1_18:1
TG 52:1	$C_{55}H_{104}O_6$	11.48	878.8177	883.7731	TG 16:0_18:0_18:1
TG 52:2	$C_{55}H_{102}O_6$	11.10	876.8020	881.7574	TG 16:0_18:1_18:1

(continued)

Table 1
(continued)

TG species	El. comp.	t_R (min)	*m/z* $[M+NH_4]^+$	*m/z* $[M+Na]^+$	TG acyl-chain annotation
TG 52:3	$C_{55}H_{100}O_6$	10.74	874.7864	879.7418	TG 16:0_18:1_18:2
TG 52:4	$C_{55}H_{98}O_6$	10.37	872.7707	877.7261	TG 16:0_18:1_18:3
TG 52:5	$C_{55}H_{96}O_6$	10.00	870.7551	875.7105	TG 16:0_18:2_18:3, TG 16:1_18:1_18:3, TG 16:1_18:2_18:2
TG 52:6	$C_{55}H_{94}O_6$	9.61	868.7394	873.6948	TG 16:0_18:3_18:3
TG 53:3	$C_{56}H_{102}O_6$	10.93	888.8020	893.7574	TG 17:1_18:1_18:1
TG 54:1	$C_{57}H_{108}O_6$	11.84	906.8490	911.8044	TG 18:0_18:0_18:1, TG 16:0_18:1_20:0
TG 54:2	$C_{57}H_{106}O_6$	11.49	904.8333	909.7887	TG 18:0_18:1_18:1
TG 54:3	$C_{57}H_{104}O_6$	11.13	902.8177	907.7731	TG 18:0_18:1_18:2
TG 54:4	$C_{57}H_{102}O_6$	10.77	900.8020	905.7574	TG 18:1_18:1_18:2, TG 18:0_18:1_18:3
TG 54:5	$C_{57}H_{100}O_6$	10.40	898.7864	903.7418	TG 18:1_18:2_18:2, TG 18:1_18:1_18:3
TG 54:6	$C_{57}H_{98}O_6$	10.02	896.7707	901.7261	TG 18:1_18:2_18:3
TG 54:7	$C_{57}H_{96}O_6$	9.86	894.7551	899.7105	TG 18:1_18:3_18:3
TG 54:8	$C_{57}H_{94}O_6$	9.22	892.7394	897.6948	TG 18:2_18:3_18:3
TG 56:1	$C_{59}H_{112}O_6$	12.05	934.8803	939.8357	TG 16:0_18:1_22:0
TG 56:2	$C_{59}H_{110}O_6$	11.86	932.8646	937.8200	TG 16:0_18:2_22:0
TG 58:1	$C_{61}H_{116}O_6$	12.16	962.9116	967.8670	TG 16:0_18:1_24:0
TG 58:2	$C_{61}H_{114}O_6$	12.06	960.8959	965.8513	TG 16:0_18:2_24:0
TG 58:3	$C_{61}H_{112}O_6$	11.89	958.8803	963.8357	TG 18:0_18:3_22:0, TG 16:0_18:3_24:0
TG 60:1	$C_{63}H_{120}O_6$	12.09	990.9429	995.8983	TG 16:0_18:1_26:0, TG 18:0_18:1_24:0
TG 60:2	$C_{63}H_{118}O_6$	12.00	988.9272	993.8826	TG 16:1_18:1_26:0, TG 18:1_18:1_24:0
TG 62:1	$C_{65}H_{124}O_6$	12.21	1018.9742	1023.9296	TG 16:0_18:1_28:0, TG 18:0_18:1_26:0
TG 62:2	$C_{65}H_{122}O_6$	12.06	1016.9585	1021.9139	TG 16:1_18:1_28:0, TG 18:1_18:1_26:0

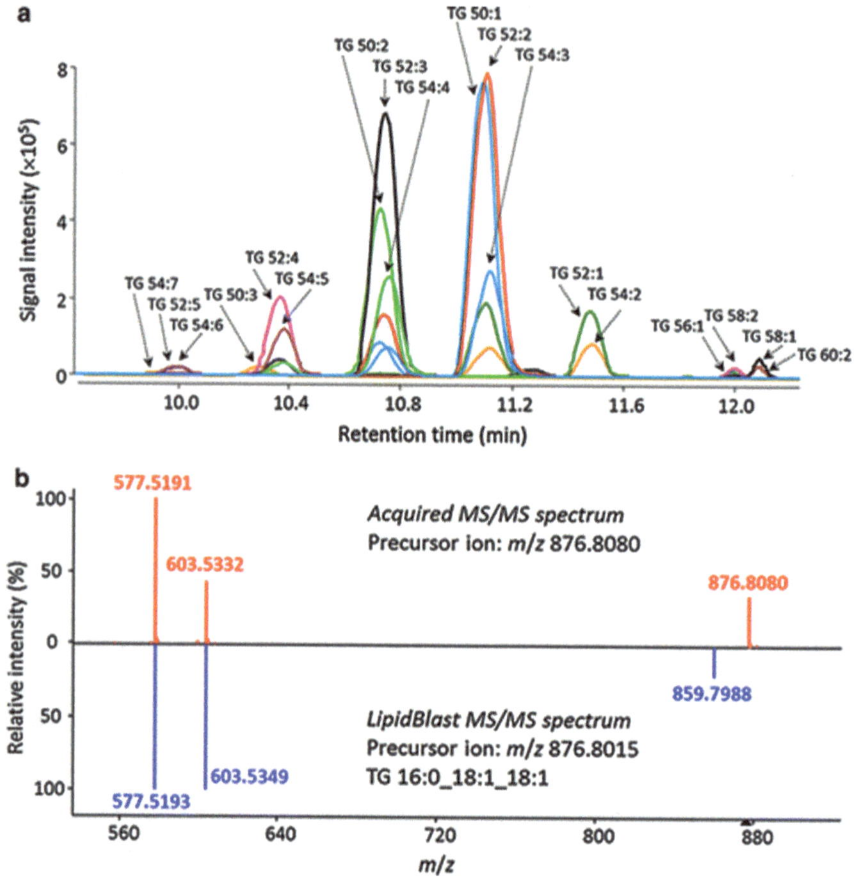

Fig. 4 (**a**) Overlay of extracted ion chromatograms of $[M+NH_4]^+$ ions of the most abundant TG in *O. externus* (UCDFST 68-934.2) lipid extract. The unannotated *m/z* traces belong to $[M+NH_4+2]^+$ ions of annotated lipids. (**b**) MS/MS spectrum of TG 52:2 with precursor ion *m/z* 876.8 acquired at a collision energy of +20 eV and MS/MS spectrum of TG 16:0_18:1_18:1 from LipidBlast in silico MS/MS library. The $[M+NH_4-17]^+$ ion (*m/z* 859.7988) was not detected using QTOFMS. For masses and peak annotations *see* Table 1. (Reproduced from [2] with permission from Springer)

4 Notes

1. Oleaginous yeasts accumulate high levels of intracellular oil upon depletion of a key nutrient, typically nitrogen [13]. Growth conditions that result in high TG include molar carbon–nitrogen ratios of 35–140, depending on the yeast species [5], and aerobic growth. Lipid concentrations are significantly lower during exponential growth than in stationary growth [5].

2. This protocol has been optimized for the Agilent 6550 iFunnel QTOFMS system. For other LC–MS systems differing in their sensitivity and linear dynamic range the protocol can be modified including (1) sample dilution and (2) volume injected.

3. The stock solution can be stored for 2 weeks. The aliquots can be stored for no more than 3 days.
4. If sampling high density cell cultures, transfer 3 μL of cell culture and add 3 μL of deionized sterile water. Swirl gently with pipette tip. If sampling cultures containing extracellular lipids (e.g., glycolipids such as sophorolipids or polyol esters of fatty acids), it is better not to swirl the culture, to avoid rupturing glycolipid droplets.
5. Nile red is carcinogenic. Dispose tips, slides, slide covers and all materials that come into contact with Nile red in biohazardous waste.
6. Some yeasts react to the Nile red solution and agglutinate, making visualization rather difficult.
7. Nile red fluorescence photobleaches in 1 or 2 s. Therefore, it is crucial to snapshot immediately after turning the fluorescent light on.
8. The maximum age of the plate that can be used in this assay is 7 days old to ensure the viability and robustness of the yeast cells.
9. Observe growth daily by taking 1-mL samples aseptically inside the biosafety cabinet and measure the optical density at 600 nm at 0, 24, 48, and 72 h to develop a growth curve. Lipid accumulation is maximum at the end of exponential and early stationary phase and is usually around 72 h after inoculation.
10. This program is set for 96-well microplate reader from Biotek Synergy 2 (Vermont, USA).
11. Nile red staining is a colorimetric method where timing and accuracy play a key role in the success of the assay. In addition, acetone evaporates quickly at room temperature. Pour the Nile red in acetone gently into a reservoir and use a multichannel pipettor for transferring and mixing the Nile red, and check briefly the level of each tip after aspirating Nile red in acetone before transferring. Transfer to the well and mix up and down. Do this step quickly enough but carefully for each well. If more than one 96 well plate is analyzed, keep the remaining Nile red in the amber bottle, and only pour enough into the reservoir for one 96-well plate. Use a new reservoir every time. This step will reduce variability of Nile red concentration due to evaporation of the solvent.
12. Dry yeast cells aerosolize very easily due to static electricity, especially in dry weather. Transferring small quantities can be difficult. Rubbing a laundry antistatic dryer sheet on the outside of the vial containing dry yeast, as well as the 2-mL screw cap before transferring the yeast prevents dry yeast cells from aerosolizing, making the transfer cleaner and easier.

13. To prepare the Folch reagent, combine two volumes of chloroform with one volume of methanol in a glass bottle and mix gently until it dissolves well. Keep the solvent at 4 °C when not in use to maintain the correct ratio. Before using, let it sit at room temperature to equilibrate.
14. Handling the 0.5 mm zirconia beads can be difficult. We recommend storing them in a container equipped with tip like dosing cap, such as those used for ketchup or mustard in a fast food restaurant.
15. Thirty seconds of bead beating at the aforementioned conditions is enough to warm up the samples considerably. To avoid lipid degradation as a consequence of heating, it is strongly recommended to remove the samples from the bead beater after every cycle and let them chill in the ice bath for at least 30 s.
16. Extracting the chloroform phase from the bottom layer can be tricky and requires practice. Use dummy samples for training to get practice before handling real samples. When handling samples, place the conical tube and the bottle from which the chloroform extract is extracted directly next to each other to avoid spilling during transfer. The chloroform extract can be stored at 4 °C for a maximum of 1 week. Equilibrate to room temperature before the next step. Vortex gently prior to transferring.
17. Before analyzing large numbers of samples, check that the detector is not saturated by monitoring the most abundant lipid species using serial dilution of a test sample. If saturation of the detector is observed either dilute samples for analysis or use a lower injection volume.
18. To enhance solubilization of ammonium formate after its addition in the mobile phase, the salt is dissolved first in a small volume of water before addition in the mobile phases (0.631 g ammonium formate/1 mL water/1 L mobile phase). Each mobile phase with modifiers is mixed, sonicated for 15 min to achieve complete dissolving of salt, mixed again, and then sonicated for another 15 min.
19. LC–ESI(+)-QTOFMS method in "discovery mode" collects both MS^1 and MS/MS spectra. Acquired MS/MS spectra can be subsequently used for structural elucidation such as automated identification of lipids using the LipidBlast in silico MS/MS library [10, 14].
20. LC–ESI(+)-QTOFMS method in "routine mode" collects MS^1 data only. This mode is preferably used for large series of samples due to better sensitivity as compared to "discovery mode."

21. Run lipid extract followed by injection of methanol for checking the carryover. The carryover effect is observed mainly for TG. Check the most intensive species such as TG 50:1 (m/z 850.786 $[M+NH_4]^+$) and TG 52:2 (m/z 876.802 $[M+NH_4]^+$) to evaluate the carryover. Carryover <0.2% for these most abundant lipid species is acceptable. Use of different wash solvents during the sample injections is highly advisable, for instance, weak wash solvent: acetonitrile–water (50:50, v/v) followed by a strong wash solvent: isopropanol. If only one solvent for washing is available, use isopropanol and also check a software option to clean the injector using valve switching (e.g., Agilent) at the different composition of mobile phase to release adsorbed lipids [15].
22. Retention times should not deviate more than ±0.3 min from the values in Table 1.
23. Information provided in this table are based on the analysis of lipid extracts of oleaginous yeasts [2].
24. TG acyl-chain annotations are based on the LipidBlast in silico MS/MS library [10, 14]. Some TG can be multiple isomers that fail to be separated chromatographically, having the same m/z and retention time and mixed MS/MS spectra. Such coelutions may occur during LC-MS-based lipidomic profiling. In this case, the particular lipid is annotated using both species (e.g., TG 54:4; TG 18:1_18:1_18:2 and TG 18:0_18:1_18:3).

Acknowledgments

L.A. Garay was funded by the National Mexican Council of Science and Technology (CONACYT) Fellowship number 291795. This work, including the efforts of T. Cajka and O. Fiehn, was funded by the National Institutes of Health [grant numbers NIH HL113452 and NIH DK097154], as well as the NIH instrument funding [grant number NIH S10-RR031630].

References

1. Ratledge C (1979) Resources conservation by novel biological processes. Part 1: Grow fats from wastes. Chem Soc Rev 8:283–296
2. Garay LA, Sitepu IR, Cajka T, Chandra I, Shi S, Lin T, German JB, Fiehn O, Boundy-Mills KL (2016) Eighteen new oleaginous yeast species. J Ind Microbiol Biotechnol 43(7):887–900
3. Sitepu IR, Sestric R, Ignatia L, Levin D, Bruce German J, Gillies LA, Almada LA, Boundy-Mills KL (2013) Manipulation of culture conditions alters lipid content and fatty acid profiles of a wide variety of known and new oleaginous yeasts species. Bioresour Technol 144:360–369
4. Ledesma-Amaro R (2015) Microbial oils: a customizable feedstock through metabolic engineering. Eur J Lipid Sci Technol 117 (2):141–144
5. Sitepu I, Garay L, Sestric R, Levin D, Block DE, German J, Boundy-Mills K (2014)

Oleaginous yeasts for biodiesel: current and future trends in biology and production. Biotechnol Adv 32(7):1336–1360. https://doi.org/10.1016/j.biotechadv.2014.08.003

6. Tsouko E, Papanikolaou S, Koutinas A (2016) Production of fuels from microbial oil using oleaginous microorganisms. In: Handbook of biofuels production. Elsevier B.V., Amsterdam, p 201
7. Oguri E, Masaki K, Naganuma T, Iefuji H (2012) Phylogenetic and biochemical characterization of the oil-producing yeast *Lipomyces starkeyi*. Antonie Van Leeuwenhoek 101 (2):359–368
8. Gujjari P, Suh S-O, Coumes K, Zhou J (2011) Characterization of oleaginous yeasts revealed two novel species: *Trichosporon cacaoliposimilis* sp. nov. and *Trichosporon oleaginousus* sp. nov. Mycologia 103(5):1110–1118. https://doi.org/10.3852/10-403
9. Cajka T, Fiehn O (2014) Comprehensive analysis of lipids in biological systems by liquid chromatography-mass spectrometry. TrAC Trends Anal Chem 61:192–206
10. Kind T, Liu K-H, Lee DY, DeFelice B, Meissen JK, Fiehn O (2013) LipidBlast in silico tandem mass spectrometry database for lipid identification. Nat Methods 10(8):755–758. https://doi.org/10.1038/nmeth.2551
11. Suutari M, Priha P, Laakso S (1993) Temperature shifts in regulation of lipids accumulated by *Lipomyces starkeyi*. J Am Oil Chem Soc 70:891–894
12. Sitepu I, Ignatia L, Franz A, Wong D, Faulina S, Tsui M, Kanti A, Boundy-Mills K (2012) An improved high-throughput Nile red fluorescence assay for estimating intracellular lipids in a variety of yeast species. J Microbiol Methods 91(2):321–328
13. Lundin H (1950) Fat synthesis by microorganisms and its possible applications in industry. J Inst Brew 56(1):17–28
14. Cajka T, Fiehn O (2017) LC–MS-based lipidomics and automated identification of lipids using the LipidBlast in-silico MS/MS library. In: Bhattacharya S (ed) Lipidomics: methods and protocols, Methods in molecular biology, vol 1609. Springer, New York. https://doi.org/10.1007/978-1-4939-6996-8_14
15. Cajka T, Garay LA, Sitepu IR, Boundy-Mills KL, Fiehn O (2016) Multiplatform mass spectrometry-based approach identifies extracellular glycolipids of the yeast Rhodotorula babjevae UCDFST 04-877. J Nat Prod 79 (10):2580–2589

Chapter 3

Fungi (Mold)-Based Lipid Production

Yan Yang, Fatemeh Heidari, and Bo Hu

Abstract

There is an increasing need for the development of alternative energy sources with a focus on reducing greenhouse gas emissions and striving toward a sustainable economy. Bioethanol and biodiesel are currently the primary choices of alternative transportation fuels. At present, biodiesel is not competitive with conventional fuel due to its high price, and the only way to compete with conventional fuel is to improve the quality, reduce the costs, and coproduce value-added products. With the high demand for lipids in the energy sector and other industrial applications, microbial lipids accumulated from microorganisms, especially oleaginous fungi and yeasts have been the important topic of many recent research studies. This chapter summarizes the current status of knowledge and technology about lipid production by oleaginous fungi and yeasts for biofuel applications and other value-added products. The chapter focuses on several aspects such as the most promising oleaginous strains, strain development, improvement of lipid production, methods and protocols to cultivate oleaginous fungi, substrate utilization, fermentation process design, and downstream processing. The feasibility and challenges during the large-scale commercial production of microbial lipids as fuel sources are also discussed. It provides an overview of microbial lipid production biorefinery and also future development directions.

Key words Single-cell oils, Oleaginous fungi, Microbial oil production, Biofuel, Biodiesel, Oil processing by-products

1 Introduction

Due to rapid depletion and associated environmental problems, fossil fuel consumption is not a sustainable solution in the long term. The reduction in dependence on fossil-based fuel is the driving force to develop biofuel, and a possible solution on this issue is the development of bioethanol and biodiesel. Current biodiesel production heavily relies on row crop-based industry, using vegetable oil and animal fat as feedstock. Microbial lipids, also referred to as single-cell oils (SCO), offer potential for sufficient production of renewable fuels. Considering the negative environmental consequences and concerns about limited supply of

Venkatesh Balan (ed.), *Microbial Lipid Production: Methods and Protocols*, Methods in Molecular Biology, vol. 1995, https://doi.org/10.1007/978-1-4939-9484-7_3,

crop-based oil resources, biodiesel from microbial lipids is friendly and sustainable, making it an attractive alternative to fossil fuels.

Oleaginous microorganisms can accumulate more than 20–25% in lipids of their dry cell biomass, mainly in the form of triacylglycerols (TAG) [1]. The accumulated lipids have similar composition and energy value to plant and animal oils, and can be used for economical biofuel production especially if it is based on inexpensive carbon source, such as waste materials, by-products and lignocellulosic substrates. Microbial lipids synthesized from oleaginous microorganisms can be used as raw materials for biofuel production including biodiesel, hydrocarbon compounds, and other complex oils. The most common biofuel produced from lipids is biodiesel via transesterification of TAG. In addition, various oleochemical products can also be generated from microbial lipids. The major process for transforming lipids into oleochemicals is the hydrolysis of TAG into glycerol and fatty acids (FA), and more products can be formed from FA, such as fatty alcohols and fatty amines [2, 3]. Some microbial lipids contain a special group of fatty acids called essential fatty acids (EFA), which are valuable as food supplements. Important EFA include gamma linoleic acid (GLA), arachidonic acid (AA), eicosapentaenoic acid (EPA), and docosahexaenoic acid (DHA) [4].

Microbial lipid biosynthesis is largely conserved among eukaryotic microorganisms, like yeast and fungi (mold). Many microalgae species can also accumulate large amount of lipids when cultured with sugar culture media. As a matter of fact, these microalgae species were commonly viewed as oleaginous fungi/yeast, until they were later on reclassified as microalgae. Heterotrophic cultivation of oleaginous microorganisms share many common features. The high cost of substrate, for instance, glucose, greatly increases the price of the microbial lipids for biodiesel production. Alternative economical and sustainable substrates have been explored, such as nongrain plants, agricultural and forest residues, lignocellulosic biomass, food wastes and other waste materials. Lignocellulosic biomass containing cellulose and hemicellulose can be potential substrates; however, oleaginous microbial species rarely can directly utilize lignocellulosic biomass [5]. The screening and selection of new oleaginous strains with desirable characteristics (e.g., high lipid content and production, capability to generate hydrolytic enzymes, wide utilization of different substrates) will provide the possibility to use abundant lignocellulosic biomass and accumulate high amount of lipids. Moreover, the improvement of oleaginous strains by metabolic engineering, the development of cultivation methods and fermentation processes, the exploitation of appropriate substrates, and the optimization of downstream processing will enhance lipid production and make microbial diesel economically competitive.

1.1 Oleaginous Fungi/Yeast Strains

1.1.1 Advantages and Limitation of Producing Lipids in Fungi/Yeast

TAG as energy storage compounds has been widely found in eukaryotic organisms such as plants, animals, algae, and fungi, but was rarely described in bacteria. Most bacteria are not oil producers and bacteria generally accumulate polyhydroxyalkanoates as storage material [6]. Only few bacterial species, belonging to the bacillus genera (*B. alcalophilus*, *B. subtilis*) and actinobacterial genera (*Mycobacterium*, *Streptomyces*, *Rhodococcus*, and *Nocardia*) accumulate relevant amounts of lipids [6, 7]. Among all the oleaginous microorganisms, the heterotrophic cultivation of oleaginous fungi (molds) and yeasts has some favorable properties for industrial biofuel production. Fungi and yeasts have a short life cycle, high growth rate, and high biomass density, and can be grown in conventional bioreactors to maximize yield and productivity. In addition, filamentous fungal cells in general are more easily harvested compared to algae and yeasts, especially when they grow in the form of pellets or mycelia [8]. Furthermore, fungal lipid production is more versatile and has much less demand on space, venue, and climate compared to current autotrophic microalgal lipid production. These characteristics make oleaginous fungi a good candidate for mass cultivation in the commercial biofuels production. Although microbial lipid content is high in oleaginous fungi and yeasts, but process optimization and isolation of organism that can grow on inexpensive sources of carbon such as industrial and agricultural by-products/wastes, food wastes and waste streams, and lignocellulosic materials would add further advantage to the production system.

1.2 Current Promising Candidates

Oleaginous fungi (filamentous fungi or molds) are defined as the fungal strains with more than 20% of lipids in their dry cell biomass. Some species can accumulate very high contents of lipids, for example *Mortierella* can store up to 80% of lipids based on dry weight [6]. Most oleaginous strains only accumulate high content of lipids in some stressed conditions, for example, *Rhodotorula* species and *Lipomyces* species, can accumulate lipids at the level of 40% and up to 70% under nutrient-limiting conditions [8, 9]. Oleaginous fungi and yeasts have been known as favorable oil-producing organisms since 1980s [10]. Lipid accumulation via oleaginous fungal strains has been extensively studied and some commercial successes have been generated, especially for the production of nutraceuticals such as polyunsaturated fatty acids (PUFA). However, research focused on the exploitation of SCO for biofuel production has just emerged in the last few years. Recent studies have primarily focused on several oleaginous yeast and fungi strains to convert various carbon sources to lipids for biofuel production, especially inexpensive carbon sources such as lignocellulosic biomass and waste materials [11–13].

Table 1 lists the oil content of several oleaginous fungi and yeast species. Among these oleaginous fungi and yeasts, some

Table 1
Oil content of some oleaginous fungi and yeast species [6, 9, 14–16]

Fungi	Oil content (% dry weight)	Yeast	Oil content (% dry weight)
Aspergillus niger	23–57	*Candida curvata*	58
Aspergillus oryzae	15–57	*Cryptococcus albidus*	65
Cunninghamella echinulata	40–57	*Lipomyces starkeyi*	64
Humicola lanuginosa	75	*Rhodotorula glutinis*	72
Mortierella alliacea	42	*Rhodotorula graminis*	36
Mortierella alpina	31	*Rhodosporidium toruloides*	76
Mortierella isabellina	60–86	*Rhizopus arrhizus*	57
Mortierella vinacea	66	*Trichosporon pullulans*	65
Mucor circinelloides	20	*Trichosporon fermentans*	62
Mucor rouxii	32	*Yarrowia lipolytica*	36

oleaginous zygomycetes fungi show more promising high oil content species. For instance, the two oleaginous fungi *M. circinelloides* and *M. isabellina* belonging to *Zygomycota* phylum have been heavily studied for their lipid accumulation for biodiesel production using various substrates [17–20]. Another oleaginous zygomycetes fungus *Cunninghamella echinulata* can accumulate up to 57% lipid content and has been studied for lipid production using starch and glycerol [21, 22]. The oleaginous fungus *Mucor circinelloides* has been used for the first commercial production of microbial lipids [17]. *M. circinelloides* represents an excellent model within *Zygomycota* phylum since the whole genome sequence and an efficient transformation procedure are readily available [23, 24]. In addition, many mucoraceous fungi including *M. circinelloides* and *M. rouxii* are known to produce high levels of lipids and GLA [25, 26]. Somashekar et al. [27] reported *M. rouxii* and *Mucor* sp. resulted in a production of 30% lipids and up to 17% GLA in lipids using glucose and peptone as carbon and nitrogen sources. Some *Mortierella* fungi also showed good capability to generate PUFA, such as AA and EPA. *M. alpina* was considered as a suitable source for the production of AA, and AA was more than 18% of dry cell mass and over 60% of the total lipids reported by Eroshin et al. [28]. A filamentous fungus *M. alliacea* Strain YN-15 was found to accumulate 7.1 g/L AA out of 19.5 g/L total lipids, and the dry cell mass was 46.1 g/L in a 50 L jar fermenter [29]. A remarkable lipid accumulation, up to 57%, as well as a considerable amount of oxalic acid production has also been reported for *Aspergillus niger* using glycerol as the sole carbon source. However, most studies have focused on lipase production using *A. niger* rather than lipid production, since it can be used in the transesterification of TAG [30].

The yeasts represent a part of the microbiota in natural ecosystems, such as soils, freshwaters and marine waters. Approximately 1500 species of yeasts belonging to over 100 genera have been described so far, and only 30 species are found to be able to accumulate more than 25% of lipids based on dry weight [24, 31]. The most deeply investigated oleaginous yeasts belong to the genera *Candida*, *Cryptococcus*, *Rhodotorula*, *Rhodosporidium*, *Lipomyces*, *Yarrowia*, *Pichia*, and *Trichosporon* [6, 32, 33]. The yeasts of Basidiomycota phylum represent a great number of oleaginous yeasts species, while other important oleaginous yeasts have been identified in the Ascomycota phylum (e.g., *Y. lipolytica* and *L. starkeyi*). One of the *Candida* genus yeast, *C. curvata* can synthetize and store significant amount of lipids [1, 34]. Yeasts of the *Cryptococcus* genus are widely distributed in nature and have been isolated from air, soil, water, plants, and animal wastes. *C. curvatus* is a yeast with good industrial potential since it requires minimal nutrients for growth, accumulates up to 60% lipids, and can grow on a very broad range of substrates [35, 36]. The red yeasts *R. glutinis* and *R. toruloides* are able to synthesize carotenoids, which serve as natural pigments, in addition to lipids [37]. Being considered as a potential pigment source, their biotechnological application is also associated with their capability to utilize glycerol and lignocellulosic materials for lipid production [38, 39]. The oleaginous yeast *L. starkeyi* showed intracellular lipids accumulation as high as 70% of their dry biomass, and can produce lipids on xylose, ethanol, and L-arabinose, or using a mixture of glucose and xylose [40]. *Y. lipolytica* is a also good candidate for SCO production [9]. It has become a model organism because it can be subjected to genetic and metabolic engineering with genomic data available [9, 24].

2 Strategies for Screening Novel Oleaginous Fungi/Yeast

The extent of lipid accumulation for oleaginous microorganisms is determined by both genetics and growth conditions. The isolation, identification and selection of appropriate oleaginous fungi and yeast strains with high growth rates and productivities are crucial steps for the overall success of biofuel production. The current studied oleaginous strains are not as efficient or robust as would be preferred for industrial applications [41]. The use of strain development techniques such as screening potential strains, directed evolution, and metabolic pathway engineering can lead to better oleaginous strains with higher biomass and lipid production than wild-type strains.

Generally, there are three different strategies available for screening and selection. The first strategy is to isolate from local environments where they are supposed to be grown on a large scale. Presumably such selected strains have evolved for growth in the local geographical, climatic and ecological conditions. The second strategy is acclimatization and directed evolution under an environment the microorganisms do not normally grow well. The studies of microbial species in some extreme conditions may result in some special strains with desired properties. The third strategy is to use genetic and metabolic engineering to modify the regulation of metabolic pathways. The genetic engineering method can improve the intrinsic properties of strains and make significant improvement, provided a transformation produced is available.

There are numerous studies demonstrating screening and selection of oleaginous fungi and yeasts. Dey et al. [42] screened two endophytic oleaginous fungi *Colletotrichum* sp. and *Alternaria* sp. with lipid content 30% and 58% respectively, and studied their growth under optimum and nutrient-stress conditions. The endophytic oleaginous fungus *Microsphaeropsis* sp. was isolated from oily wood-stem and used for lipid production from wheat straw hemicellulose hydrolysate [43]. Other studies reported oleaginous fungus *Aspergillus* sp. from soil samples and accumulated 22% lipid using corncob waste liquor as a substrate, which was used for biodiesel production [44]. Kitcha and Cheirsilp [45] screened multiple oleaginous yeast strains such as *Trichosporonoides spathulata* with more than 40% lipid from soils and palm oil mill wastes, and optimized lipid production with crude glycerol. Tanimura et al. [46] selected *Cryptococcus* sp. with high lipid content (>60% lipid) and productivity from hundreds of yeast isolates, and studied their properties for biodiesel production.

3 Methods to Improve Lipid Production

3.1 Cultivation Conditions

Accumulation of lipids in the fungal cells is highly dependent on cultivation conditions. Different oleaginous species of fungi/yeasts have their own characteristics for optimal lipid production so that it is feasible to change the cultivation conditions for the synthesis of lipids with varied fatty acid composition to meet different industrial needs. The theoretical yield for converting sugars to lipids is 32% wt/wt from glucose and 34% wt/wt from xylose [47], and most of oleaginous cell cultures can reach the yield of 20–25% wt/wt [48], while some sugars have to be diverted to support the cell growth and metabolism. This conversion efficiency relies on the nature of microorganism, but the cultivation conditions also greatly affect microbial lipid content and lipid composition of fungi and yeasts [27, 31, 49].

Lipid accumulation by oleaginous fungi and yeasts mostly happens when a nutrient in the medium becomes limited, and the excessive carbon source is converted into storage of TAG. The limited supply of nitrogen, phosphorus, sulfur, iron, or zinc was found to trigger lipid accumulation in oleaginous fungi and yeasts [50, 51]. However, nitrogen limitation is the most effective condition for most oleaginous fungi and yeast species in lipid accumulation, leading to the highest substrate/lipid conversion and lipid content within cells [24, 52]. Thus, nitrogen limitation condition is widely used to induce lipogenesis and the selection of appropriate C:N ratio is critical to maximize fungal/yeasts lipid production. The effect of the C:N ratio on lipid synthesis has been investigated for different oleaginous fungi and yeasts, such as many oleaginous species of *Mortierella*, *Yarrowia*, *Rhodotorula*, and *Candida* [53, 54]. An optimally high C:N molar ratio for lipid accumulation is usually at the range of 65 to near 100 [41, 48].

Besides C:N ratio, many other nutritional and environmental factors influence cell growth and lipid production in fungi and yeasts, including carbon sources, nitrogen sources, and other essential macronutrients and micronutrients in the growth medium, temperature, pH level, and dissolved oxygen etc. Fatty acid composition is also affected by cultural medium and cultivation conditions. The main fatty acids in oleaginous fungi and yeasts are the palmitic (C16:0), palmitoleic (C16:1), stearic (C18:0), oleic (C18:1), and linoleic (C18:2) acids [6, 55]. Somashekar et al. [27] studied the medium composition for the lipid accumulation of *Mucor* species and found that cultivation using different carbon sources leads to different lipid production, and lactose was a poor promoter for biomass and lipid production. Athenstaedt et al. [56] got a similar conclusion that changing of carbon source modifies the lipid and protein composition in *Y. lipolytica*. Meanwhile, nitrogen sources are found to affect the total lipid production and lipid compositions, and the selectivity of nitrogen sources varied among different fungi and yeast species [27]. As an example, switch of nitrogen source from nitrate to urea diminished the accumulation of unsaturated fatty acids in *Ustilago maydis* [57]. Growth temperature of oleaginous microorganisms influences the production and fatty acid composition of accumulated lipids [32]. Culture of *Sporobolomyces roseus* at different temperature obtained the lowest lipid production of 0.96 g/L at 30 °C and the highest lipid production of 2.35 g/L at 14 °C. Meanwhile, the C18:1, C18:2, and C18:3 fatty acids represented 60%, 10%, and 1%, respectively, of the total fatty acids at 30 °C. These percentages switched to 49.2%, 15%, and 9%, respectively, at a lower growth temperature (between 6 and 22 °C). Similar result was obtained on the study of lipid accumulation by *Metschnikowia pulcherrima*, in which lower temperature triggered high levels of oil production [58]. Initial pH is another critical parameter in lipid accumulation. Angerbauer et al.

[59] studied the influence of the pH value (from 5 to 7) on the lipid production of *L. starkeyi* in a basal medium (C:N ratio of 100 g/g). The highest lipid production (56% biomass DW; 7.5 g/L) was obtained at pH 5, while the highest biomass was obtained at pH 6.5 (15.12 g/L). The lipid accumulation was not significantly modified for the pH values that ranged from 5.5 to 6.5 but decreased dramatically (7% biomass DW, 1.1 g/L) at pH 7. It is considered that optimal growth temperature for oleaginous fungi is generally 20–28 °C, and 25–30 °C for yeasts, and optimal growth pH ranges are typically 5–7 for fungi and yeasts [32, 47].

3.2 Metabolic Engineering

Metabolic engineering is another method that can efficiently improve the overall lipid production. Improvement of lipid production through the optimization of nutritional and environmental factors encounters several limitations, mainly due to the genetics of microorganisms. Thus, research is currently increasingly directed to the genetic modifications to enhance lipid synthesis by oleaginous microorganisms. Extensive studies have been carried out on this aspect, and the development of genome data and genetic tools provided the possibility to increase the yield of stored lipids by metabolic engineering. The most frequent studied model strains are *Y. lipolytica* and *M. circinelloides*, and many other oleaginous microorganisms have been used in the study of metabolic pathway engineering with the availability of genome data and genetic tools [9, 23, 24].

The ability of oleaginous microorganisms to accumulate high amounts of lipid depends mostly on the regulation the biosynthetic pathway and the supply of the precursors (e.g., acetyl-CoA, malonyl-CoA, and glycerol-3-phosphate) and the cofactor NADPH. Regulation method can be classified into five approaches: (1) overexpressing enzymes of FA biosynthesis pathway; (2) overexpressing enzymes of TAG biosynthesis pathway; (3) regulation of related TAG biosynthesis bypasses; (4) blocking competing pathways; and (5) multigene transgenic methods [60].

Overexpression of enzyme(s) is one of the most common strategies, which aims to increase the expression of rate-limiting enzyme and increase lipid production. It was reported that overexpression of diacylglycerol acyltransferase (DGA1) in *Y. lipolytica* gave a fourfold increase in lipid production over control, and the overexpression of acetyl-CoA carboxylase (ACC1) in *Y. lipolytica* gave a twofold increase over the control strain [61]. Regulation of related TAG biosynthesis bypasses focuses on some enzymes that are not directly involved but influence the rate of lipid production. Regulation of these enzymes will mainly affect the content of essential metabolites for lipid synthesis. For example, the overexpression of malic enzyme in *M. circinelloides*, which is involved in NADPH formation and pyruvate generation from malate, leads to a

2.5-fold increase of lipid production over control [62]. Blocking competing pathways are aimed to direct the metabolic flux to TAG biosynthesis rather than other pathways. Deletion of GUT2 gene, which codes for glycerol-3-phosphate dehydrogenase isomer in *Y. lipolytica*, was also demonstrated to have a threefold increase in lipid accumulation compared to the wild-type strain [63]. Multi-gene transgenic methods are mostly related to the manipulation of more than one key enzyme in lipid synthesis pathways. Besides the lipid production increased by the overexpression of DGA1 and ACC1 described above, the co-overexpression of these two genes resulted in an almost fivefold increase than control [61].

There are also some studies which have proven the success of recombination oleaginous microorganisms for lipid production. For example, fatty acid ethyl esters (FAEEs) and fatty acid isoamyl esters (FAIEs) production by heterologous expression of bacterial acyltransferase in *S. cerevisiae* [64], and FAEEs production by heterologous expression of *Zymomonas mobilis* pyruvate decarboxylase, alcohol dehydrogenase, and unspecific acyltransferase in *E. coli* [65].

4 Methods and Substrates for Fungi/Yeast Lipid Production

4.1 Cultivation Methods and Fermentation Processes

In order to obtain a higher microbial lipid production rate, various cultivation methods or modes have been developed to culture oleaginous microorganisms. Mostly studied cultivation modes include batch, fed-batch, and continuous mode. Besides these widely studies methods, solid-state fermentation and pelletized submerged fermentation have also been investigated.

Batch cultivation is a partially closed system in which most of the materials required are loaded into the reactor in the first place. In this mode of operation, conditions are continuously changing with time, and the reactor is an unsteady-state system. Culture medium in batch cultivation always has a high initial C:N ratio, and the change of nitrogen concentration determines the passage from a balanced growth phase to a lipid accumulation phase. During the growth phase, the carbon source is managed for cell growth and consequently lipid contents are low. As nitrogen concentration becomes limited, cell growth ceases and microbial metabolism shifts to lipid accumulation [66]. Therefore, the initial C:N ratio of growth medium in batch cultures is vital in determining the bioprocess performance and affects both the amount of biomass produced and lipid content within cells.

It was reported that batch culture of *M. isabellina* using corn stover hydrolysate can achieve a lipid production of 6.9 g/L in a 7.5 L bioreactor [18], culture of *L. starkeyi* using glucose and xylose achieved a lipid production of 12.6 g/L in flask cultures [40], and the culture of *C. echinulata* using starch obtained a lipid production of 7.65 g/L also in flasks [21]. Xu et al. [67] showed that both biomass (26.7 g/L) and lipid (18.7 g/L) production of *R. toruloides* was improved using crude glycerol instead of glucose (biomass: 16.7 g/L, lipids: 11.2 g/L) during one-stage batch fermentation. *T. fermentans* was reported to have a biomass of 28.1 g/L and a lipid content of 62.4% with peptone as nitrogen source, glucose as carbon source in a batch culture [68].

In fed-batch processes, cells are grown under a batch mode for some time, usually until close to the end of the exponential growth phase, and then fed with a solution of substrates without the removal of culture fluid. The control of nutrient concentrations in reactor can help to monitor and control the specific growth rate and carbon utilization, and also maintain the oleaginous microorganisms at a desired specific growth phase. Fed-batch cultivation has been proved to effectively increase both the cell density and lipid contents of oleaginous microorganisms. *M. alpina* was cultivated in fed-batch mode using glucose and nitrate as carbon and nitrogen sources, and a dry cell concentration of 37 g/L and a lipid production of 14.2 g/L were achieved [69]. The fed-batch cultivation of oleaginous yeast *C. curvatus* in 30 L stirred-tank fermenter has reached a 104.1 g/L dry cell concentration and a 86.1 g/L lipid production in about 8 days, which was higher than the dry cell biomass (51.8 g/L) and lipid production (33 g/L) under the batch mode [70]. And the fermentation of oleaginous yeast *R. toruloides* using flask fed-batch cultivation has reached a 151.5 g/L dry cell concentration with a lipid content of 48.0% (w/w) in 25 days, a pilot-scale fed-batch cultures in 15 L stirred-tank bioreactor also achieved a 106.5 g/L dry cell concentration with a lipid content of 67.5% (w/w) in 5.5 days [14].

Unlike batch culture, continuous culture feeds the microorganisms with fresh nutrients, and at the same time removing spent medium plus cells from the system. At the steady state when high cell yield occurs, the microbial growth, assimilation of C and N sources and withdrawal of nutrients are maintained at constant rates. Continuous culture usually results in a high biomass and end-product productivity in the fermentation process. The constant C:N ratio in reactor will help to reduce the loss of cell viability and acid production [9]. Besides the C:N ratio in growth medium, microbial growth and nutrient assimilation both largely depend on the dilution rate. It was reported that as dilution rate increased in continuous cultivation of *C. curvata*, specific lipid accumulation rate increased and nonlipid cell accumulation decreased [71]. This result indicates that the microbial metabolism is partially controlled by dilution rate. A continuous culture of *C. curvatus* was operated to produce lipids from acetic acid and at steady state, a cell mass of 26.7 g/L with up to 53% of lipid content was achieved with optimal dilution rate and C:N ratio [72]. Another report mentioned that lipid production of *Y. lipolytica* was favored at low specific dilution rates and highly aerated condition, and a high lipid production (3.5 g/L) with high lipid content (43% w/w) can be achieved [73].

Submerged culture is the major cultivation process used for lipid production from lignocellulosic biomass, but researchers also explore solid-state fermentation using oleaginous fungi. This is a method to directly use lignocellulosic materials as feedstocks to reduce the pretreatment cost in lipid production. Meanwhile, solid-state fermentation has drawbacks in areas such as heat and mass transfer, scale-up, and cell and lipid harvest. Zhang and Hu [5]. conducted such fermentation using *M. isabellina* on soybean hulls at around 40–90% moisture content, but an overall low efficiency was obtained due to the lack of cellulase for degrading cellulose to provide sugars. Other attempts to carry out lipid production in solid-state fermentation are also described: Lin et al. [15] performed solid-state fermentation using a cellulolytic strain of *A. oryzae* on wheat straw. This strain can directly convert cellulose into lipids in a low-cost fermentation system. Solid-state fermentation of sweet potato was also carried out with *A. niger* and *A. oryzae* fungi, and the total lipid content increased from 1.93% to 3.17% and 8.71% respectively [74]. Peng and Chen [75] performed solid-state fermentation using a substrate consisting of steam-exploded wheat straw and wheat bran with an isolated *Microsphaeropsis* sp. fungus, and the lipid yield was 80 mg/g, and the lipid content of the dry fermented mass was 10.2%. Economou et al. [76] performed semisolid-state fermentation using *M. isabellina* on sweet sorghum, and the highest oil efficiency of 11 g/100 g dry weight of substrate was obtained at 92% moisture content.

Another novel method is pelletized submerged fermentation, in which filamentous microorganisms aggregate in medium and grow as pellets/granules. As a commercial fungal strain to produce cellulase, *A. oryzae* naturally forms homogenous, stable and relatively large cell pellets during cell growth. The major benefit for pelletized submerged fermentation is easy harvest of biomass after lipid production, which reduces the cost in biomass harvest and finally decreases the cost of biodiesel. One approach for pelletized submerged fermentation is to use $CaCO_3$ powder to induce the fungal pelletization [77], and another approach is to conduct pH adjustment during cell growth [8].

The two dominant processes for bioconversion of lignocellulose are simultaneous saccharification and fermentation (SSF) and separate hydrolysis and fermentation (SHF). SSF combines enzymatic hydrolysis and fermentation steps and improves the yield of hydrolysis. SSF also reduces the contamination risk and capital cost of lipid production. However, SSF are usually set at temperature and pH suitable for fermentation (around 30 °C, pH > 5), which is normally lower than the temperature required for hydrolysis. This will affect the release of sugar from lignocellulosic biomass, decrease lipid production, and leave more unhydrolyzed solids to reduce lipid recovery efficiency [41]. Currently, most studies still focus on SHF, in which both hydrolysis and fermentation can be controlled and optimized. Culture of *M. isabellina* using SHF in glucose medium obtained a lipid production of 10.2 g/L in flasks [18]. A few studies use SSF process, for example, Liu et al. [78] cultured *T. cutaneum* using pretreated corn stover and achieved a lipid production of 3.23 g/L in a 50 L bioreactor.

4.2 Utilization of Different Substrates

Oleaginous fungi and yeasts are able to utilize a variety of carbon sources. Glucose is the most commonly used in the research of fungal growth and lipid production, and other carbon sources such as fructose, xylose [79], lactose [80], arabinose, mannose [81], pectin, starch [82], and ethanol have also been used. In order to make biodiesel produced from microbial lipids competitive to the current plant oil-based biodiesel, cheap and abundant feedstock need to be explored to reduce the costs. Carbons sources obtained from lignocellulosic biomass, by-products, or surplus have been predominantly investigated to address this question. Besides lignocellulosic biomass, some waste materials such as agricultural residues, forest residues, food waste, municipal wastes, and animal wastes can also be utilized for the production of lignocellulosic-based microbial lipids. Cellulose and hemicellulose are the main target compounds in these materials to release sugar for the growth of oleaginous microorganisms, whereas pretreatment and hydrolysis processes are usually required before fermentation. Besides the relatively high cost of the pretreatment and hydrolysis processes, the difficulty to use these materials for lipid production is the inhibitory compounds (inhibitors) released during these steps such as furfural, 5-hydroxymethylfurfural, and water soluble lignin [83]. Detoxification is an essential step on efficient utilization of these hydrolysates. Many oleaginous fungi and yeasts have been studied for lipid accumulation on different kinds of substrates, such as sweet sorghum juice [76], sorghum bagasse hydrolysates [84], potato starch [85], cassava starch [86], tomato waste [87], sugar cane molasses [68], whey permeate [88, 89], acetate [90], lignocellulosic wastes from palm oil mill [45], crude glycerol [91], corncob hydrolysate [92], corncob waste liquor [44], sewage sludge [59], starch wastewater [93], Jerusalem artichoke hydrolysate [92], sugar cane bagasse hydrolysate [94], hemicellulose sugars from wheat straw [95], wheat straw [15], rice bran [96], rice hulls hydrolysate [97], and rice straw hydrolysate [82] (Table 2).

4.3 Potential of Lignocellulosic Biomass as Carbon Sources

Lignocellulosic biomass is a biological material that is composed mainly of cellulose (35–50%, dry weight basis), hemicelluloses (20–35%), and lignin (10–25%) [105]. As one of the most abundant materials in the world, it serves as a feedstock for the traditional paper industry and is used for production of chemicals and

Table 2
Examples of fungi and yeast cultivation for oil accumulation on various substrates

Fungi/yeast species	Substrates	Biomass (g/L)	Biomass productivity (g/L/day)	Lipid content (% dry weight)	Lipid (g/L)	Lipid productivity (g/L/day)	Lipid yield (g/g substrate)	Culture mode	Reference
Mortierella isabellina	Glucose	11.4–22.3	1.32–4.80	40.0–45.0	5.0–10.2	0.52–2.11	0.12	Batch	[18]
Mortierella isabellina	Xylose	9.8–21.6	1.11–3.36	37.0–43.0	3.8–8.8	0.38–1.27	0.10	Batch	[18]
Mortierella isabellina	Corn stover hydrolysate	10.9–14.1	2.84–4.5	29.5–38.4	2.48–4.82	0.65–1.61	–	Batch	[18]
Mortierella isabellina	Enzymatic hydrolysate slurry	–	–	24.8	3.2	0.87	0.07	Batch	[18]
Trichosporon cutaneum	Glucose	22.9	4.58	52.4	12.0	2.4	0.20	Batch	[98]
Trichosporon cutaneum	Xylose	21.2	4.24	46.5	9.9	1.98	0.16	Batch	[98]
Trichosporon fermentans	Glucose	28.1	4.0	62.4	17.5	2.50	0.18	Batch	[67]
Trichosporon fermentans	Molasses + glucose	27.9	3.99	52.7	14.7	2.09	0.13	Batch	[67]
Trichosporon fermentans	Rice straw hydrolysate	28.6	3.58	40.1	11.5	1.44	–	Batch	[82]
Yarrowia lipolytica	Glucose	9.3	0.30	24.7	2.3	1.78	0.12	Continuous	[99]
Yarrowia lipolytica	Rice bran hydrolysate	10.8	–	48.0	5.16	–	0.17	Batch	[100]
Yarrowia lipolytica	Sugarcane bagasse hydrolysate	11.4	3.01	58.5	6.68	1.76	0.33	Batch	[94]
Yarrowia lipolytica	Glycerol	4.7	2.26	22.30	1.05	0.50	0.04	Repeated-batch	[101]

(continued)

Table 2
(continued)

Fungi/yeast species	Substrates	Biomass (g/L)	Biomass productivity (g/L/day)	Lipid content (% dry weight)	Lipid (g/L)	Lipid productivity (g/L/day)	Lipid yield (g/g substrate)	Culture mode	Reference
Rhodosporidium toruloides	Glucose	106.5	19.07	67.5	71.89	12.96	–	Fed-batch	[14]
Cryptococcus curvatus	Glucose	104.1	13.50	82.7	86.09	11.28	–	Fed-batch	[69]
Apiotrichum curvatum	Whey	85.0	29.1	35.0	29.75	9.60	–	Fed-batch	[87]
Rhodotorula glutinis	Starch wastewater	60.0	24.0	30.0	18.0	7.20	–	Batch	[93]
Lipomyces starkeyi	Cellubiose	28.0	6.43	50.0	14.0	3.22	0.20	Batch	[102]
Lipomyces starkeyi	Cellubiose + xylose	31.5	5.24	55.0	17.3	2.88	0.19	Batch	[102]
Mucor circinelloides	Stillage + glycerol	20.0	10.0	46.0	9.2	4.60	–	Batch	[103]
Lipomyces tetrasporus, Lipomyces kononenkoae, Rhodosporidium toruloides	Lignocellulosic hydrolyzates	42.6–54.3	–	50–65	25–30 g/L	2.88–5.28	0.13–0.15	Two-stage batch	[104]

biomaterials/biopolymers. In recent years, more studies have recognized the potential to use lignocellulosic biomass as an alternative feedstock for the biofuel production. The fermentable sugars in cellulose and hemicellulose components can be suitable carbon source for microbial growth. However, these sugars are not easily accessible. Together with lignin, cellulose and hemicellulose comprise a complex structure with cellulose fibers embedded in a hemicellulose and lignin polysaccharide matrix. Cellulose crystallinity, accessible surface area, lignin seal, and the heterogeneous character of biomass particles all contribute to the recalcitrance of lignocellulosic biomass and prohibit the utilization of sugars [106]. In this case, either pretreatment of lignocellulosic biomass or a high activity of cellulase is needed to break the structure and release the polysaccharide for its utilization.

The bioconversion of lignocellulosic biomass into lipids by oleaginous microorganisms generally involves biomass pretreatment, enzymatic hydrolysis, and fermentation of sugars into lipids. This oleaginous cell cultivation on lignocelluloses shares many similarities with lignocellulosic ethanol production. Pretreatment is the key step to disrupt the complex lignocellulose structure and make cellulose and hemicellulose more accessible to cellulase enzymes for hydrolysis. Acid, alkaline, or ammonia fiber explosion (AFEX) pretreatment methods have been successfully applied in industrial conditions. The major sugars hydrolyzed from lignocellulosic biomass are glucose and xylose, with some amount of minor sugars such as arabinose, mannose, or galactose. Thus the oleaginous strains that can utilize both glucose and xylose are preferred, and the ability to utilize those minor sugars is also desirable [107]. Separate hydrolysis and fermentation (SHF) and simultaneous saccharification and fermentation (SSF) are the two common fermentation strategies in lignocellulose-based lipid conversion. In SHF, the pretreated biomass was enzymatically hydrolyzed, and then the hydrolysate was used as substrate for microbial cultivation. SSF combines cellulose hydrolysis with microbial lipid fermentation into an integrated process. SSF can prevent the inhibition of enzymes by sugar and improve the yield of enzymatic hydrolysis, thus improve lipid yield. It also renders some difficulties and challenges due to the higher working temperature of cellulase, compared to the lower temperatures where most fungi and algae strains grow.

Recent studies confirmed the conversion of cellulose and hemicellulose hydrolysate into lipids by oleaginous fungi, yeast, or algae strains [12, 13]. Both C5 and C6 monomeric sugars become available after hydrolysis, as well as various inhibitors. Selecting the strains that can utilize both C5 and C6 sugars and also high tolerance to lignocellulose degradation compounds will be an important factor to the success of lignocellulosic-based biofuel production.

4.4 Mass Cultivation of Oleaginous Fungi/ Yeast for Commercial Applications

The high cell density during the cultivation of oleaginous fungi and yeasts is one of the most important prerequisites for the mass cultivation in the commercial production of biofuels and coproducts. For example, some oleaginous species such as *C. curvatus* and *R. toruloides* [35, 69] can achieve more than 100 g/L cell biomass with more than 60% lipid content, which is attractive in the industrial production of microbial lipids. The scaling-up of lipid production can be realized in large-scale fermenters or stirred tank bioreactors. Batch, fed-batch and continuous processes are also commonly used cultivation methods in the mass production of oleaginous fungi. Researchers have attempted to increase lipid production and yield at the industrial level. Recently, Lorenz et al. [108] developed a two-stage fed-batch process for lipid generation in *R. glutinis*. By manipulating different factors, an industrially applicable process was established with biomass concentration of 106 g/L combined with a high lipid content of 63% and a lipid yield of 0.18 ± 0.02 g/g using sucrose as carbon source in 84 h. Similarly, Lin et al. [109] performed a two-stage fermentation of *L. starkeyi* with fed-batch process using nonsterile glucose solution as carbon source, and the biomass, lipid content and lipid productivity reached 104.6 g/L, 64.9% and 1.6 g/L/h respectively in 40 h.

Another prerequisite for the commercial heterotrophic cultivation of oleaginous fungi in bioreactors is that the fungal strain has the ability to grow in an inexpensive medium. Besides the conventional sugar substrate (such as glucose, xylose, fructose, and sucrose) and sugar-containing materials (such as molasses, starch wastewater), the lignocellulosic materials are increasing used as feedstock to lower the costs and increase sustainability and economic viability. As mentioned above, SHF and SSF are widely used in the lignocellulosic conversion; nonetheless, consolidated bioprocessing (CBP) is another process that has been attempted for lignocellulosic lipid production [20]. In CBP, biological conversion is consolidated into a single-scheme process that comprises enzyme production, cellulose hydrolysis, and fermentation. Although the current lipid yield from CBP is still low [20], it shows great potential as an ultimate low-cost process for biofuel production [110].

Other than the cultivation methods and process design during the production process, the supply of feedstock also comprises an important component of commercial production. Feedstock for large-scale lipid production must be available in large quantities and easy accessibility with low costs. There is an inherent dilemma between the distributed feedstock production system from marginal land and the large-scale conversion in the traditionally compact refineries. The biofuel production system needs to be compatible with the distributed biomass production system to maximize the usage of marginal lands. Second, future biofuels need to be "drop-in" and compatible with the current energy infrastructure. Numerous projects have been funded by the US Department of Energy (DOE) and the US Department of Agriculture (USDA), as well as the National Science Foundation (NSF) to study the whole process for generating liquid biofuel products from non-food source lipids. For example, projects funded from the Biomass Research and Development Initiative (BRDI) program under USDA and DOE are focusing on various topics such as feedstock development, cell biomass harvesting, biological and chemical conversion, and life cycle assessment.

5 Heterotrophic Microalgae Cultivation for Lipids Production

The lipid production of oleaginous fungi and yeasts is usually compared to the lipid production of oleaginous microalgae under heterotrophic condition. Microalgae use carbon dioxide as carbon source and sunlight as energy sources in most cases to produce microbial lipids as most microalgae are obligate autotrophic organisms. By changing the culture conditions or using genetic engineering, some autotrophic microalgae can be converted to heterotrophic ones, which utilize organic carbon source for the growth [111]. The carbon source could be sugars, starch, acetate, glycerol, and waste materials. Many heterotrophic microalgae exhibit rapid growth and high productivity, and many species can be induced to accumulate substantial quantities of lipids, often greater than 60% of dry biomass [112]. Compared to autotrophic microalgae cultivation, the heterotrophic cultures have several major limitations: limited number of microalgal species can grow heterotrophically; increasing energy expenses and costs by adding an organic substrate; and inability to produce light-induced metabolites [113]. However, the heterotrophic cultivation of microalgae can achieve significant higher biomass density and lipid productivity in comparison to the autotrophic mode [114, 115], and it can be adapted to generate a wide variety of microalgal metabolites at different scales from lab bench to industrial scale [116–118]. Heterotrophic cultivation of microalgae has higher biomass production with no light limitation, a high degree of process control, and lower costs for harvesting [119]. In most cases, heterotrophic cultivation is far cheaper, simpler to construct facilities, and easier to maintain on a large scale. These advantages significantly reduce the operation costs in most processes with heterotrophic culture, and will fit the need for industrial biofuel and bioproducts production. Since most of these unique microalgae strains have been historically identified as fungi/yeast. Heterotrophic microalgae can reach attractive high cell density under specific culture conditions, therefore, sharing many same features as fungal lipid accumulation. Thus some current studies about heterotrophic microalgae are also included in this chapter.

Table 3
Oil content of some oleaginous microalgae [120, 121]

Microalgae	Oil content (% dry weight)	Microalgae	Oil content (% dry weight)
Botryococcus braunii	25–75	*Monallanthus salina*	>20
Cylindrotheca sp.	16–37	*Nannochloris* sp.	20–35
Chlorella vulgaris	14–22	*Nannochloropsis* sp.	31–68
Chlorella protothecoides	55.2	*Neochloris oleoabundans*	35–54
Chlamydomonas reinhardtii	21	*Nitzschia* sp.	45–47
Crypthecodinium cohnii	15–30	*Phaeodactylum tricornutum*	20–30
Dunaliella primolecta	23	*Scenedesmus obliquus*	14–22
Euglena gracilis	14–20	*Schizochytrium* sp.	50–77
Isochrysis sp.	25–33	*Tetraselmis sueica*	15–23

5.1 Current Promising Candidates

Several algal species have the capability to accumulate a high content of lipids in their cell biomass. Table 3 lists the oil content of several oleaginous algae species. *Chlorella vulgaris* is a commercially important green microalga due to its high photosynthetic efficiency. *C. protothecoides* is another single-cell green microalga, and heterotrophic growth of *C. protothecoides* supplied with acetate, glucose, or other organic compounds as carbon source, results in high biomass and high content of lipid in cells [114]. The heterotrophic microalgae *Schizochytrium* sp. can accumulate 50–77% lipids with up to 48.95% docosahexaenoic acid (DHA) of its total lipids [122], and it has been used for commercial production of docosahexaenoic acid (DHA). Yang et al. [123] isolated and screened *Schizochytrium* strains for DHA production from macroalgae, fallen leaves of mangrove trees and mud from mangrove ecosystem. DHA is also commercially manufactured from another heterotrophic microalgae *Crypthecodinium cohnii.* In addition, some microalgae have the capability to accumulate up to 100 g/L dry biomass in the continuous heterotrophic cultivation (e.g., *Schizochytrium* sp., *C. cohnii*, and *Chlorella* sp.) [124, 125], which also makes these algae attractive for fuel production.

5.2 Cultivation of Heterotrophic Microalgae for Lipid Production

The mass cultivation of heterotrophic oleaginous microalgae is feasible since the heterotrophic algae growth is independent of light, and allows for much simpler scale-up possibilities since smaller reactor surface to volume ratio may be used than autotrophic because light intensity does not have to be optimized. The microalgal species which are currently attracting commercial interest can grow heterotrophically and perform efficiently in conventional bioreactors in a similar manner to bacteria or yeast [126]. Li et al.

[127] outlined the feasibility for large-scale biodiesel production based on heterotrophic cultivation of *Chlorella protothecoides* in different scales of bioreactors. Also, several green microalgae (*Chlorella* sp., *Dunaliella salina*, *C. cohnii*, *Schizochytrium* sp.) and cyanobacteria (*Spirulina* sp.) have been commercially produced at large scale as food additives for the human nutrition application [128, 129]. The cost efficiency of large-scale systems is dependent on the culture stability, nutrient source sustainability, the level of biomass and lipid productivity, and the extraction and conversion of lipids to biofuel products. The algal biodiesel has to be made to be competitive compete with the current diesel fuel. The cost of feed stock or carbon source required for the production of microbial lipids accounts for 50–60% of the total costs of the biodiesel [130]. Glucose is the carbon source most widely employed for the growth of all oleaginous microorganisms including microalgae, but other than glucose, carbon sources such as xylose, lactose, galactose, fructose, sucrose, mannose, acetate, ethanol, and glycerol have also been investigated for certain microalgae species [131]. Some researchers have focused on finding cheaper less expensive organic carbon sources such as agricultural/forest residues, lignocellulosic biomass, food wastes, municipal wastes, and animal wastes, similar to the efforts made in fungal oil accumulation processes. Xu et al. [114] cultivated the alga *C. protothecoides* in a heterotrophic condition with inexpensive corn powder hydrolysate as the sole food source and resulted in high biomass (2.02 g/L/day) and lipid (0.93 g/L/day) productivities. Li et al. (2011) produced high oil content and fast growing algae *C. pyrenoidosa* by using low-cost rice straw hydrolysate and achieved 1.10 g/L/day biomass production and 0.62 g/L/day lipid production.

Lists of several substrates used for lipid production by heterotrophic microalgae are provided in Table 4.

6 Downstream Processing of Microbial Lipid Production

Downstream processing costs remain a major obstacle to economically produced microbial lipids. Among all downstream processes/steps such as lipid extraction, conversion of lipids to biofuels, and coproducts utilization, the extraction of lipids from microbial biomass is most important, because efficiency of cell disruption directly influences subsequent downstream operations and overall extraction efficiencies.

6.1 Extraction of Lipids

Extraction process is followed after lipid production with the purpose to separate lipids from other constituents, and it is one of the most costly processes determining the economic sustainability of whole biofuel production process. When using oleaginous microorganisms for lipid production, drying and cell disruption is often

Table 4
Examples of heterotrophic microalgae cultivation for oil accumulation on various substrates

Microalgae	Substrates	Biomass (g/L)	Biomass productivity (g/L/day)	Lipid content (% dry weight)	Lipid (g/L)	Lipid productivity (g/L/day)	Lipid yield (g/g substrate)	Culture mode	Reference
Chlorella protothecoides	Glucose	12.8–15.5	1.67–2.02	44.3–48.7	6.24–7.15	0.73–0.81	–	Fed-batch	[127]
Chlorella protothecoides	Glucose	3.2	0.55	57.8	1.85	0.32	0.12	Batch	[132]
Chlorella protothecoides	Glucose	16.8–51.2	2.19–7.31	50.3–55.2	9.27–25.75	1.21–3.68	–	Fed-batch	[132]
Chlorella protothecoides	Whey permeate hydrolysate	9.1	1.30	42.0	3.82	0.55	–	Batch	[133]
Chlorella protothecoides	Whey permeate hydrolysate	17.2	1.72	20.5	3.53	0.35	–	Fed-batch	[133]
Chlorella protothecoides	Whey permeate	7.3	0.73	49.9	3.64	0.36	–	SSF with batch mode	[133]
Chlorella protothecoides	Waste molasses hydrolysate	70.9	9.56	57.6	40.8	5.50	–	Fed-batch	[134]
Chlorella vulgaris	Glucose	1.21	0.15	23.0	0.28	0.035	0.028	Batch	[131]
Chlorella vulgaris	Acetate	0.9–0.99	0.08–0.09	31.0–36.0	0.32–0.35	0.027–0.029	0.032–0.035	Batch	[131]
Chlorella vulgaris	Glycerol	0.66–0.72	0.09–0.10	22.0–34.0	0.15–0.22	0.022–0.031	0.011–0.015	Batch	[131]
Chlorella vulgaris	Glucose	0.8	0.13	15.8	0.13	0.021	–	Batch	[135]
Chlorella zofingiensis	Glucose	9.7	0.69	51.1	4.96	0.35	0.17	Batch	[115]

Chlorella sorokiniana	Glucose	3.2	0.46	56.0	1.80	0.26	–	Batch	[136]
Chlorella pyrenoidosa	Rice straw hydrolysate	2.83	1.42	56.3	1.59	0.80	0.12	Batch	[137]
Crypthecodinium cohnii	Glucose	26.0	5.20	15.0	3.80	0.76	–	Fed-batch	[124]
Crypthecodinium cohnii	Acetic acid	109.0	6.54	56.0	61.0	3.66	–	Fed-batch	[124]
Crypthecodinium cohnii	Ethanol	83.0	9.05	42.0	35.0	3.82	–	Fed-batch	[138]
Schizochytrium sp.	Glucose	71.0	12.91	49.9	35.17	6.39	–	Fed-batch	[122]
Schizochytrium limacinum	Maize starch hydrolysate	85.27	21.32	52.4	44.68	11.17	–	Fed-batch	[139]
Scenedesmus sp.	Glucose	3.46	0.58	43.4	1.50	0.25	–	Batch	[140]
Nitzschia laevis	Glucose	–	2.80	14.7	–	0.41	–	Continuous	[141]
Monoraphidium sp.	Glucose	3.96	0.40	37.6	1.49	0.15	–	Batch	[142]

required before lipid extraction to recover intracellular products. After extraction, lipids can go through transesterification process to produce biodiesel, or transfer into value-added products through other process like thermochemical conversion.

Based on the structure, texture, sensitivities, and lipid content differences between microbial, plant, and animal tissues, different extraction strategies have been developed. The current industrial methods include solvent extraction and pressing. These methods can be readily applied to microbial lipids extraction, as they are efficient and cost effective. There are also new types of extraction methods like supercritical fluid extraction, high pressure homogenization, microwave-assisted extraction, and ultrasonic extraction that have been developed to improve extraction efficiency [143]. Except for supercritical fluid extraction, the other three methods are actually pretreatment processes that need to combine with other extraction methods for increased efficiency.

Solvent extraction is probably the most widely used method in lab scale lipid analysis, which transfers lipids from water solution into organic solvents based on their different relative solubilities in these two immiscible liquids. In practical, different lipids have different polarities and it is impossible to select a single organic solvent to extract them all. In addition, solvents must both be readily dissolve the lipids and also overcome the interactions between the lipids and the tissue matrix. In this situation a mixture of organic solvent is usually applied. The greatest improvement in lipid and wax extraction was made in 1957 when Folch et al. [144] described their classic extraction procedure. This procedure remains the most commonly used by researchers around the world. Another commonly used extraction method is Soxhlet method, in which organic solvent are repeated washing dried biomass in Soxhlet extractor. Although this method has some limitations like low extraction of polar lipids and hazards of boiling organic solvents, it also offers several advantages like low cost and less solvent use, and has been applied in some large-scale biodiesel research.

Pressing is a mechanic method which uses pressure to squeeze lipids out from lipid-containing biomass. It may be the oldest method for oil extraction, requires simple and sturdy equipment, and is easy to operate. The extraction efficiency is highly dependent on the raw material used in pressing, varies from low limit of 6.1% obtained from hazelnut seeds [145] to 86–92% obtained from rapeseed [146]. When using microalgae as raw biomass, the extraction efficiency can reach up to 75% [147]. Pressing process can be adapted quickly to process different kinds of oilseeds, and chemical-free protein-rich cake can also be obtained for other purpose like feeding material [148].

Another extraction technology that draws attention is called supercritical fluid extraction, which uses supercritical fluid solvent in extraction. Compared with solvent extraction, it eliminates the use of organic solvents, which reduces the problems of solvent storage, disposal, and environmental concerns. The most commonly used supercritical solvent is carbon dioxide, which does not lead to contamination or thermal degradation of target compounds. During extraction process, CO_2 is compressed beyond its critical temperature of 31 °C and critical pressure of 74 bar. At this state no surface tension is present in supercritical fluids and viscosities are much lower than in liquids, so the supercritical fluids have much higher diffusion coefficients than liquids and can penetrate into small pores that are inaccessible to liquid. Research has shown that the lipid extraction has the ability to reach more than 90% of theoretical value in a short time [149]. Some parameters are generally considered to influence the extraction yields, including pressure [150], moisture content [149] and pretreatment of sample [151]. Like pressing method, the supercritical fluid extraction method is able to produce chemical-free residue as feeding material or fertilizer. Currently supercritical fluid extraction remains at laboratory research stage, factors like high capital investment and high pressure requirements limit its use in oil industry.

6.2 Conversion of Lipids to Biofuels

6.2.1 Transesterification to Biodiesel

Triglycerides or TAG are the major components of accumulated microbial lipids and the main material for biodiesel production. After extraction, TAG has high viscosity, low volatility and polyunsaturated characteristics, which prohibits its direct use as liquid biofuel. The most widely studied and applied technique that improves biofuel properties is transesterification. In this process, long-chain fatty acids are exchanged from TAG by methanol or ethanol, generating glycerol and FAME or FAEE. This reaction is reversible in nature, so excess alcohol is generally used to shift the equilibrium toward the product side. There are multiple choices of catalysts in transesterification, such as acid [152], alkaline [153, 154], or enzyme [127]. Alkaline catalysts like NaOH and KOH can achieve a high conversion yield in a short time under low temperature and atmospheric pressure. However, they cause the free fatty acids to produce soap, and are not ideal for microbial biodiesel production since high free fatty acid content is normally found in microbial oils. Compared to alkaline catalysts, acid catalysts (normally H_2SO_4 or HCl) require larger response time and higher temperature than alkaline catalysts, and are not sensitive to free fatty acid content. They are typically used when free fatty acids content is higher than 1%. Enzymes can also be used in the transesterification process; a 98% biodiesel conversion yield has been reported using immobilized lipase from *Candidiasis* sp. and oil extracted from microalgae [127]. Lipases from other species are also reported to have a high biodiesel conversion yield [155], revealing that lipase could be an effective catalyst in transesterification.

To improve the transesterification process, other types of technologies including microwave and ultrasonic technology have also been applied to increase efficiency. For microwave technology, the main mechanism is to let methanol absorb microwave radiation, redirect its dipole rotation and allow the destruction of methanol--lipid interface [156]. It has been proved that this technology can efficiently accelerate the reaction to greatly reduce the reaction time [157]. One major flaw is that KOH is known as the most suitable catalyst when using microwave irradiation in transesterification [158, 159], and that could be a concern because of the high free fatty acid in microbial oils. Microwave assisted transesterification without catalysts has also been investigated, in which higher temperature, larger amount of solvents and longer reaction times were required [160]. For ultrasonic technology, it uses high frequency sound wave to compress and stretch the molecular spacing of media, enhance the mass transfer rate and initiate the transesterification reaction [161]. It has a similar function as microwave radiation to accelerate reaction, and a biodiesel yield of also 100% in 10–20 min using continuous ultrasonic was obtained [161].

Besides the two step extraction–transesterification process, one-step transesterification has also been investigated. This process is also called as direct transesterification, because it combines oil extraction and transesterification in one step and directly converts oil-rich biomass to biodiesel. Instead of using refined lipids as feedstock, using biomass as feedstock eliminates the lipid extraction and purification process, which could be a useful method to decrease biodiesel production cost [162]. In addition, direct transesterification could utilize membrane phospholipids which are hard to be extracted from dried biomass, thus increase the biodiesel production yield [163]. Solvents like methanol that could be simultaneously used as extraction and transesterification reagent would be desirable to apply in direct transesterification process, and the use of additional solvent like hexane or chloroform helps to enhance both extraction and esterification efficiency [164]. Different reports claim that important parameters affect this process includes catalysts, alcohol type and alcohol–lipids molar ratio, reaction time and temperature, cosolvent, and water content [163, 165–167].

The novel technologies including supercritical fluid, microwave irradiation and ultrasonic have also been applied in direct transesterification to enhance the efficiency of direct transesterification. Supercritical fluid was used to extract lipid from biomass. When applied in direct transesterification, the fluid has been changed from CO_2 to methanol in order to conduct extraction and transesterification process at same time. This process does not need catalysts, and can reach 85.8% conversion yield with relatively high moisture content [156]. Microwave irradiation and ultrasonic are both found to be assisted in lipid extraction. They can break cell wall structure either by microwave penetration [159] or cavitation bubbles produced by ultrasound [168]. These two processes were also proved to have higher conversion yield than transesterification process using conventional heating [168, 169]. Although all these novel technologies enhance the biodiesel conversion yield or reaction rate, for commercialization the production cost needs to decrease.

6.2.2 Thermochemical Conversion to Biofuels

Besides transesterification processes discussed above, thermochemical conversion is another method to utilize organic materials like microbial biomass for fuel production. This method uses thermal decomposition to treat organic compounds to generate fuel products. The easiest and most direct method is combustion. In this process biomass is burnt with the presence of air to generate heat, mechanical power of electricity. However, combustion is feasible only for biomass with less than 50% moisture content [170]. Meanwhile, the heat produced must be used immediately which makes the combustion method less attractive. More widely researched thermochemical conversion processes include gasification, liquefaction, and pyrolysis. In these methods, either the whole biomass or the residual fraction after extraction of value-added products can be used, which undoubtedly increase financial feasibility. Among these conversion methods, gasification has the highest temperature requirement, and pyrolysis has received special attention based on its efficient utilization of various types of biomass [170].

In gasification, biomass is partial oxidized with lean oxygen at a high temperature of 800–900 °C [170]. Syngas, which is generated by reaction between biomass and oxygen or water (steam), is a gas mixture of CO, H_2, CO_2, N, and CH_4 [171]. The proportion of syngas is determined by the use of air, oxygen, or steam as the gasification medium [172]. Syngas can be produced by a variety of biomass and can be burnt directly or used as fuel for gas engines or gas turbines [172]. At current stage, most research uses microalgae as feedstock to study their gasification characteristics, and a marginal positive energy balance was obtained when using algae biomass as feedstock for gasification at 1000 °C [173]. However, as mentioned above, the cost associated with harvesting and downstream processing of microalgae is a huge barrier to commercialization of this technology. Using microalgae cocultured with fungi can create stable pellets, resulting in clear media. Developing novel inexpensive methods which are environmentally friendly are practical alternatives to currently available technologies such as filtration, centrifugation, and chemical flocculation [174].

In liquefaction, biomass material is converted into liquid fuel at a medium temperature (300–350 °C), high pressure (5–20 MPa) environment with presence of hydrogen [170]. Research related to liquefaction of microalgae has significantly increased in the last years, which indicates a raising interest for this conversion technology [175]. When using algae as feedstock, a maximum yield of 64% dry weight of oil is obtained with high heating value product and positive energy balance [176]. The major flaw of this process is that it generates a tarry lump product, which is difficult to handle [177], so reactors for liquefaction and fuel-feed systems are complex and expensive [172].

Pyrolysis is another thermochemical method to convert biomass into fuel products. Compared to gasification and liquefaction, pyrolysis is conducted at a medium to high temperature (350–700 °C) in the absence of air or oxygen [170]. The products of pyrolysis include bio-oil, syngas, and charcoal, and the composition of products can be modified by adjusting reaction parameters. Pyrolysis is usually divided into fast and slow pyrolysis. Fast pyrolysis requires higher temperature and short reaction time to mainly generate liquid fuels. Slow pyrolysis uses lower temperature and longer reaction time, and enhances the production of tars and chars [178]. Pyrolysis is proven to produce nonoxygenated liquid hydrocarbon mixtures as diesel fuel additive from triglycerides [179]. It can also achieve a high biomass-to-liquid conversion ratio, which makes this process appropriate to be scaled up to produce liquid fuel in a large scale and partially replace petroleum fuel [180].

6.2.3 Biochemical Conversion to Biofuels

Biochemical conversion is another method to utilize organic compounds like microorganisms for fuel production. This method use biological process to convert biomass into fuels, and major process using this method includes anaerobic digestion, alcoholic fermentation and photobiological hydrogen production.

Anaerobic digestion converts organic wastes into biogas, in which methane and carbon dioxide are the primarily components. This process occurs in three sequential stages of bacterial hydrolysis, acidogenesis, and methanogenesis. Organic compounds are broken down into soluble derivatives like sugars and amino acids through bacterial hydrolysis, and then converted into acetic acid, volatile fatty acids, carbon dioxide, hydrogen, and ammonia. In methanogenesis process, these components are further converted into methane and carbon dioxide [181]. Anaerobic digestion process is appropriate for high moisture content (80–90% moisture) organic wastes [172], which makes wet oleaginous microorganisms biomass a suitable substrate. Besides energy recovery from methane gas, the anaerobic digestion process also leaves a nutrient rich waste product that can be recycled into new growth medium or served as fertilizer in cropland [182]. Alcoholic fermentation is the conversion of biomass materials into ethanol [172]. This process uses sugars, starch or cellulose in microorganisms, and converts them into ethanol by anaerobic fermentation using yeast. Compared with anaerobic digestion, this process may leave high amounts of solid residue with significant amount of lipids and protein, which can be used as substrate of gasification or as feeding material.

Photobiological hydrogen production can only applied on algae, which is a process for microalgae to produce hydrogen under certain conditions in a closed photobioreactor. During photosynthesis, microalgae convert water molecules into hydrogen ions (H^+) and oxygen; the hydrogen ions are then subsequently converted by hydrogenase enzymes into H_2 under anaerobic conditions [181]. The first approach is a two-stage process starting with the photosynthetically growth of microalgae. After this cultivation, algae are deprived of sulfur and thereby anaerobic conditions are induced to stimulate consistent hydrogen production [183]. The second approach involves the simultaneous production of photosynthetic O_2 and H_2 gas. Electrons that are released upon photosynthetic H_2O oxidation directly feed into the hydrogenase-mediated H_2-evolution process, without involving intermediate CO_2 fixation and energy storage as cellular metabolites [184]. Currently, the single-stage mechanism has encountered several limitations, including the suppression of H_2 production in the presence of O_2, and the actual yield of H_2 from two-stage process can only reach to about 10% of the theoretical maximum value, and an actual yield is around 1 mol H_2 per m^2 culture area per day [183].

6.3 Using of By-Products

Fatty acids (straight chain hydrocarbons) are the desired product for biodiesel production, and the prevailing fatty acids in oleaginous microorganisms are palmitic acid, oleic acid and linoleic acid [185]. However, they are not the only lipids produced in cell, and oleaginous microorganism production can also provide many valuable by-products like lubricants, emulsifiers, biopolymers, plasticizers, nutraceuticals, pharmaceuticals, animal feed, and fertilizers [41, 186]. Lipids discovered in cell also include steryl esters, free sterols, glycerophospholipids, sphingolipids, glycolipids, terpenoids, carotenoids, hydrocarbons long chain alcohols (fatty alcohols), waxes, polyprenols, isoprenoid quinones, and others [187]. Some of these components are also feasible and economical to produce for other types of biofuel in addition to biodiesel. For example, *Botryococcus braunii* is regarded as a potential source of renewable fuel because of its ability to produce large amounts of hydrocarbons [188, 189]. Depending on the strain and growth conditions, up to 75% of algal dry mass can be in the form of hydrocarbons [188]. *Gliocladium roseum*, an endophytic fungus, was shown to produce a series of volatile hydrocarbons and hydrocarbon derivatives on an oatmeal-based agar and pure cellulose-based agar under limited oxygen conditions [190]. Many of these hydrocarbons are major substances in diesel fuel, like octane, 1-octene, 2-methyl heptane, hexadecane, 4-methyl undecane, 3-methyl nonane, and 1,3-dimethyl benzene [190].

A variety of high value lipid compounds can also be obtained oleaginous microorganisms, and the most well-studied ones are PUFA (as dietary supplements), terpenoids (as pharmaceuticals), and carotenoids such as astaxanthin (as dietary supplements and food colorant) [41, 191, 192]. The most important PUFAs omega-3 and omega-6 such as alphalinolenic acid (ALA), linoleic acid (LA), EPA, DHA, GLA, and AA are used as nutraceuticals. They are important structural components of the phospholipid cell membranes and have multiple physiological functions [193, 194]. A few processes have focused on commercial manufacture of PUFA using heterotrophic growth of oleaginous fungi and microalgae [26, 28, 195, 196], which can serve as a sustainable alternative of PUFA from fish oil or plant oil. Terpenoids are a large group of natural organic compounds that are similar to terpenes, and have been proposed for biofuel because of their hydrocarbon-like structure. In reality, the yield of terpenoids with the microbial production system is very low [197] and it requires the input of glucose or sucrose which undoubtedly increases cost. Currently, the production of terpenoids is more suitable as high value drugs for the pharmaceutical industry, and they are under investigation for antibacterial, antineoplastic, and other pharmaceutical applications. Carotenoids are organic pigments found in bacteria, fungi, yeast, and photosynthetic organisms, and also in eukaryotes [198]. They are substantially hydrophobic antioxidants that are widely utilized in medicine as antioxidants or in food chemistry as colorants, and humans are mostly incapable of synthesizing carotenoids. Astaxanthin is a keto-carotenoid found in microalgae, yeast and some fish (salmon, trout). It is the most powerful natural antioxidant and widely applied as a nutraceutical, pharmaceutical, safe colorant, and aquaculture feed additive [199]. These high value lipids will be more suitable for pharmaceutical or nutraceutical use rather than converted into biodiesel.

Besides the various lipid compounds produced in cell, oleaginous microorganisms also have significant amount of protein, carbohydrates and other types of nutrients, which can be used as by-products from biofuel industry. Based on different biodiesel production methods, various types of by-products can be obtained for specific purpose. Nonsolvent extraction method like pressing or supercritical fluid extraction methods produce chemical-free residue. They can be served both as animal feed or farmland fertilizer, especially in the case that algae are the natural food source of many important aquaculture species such as molluscs, shrimps, and fish [200]. In the case of oleaginous fungi and yeasts, the microorganism cake after lipid extraction could be sold as animal feed or be further processed to produce amino acids or peptides, which have wide applications like biomaterials, bioplastic, and biofoam [201]. When using pyrolysis for biodiesel conversion, solid charcoal residue "biochar" is generated together with bio-oil and syngas. It can be used as biofertilizer in agricultural for carbon sequestration [202–204]. Glycerol generated from transesterification processes could be easily transformed into a list of value-added products

[205], or can be directly use as carbon source for heterotrophic growth of oleaginous microorganisms.

In another way, the biomass residues after lipid extraction can be used as feedstock to generate various types of biofuel. Microalgae meal has been explored to produce ethanol [206, 207] and biogas [181, 208] from fermentation process, and the same method could be applied on fungal biomass [209]. Modification of the fermentation conditions will help to obtain other types of energy like H_2 [184]. Moreover, a few lab-scale heterotrophic growth of fungi, yeast, and microalgae have been established for the enrichment of specific compounds like pigments and antioxidants [128, 200, 210, 211]. Some recent studies also suggested that besides lipid-rich biomass, microalgae with low lipid content could also be used to produce a hydrocarbon rich fuel as refinery feedstock. The efficient use of these by-products can improve biofuel industrial value, leading to significant positive impact to reduce the production cost of biodiesel and make it more economically feasible.

7 Main Related Technical Challenges

Oleaginous microorganisms have become promising candidates for biofuel production. They can avoid competing with food crops for arable lands if lignocellulosic and other types of waste materials are used for cell cultivation, and can reach a very high production rate compared to crops. In general, high growth rates, reasonable growth densities and high oil contents are the main reasons for microbial lipid to be attractive as biodiesel feedstocks. Correspondingly, the main technical challenges for the economically feasible biofuel include: biomass productivity, cellular lipid content, overall lipid productivity and yield, and substrates including carbon sources and other nutrients.

In early explorations, expensive carbon sources such as starch are the substrates of interest, but the microbial production of TAGs from starch was not profitable compare to low-cost plant oils and fats. For profitability, the relatively low-cost biodiesel end-product will require low-cost substrates, highly efficient conversion system and the capability to yield valuable products [41]. Heterotrophic fermentation using lignocellulosic sources of carbohydrates may best fit these requirements. Current cultures of oleaginous strains on lignocellulosic hydrolysates are facing several challenges such as low lipid content, low lipid productivity and low lipid yield, which are caused by several technical problems like limited sugar concentrations, low control on C:N ratio and low tolerance of inhibitor generated from pretreatment [41]. Besides carbon source, the major nutrients required for biomass growth like nitrogen, phosphorous, iron and sulfur are also crucial. Stable supply and management of nutrients is also important in the consistent accumulation of intercellular lipids. Once the culture of oleaginous microorganisms finished, there is also technical requirements on low water use, high efficiency harvest, high efficiency oil extraction, and downstream processing. The capital cost of required equipment and energy required to extract the oil are relatively high, and algal-based biofuel price is estimates to range from $300 to $2600 per barrel based on current technology [212]. Obviously, it is needed to overcome these technical hurdles and make the microbial-based biodiesel economical feasible.

The crude microbial oil is chemically similar to crude fossil fuel oil, and the engineering challenges on oil conversion are mostly well managed by current petroleum companies [179]. This inspires researchers to build collaborations between microbial lipid production companies and major oil companies for maximizing downstream processing efficiencies. The potential to yield separable, valued coproducts would be a bonus for the profitability of microbial-based biodiesel. A recent popular concept is SCO biorefinery, which integrates the conversion process of biomass into biofuels with production of coproducts like oleochemicals and power. With difference choices of industrial process discussed above, various types of biofuel refinery can be designed. Furthermore, biorefinery process can be more economical and efficient when integrated with industries dealing with large amounts of biomass, such as pulp and paper mill, agricultural residues, and municipal solid waste treatment. It has been considered as a way to maximize the biofuel production and coproducts while reducing industrial waste disposal.

8 Conclusions

Microbial lipids offer potential for sufficient production of renewable fuels. In response to the demand and increasing price of edible oils used in the production of first generation biofuels, many studies have considered the lipid synthesized by oleaginous fungi and yeasts as a potentially economically viable process. At current stage, heterotrophic cultivation of fungi, yeast, and algae can accumulate a significant amount of microbial oil for downstream process. In order to make biofuels economically competitive with petroleum fuels, underutilized substrates such as organic waste and lignocellulosic biomass are under investigation to provide inexpensive carbon sources for lipid accumulation. The utilization of microbial lipid depends on optimized cell cultivation systems, while efficient oil extraction and biofuel production systems are also essential components. Utilization of by-products can be an effective component to make biofuel more economic feasible.

The exploration and screening of the natural biodiversity with a great number of native species is a promising strategy to identify novel oleaginous species that are capable of assimilating and accumulating lipids from agro-industrial residues, particularly the lignocellulosic biomass and by-products/waste materials. Based on the contents discussed in this chapter, the future research on microbial lipid biofuel will mainly focus on two aspects: increase the production rate and decrease the cost of biofuel production from microbial lipids. Genetic modification of oleaginous microorganisms and further improvement of pretreatment technologies, and process design and the development of technologies for valuable coproduct generation are focused in this review. These research works will undoubtedly open a possibility of industrial application of biofuel production.

References

1. Ratledge C (1991) Microorganisms for lipids. Acta Biotechnol 11(5):429–438
2. Biermann U, Friedt W, Lang S, Lühs W, Machmüller G, Metzger JO, Rüsch Gen Klaas M, Schäfer HJ, Schneider MP (2000) New syntheses with oils and fats as renewable raw materials for the chemical industry. Angew Chem Int Ed Eng 39:2206–2224
3. Corma A, Iborra S, Velty A (2007) Chemical routes for the transformation of biomass into chemicals. Chem Rev 107:2411–2502
4. Sakuradani E (2010) Advances in the production of various polyunsaturated fatty acids through oleaginous fungus Mortierella alpina breeding. Biosci Biotechnol Biochem 74:908–917
5. Zhang J, Hu B (2012) Solid-state fermentation of Mortierella isabellina for lipid production from soybean hull. Appl Biochem Biotechnol 166:1034–1046
6. Meng X, Yang J, Xu X, Zhang L, Nie Q, Mian M (2009) Biodiesel production from oleaginous microorganisms. Renew Energy 34:1–5
7. Patnayak S, Sree A (2005) Screening of bacterial associates of marine sponges for single cell oil and PUFA. Lett Appl Microbiol 40:358–363
8. Xia C, Zhang J, Zhang W, Hu B (2011) A new cultivation method for microbial oil production: cell pelletization and lipid accumulation by Mucor circinelloides. Biotechnol Biofuels 4:15
9. Beopoulos A, Cescut J, Haddouche R, Uribelarrea JL, Molina-Jouve C, Nicaud JM (2009) *Yarrowia lipolytica* as a model for bio-oil production. Prog Lipid Res 48:375–387
10. Ratledge C (1993) Single cell oils—have they a biotechnological future? Trends Biotechnol 11:278–284
11. Papanikolaou S, Muniglia L, Chevalot I, Aggelis G, Marc I (2003) Accumulation of a cocoa-butter-like lipid by Yarrowia lipolytica cultivated on agro-industrial residues. Curr Microbiol 46:124–130
12. Hu C, Zhao X, Zhao J, Wu S, Zhao ZK (2009) Effects of biomass hydrolysis by-products on oleaginous yeast Rhodosporidium toruloides. Bioresour Technol 100 (20):4843–4847
13. Ruan Z, Zanotti M, Zhong Y, Liao W, Ducey C, Liu Y (2012) Co-hydrolysis of lignocellulosic biomass for microbial lipid accumulation. Biotechnol Bioeng 110:1039–1049
14. Li Y, Zhao Z, Bai F (2007) High-density cultivation of oleaginous yeast *Rhodosporidium toruloides* Y4 in fed-batch culture. Enzyme Microb Technol 41:312–317
15. Lin H, Cheng W, Ding HT, Chen XJ, Zhou QF, Zhao YH (2010) Direct microbial conversion of wheat straw into lipid by a cellulolytic fungus of Aspergillus oryzae A-4 in solid-state fermentation. Bioresour Technol 101:7556–7562
16. Vicente G, Bautista LF, Gutierrez FJ, Rodriguez R, Martinez V, Rodriguez-Frometa RA, Rui-Vazquez RM, Torres-Martinez S, Garre V (2010) Direct transformation of fungal biomass from submerged cultures into biodiesel. Energy Fuel 24:3173–3178
17. Ratledge C (2004) Fatty acid biosynthesis in microorganisms being used for single cell oil production. Biochimie 86:807–815

18. Ruan Z, Zanotti M, Wang X, Ducey C, Liu Y (2012) Evaluation of lipid accumulation from lignocellulosic sugars by Mortierella isabellina for biodiesel production. Bioresour Technol 110:198–205
19. Zeng J, Zheng Y, Yu X, Yu L, Gao D, Chen S (2013) Lignocellulosic biomass as a carbohydrate source for lipid production by *Mortierella isabellina*. Bioresour Technol 128:385–391
20. Wei H, Wang W, Yarbrough JM, Baker JO, Laurens L, Wychen SV, Chen X, Taylot LE II, Xu Q, Himmel ME, Zhang M (2013) Genomic, proteomic, and biochemical analyses of oleaginous Mucor circinelloides: evaluating its capability in utilizing cellulolytic substrates for lipid production. PLoS One 8:E71068
21. Chen H, Liu T (1997) Inoculum effects on the production of γ-linolenic acid by the shake culture of Cunninghamella echinulata CCRC31840. Enzyme Microb Technol 21:137–142
22. Fakas S, Papanikolaou S, Batsos A, Galiotou-Panoyotou M, Mallouchos A, Aggelis G (2009) Evaluating renewable carbon sources as substrates for single cell oil production by Cunninghamella echinulata and Mortierella isabellina. Biomass Bioenergy 33:573–580
23. Gutiérrez A, López-García S, Garre V (2011) High reliability transformation of the basal fungus Mucor circinelloides by electroporation. J Microbiol Methods 84:442–446
24. Rossi M, Amaretti A, Raimondi S, Leonardi A (2011) Getting lipids for biodiesel production from oleaginous fungi. In: Biodiesel: feedstocks and processing technologies. Intech. www.intechopen.com
25. Du Preez J, Immelman M, Kock JLF, Kilian SG (1995) Production of gamma-linolenic acid by Mucor circinelloides and Mucor rouxii with acetic acid as carbon substrate. Biotechnol Lett 17(9):933–938
26. Mamatha S, Ravi R, Venkateswaran G (2008) Medium optimization of gamma linolenic acid production in Mucor rouxii CFR-G15 using RSM. Food Bioprocess Technol 1:405–409
27. Somashekar D, Venkateshwaran G, Sambaiah K, Lokesh BR (2003) Effect of culture conditions on lipid and gamma-linolenic acid production by mucoraceous fungi. Process Biochem 38(12):1719–1724
28. Eroshin V, Dediukhina EG, Satrutdinov AD, Chistiakova TI (2000) Arachidonic acid production by Mortierella alpina with growth-coupled lipid synthesis. Process Biochem 35:1171–1175
29. Aki T, Nagahata Y, Ishihara K, Tanaka Y, Morinaga T, Higashiyama K, Akimoto K, Fujikawa S, Kawamoto S, Shigeta S, Ono K, Suzuki O (2001) Production of arachidonic acid by filamentous fungus, Mortierella alliacea strain YN-15. J Am Oil Chem Soc 78:5999–5604
30. Adachi D, Hama S, Numata T, Nakashima K, Ogino C, Fukuda H, Kondo A (2011) Development of an Aspergillus oryzae whole-cell biocatalyst coexpressing triglyceride and partial glyceride lipases for biodiesel production. Bioresour Technol 102(12):6723–6729
31. Beopoulos A, Chardot T, Nicaud J (2009) Yarrowia lipolytica: a model and a tool to understand the mechanisms implicated in lipid accumulation. Biochimie 91:692–696
32. Ageitos JM, Vallejo JA, Veiga-Crespo P, Villa TG (2011) Oily yeasts as oleaginous cell factories. Appl Microbiol Biotechnol 90:1219–1227
33. Li Q, Du W, Liu D (2008) Perspectives of microbial oils for biodiesel production. Appl Microbiol Biotechnol 80:749–756
34. Holdsworth J, Ratledge C (1991) Triacylglycerol synthesis in the oleaginous yeast Candida curvata. Lipids 26:111–118
35. Meesters P, Huijberts G, Eggink G (1996) High cell density cultivation of the lipid accumulating yeast Cryptococcus curvatus using glycerol as a carbon source. Appl Microbiol Biotechnol 45:575–579
36. Zhang L, Tang Y, Guo ZP, Ding ZY, Shi GY (2011) Improving the ethanol yield by reducing glycerol formation using cofactor regulation in Saccharomyces cerevisiae. Biotechnol Lett 33:1375–1380
37. Wang S, Sun JS, Han BZ, Wu XZ (2007) Optimization of β-carotene production by Rhodotorula glutinis using high hydrostatic pressure and response surface methodology. J Food Sci 72:M325–M329
38. Easterling ER, French WT, Harnandez R, Licha M (2009) The effect of glycerol as a sole and secondary substrate on the growth and fatty acid composition of Rhodotorula glutinis. Bioresour Technol 100:356–361
39. Yu X, Zheng Y, Dorgan KM, Chen S (2011) Oil production by oleaginous yeasts using the hydrolysate from pretreatment of wheat straw with dilute sulfuric acid. Bioresour Technol 102:6134–6140
40. Zhao X, Kong X, Hua Y, Feng B, Zhao Z (2008) Medium optimization for lipid production through co-fermentation of glucose and xylose by the oleaginous yeast Lipomyces

starkeyi. Eur J Lipid Sci Technol 110 (5):405–412

41. Jin M, Slininger PJ, Dien BS, Waghmode S, Moser BR, Orjuela A, Sousa Lda C, Balan V (2015) Microbial lipid-based lignocellulosic biorefinery: feasibility and challenges. Trends Biotechnol 33(1):43–54
42. Dey P, Banerjee J, Maiti M (2011) Comparative lipid profiling of two endophytic fungal isolates – Colletotrichum sp. and Alternaria sp. having potential utilities as biodiesel feedstock. Bioresour Technol 102:5815–5823
43. Peng X, Chen H (2007) Microbial oil accumulation and cellulase secretion of the endophytic fungi from oleaginous plants. Ann Microbiol 57:239–242
44. Subhash VG, Venkata Mohan S (2011) Biodiesel production from isolated oleaginous fungi Aspergillus sp. using corncob waste liquor as a substrate. Bioresour Technol 102:9286–9290
45. Kitcha S, Cheirsilp B (2014) Bioconversion of lignocellulosic palm byproducts into enzymes and lipid by newly isolated oleaginous fungi. Biochem Eng J 88:95–100
46. Tanimura A, Takashima M, Sugita T, Endoh R, Kikukawa M, Yamaguchi S, Sakuradani E, Ogawa J, Shima J (2014) Selection of oleaginous yeasts with high lipid productivity for practical biodiesel production. Bioresour Technol 153:230–235
47. Papanikolaou S, Aggelis G (2011) Lipids of oleaginous yeasts. Part I: Biochemistry of single cell oil production. Eur J Lipid Sci Technol 113:1031–1051
48. Weete J (1980) Sphingolipids. In: Lipid biochemistry of fungi and other organisms. Springer US, New York, pp 180–195
49. Sitepu I, Sestric R, Ignatia L, Levin D, German JB, Gillies LA, Almada LA, Boundy-Mills KL (2013) Manipulation of culture conditions alters lipid content and fatty acid profiles of a wide variety of known and new oleaginous yeast species. Bioresour Technol 144:360–369
50. Wu S, Hu C, Jin G, Zhao X, Zhao ZK (2010) Phosphate-limitation mediated lipid production by Rhodosporidium toruloides. Bioresour Technol 101(15):6124–6129
51. Zhao X, Peng F, Du W, Liu C, Liu D (2012) Effects of some inhibitors on the growth and lipid accumulation of oleaginous yeast Rhodosporidium toruloides and preparation of biodiesel by enzymatic transesterification of the lipid. Bioprocess Biosyst Eng 35:993–1004
52. Wynn JP, Hamid AA, Li Y, Ratledge C (2001) Biochemical events leading to the diversion of carbon into storage lipids in the oleaginous fungi Mucor circinelloides and Mortierella alpina. Microbiology 147:2857–2864
53. Jang H, Lin Y, Yang S (2005) Effect of culture media and conditions on polyunsaturated fatty acids production by Mortierella alpina. Bioresour Technol 96:1633–1644
54. Granger L-M, Perlot P, Goma G, Pareilleux A (1992) Kinetics of growth and fatty acid production of Rhodotorula glutinis. Appl Microbiol Biotechnol 37(1):13–17
55. Uemura H (2012) Synthesis and production of unsaturated and polyunsaturated fatty acids in yeast: current state and perspectives. Appl Microbiol Biotechnol 95:1–12
56. Athenstaedt K, Jolivet P, Boulard C, Zivy M, Negroni L, Nicaud JM, Chardot T (2006) Lipid particle composition of the yeast Yarrowia lipolytica depends on the carbon source. Proteomics 6:1450–1459
57. Zavala-Moreno A, Arreguin-Espinosa R, Pardo JP, Romero-Aguilar L, Guerra-Sanchez G (2014) Nitrogen source affects glycolipid production and lipid accumulation in the phytopathogen fungus *Ustilago maydis*. Adv Microbiol 04:934–944
58. Santamauro F, Whiffm FM, Scott RJ, Chuck CJ (2014) Low-cost lipid production by an oleaginous yeast cultured in non-sterile conditions using model waste resources. Biotechnol Biofuels 110:198–205
59. Angerbauer C, Siebenhofer M, Mittelbach M, Guebitz GM (2008) Conversion of sewage sludge into lipids by Lipomyces starkeyi for biodiesel production. Bioresour Technol 99:3051–3056
60. Liang M, Jiang J (2013) Advancing oleaginous microorganisms to produce lipid via metabolic engineering technology. Prog Lipid Res 52:395–408
61. Tai M, Stephanopoulos G (2013) Engineering the push and pull of lipid biosynthesis in oleaginous yeast Yarrowia lipolytica for biofuel production. Metab Eng 15:1–9
62. Zhang Y, Adams I, Ratledge C (2007) Malic enzyme: the controlling activity for lipid production? Overexpression of malic enzyme in Mucor circinelloides leads to a 2.5-fold increase in lipid accumulation. Microbiology 53:2013–2025
63. Beopoulos A, Mrozova Z, Thevenieau F, Le Dall MT, Hapala I, Papanikolaou S, Chardot T, Nicaud JM (2008) Control of lipid accumulation in the yeast Yarrowia

lipolytica. Appl Environ Microbiol 74:7779–7789

64. Kalscheuer R, Luftmann H, Steinbuchel A (2004) Synthesis of novel lipids in Saccharomyces cerevisiae by heterologous expression of an unspecific bacterial acyltransferase. Appl Environ Microbiol 70:7119–7125
65. Kalscheuer R (2006) Microdiesel: Escherichia coli engineered for fuel production. Microbiology 152:2529–2536
66. Vazquez-Duhalt R, Greppin H (1987) Growth and production of cell constituents in batch cultures of Botryococcus sudeticus. Phytochemistry 26:885–889
67. Xu J, Zhao X, Wang W, Du W, Liu D (2012) Microbial conversion of biodiesel byproduct glycerol to triacylglycerols by oleaginous yeast Rhodosporidium toruloides and the individual effect of some impurities on lipid production. Biochem Eng J 65:30–36
68. Zhu L, Zong M, Wu H (2008) Efficient lipid production with *Trichosporon fermentans* and its use for biodiesel preparation. Bioresour Technol 99:7881–7885
69. Zhu M, Yu L-J, Li W, Zhou P-P, Li C-Y (2006) Optimization of arachidonic acid production by fed-batch culture of Mortierella alpina based on dynamic analysis. Enzyme Microb Technol 38:735–740
70. Zhang J, Fang X, Zhu X-L, Yan L, Xu H-P, Zhao B-F, Chen L, Zhang X-D (2011) Microbial lipid production by the oleaginous yeast Cryptococcus curvatus O3 grown in fed-batch culture. Biomass Bioenergy 35 (5):1906–1911
71. Brown BD, Hsu KH, Hammond EG, Glatz B (1989) A relationship between growth and lipid accumulation in Candida curvata D. J Ferment Bioeng 68:344–352
72. Béligon V, Poughon L, Christophe G, Lebert A, Larroche C, Fontanille C (2016) Validation of a predictive model for fed-batch and continuous lipids production processes from acetic acid using the oleaginous yeast Cryptococcus curvatus. Biochem Eng J 111:117–128
73. Papanikolaou S, Aggelis G (2002) Lipid production by Yarrowia lipolytica growing on industrial glycerol in a single-stage continuous culture. Bioresour Technol 82:43–49
74. Abu O (2000) Changes in lipid, fatty acids and protein composition of sweet potato (Ipomoea batatas) after solid-state fungal fermentation. Bioresour Technol 72:189–192
75. Peng X, Chen H (2008) Single cell oil production in solid-state fermentation by Microsphaeropsis sp. from steam-exploded wheat straw mixed with wheat bran. Bioresour Technol 99:3885–3889
76. Economou CN, Makri A, Aggelis G, Pavlou S, Vayenas DV (2010) Semi-solid state fermentation of sweet sorghum for the biotechnological production of single cell oil. Bioresour Technol 101:1385–1388
77. Liao W, Liu Y, Chen S (2007) Studying pellet formation of a filamentous fungus Rhizopus oryzae to enhance organic acid production. Appl Biochem Biotechnol 137:689–701
78. Liu W, Wang Y, Yu Z, Bao J (2012) Simultaneous saccharification and microbial lipid fermentation of corn stover by oleaginous yeast Trichosporon cutaneum. Bioresour Technol 118:13–18
79. Heredia L, Ratledge C (1988) Simultaneous utilization of glucose and xylose by Candida curvata D in continuous culture. Biotechnol Lett 10(1):25–30
80. Daniel HJ, Otto RT, Binder M, Reuss M, Syldatk C (1999) Production of sophorolipids from whey: development of a two-stage process with Cryptococcus curvatus ATCC 20509 and Candida bombicola ATCC 22214 using deproteinized whey concentrates as substrates. Appl Microbiol Biotechnol 51:40–45
81. Hansson L, Dostlek M (1986) Influence of cultivation conditions on lipid production by Cryptococcus albidus. Appl Microbiol Biotechnol 24:12–18
82. Papanikolaou S, Galiotou-Panayotou M, Fakas S, Komaitis M, Aggelis G (2007) Lipid production by oleaginous Mucorales cultivated on renewable carbon sources. Eur J Lipid Sci Technol 109:1060–1070
83. Huang C, Zong M-H, Wu H, Liu Q-P (2009) Microbial oil production from rice straw hydrolysate by Trichosporon fermentans. Bioresour Technol 100:4535–4538
84. Liang Y, Tang T, Siddaramu T, Choudhary R, Umagiliyage AL (2012) Lipid production from sweet sorghum bagasse through yeast fermentation. Renew Energy 40:130–136
85. Wild R, Patil S, Popovic M, Zappi M, Dufreche S, Bajpai R (2010) Lipids from Lipomyces starkeyi. Food Technol Biotechnol 48:329–335
86. Wang Q, Guo F-J, Rong Y-J, Chi Z-M (2012) Lipid production from hydrolysate of cassava starch by Rhodosporidium toruloides 21167 for biodiesel making. Renew Energy 46:164–168
87. Fakas S, Certik M, Papanikolaou S, Aggelis G, Komaitis M, Galiotou-Panayotou M (2008) Gama-linolenic acid production by

Cunninghamella echinulata growing on complex organic nitrogen sources. Bioresour Technol 99:5986–5990

88. Ykema A, Verbree EC, Kater MM, Smit H (1988) Optimization of lipid production in the oleaginous yeast Apiotrichum curvatum in whey permeate. Appl Microbiol Biotechnol 29:211–218
89. Akhtar P, Gray J, Asghar A (1998) Synthesis of lipids by certain yeast strains grown on whey permeate. J Food Lipids 5:283–297
90. Christophe G, Deo JL, Kumar V, Nouaille R, Fontanille P, Larroche C (2011) Production of oils from acetic acid by the oleaginous yeast Cryptococcus curvatus. Appl Biochem Biotechnol 167:1270–1279
91. Athalye S, Garcia R, Wen Z (2009) Use of biodiesel-derived crude glycerol for producing eicosapentaenoic acid (EPA) by the fungus Pythium irregulare. J Agric Food Chem 57:2739–2744
92. Chang Y-H, Chang K-S, Lee C-F, Hsu C-L, Huang C-W, Jang H-D (2015) Microbial lipid production by oleaginous yeast Cryptococcus sp. in the batch cultures using corncob hydrolysate as carbon source. Biomass Bioenergy 72:95–103
93. Xue F, Gao B, Zhu Y, Zhang Z, Feng W, Tan T (2010) Pilot-scale production of microbial lipid using starch wastewater as raw material. Bioresour Technol 101:6092–6095
94. Tsigie YA, Wang CY, Truong CT, Ju YH (2011) Lipid production from Yarrowia lipolytica Po1g grown in sugarcane bagasse hydrolysate. Bioresour Technol 102:9216–9222
95. Xiaowei P, Hongzhang C (2012) Hemicellulose sugar recovery from steam-exploded wheat straw for microbial oil production. Process Biochem 47:209–215
96. Oliveira Mdos S, Feddern V, Kupski L, Cipolatti EP, Badiale-Furlong E, de Souza-Soares LA (2011) Changes in lipid, fatty acids and phospholipids composition of whole rice bran after solid-state fungal fermentation. Bioresour Technol 102:8335–8338
97. Economou C, Aggelis G, Pavlou S, Vayenas DV (2011) Single cell oil production from rice hulls hydrolysate. Bioresour Technol 102:9737–9742
98. Hu C, Wu S, Wang Q, Jin G, Shen H, Zhao ZK (2011) Simultaneous utilization of glucose and xylose for lipid production by Trichosporon cutaneum. Biotechnol Biofuels 4:25
99. Aggelis G, Komaitis M (1999) Enhancement of single cell oil production by Yarrowia lipolytica growing in the presence of Teucrium polium L. aqueous extract. Biotechnol Lett 21:747–749
100. Tsigie YA, Wang C-Y, Kasim NS, Diem Q-D, Huynh L-H, Ho Q-P, Truong C-T, Ju Y-H (2012) Oil production from Yarrowia lipolytica Po1g using rice bran hydrolysate. J Biomed Biotechnol 2012:1–10
101. Makri A, Fakas S, Aggelis G (2010) Metabolic activities of biotechnological interest in Yarrowia lipolytica grown on glycerol in repeated batch cultures. Bioresour Technol 101:2351–2358
102. Gong Z, Wang Q, Shen H, Hu C, Jin G, Zhao ZK (2012) Co-fermentation of cellobiose and xylose by Lipomyces starkeyi for lipid production. Bioresour Technol 117:20–24
103. Mitra D, Rasmussen ML, Chand P, Chintareddy VR, Yao L, Grewell D, Verkade JG, Wang T, van Leeuwen JH (2012) Value-added oil and animal feed production from corn-ethanol stillage using the oleaginous fungus Mucor circinelloides. Bioresour Technol 107:368–375
104. Slininger PJ, Dien BS, Kurtzman CP, Moser BR, Bakota EL, Thompson SR, O'Bryan PJ, Cotta MA, Balan V, Jin M, Sousa Lda C, Dale BE (2016) Comparative lipid production by oleaginous yeasts in hydrolyzates of lignocellulosic biomass and process strategy for high titers. Biotechnol Bioeng 113:1676–1690
105. Sun Y, Cheng J (2002) Hydrolysis of lignocellulosic materials for ethanol production: a review. Bioresour Technol 83:1–11
106. Chang V, Holtzapple M (2000) Fundamental factors affecting biomass enzymatic reactivity. Appl Biochem Biotechnol 84–86:5–38
107. Huang C, Chen X-F, Xiong L, Chen X-D, Ma L-L, Chen Y (2013) Single cell oil production from low-cost substrates: the possibility and potential of its industrialization. Biotechnol Adv 31:129–139
108. Lorenz E, Runge D, Marbà-Ardébol AM, Schmacht M, Stahl U, Senz M (2017) Systematic development of a two-stage fed-batch process for lipid accumulation in Rhodotorula glutinis. J Biotechnol 246:4–15
109. Lin J, Shen H, Tan H, Zhao X, Wu S, Hu C, Zhao ZK (2011) Lipid production by Lipomyces starkeyi cells in glucose solution without auxiliary nutrients. J Biotechnol 152:184–188
110. Lynd L (1996) Overview and evaluation of fuel ethanol from cellulosic biomass: technology, economics, the environment, and policy. Annu Rev Energy Environ 21:403–465

111. Li Y, Horsman M, Wu N, Lan CQ, Dubois-Calero N (2008) Biofuels from microalgae. Biotechnol Prog 24(4):815–820
112. Sheehan J, Dunahay T, Benemann J, Roessler P (1998) Look back at the U.S. department of energy's aquatic species program: biodiesel from algae; close-out report. Office of Scientific and Technical Information (OSTI). Prepared for: U.S. Department of Energy's Office of Fuels Development, pp 1–328
113. Chen F (1996) High cell density culture of microalgae in heterotrophic growth. Trends Biotechnol 14:421–426
114. Xu H, Miao X, Wu Q (2006) High quality biodiesel production from a microalga Chlorella protothecoides by heterotrophic growth in fermenters. J Biotechnol 126:499–507
115. Liu J, Huang J, Sun Z, Zhong Y, Jiang Y, Cheng F (2010) Differential lipid and fatty acid profiles of photoautotrophic and heterotrophic chlorella zofingiensis: assessment of algal oils for biodiesel production. Bioresour Technol 102:106–110
116. Apt K, Behrens P (1999) Commercial developments in microalgal biotechnology. J Phycol 35:215–226
117. Wen Z, Chen F (2001) Optimization of nitrogen sources for heterotrophic production of eicosapentaenoic acid by the diatom Nitzschia laevis. Enzyme Microb Technol 29:341–347
118. Sansawa H, Endo H (2004) Production of intracellular phytochemicals in chlorella under heterotrophic conditions. J Biosci Bioeng 98:437–444
119. Barclay W, Meager K, Abril J (1994) Heterotrophic production of long chain omega-3 fatty acids utilizing algae and algae-like microorganisms. J Appl Phycol 6:123–129
120. Subramaniam R, Dufreche S, Zappi M, Bajpai R (2010) Microbial lipids from renewable resources: production and characterization. J Ind Microbiol Biotechnol 37:1271–1287
121. Heredia-Arroyo T, Wei W, Ruan R, Hu B (2011) Mixotrophic cultivation of Chlorella vulgaris and its potential application for the oil accumulation from non-sugar materials. Biomass Bioenergy 35:2245–2253
122. Ren LJ, Ji XJ, Huang H, Qu L, Feng Y, Tong QQ, Ouyang PK (2010) Development of a stepwise aeration control strategy for efficient docosahexaenoic acid production by Schizochytrium sp. Appl Microbiol Biotechnol 87:1649–1656
123. Yang HL, Lu CK, Chen SF, Chen YM, Chen YM (2010) Isolation and characterization of Taiwanese heterotrophic microalgae: screening of strains for docosahexaenoic acid (DHA) production. Mar Biotechnol 12:173–185
124. De Swaaf M, Sijtsma L, Pronk J (2003) High-cell-density fed-batch cultivation of the docosahexaenoic acid producing marine alga Crypthecodinium cohnii. Biotechnol Bioeng 81:666–672
125. Wu Z, Shi X (2007) Optimization for high-density cultivation of heterotrophic chlorella based on a hybrid neural network model. Lett Appl Microbiol 44:13–18
126. Riesenberg D, Guthke R (1999) High-cell-density cultivation of microorganisms. Appl Microbiol Biotechnol 51:422–430
127. Li X, Xu H, Wu Q (2007) Large-scale biodiesel production from microalga Chlorella protothecoides through heterotrophic cultivation in bioreactors. Biotechnol Bioeng 98:764–771
128. Pulz O, Gross W (2004) Valuable products from biotechnology of microalgae. Appl Microbiol Biotechnol 65:635–648
129. Brennan L, Owende P (2010) Biofuels from microalgae—a review of technologies for production, processing, and extractions of biofuels and co-products. Renew Sust Energ Rev 14:557–577
130. Hughes E, Benemann J (1997) Biological fossil CO_2 mitigation. Energy Convers Manag 38:S467–S473
131. Liang Y, Sarkany N, Cui Y (2009) Biomass and lipid productivities of Chlorella vulgaris under autotrophic, heterotrophic and mixotrophic growth conditions. Biotechnol Lett 31:1043–1049
132. Xiong W, Li X, Xiang J, Wu Q (2008) High-density fermentation of microalga Chlorella protothecoides in bioreactor for microbiodiesel production. Appl Microbiol Biotechnol 78:29–36
133. Espinosa-Gonzalez I, Parashar A, Bressler D (2014) Heterotrophic growth and lipid accumulation of Chlorella protothecoides in whey permeate, a dairy by-product stream, for biofuel production. Bioresour Technol 155:170–176
134. Yan D, Lu Y, Chen YF, Wu Q (2011) Waste molasses alone displaces glucose-based medium for microalgal fermentation towards cost-saving biodiesel production. Bioresour Technol 102:6487–6493
135. Kim D, Hur S (2013) Growth and fatty acid composition of three heterotrophic chlorella species. Algae 28:101–109
136. Wan MX, Wang RM, Xia JL, Rosenberg JN, Nie ZY, Kobayashi N, Oyler GA, Betenbaugh MJ (2012) Physiological evaluation of a new

Chlorella sorokiniana isolate for its biomass production and lipid accumulation in photoautotrophic and heterotrophic cultures. Biotechnol Bioeng 109:1958–1964

137. Li P, Miao X, Li R, Zhong J (2011) In situ biodiesel production from fast-growing and high oil content Chlorella pyrenoidosa in rice straw hydrolysate. J Biomed Biotechnol 2011:1–8

138. De Swaaf M, Pronk J, Sijtsma L (2003) Fed-batch cultivation of the docosahexaenoic-acid-producing marine alga Crypthecodinium cohnii on ethanol. Appl Microbiol Biotechnol 61:40–43

139. Song X, Zang X, Zhang X (2015) Production of high docosahexaenoic acid by Schizochytrium sp. using low-cost raw materials from food industry. J Oleo Sci 64:197–204

140. Ren H-Y, Liu B-F, Ma C, Zhao L, Ren N-Q (2013) A new lipid-rich microalga Scenedesmus sp. strain r-16 isolated using Nile red staining: effects of carbon and nitrogen sources and initial pH on the biomass and lipid production. Biotechnol Biofuels 6:143

141. Wen Z, Chen F (2002) Continuous cultivation of the diatom Nitzschia laevis for eicosapentaenoic acid production: physiological study and process optimization. Biotechnol Prog 18:21–28

142. Yu X, Zhao P, He C, Li J, Tang X, Zhou J, Huang Z (2012) Isolation of a novel strain of Monoraphidium sp. and characterization of its potential application as biodiesel feedstock. Bioresour Technol 121:256–262

143. Kim J, Yoo G, Lee H, Lim J, Kim K, Kim CW, Park MS, Yang J-W (2013) Methods of downstream processing for the production of biodiesel from microalgae. Biotechnol Adv 31:862–876

144. Folch J, Lees M, Stanley G (1957) A simple method for the isolation and purification of total lipids from animal tissue. J Biol Chem 226:497–509

145. Uquiche E, Jerez M, Ortiz J (2008) Effect of pretreatment with microwaves on mechanical extraction yield and quality of vegetable oil from Chilean hazelnuts (Gevuina avellana Mol). Innov Food Sci Emerg Technol 9:495–500

146. Singh J, Bargale P (2000) Development of a small capacity double stage compression screw press for oil expression. J Food Eng 43:75–82

147. Topare N, Raut SJ, Genge CV, Khedkar SV, Chavan YPO, Bhagat SL (2011) Extraction of oil from algae by solvent extraction and oil expeller method. Int J Chem Sci 9 (4):1746–1750

148. Pradhan RC, Mishra S, Naik SN, Bhatnagar N, Vijay VK (2011) Oil expression from Jatropha seeds using a screw press expeller. Biosyst Eng 109:158–166

149. King J (2002) Supercritical fluid extraction: present status and prospects. Grasas Aceites 53:8–21

150. Salgin U, Doker O, Calimli A (2006) Extraction of sunflower oil with supercritical CO_2: experiments and modeling. J Supercrit Fluids 38:326–331

151. Del Valle J, Germain JC, Uquiche E, Zetzl C, Brunner G (2006) Microstructural effects on internal mass transfer of lipids in prepressed and flaked vegetable substrates. J Supercrit Fluids 37:178–190

152. Krohn B, McNeff CV, Yan B, Nowlan D (2011) Production of algae-based biodiesel using the continuous catalytic Mcgyan® process. Bioresour Technol 102:94–100

153. Umdu E, Tuncer M, Seker E (2009) Transesterification of Nannochloropsis oculata microalga's lipid to biodiesel on Al_2O_3 supported CaO and MgO catalysts. Bioresour Technol 100:2828–2831

154. Vijayaraghavan K, Hemanathan K (2009) Biodiesel production from freshwater algae. Energy Fuel 23:5448–5453

155. Tran D, Yeh K-L, Chen C-L, Chang J-S (2012) Enzymatic transesterification of microalgal oil from Chlorella vulgaris Esp-31 for biodiesel synthesis using immobilized Burkholderia lipase. Bioresour Technol 108:119–127

156. Patil P, Gude VG, Mannarswamy A, Deng S, Cooke P, Munson-McGee S, Rhodes I, Lammers P, Nirmalakhandan N (2011) Optimization of direct conversion of wet algae to biodiesel under supercritical methanol conditions. Bioresour Technol 102:118–122

157. Rathana Y, Roces SA, Bacani FT, Tan RR, Kubouchi M, Yimsiri P (2010) Microwave-enhanced alkali catalyzed transesterification of kenaf seed oil. Int J Chem React Eng 8. https://doi.org/10.2202/1542-6580.2324

158. Refaat A (2009) Different techniques for the production of biodiesel from waste vegetable oil. Int J Environ Sci Technol 7:183–213

159. Patil P, Gude VG, Mannarswamy A, Cooke P, Munson-McGee S, Nirmalakhandan N, Lammers P, Deng S (2011) Optimization of microwave-assisted transesterification of dry algal biomass using response surface methodology. Bioresour Technol 102:1399–1405

160. Geuens J, Kremsner JM, Nebel BA, Schober S, Dommisse RA, Mittelbach M, Tavernier S, Kappe CO, Maes BUW (2007) Microwave-assisted catalyst-free transesterification of triglycerides with 1-butanol under supercritical conditions. Energy Fuel 22:643–645
161. Ji J, Wang J, Li Y, Yu Y, Xu Z (2006) Preparation of biodiesel with the help of ultrasonic and hydrodynamic cavitation. Ultrasonics 44: E411–E414
162. Amaro H, Guedes A, Malcata F (2011) Advances and perspectives in using microalgae to produce biodiesel. Appl Energy 88:3402–3410
163. Wahlen B, Willis R, Seefeldt L (2011) Biodiesel production by simultaneous extraction and conversion of total lipids from microalgae, cyanobacteria, and wild mixed-cultures. Bioresour Technol 102:2724–2730
164. Cao H, Zhang Z, Wu X, Miao X (2013) Direct biodiesel production from wet microalgae biomass of Chlorella pyrenoidosa through in situ transesterification. Biomed Res Int 2013:1–6
165. Johnson M, Wen Z (2009) Production of biodiesel fuel from the microalga Schizochytrium limacinum by direct transesterification of algal biomass. Energy Fuel 23:5179–5183
166. Ehimen E, Sun Z, Carrington C (2010) Variables affecting the in situ transesterification of microalgae lipids. Fuel 89:677–684
167. Özgül-Yücel S, Türkay S (2002) Variables affecting the yields of methyl esters derived from in situ esterification of rice bran oil. J Am Oil Chem Soc 79:611–614
168. Ehimen E, Sun Z, Carrington G (2012) Use of ultrasound and co-solvents to improve the in-situ transesterification of microalgae biomass. Procedia Environ Sci 15:47–55
169. Cheng J, Yu T, Li T, Zhou J, Cen K (2013) Using wet microalgae for direct biodiesel production via microwave irradiation. Bioresour Technol 131:531–535
170. Goyal H, Seal D, Saxena R (2008) Bio-fuels from thermochemical conversion of renewable resources: a review. Renew Sust Energy Rev 12:504–517
171. Demirbaş A (2001) Biomass resource facilities and biomass conversion processing for fuels and chemicals. Energy Convers Manag 42:1357–1378
172. Mckendry P (2002) Energy production from biomass (part 2): conversion technologies. Bioresour Technol 83:47–54
173. Hirano A, Hon-Nami K, Hunito S, Hada M, Ogushi Y (1998) Temperature effect on continuous gasification of microalgal biomass: theoretical yield of methanol production and its energy balance. Catal Today 45:399–404
174. Mackay S, Gomes E, Holliger C, Bauer R, Schwitzguébel JP (2015) Harvesting of Chlorella sorokiniana by co-culture with the filamentous fungus Isaria fumosorosea: a potential sustainable feedstock for hydrothermal gasification. Bioresour Technol 185:353–361
175. López Barreiro D, Prins W, Ronsse F, Brilman W (2013) Hydrothermal liquefaction (HTL) of microalgae for biofuel production: state of the art review and future prospects. Biomass Bioenergy 53:113–127
176. Dote Y, Sawayama S, Innoue S, Minowa T, Yokoyama S-Y (1994) Recovery of liquid fuel from hydrocarbon-rich microalgae by thermochemical liquefaction. Fuel 73:1855–1857
177. Demirbaş A (2000) Mechanisms of liquefaction and pyrolysis reactions of biomass. Energy Convers Manag 41:633–646
178. Jena U, Das K (2011) Comparative evaluation of thermochemical liquefaction and pyrolysis for bio-oil production from microalgae. Energy Fuel 25:5472–5482
179. Maher K, Bressler D (2007) Pyrolysis of triglyceride materials for the production of renewable fuels and chemicals. Bioresour Technol 98:2351–2368
180. Demirbaş A (2006) Oily products from mosses and algae via pyrolysis. Energy Sources Part A 28:933–940
181. Cantrell KB, Ducey T, Ro KS, Hunt PG (2008) Livestock waste-to-bioenergy generation opportunities. Bioresour Technol 99:7941–7953
182. Phang SM, Miah MS, Yeoh BG, Hashim MA (2000) Spirulina cultivation in digested sago starch factory wastewater. J Appl Phycol 12 (3–5):395–400
183. Melis A, Happe T (2001) Hydrogen production. Green algae as a source of energy. Plant Physiol 127:740–748
184. Ghirardi M (2000) Microalgae: a green source of renewable H_2. Trends Biotechnol 18:506–511
185. Feofilova E, Sergeeva Y, Ivashechkin A (2010) Biodiesel-fuel: content, production, producers, contemporary biotechnology (review). Appl Biochem Microbiol 46:369–378
186. Dahiya A (2015) Algae biomass cultivation for advanced biofuel production. In: Bioenergy. Elsevier, Amsterdam, pp 219–238

187. Sitepu IR, Garay LA, Sestric R, Levin D, Block DE, German JB, Boundy-Mills KL (2014) Oleaginous yeasts for biodiesel: current and future trends in biology and production. Biotechnol Adv 32:1336–1360
188. Banerjee A, Sharma R, Chisti Y, Banerjee UC (2002) Botryococcus braunii: a renewable source of hydrocarbons and other chemicals. Crit Rev Biotechnol 22:245–279
189. Metzger P, Largeau C (2004) Botryococcus braunii: a rich source for hydrocarbons and related ether lipids. Appl Microbiol Biotechnol 66:486–496
190. Strobel GA, Knighton B, Kluck K, Ren Y, Livinghouse T, Griffin M, Spakowicz D, Sears J (2008) The production of myco-diesel hydrocarbons and their derivatives by the endophytic fungus Gliocladium roseum (NRRL 50072). Microbiology 154:3319–3328
191. Chen G, Chen F (2006) Growing phototrophic cells without light. Biotechnol Lett 28:607–616
192. Williams P, Laurens LML (2010) Microalgae as biodiesel & biomass feedstocks: review and analysis of the biochemistry, energetics & economics. Energy Environ Sci 3:554
193. Simopoulos A (1999) Essential fatty acids in health and chronic disease. Am J Clin Nutr 70:560S–569S
194. Simopoulos A (2002) Omega-3 fatty acids in inflammation and autoimmune diseases. J Am Coll Nutr 21:495–505
195. Behrens P (2005) Photobioreactors and fermentors, in algal culturing techniques. The light and dark sides of growing algae. In: Anderson RA (ed) Algal culturing techniques. Elsevier Academic Press, Burlington, pp 189–203
196. Mendes A, Reis A, Vasconcelos R, Guerra P (2008) Crypthecodinium cohnii with emphasis on DHA production: a review. J Appl Phycol 21:199–214
197. Martin VJ, Pitera DJ, Withers ST, Newman JD, Keasling JD (2003) Engineering a mevalonate pathway in Escherichia coli for production of terpenoids. Nat Biotechnol 21:796–802
198. Cohen G (2011) Microbial biochemistry. Springer, Netherlands
199. Guerin M, Huntley M, Olaizola M (2003) Haematococcus astaxanthin: applications for human health and nutrition. Trends Biotechnol 21:210–216
200. Spolaore P, Joannis-Cassan C, Duran E, Isambert A (2006) Commercial applications of microalgae. J Biosci Bioeng 101:87–96
201. Könst P, Franssen MCR, Scott EL, Sanders JPM (2011) Stabilization and immobilization of Trypanosoma brucei ornithine decarboxylase for the biobased production of 1,4-diaminobutane. Green Chem 13:1167
202. Marris E (2006) Black is the new green. Nature 442:624–626
203. Lehmann J, Gaunt J, Rondon M (2006) Biochar sequestration in terrestrial ecosystems – a review. Mitig Adapt Strat Gl 11(2):403–427
204. Lal R (2008) Black and buried carbons' impacts on soil quality and ecosystem services. Soil Tillage Res 99:1–3
205. Tan H, Aziz A, Aroua M (2013) Glycerol production and its applications as a raw material: a review. Renew Sust Energy Rev 27:118–127
206. Hirano A, Ueda R, Hirayama S, Ogushi Y (1997) CO_2 fixation and ethanol production with microalgal photosynthesis and intracellular anaerobic fermentation. Energy 22:137–142
207. Ueno Y, Kurano N, Miyachi S (1998) Ethanol production by dark fermentation in the marine green alga, Chlorococcum littorale. J Ferment Bioeng 86:38–43
208. Sialve B, Bernet N, Bernard O (2009) Anaerobic digestion of microalgae as a necessary step to make microalgal biodiesel sustainable. Biotechnol Adv 27(4):409–416
209. Asachi R, Karimi K, Taherzadeh M (2011) Ethanol production by mucor indicus using the fungal autolysate as a nutrient supplement. In: Proceedings of the world renewable energy congress. Sweden Linköping University Electronic Press, Linköping, Sweden
210. Raja R, Hemaiswarya S, Kumar NA, Sridhar S, Rengasamy R (2008) A perspective on the biotechnological potential of microalgae. Crit Rev Microbiol 34:77–88
211. Donot F, Fontana A, Baccou JC, Strub C, Schorr-Galindo S (2014) Single cell oils (SCOs) from oleaginous yeasts and moulds: production and genetics. Biomass Bioenergy 68:135–150
212. Hannon M, Gimpel J, Tran M, Rasala B, Mayfield S (2010) Biofuels from algae: challenges and potential. Biofuels 1:763–784

Chapter 4

Qualitative and Quantitative Estimation of Bacterial Lipid Production

Gangatharan Muralitharan, Manickam Gayathri, and Sumathy Shunmugam

Abstract

An ever increasing energy demand and fast depletion of fossil fuels have led to increased consideration of bacterial lipids as a renewable biofuel source. Many methods are available for both physical and chemical extraction of bacterial lipids. The method of choice will depend on the nature of sample to be analyzed, combinations of solvent systems preferred, content and quality of the lipid to be analyzed, types of equipment available, and time of the extraction procedures employed. Here we describe the most reliable, routine method of extracting bacterial lipids and evaluating the growth kinetic parameters like biomass and lipid productivity and lipid content. We also describe the method of comparing bacterial fatty acid methyl ester peaks with standard peaks for analysis.

Key words Bacteria, Cyanobacteria, Growth kinetic parameters, Lipid extraction, Fatty acid methyl ester (FAME), Transesterification

1 Introduction

The identification of oleaginous microbes starts with identification of their lipid production potential. Bacteria including photosynthetic cyanobacteria along with microalgae are considered promising sources of commercial lipids. The effective estimation of bacterial lipid consists of mechanical, physical, chemical, or enzymatic methods of lipid extraction from biomass followed by transesterification of fatty acids into corresponding methyl esters called fatty acid methyl esters (FAMEs) and glycerol through saponification with methanol using an alkali as a catalyst. Fatty acids are the important nonpolar constituents of the bacterial lipid. FAMEs are the major constituent of biodiesel. This reversible reaction of transesterification is depicted in Eq. 1.

$$\text{Triglyceride} + 3CH_3OH \leftrightarrow C_3H_8O_3\ \text{(Glycerol)} + 3\text{FAME} \quad (1)$$

Venkatesh Balan (ed.), *Microbial Lipid Production: Methods and Protocols*, Methods in Molecular Biology, vol. 1995, https://doi.org/10.1007/978-1-4939-9484-7_4,

The most widely used classical methods of lipid extraction [1, 2] involve the use of methanol and chloroform as solvents. Other less toxic and less effective solvents like ethanol, isopropanol [3], butanol, methyl *tert*-butyl ether (MTBE) [4], acetic acid esters, hexane, 2-ethoxyethanol (2-EE) [5], and various combinations of these have been attempted [6]. The choice of a proper solvent system influences the efficiency of lipid extraction. Many solvent extraction methods are being combined with some mechanical and physicochemical approaches for increased recovery of microbial lipids. When evaluating microbes for lipid production, growth kinetic parameters like biomass and lipid productivity along with lipid content should be taken into account.

This protocol describes how to calculate the growth kinetic parameters, extract the lipids using the method of Folch et al. [1] (Fig. 1), and transesterify the triacylglycerols into FAMEs (Fig. 1) for both bacterial and cyanobacterial biomass. Direct transesterification to FAMEs without extraction is also described. Once the FAME profile is generated by gas chromatography, the steps for comparing the peaks with the known standard peaks are enlisted.

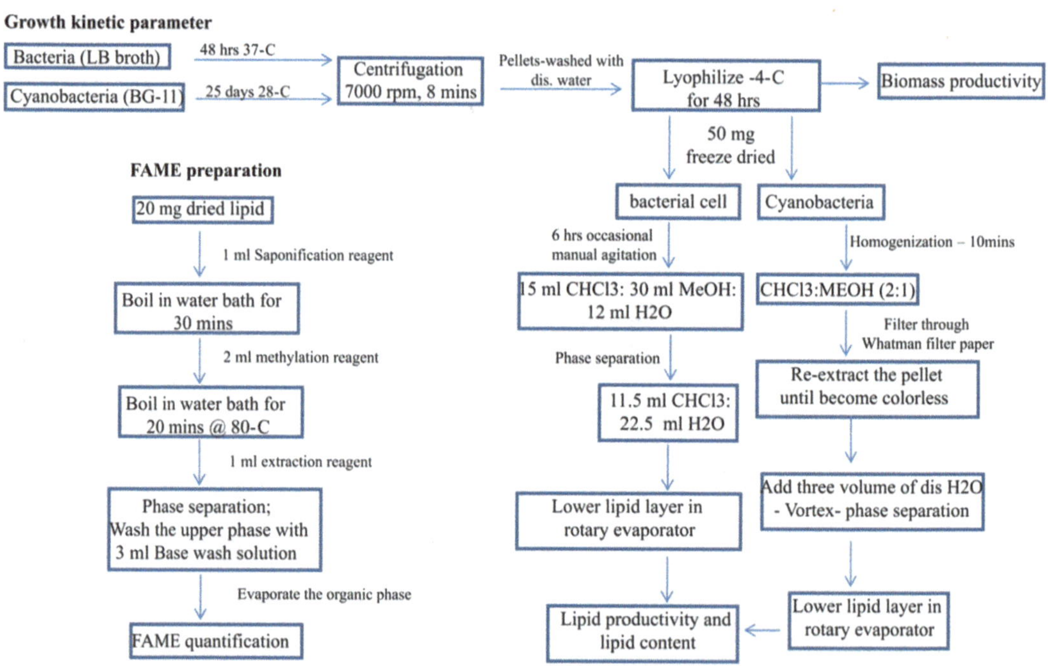

Fig. 1 Schematic workflow of growth kinetic parameters estimation and FAME preparation of cyanobacterial and bacterial cells

2 Materials

Prepare all media and solutions in ultrapure water (deionized water purified to attain a sensitivity of 18 MΩ-cm at 25 °C) and analytical grade solvents.

2.1 Growth Medium

2.1.1 For Bacteria

1. LB (Luria–Bertani) medium: Add about 100 mL of water to a 1 L graduated glass beaker. Weigh 10 g tryptone, 5 g yeast extract, and 10 g NaCl and transfer to the beaker. Add water to a volume of 900 mL. Mix thoroughly until the solutes have dissolved. Adjust the pH to 7.0 with 5 N NaOH (~0.2 mL). Make up to 1 L with water. Sterilize by autoclaving for 20 min at 15 psi (1.05 kg/cm^2) on liquid cycle. For solid medium, add 15 g/L agar just before autoclaving.

2.1.2 For Cyanobacteria

1. BG11 (Blue-green) medium [7]: Make stock solutions of the following components by dissolving each of them separately in individual (100 mL) screw-capped graduated reagent bottles labeled numbers 1 through 8. Add about 100 mL of water to each bottle. Then add to the appropriate numbered bottle: (1) 75 g $NaNO_3$ (*see* **Note 1**), (2) 2 g K_2HPO_4, (3) 3.75 g $MgSO_4{\cdot}7H_2O$, (4) 1.8 g $CaCl_2{\cdot}2H_2O$, (5) 0.3 g citric acid, (6) 0.3 g ammonium ferric citrate green, (7) 0.05 g $EDTANa_2$, (8) 1 g Na_2CO_3, (9) trace metal solution containing 1.43 g H_3BO_3, 0.905 g $MnCl_2{\cdot}4H_2O$, 0.11 g $ZnSO_4{\cdot}7H_2O$, 0.195 g $Na_2MoO_4{\cdot}2H_2O$, 0.04 g $CuSO_4{\cdot}5H_2O$ and 0.025 g Co $(NO_3)_2{\cdot}6H_2O$. Mix thoroughly until the solutes have dissolved. Add water to a final volume of 500 mL in each bottle. Screw-cap them and store at 4 °C.

   Make a working solution by mixing 10 mL each from stock solutions (1) to (8) and 1 mL of trace metal solution, *i.e.* stock solution (9). Adjust pH to 7.1 with 1 M NaOH or HCl. Bring the volume up to 1 L with water. Sterilize by autoclaving for 20 min at 15 psi (1.05 kg/cm^2) on liquid cycle (*see* **Note 2**).

2.2 Lipid Extraction

1. 50 mg freeze-dried bacterial/cyanobacterial biomass.
2. Preweighed glass vials.
3. Chloroform:methanol:water (1:2:0.8, v/v) for bacteria, chloroform:methanol (2:1, v/v) for cyanobacteria.
4. Separation funnel.
5. Glass funnel.
6. Whatman No. 1 filter paper.
7. Mortar and pestle.
8. Test tubes.
9. Spatula.

2.3 Preparation of FAMEs

1. Saponification reagent: 15 g sodium hydroxide in 100 mL of methanol–water (1:1, v/v).
2. Methylation reagent: methanol:6 N HCl (1:1.18, v/v).
3. Extraction solvent: distilled hexane:anhydrous diethyl ether (1:1, v/v).
4. Methanol.
5. Screw-cap test tube.
6. Teflon tape.
7. Base wash solution—1.2% NaOH (w/v).
8. Glass vials.

2.4 Direct Transesterification

2.4.1 For Bacteria

1. 10 mg freeze-dried bacterial biomass.
2. Chloroform.
3. Methanol:sulfuric acid at the concentration of 85:15 (v/v).
4. Ice-cold water.
5. Screw-cap test tube.
6. Glass vial.
7. Teflon tape.

2.4.2 For Cyanobacteria

1. 30 mg of freeze-dried cyanobacterial biomass.
2. Direct transesterification reagent: methanol:concentrated sulfuric acid:chloroform (4.25:0.75:5, v/v).
3. Screw-cap test tubes.
4. Teflon tape.
5. Glass vials.

3 Methods

1. (a) Inoculate 50 mL of LB broth in 250 mL Erlenmeyer flask with a single colony of bacteria from LB agar plate. Incubate the culture at 37 °C for 48 h.
   (b) Inoculate 180 mL of BG11 medium in 500 mL Erlenmeyer flask with 20 mL of cyanobacterial biomass at 0.8 A_{600}. Incubate the culture at 28 ± 2 °C, 14/10 h light/dark cycle, with the light intensity of 50 μE/m^2/s under static conditions (*see* **Note 3**) for 25 days.
2. Harvest the bacterial/cyanobacterial cells by centrifugation at 3300 × *g* for 8 min. Discard the supernatant and wash the pellets twice with sterile distilled water.
3. Lyophilize at −40 °C for 48 h and calculate the dry weights of bacterial/cyanobacterial cells by gravimetry. Biomass

productivity (P_{dwt}) in g/L/day is calculated from the formula P_{dwt} = (biomass concentration in g/L/number of incubation days) (*see* **Note 4**).

3.1 Lipid Extraction from Bacteria [8]

1. To 50 mg of freeze-dried bacterial cells, add 15 mL of chloroform, 30 mL of methanol, and 12 mL of water in a 500 mL separatory funnel to achieve 1:2:0.8 (v/v) (*see* **Notes 5–7**).
2. Allow the samples to stand for 6 h with occasional manual agitation (*see* **Note 8**).
3. For phase separation, add 11.5 mL of chloroform and 22.5 mL of water to achieve a final concentration of 1:1:0.9 (v/v) of chloroform, methanol, and water, respectively.
4. Recover and evaporate the bottom layer, which is the chloroform phase, in preweighed glass vials.
5. Calculate the lipid content (L_C) by gravimetry (*see* **Note 9**). L_C (% dwt) = (lipid weight (mg)/biomass taken for lipid extraction) × 100%.
6. Calculate the lipid productivity (L_P) in mg/L/day using the formula: $L_P = P_{dwt} \times L_C$.

3.2 Lipid Extraction from Cyanobacteria [1]

1. Weight then homogenize approximately 50 mg of freeze-dried cyanobacterial biomass using chloroform:methanol (2:1, v/v) (~1 mL) in a mortar and pestle for 10 min (*see* **Notes 5, 6,** and **10**). Filter the slurry through Whatman No. 1 filter paper fitted in a glass funnel into a 50 mL test tube (*see* **Note 11**). Reextract the pellet until it becomes colorless (*see* **Note 12**).
2. Add three volumes of distilled water with the filtrate to remove the water soluble impurities.
3. Vortex the filtrate, then allow it to stand (until clear phases appear) for phase separation. After phase separation, transfer the lower lipid layer using a pipette to a preweighed glass vials and allow it to dry in rotary evaporator.
4. Weigh the glass vials and calculate the lipid content (L_C) and lipid productivity (L_P) (as described before).

3.3 Preparation of FAMEs

1. To 20 mg of weighted, dried lipid, add 1 mL of saponification reagent and boil it for 30 min (*see* **Note 13**).
2. Add 2 mL of methylation reagent and boil it in a water bath for 20 min at 80 °C.
3. Cool the extracts and add 1 mL of extraction solvent and mix it thoroughly.
4. Allow the mixture for phase separation. Discard the lower aqueous phase and wash the upper organic phase with 3 mL of base wash solution.
5. Evaporate the organic phase in glass vials. After drying, dissolve it in 10 μL of methanol for GC/MS quantification of FAMEs.

3.3.1 Direct Transesterification to FAMEs in Bacterial Cells [9]

1. Weigh approximately 10 mg of freeze-dried bacterial cells in preweighed Teflon-stoppered glass vials and add 1 mL of chloroform and 1 mL of methanol:sulfuric acid (85:15, v/v).
2. Incubate the vials containing the mixture in a water bath at 105 °C for 2 h. (*see* **Note 13**).
3. Cool the mixture to room temperature and incubate on ice for 10 min.
4. Add 0.5 mL of ice-cold water and vortex for 1 min for thorough mixing.
5. Centrifuge at 2000 × *g* and transfer the bottom organic phase in clean glass tube using a pipette. Evaporate the organic phase and dissolve it with 10 μL of methanol for GC/MS quantification of FAMEs.

3.3.2 Direct Transesterification to FAMEs in Cyanobacterial Cells [10]

1. In a preweighed clean glass vial, weigh approximately 30 mg of freeze-dried cyanobacterial samples.
2. Add 10 mL of direct transesterification reagent.
3. Place the mixture in a water bath at 90 °C for 90 min (*see* **Notes 13–15**).
4. Cool the mixture to room temperature and add 1 mL of distilled water and vortex for 5 min.
5. Allow it to stand for phase separation (*see* **Note 16**). Transfer the lower phase containing FAMEs to a clean glass vial. Repeat **step 3** for further FAME extraction from the original sample.
6. Evaporate the FAME sample by rotary evaporator and dissolve the dried mixture in methanol and analyze through GC/MS.

3.4 Parameter for Gas Chromatography

The chromatographic settings for FAME analysis are as follows:

1. *Column*: SP-2560 column (Supelco, USA) (100 m × 0.25 mm I.D. × 0.20–0.25 μm film thickness).
2. *Detector*: Flame ionization detector (FID).
3. *Carrier gas*: (1) Helium with a flow rate of 2 mL/min.
4. *Injection volume*: 1–2 μL.
5. Temperature program: Initial oven temperature—140 °C for 5 min, ramping temperature—5 °C/min to 180 °C (8 min).
6. *Injector and detector temperature*: 200 and 250 °C.
7. *Split ratio*: 100:1.
8. *Run time*: 55 min per sample.
9. *Standard*: Supelco FAME mix C4–C24 (Bellefonte, PA, USA).

3.4.1 Comparison of FAME Peak with Standard Peak

1. The standard GC peak (Fig. 1) and area percentage of fatty acid from C4–C22 (Table 1) are used for the determination of fatty acids in the samples.

Table 1
GC peak of the standard and sample with retention time, area percentage, and corresponding fatty acids

Peak	Retention time	Area percentage	Fatty acid (standard)	Sample peak	Retention time	Area percentage	Fatty acid in sample
1	9.650	0.434	C4:0				
2	10.039	3.7247	C6:0				
3	11.106	4.0800	C8:0	1	11.425	3.3166	C8:0
4	12.887	4.3378	C10:0	2	13.171	0.3341	C10:0
5	14.092	2.0577	C11:0	3	14.637	0.117	C11:0
6	15.483	4.5035	C12:0	4	15.718	0.6577	C12:0
7	17.037	2.1495	C13:0	5	17.215	0.1905	C13:0
8	18.702	4.5664	C14:0	6	18.561	0.2606	C14:0
9	20.173	2.0428	C14:1	7	20.008	2.2615	C14:1
10	20.411	2.2180	C15:0	8	20.485	0.0862	C15:0
11	21.889	2.0802	C15:1				
12	22.151	7.2578	C16:0	9	21.991	23.4727	C16:0
13	23.362	2.0731	C16:1	10	23.171	4.2869	C16:1
14	23.831	2.2606	C17:0	11	23.617	3.887	C17:0
15	25.027	2.2271	C17:1	12	25.275	1.2331	C17:1
16	25.513	4.7938	C18:0	13	25.813	1.2393	C18:0
17	26.199	2.1650	C18:1n9t	14	26.001	3.0471	C18:1n9t
18	26.523	4.6507	C18:1n9c	15	26.430	0.7182	C180:1n9c
19	27.316	1.8571	C18:2n6t	16	27.219	0.3159	C18:2n6t
20	28.024	1.8138	C18:2n6c	17	28.143	3.6240	C18:2n6c
21	28.679	4.8729	C20:0	18	28.932	1.2077	C20:0
22	29.146	1.5494	C18:3n6	19	29.157	0.8825	C18:3n6
23	29.571	2.2377	C20:1n9	20	29.520	3.3360	C20:1n9
24	29.747	1.5795	C18:3n3	21	29.773	2.2755	C18:3n3
25	30.145	2.3276	C21:0	22	30.126	0.6716	C21:0
26	30.998	1.8362	C20:2	23	30.823	1.337	C20:2
27	31.669	4.8663	C22:0	24	31.909	1.2321	C22:0
28	32.096	1.5315	C20:3n6	25	33.193	0.5212	C23:0
29	32.545	2.1803	C22:1n9				
30	32.706	1.4579	C20:3n3				

(continued)

Table 1
(continued)

Peak	Retention time	Area percentage	Fatty acid (standard)	Sample peak	Retention time	Area percentage	Fatty acid in sample
31	32.985	1.0102	C20:4n6				
32	33.138	2.4107	C23:0				
33	34.019	1.8620	C22:2				
34	34.776	4.6740	C24:0				
35	34.921	1.2381	C20:5n3				
36	35.712	2.2775	C24:1				
37	39.695	0.7943	C22:6n3				
Total		100.0000					

2. Mark the fatty acid C4:0 to C22:6n3 in standard peak (Fig. 1 and Table 1).
3. Compare the standard peak (Fig. 2) and sample peak (Fig. 3), and correlate the retention times.
4. Calculate the area percentage of each fatty acid to generate the fatty acid profile of the sample (*see* **Note 17**).

4 Notes

1. For preparing BG11N_0 medium, the stock solution containing nitrogen sources should be omitted. BG11N_0 medium is used for growing heterocystous cyanobacterial strains that are capable of fixing N_2 with their specialized heterocystous cells.
2. To reduce precipitation of BG11 medium during autoclaving, autoclave stocks (5) 0.3 g citric acid, and (6) 0.3 g ammonium ferric citrate green, separately in 200 mL water and then add 10 mL of each stock to the autoclaved medium aseptically when cooled.
3. Shake the cultures manually by hand for 1 min on alternate days for better growth.
4. Calculate the biomass concentration in g/L using the following formula: Biomass concentration = (dry weight in grams/mL of culture grown) × 1000 mL.
5. The solvent used during extraction are flammable. So naked flames should be excluded from the work bench. All procedures involving solvent should be carried out inside a fume hood.

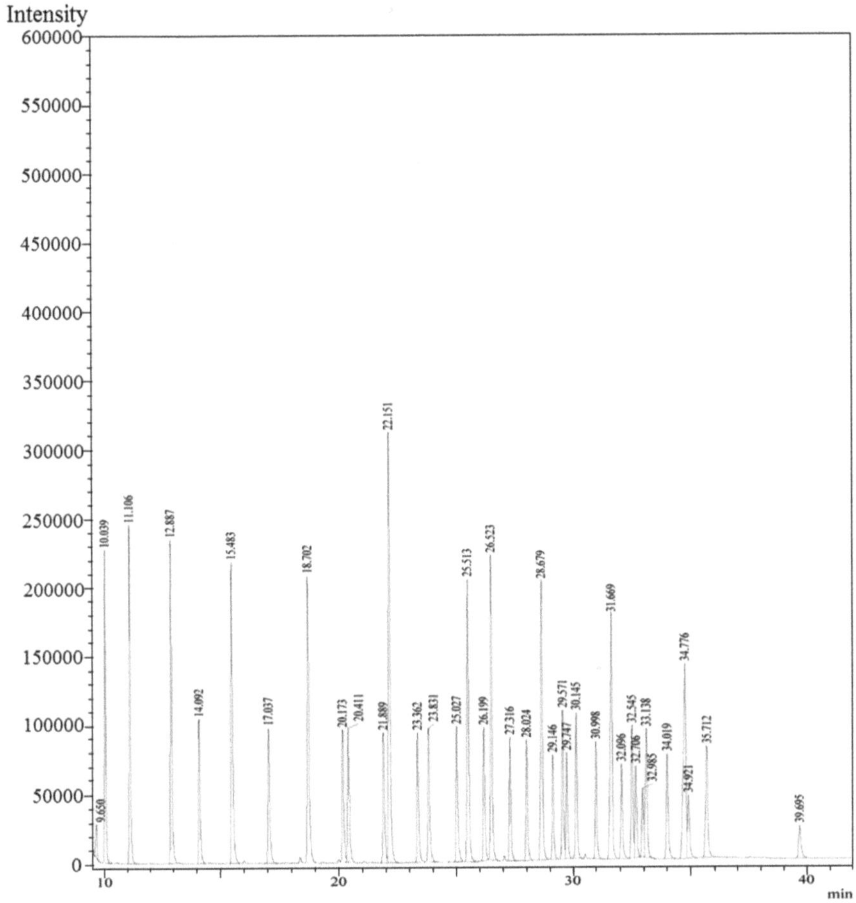

Fig. 2 Gas chromatogram showing the standard fatty acid peaks at various retention times

6. Clean the small spills with absorbent material and dispose it properly.
7. Shake for 15 s after adding each solvent.
8. Agitate every 15 min for better extraction.
9. Subtract the glass vial weight from the weight after evaporating the chloroform phase.
10. Add a similar volume (~1 mL) of the solvent when it is evaporated during homogenization.
11. Make a paper cone with Whatman No. 1 filter paper and place inside the glass funnel for filtering.
12. Remove the pellet from the filter paper using a spatula and repeat the homogenization step with ~1 mL of the solvent system. Repeat this step five times during which the pellet will become colorless.
13. FAME preparation should be carried out in screw-capped glass tubes sealed with Teflon tape to avoid evaporation. Reseal the

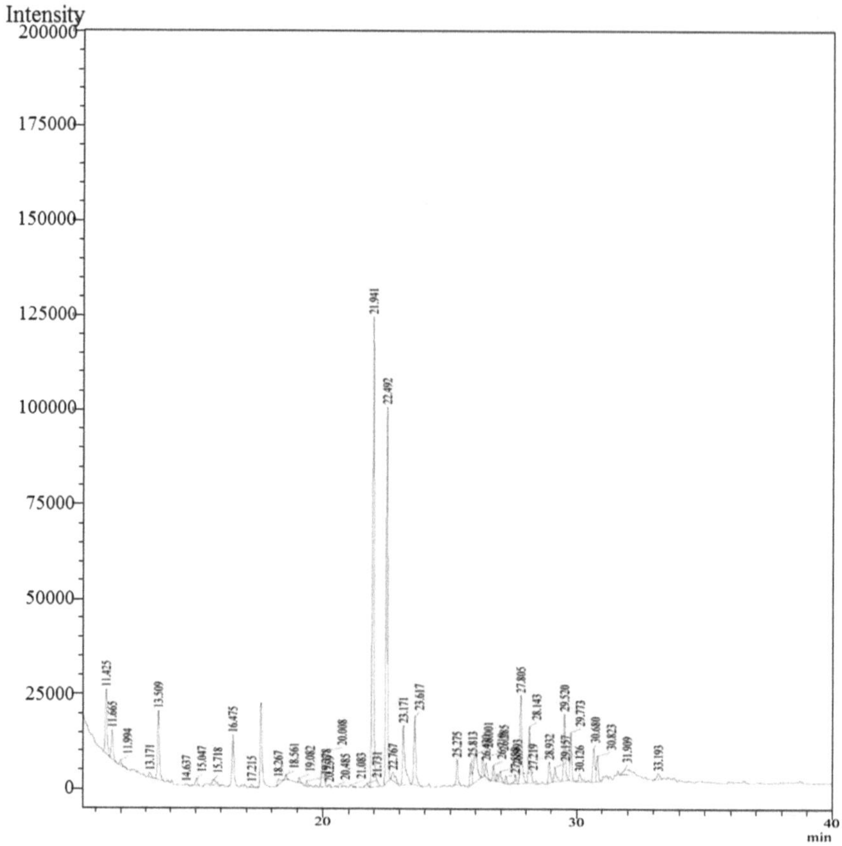

Fig. 3 Gas chromatogram of the sample showing the fatty acid peaks at various retention times

tube with Teflon tape if any hole is formed during boiling. Wear a mask and gloves since the extract in the tube may spill out during heat and the vapors may cause irritation.

14. Safety glasses, goggles, or face shield are recommended to protect eyes from splashing. PVC gloves are recommended to prevent skin contact.
15. For cyanobacteria extend the heating step for another 30–45 min.
16. Repeat the **steps 3–5** until the oil-like droplets are formed in the lower phase.
17. For example: If the retention time is 11.106 min for the C8:0 fatty acid in the standard, the peak with the same retention time in the sample indicates the same fatty acid. Note the peak area of the C8:0 fatty acid in the sample as a percentage of the total summed peak areas. Similarly determine the area percentage of each peak for each fatty acid in the sample chromatogram to generate the complete fatty acid compositional profile. If there

is no peak at the same retention time (11.106) in the sample chromatogram, the C8:0 fatty acid can be listed as "not detected."

Acknowledgments

M.G. acknowledges Bharathidasan University authorities for the University Research Fellowship (05441/URF/K7/2013 dated 04.07.2013). The authors are thankful to DST-FIST program (SR/FIST/LSI/-013/2012 dated 13.08.2012) for instrument facilities.

References

1. Folch J, Lees M, Sloane-Stanley GH (1957) A simple method for the isolation and purification of total lipids from animal tissues. J Biol Chem 226(1):497–509
2. Bligh EG, Dyer WJ (1959) A rapid method of total lipid extraction and purification. Can J Biochem Physiol 37(8):911–917
3. Jensen S, Häggberg L, Jörundsdóttir H, Odham G (2003) A quantitative lipid extraction method for residue analysis of fish involving non halogenated solvents. J Agric Food Chem 51(19):5607–5611
4. Matyash V, Liebisch G, Kurzchalia TV, Shevchenko A, Schwudke D (2008) Lipid extraction by methyl-tert-butylether for high-throughput lipidomics. J Lipid Res 49:1137–1146
5. Jones J, Manning S, Montoya M, Keller K, Poenie M (2012) Extraction of algal lipids and their analysis by HPLC and mass spectrometry. J Am Oil Chem Soc 89:1371–1381
6. Sheng J, Vannela R, Rittmann BE (2011) Evaluation of methods to extract and quantify lipids from *Synechocystis* PCC 6803. Bioresour Technol 102:1697–1703
7. Stanier RY, Kunisawa R, Mandel M, Cohen-Bazire G (1971) Purification and properties of unicellular blue-green algae (Order Chroococcales). Bacteriol Rev 35:171–205
8. Cea M, Sangaletti-Gerhard N, Acuña P, Fuentes I, Jorquera M, Godoy K, Osses F, Navia R (2015) Screening transesterifiable lipid accumulating bacteria from sewage sludge for biodiesel production. Biotechnol Rep 8:116–123
9. Bhatia SK, Yi DH, Kim YH, Kim HJ, Seo HM, Lee JH, Kim JH, Jeon JM, Jang KS, Kim YG, Yang YH (2015) Development of semi-synthetic microbial consortia of *Streptomyces coelicolor* for increased production of biodiesel (fatty acid methyl esters). Fuel 159:189–196
10. Indarti E, Majid MIA, Hashim R, Chong A (2005) Direct FAME synthesis for rapid total lipid analysis from fish oil and cod liver oil. J Food Compos Anal 18(2):161–170

Chapter 5

Rhodococcus and *Yarrowia*-Based Lipid Production Using Lignin-Containing Industrial Residues

Rosemary K. Le, Kristina M. Mahan, and Arthur J. Ragauskas

Abstract

Improvement in biorefining technologies coupled with development of novel fermentation strategies and analysis will be paramount in establishing supplementary and sustainable biofuel pathways. Oleaginous microorganisms that are capable of accumulating triacylglycerides (TAGs) and fatty acid methyl esters (FAMEs), such as *Rhodococcus* and *Yarrowia* species, can be used to produce second-generation biofuels from non-food competing carbon sources. These "microbiorefineries" provide a pathway to upgrade agricultural and industrial waste streams to fungible fuels or precursors to chemicals and materials. Here we provide a general overview on cultivating *Rhodococcus* and *Yarrowia* on agro-waste/industrial biomass pretreatment waste streams to produce single-cell oils/lipids and preparing samples for FAME detection.

Key words Yeast, *Rhodococcus*, *Yarrowia*, Fermentation, Agro-waste, Lipid production

1 Introduction

The use of microbial autotrophs, such as algae, and heterotrophs, like fungi and bacteria, that are capable of accumulating intracellular lipids can be employed as single-cell oil microfactories; which is a rapidly expanding area of interest [1, 2]. Yeast species present an option for such microfactories with their ability to produce lipids or triacylglycerides (TAGs), storing them at 20–70% of their cell weight, depending on the culture conditions [3]. It has been previously reported that lipid accumulation in oleaginous yeasts such as *Cryptococcus albidus*, *Cryptococcus curvatus*, *Lipomyces starkeyi*, *Lipomyces tetrasporus*, *Rhodotorula glutinis*, and *Trichosporon pullulans* can reach beyond 65% of the cell dry weight, under specific conditions [4]. *Yarrowia lipolytica* is a model oleaginous yeast strain, also considered a "biotechnological chassis" that has been used extensively for studies in heterologous protein synthesis,

Venkatesh Balan (ed.), *Microbial Lipid Production: Methods and Protocols*, Methods in Molecular Biology, vol. 1995, https://doi.org/10.1007/978-1-4939-9484-7_5, © Springer Science+Business Media, LLC, part of Springer Nature 2019

lipid accumulation, and catabolic processing [5–7]. It can degrade a wide variety of substrates such as citrate, isocitrate, alpha-ketoglutarate, pyruvate, succinate, polyalcohols, carotenoids, alkanes, and aromatics.

In previous years, the gram-positive, oleaginous, soil bacteria *Rhodococcus opacus* have been studied extensively for their ability to accumulate intracellular lipids >20%, based on cell dry weight (CDW) [8–14]. For example, the *R. opacus* PD 630 strain exhibited oleaginicity above 80% CDW, when glucose was utilized as a carbon source under nitrogen-limited conditions [15, 16]. When glucose was not used, the wild-type species have also demonstrated the ability to effectively degrade aromatic compounds commonly found in lignocellulosic biomass.

While algae- and yeast-to-biofuel processes are prominent and continuously being refined, oil-producing bacteria that also boast a broader threshold of viable nutrients and tolerable growth conditions represent an area of research with considerable potential for advancement [17–20]. The genera of bacteria used in experimental work commonly include *Acinetobacter*, *Mycobacterium*, *Streptomyces*, and more recently *Rhodococcus* [21]. *R. opacus* and *R. jostii* have extensively been studied in the last few years and are well noted for their ability to accumulate oleaginous extents of intracellular single cell oils based on cell dry weight (CDW) >20%. Beyond glucose, these wild-type species have also demonstrated the effective degradation of aromatic material common in lignocellulosic biomass, specifically, substituted aryl species found in lignin. The biochemical degradation of aryl units into lipids is made feasible via enzyme-mediated sequences such as the β-ketoadipate pathway (β-KAP) [22].

Yarrowia lipolytica, the archetypical oleaginous yeast of the *Yarrowia* clade owing to the comprehensive information provided by genomics, systems biology, genetic engineering, and transcriptomic data accumulated in recent years, could be modified to be effective in producing valuable lipids. Bioconversion with *Y. lipolytica* is considered as a promising sustainable biocommodity production pathway for biodiesel due to the vegetable oil-like profile of fatty acids that it can amass.

TAGs and fatty acid methyl esters (FAMEs) originating from oleaginous yeasts and *Rhodococcus* species can be used to generate second generation biofuels from nonfood plant sources, which provides a pathway to upgrade agricultural and industrial waste streams to fungible fuels or precursors to chemicals and materials. The capability of microbes to selectively create chemicals with special features point to mechanistic controls that, in certain cases, can be harnessed to generate targets or new molecules that could give rise to unique materials or fuels.

Here we provide a general overview on cultivating *Rhodococcus* and *Yarrowia* on agro-waste/industrial waste streams to produce single-cell oils/lipids and preparing samples for FAMEs detection. The materials and methods presented can be adapted to suit the requirements for other oleaginous organisms.

2 Materials

Prepare all solutions with deionized water and biological grade or analytical grade reagents. Store all media at 4 °C, unless otherwise specified. Several strains of *Rhodococcus* have been studied, but the focus of this chapter is on *Rhodococcus opacus* (DSMZ 1069, referred to as DSM 1069) and *Rhodococcus wratislaviensis* (DSMZ 44193, referred to as PD 630) obtained from the German Collection of Microorganisms and Cell Cultures (DSMZ, www.dsmz.de). Several strains of *Yarrowia* have also been studied and are widely available in culture collections.

2.1 Strain Preservation

1. Complex culture media (as shown in Subheading 2.2).
2. 250 mL baffled flask, 15 or 50 mL culture tubes.
3. Shaking incubator.
4. Glycerol.
5. Water.
6. −20 °C and/or −80 °C freezer.
7. 2 mL cryovials.
8. Freeze dryer.

2.2 Complex Culture Media

2.2.1 For Rhodococcus

1. For inoculum (50 mL) fermentation: 250 mL baffled flask with cap.
2. For small-scale (150 mL) fermentation: 500 mL baffled flask with cap.
3. *R. opacus* DSM 1069 liquid media (DSMZ Medium 65, GYM Streptomyces Medium):

• Glucose	4.0 g/L
• Yeast extract	4.0 g/L
• Malt extract	10.0 g/L

Make up to 1 L with water. Autoclave for 15 min. Let cool to room temperature before use (*see* **Notes 1** and **2**).

4. *R. opacus* DSM 1069 agar plates:

• Glucose	4.0 g/L
• Yeast extract	4.0 g/L
• Malt extract	10.0 g/L
• Agar	12.0 g/L

Make up to 1 L with water. Autoclave for 15 min. Let cool until able to handle with heat resistant gloves and pour ~20 mL into sterile 100 mm (D) × 15 mm (h) petri dishes in a sterile hood.

5. *R. opacus* DSM PD 630 liquid media (DSMZ Medium 535):

• Trypticase soy broth	30.0 g/L

Make up to 1 L with water. Autoclave for 15 min. Let cool to room temperature before use.

6. *R. opacus* DSM PD 630 agar plates:

• Trypticase soy broth	30.0 g/L
• Agar	15.0 g/L

Make up to 1 L with water. Autoclave for 15 min. Let cool until able to handle with heat-resistant gloves and pour ~20 mL into sterile 100 mm (D) × 15 mm (h) petri dishes in a sterile hood.

2.2.2 For Yarrowia

1. Shake flasks (baffled or unbaffled).
2. Fermenters (batch-fed or continuous).
3. Yeast extract—peptone—glucose—medium (YPD):

• Yeast extract	10.0 g/L
• Bacto peptone	20.0 g/L
• Glucose	20.0 g/L

4. Yeast extract–malt extract–peptone–glucose Medium (YMG):

• Yeast extract	3.0 g/L
• Malt extract	3.0 g/L
• Bacto peptone	5.0 g/L
• Glucose	10.0 g/L

5. For agar plates: make up complex culture media as described above and amend with agar.

• Agar	15.0–20.0 g/L

2.3 Minimal Media

2.3.1 For Rhodococcus

1. *R. opacus* DSM 1069 minimal media (1×) [9]:

• KH_2PO_4	0.40 g/L
• K_2HPO_4	1.60 g/L
• $MgSO_4{\cdot}7H_2O$	0.20 g/L
• $FeCl_3$	0.03 g/L
• $MnSO_4{\cdot}H_2O$	0.50 mg/L
• $CuSO_4{\cdot}5H_2O$	1.00 mg/L
• $ZnSO_4{\cdot}7H_2O$	1.00 mg/L
• $CaCl_2$	0.50 mg/L
• KCl	0.10 mg/L
• H_3BO_3	0.50 mg/L

2. Make 20× stock for ease of weighing out salts. Store at 4 °C (*see* **Note 3**):

• KH_2PO_4	8.0 g/L
• K_2HPO_4.	32.0 g/L
• $MgSO_4{\cdot}7H_2O$	4.0 g/L
• $FeCl_3$	0.6 g/L
• $MnSO_4{\cdot}H_2O$	10 mg/L
• $CuSO_4{\cdot}5H_2O$	20.0 mg/L
• $ZnSO_4{\cdot}7H_2O$	20.0 mg/L
• $CaCl_2$	10 mg/L
• KCl	2.0 mg/L
• H_3BO_3	10.0 mg/L

Make up to 1 L with water. Store at 4 °C.

3. *R. opacus* PD 630 minimal media (1×) [23]:

• $Na_2HPO_4{\cdot}12H_2O$	9.00 g/L
• KH_2PO_4	1.50 g/L
• $MgSO_4{\cdot}7H_2O$	0.20 g/L

(continued)

• $FeNH_4$ citrate	1.20 mg/L
• $CaCl_2$	20.0 mg/L
• Hoagland solution (Sigma, H2395)	2.00 mL
• $NaHCO_3$	0.50 g/L

4. Make 20× stock for ease of weighing out salts. Store at 4 °C (*see* **Note 3**):

• $Na_2HPO_4{\cdot}12H_2O$	180.0 g/L
• KH_2PO_4	30.0 g/L
• $MgSO_4{\cdot}7H_2O$	4.0 g/L
• $FeNH_4$ citrate	24 mg/L
• $CaCl_2$	0.4 g/L
• Hoagland solution	40.0 mL
• $NaHCO_3$	10.0 g/L

Make up to 1 L with water. Store at 4 °C.

5. The nitrogen source for minimal media for DSM 1069 is NH_4NO_3 and the source for PD 630 is NH_4Cl. Make 20% (w/v) solution to be diluted to 0.1% or 0.05% final for use. For example, 2.0 g NH_4NO_3 or NH_4Cl made up to 10 mL volume with water. Store at 4 °C.

2.3.2 For *Yarrowia*

1. Nitrogen-limited medium (NLM) [24]:

• Glucose	70.0 g/L
• $(NH_4)_2SO_4$	0.1 g/L
• Yeast extract	0.75 g/L
• KH_2PO_4	1.5 g/L
• $MgSO_4{\cdot}7H_2O$	1.5 g/L

2. Yeast Nitrogen Base Medium (YNB) (DIFCO).

• YNB	6.7 g/L

Different concentrations of carbon sources are recommended (*see* **Note 4**):

• Glucose	10.0–20.0 g/L (up to 100.0 g/L)
• Sodium acetate	4.0 g/L

3. Solution of trace elements (mg in 100 mL):

• $Ca(NO_3)_2{\cdot}4H_2O$	2.0 g
• $FeCl_3{\cdot}6H_2O$	200 mg
• H_3BO_3	50 mg
• $CuSO_4{\cdot}5H_2O$	10 mg
• $MnSO_4{\cdot}4H_2O$	40 mg
• $ZnSO_4{\cdot}7H_2O$	40 mg
• Na_2MoO_4	20 mg
• $CoCl_2$	10 mg
• KI	10 mg

4. Minimal Medium with Thiamine (MMT).

   Some strains of *Y. lipolytica* grow slowly in YNB (*see* **Note 5**), therefore this medium is recommended to obtain higher growth rates.

• $NH_4H_2PO_4$	5.0 g/L
• KH_2PO_4	2.5 g/L
• $MgSO_4{\cdot}7H_2O$	1.0 g/L
• Thiamine–HCl (Vitamin B1)	0.3 g/L (*see* **Note 6**)

5. S2 Minimal Medium pH 6.4 [25] (*see* **Note 7**).

• $MgSO_4{\cdot}7H_2O$	1.5 g/L
• Glucose	30.0 g/L
• $(NH_4)_2SO_4$	0.5 g/L
• Yeast extract	0.5 g/L

2.4 Optical Density

1. UV-Vis spectrophotometer.
2. Cuvettes.

2.5 Serial Dilutions and Plating (SDP)

1. Petri dishes.
2. Screw-cap glass culture tubes.
3. 0.9% Physiological NaCl solution: 9.0 g/L.

Add 9.9 mL salt solution to each glass tube. Autoclave and store at room temperature.

2.6 Cell Dry Weight

1. 10 or 50 mL culture tubes.
2. 0.45 μm polyethersulfone syringe filters.

3. Centrifuge.
4. Desiccator.
5. Oven or freeze dryer.

2.7 Methanolysis/Transesterification

2.7.1 For Rhodococcus

1. Screw-cap culture tubes with PTFE lined phenolic caps.
2. Autosampler vials.
3. For one sample:

• Freeze-dried cells (*see* **Note 8**)	6–10 mg
• $CHCl_3$	1.0 mL
• MeOH	0.85 mL
• 1 M H_2SO_4	0.15 mL
• H_2O	0.5 mL

2.7.2 For Yarrowia

1. Screw-cap culture tubes with PTFE lined phenolic caps.
2. 2 mL autosampler vials.
3. For one sample:

(a) Freeze-dried cells	6–10 mg
(b) 0.5 N CH_3NaO	0.5 mL
(c) H_2SO_4 (98% purity)	0.04 mL
(d) C_6H_{14}	0.5 mL
(e) H_2O	0.5 mL

If cells cannot be effectively separated from the substrate, culture broth can be used; but only in the case where minimal media is used without the inherent presence of lipids in complex culture media.

3 Methods

3.1 Strain Preservation

3.1.1 For Rhodococcus

1. Prepare a 50% glycerol stock by mixing equal parts 100% glycerol and water, sterilize by autoclaving or sterile filter into a sterile vessel.
2. Add 50 mL of complex culture media for the strain of interest to a 250 mL baffled flask (or adjust volume if using smaller culture tubes) and sterilize by autoclaving at 121 °C for 15 min. After cooling, aseptically inoculate the flask with a sterile loop of *R. opacus* cells from a plate. Incubate the flask with shaking at 28 °C, 150 rpm, overnight or until OD_{600} ~ 0.5–1.0.

3. Aliquot 0.5 mL of *R. opacus* culture at OD_{600} ~ 0.5–1.0 with 0.5 mL of 50% glycerol into sterile 2 mL cryovials.
4. Store the cryovials at −80 °C.

3.1.2 For Yarrowia

Y. lipolytica requires oxygen, therefore strains cannot be stored long (>1 month) in sealed plastic bags or anaerobic tubes.

1. Conservation with glycerol: A culture of *Y. lipolytica* cells from the mid logarithmic growth phase is mixed with sterilized glycerol at a final concentration of 25–50% in a cryogenic tube. Tubes are stored at −80 °C.
2. Conservation using liquid nitrogen and freeze drying: After subcloning on solid YPD, one colony is transferred to 2 mL YM liquid medium for 24 h at 28 °C under aeration; 25 mL of fresh YM medium are inoculated with 0.5 mL culture, and grown at 28 °C under aeration for 24 h. Cells from this culture will be preserved by freezing and freeze drying.
3. Preservation by liquid nitrogen [26]: 10 mL of the culture are mixed with 2.5 mL of fresh YM medium containing 50% of glycerol (final glycerol concentration 10%). The mixture is dispensed in ten cryotubes and chilled at 4 °C for 1 h. The tubes are frozen in the Nalgene cryobox at −80 °C in order to obtain a temperature decrease of −1 °C/min. The tubes are then preserved in liquid nitrogen.
4. Preservation by freeze drying [27]: 8 mL of the culture is centrifuged at 4000 × *g* for 5 min and cells are resuspended in 2 mL of the following mixture: 10% skim milk, 5% honey, and monosodium glutamate 5%. The suspension is chilled at 4 °C and 0.1–0.15 mL of cell suspension is dispensed in glass tubes. After being capped with cotton, the tubes are placed in a −20 or −80 °C freezer until frozen solid. Freeze drying can be performed in a Virtis freeze dryer (below −70 °C and 200 mTorr) until all frozen liquid crystals have been sublimated (~24 h). The temperature at the beginning is about −40 °C and rises to 20 °C at the end. The tubes are sealed under vacuum and stored at 4 °C in the dark [28].

3.2 Cultivation

Fermentations can be carried out in shake flasks (baffled, for *Rhodococcus*, and/or unbaffled, for yeast) or in large stirred-tank fermentors in batch-fed or continuous cultures [25]. Strains can also be grown in 96-well microtiter plates [29].

3.2.1 For Rhodococcus

1. For inoculum (50 mL) make up 50 mL of complex culture media in a 250 mL baffled flask with cap. (Adjust initial volume as appropriate for fermentation scale.)
2. Using aseptic technique inoculate 50 mL (adjust volume for initial adaptation phase or number of samples needed for

downstream inoculation) with a loop of cells from the agar plate. Incubate at 28 °C, 150 rpm, overnight or until OD_{600} ~0.6. Optical density is measured using a spectrophotometer at 600 nm (*see* **Note 9**).

3. Cells are centrifuged at 4000 × *g* in sterile tubes for 5 min to form a pellet. The supernatant is removed and the pellet is gently resuspended in sterile minimal media containing 0.1% nitrogen, centrifuging and washing two times as previously stated, to remove the complex culture media. Resuspend the cell pellet in a volume such that the OD_{600} value is approximately 10× or more, and a small volume can be used to inoculate the preculture phase broth to an initial OD_{600} ~ 0.6.
4. Prepare the initial preculture broth with the effluent or substrate of interest at the conditions desired for the study containing 0.1% nitrogen and autoclave (*see* **Note 10**).

   Inoculate the initial preculture broth at an OD_{600} ~ 0.6.
5. After 24 h and if the preculture grows successfully, centrifuge the initial adaption broth as before and resuspend the pellet in a sterile study media (at the same solids % you used for your initial adaptation) now containing 0.05% nitrogen and transfer to the full-scale vessel, typically 150 mL in a 500 mL baffled flask for bench-scale fermentation (*see* **Note 11**). This is the start of full-scale fermentation (*see* **Note 12**). Incubate flasks at 28 °C, 150 rpm.

3.2.2 For Yarrowia

1. Temperature: Most strains of *Y. lipolytica* are capable of growing in temperatures up to 34 °C, but there are some strains which are adapted to higher temperatures [30, 31]. The recommended temperature for growth is 25–30 °C. Induction of sporulation is highest at 23 °C. Conjugation and sporulation frequencies are highest at 23–28 °C, but decrease strongly above 30 °C.
2. pH Values of the growth media: Most strains of *Y. lipolytica* tolerate low pH values well, down to pH 3, but growth is reduced above pH 7 and stops above pH 8. Therefore, addition of buffer with a pH lower than pH 7 or low pH values (pH 3.5–4) of the medium at the beginning of the cultivation are recommended for utilization of carbon sources such as acetate.
3. Aeration: *Y. lipolytica* is an obligatory aerobe and cannot grow or ferment without oxygen. Low oxygen pressure strongly reduces the growth rate. Therefore, vigorous shaking of batch cultures or high oxygen supply in fermentors is recommended. However, development of foam may occur, especially during cultivation in complete media like YEPD. This has to be inhibited, because *Y. lipolytica* cells migrate rapidly into the foam,

thus reducing the yield. Addition of small amounts of oil or antifoam efficiently reduces the formation of foam.

4. Shake flask fermentations:
   - Inoculate a single colony in 2 mL defined medium: 6.7 g/L yeast nitrogen base without amino acids, complete and 20 g/L glucose or YPD.
   - Harvest after 24 h by centrifugation at 8000 × *g* and wash twice with low-nitrogen defined medium (1.7 g/L yeast nitrogen base without amino acids and ammonium sulfate, 1.1 g/L ammonium sulfate and 50 g/L glucose).
   - Wash cells used to inoculate 50 mL low-nitrogen defined medium in 250 mL shake flasks at OD_{600} ~ 0.01–0.05 and grown at 30 °C for 120 h (220–250 rpm).
   - 2–10 mL of cell suspension is sampled every 24 h (from 48 to 120 h) for OD_{600} or SDP (*see* **Notes 13** and **14**), cell dry weight, extracellular metabolites, and lipid measurements [32] 2–10 mL of cell suspension is sampled every 24 h (from 48 to 120 h) for OD600, cell dry weight, extracellular metabolites, and lipid measurements [32].
5. Batch mode bioreactor fermentations [25]:
   - Inoculate a single colony in 2 mL defined medium: 6.7 g/L yeast nitrogen base without amino acids, complete and 20 g/L glucose or YPD.
   - Harvest the seed culture by centrifugation at 4000 × *g* and wash twice with sterile fermentation medium.
   - Inoculate into the bioreactor for a starting concentration OD_{600} of ~0.05.
   - 1 L S2 medium in a 1.5 L bioreactor (or 1.6 L working volume in a 3 L bioreactor) at 30 °C with an aeration rate of 1 vvm (volume air [L]/volume media [L]/min).

3.3 Cell Dry Weight

1. Weigh an empty, sterile 50 mL conical centrifuge tube (polypropylene) and record the mass. Aseptically transfer 10 mL of fermentation broth to the 50 mL tube. This will be used for cell dry weight (CDW) analysis. Centrifuge the broth at 4000 × *g* for 5 min to pellet the cells. Save the initial supernatant for analysis and wash the pellet with water and centrifuge as before, repeating twice to remove residual fermentation broth. Freeze dry for analysis following procedure in Subheading 3.1, **step 4** (for *Yarrowia*).
2. After freeze drying the cell pellet, reweigh the tube and record the mass of the cell pellet, by subtracting the weight of the empty tube. The freeze-dried cell pellet can be used for lipid analysis. *See* Subheading 3.4.

3. Alternatively, the washed cell pellet can be resuspended in a small amount of water and poured onto an aluminum pan, previously weighed and mass recorded, and reweighing after the liquid has been removed and the dried cells weighed, recording the mass and subtracting the mass of the empty pan.

Due to hyphal growth of *Y. lipolytica* in YPD and YM media, maximum growth rates must be determined using a dry weight method.

1. 5–10 mL cultures are collected every few hours and filtered using 0.45 μm polyethersulfone filters, which can be dried using a microwave and subsequently placed in a desiccator for 2 days [33]. 5–10 mL cultures are collected every few hours and filtered using 0.45 μm polyethersulfone filters, which can be dried using a microwave and subsequently placed in a desiccator for 2 days [33].
2. Instead of filtering, the culture broth can also be harvested by centrifugation at 10,000–18,000 × *g*, washed (with water, salt solution, or buffer), and dried for 24 h at 105 °C until constant weight in preweighed tubes [25].

3.4 Transesterification for Gas Chromatography-- Mass Spectrometry (GC-MS or GC-FID) Analysis

3.4.1 For Rhodococcus

GC-MS is used to determine lipid (fatty acid methyl ester) content of the cells.

1. Approximately 10 mg of freeze-dried cell–broth mixture is mixed with 1.0 mL chloroform, 0.85 mL methanol, and 0.15 mL 1 M sulfuric acid. This is incubated at 100 °C for 2 h and 20 min in a sand bath or water bath (*see* **Note 11**). Subsequently, 0.5 mL of distilled water is added to each sample, vortexed for approximately 1 min, and the lower phase (~0.6 mL), containing the fatty acid methyl esters is transferred to 2 mL autosampler vials. Samples are stored at −20 °C until ready for processing in an explosion-proof freezer.
2. Run samples on GC-MS or GC-FID using previously documented literature as reference for column/method conditions. Example conditions for FAMEs detection using Agilent 7890A GC system equipped with FID and a Supelco SP-2560 column:

   Using helium as the carrier gas at flow rate of 19.7 cm/s, split inject 2 μL samples at 100:1 ratio; set initial oven temperature to 140 °C and hold for 5 min. Then ramp by 4 °C/min until 240 °C is reached, and hold at the final temperature for 15 min. HP-5MS columns can also be used for FAMEs detection.
3. Generate a calibration curve using an appropriate standard (ex. Supelco® 37 Component FAME Mix, Sigma Cat. No.: 47885-U SUPELCO when testing for FAMEs).

3.4.2 For Yarrowia

1. Prior to chemical derivatization, cells at desired sampling times can be pelleted by centrifugation and freeze-dried or frozen. Cells can either be saponified from a fresh/frozen cell pellet [34] or 20 mg of freeze-dried cells can be derivatized [35, 36].
2. For saponification: Approximately 1 mL of cell suspension (depending on cell density) will yield ~2 mg of dry cell weight (DCW). Centrifuge cell suspension at 18,000 × *g* for 10 min and carefully remove the supernatant. Cells can be frozen at −20 °C before derivatization. At a minimum, 2 mg of should be used. For lipid analysis, add an internal standard (*see* **Note 16**). Lipid transesterification is initiated by the addition of 500 μL of 0.5 N sodium methoxide (20 g/L sodium hydroxide in anhydrous methanol). The cells are then vortexed at 1200 × *g* for 60 min at room temperature. The mixture is then neutralized with 40 μL sulfuric acid (98% purity) (*see* **Note 17**). The resulting fatty acid methyl esters (FAMEs) are extracted by 500 μL hexane, followed by vortexing for 30 min again at 1200 × *g*. Centrifuge the samples at 8000 × *g* for 1 min. The top hexane layer can be carefully removed and analyzed by a gas chromatography–flame ionization detector (GC-FID) to quantify fatty acids and lipid titer.
3. For transesterification derivatization: Pellet cell suspension via centrifugation as described above and freeze-dry. Use approximately 20 mg of freeze-dried cells for lipid content and fatty acid analysis. Add an internal standard (*see* **Note 18**) to samples of freeze-dried cells. The cell lipids are extracted by the addition of 0.5 N hydrochloric acid in methanol (*see* **Note 19**) at 85 °C for 90 min. After transesterification, the lipid-soluble components are separated water-soluble components by two-phase liquid extraction. Water and isooctane are added and the isooctane fraction can subsequently be analyzed via GC-FID.

4 Notes

1. DSMZ recommends the addition of 2.0 g of calcium carbonate when making agar plates. We exclude this in our agar as it interferes with counting colonies for serial dilute plating due to poor dissolution/incorporation into the media.
2. When preparing liquids to be autoclaved, never fill the autoclavable vessel more than 2/3 full to prevent "boil-over." Loosen caps or use vented closures before autoclaving to prevent excess pressure build up. Place items to be autoclaved in secondary containment.
3. After storing 20× minimal media at 4 °C, salts may form crystals or precipitate. Place bottle in water bath at 60 °C,

shake to mix occasionally, until salts have dissolved before making 1× solution.

4. Higher concentrations reduce growth, toxic above 10 g/L, addition of 0.05 M phosphate buffer of pH 5 is recommended; n-Alkanes, fats, and fatty acids have no toxic effects and can therefore be also used at high concentrations. Emulsifiers like Brij 58 or Tween can be added but are not necessary because *Y. lipolytica* secretes very efficient bioemulsifiers.
5. Addition of carbon sources as recommended for YNB. The pH value should be adjusted at 6 for most carbon sources, but at pH 4 for acetate or fatty acids (or strongly buffered at pH 6 with phosphate buffer). Sometimes, problems occur with strains harboring multiple amino acid auxotrophies: use of 0.1% glutamine as N-source is recommended in such cases.
6. Thiamine hydrochloride can be sterilized by autoclaving when in solution at pH 3.5. Solutions above pH 5 are heat sensitive and should be filter sterilized into sterile media. Store thiamine hydrochloride at room temperature.
7. S2 is an efficient, cost effective, and scalable minimal culture medium for maximum lipid production. Phosphate limiting conditions produce high lipid content. Alternate carbon sources including fructose, sucrose, or glycerol can be used to produce different FAME profiles of oils produced from *Y. lipolytica*. Strains do not grow well and are unable to produce lipids when glucose has been replaced by maltose, xylose, trehalose, L-malate, or citrate.
8. Record the mass of freeze-dried cells that is used for methanolysis/transesterification so the total lipids can be related to the initial mass to determine cell oleaginicity.
9. At this optical density, the cells are in the exponential phase [9] of growth and are more amenable to acclimation.
10. It is typically desired to maximize the amount of the carbohydrate source in the media such that the cells have excess material to metabolize. At a minimum, the broth should contain 40 mg solids/mL, if possible. For lignocellulosic material, it has been determined that a lignin content of less than 1.5% w/v is ideal for lipid production in *Rhodococcus* [37].
11. The nitrogen-limited conditions cause the cells to change their metabolism and reroute energy into storing fats rather than any other cell function (e.g., cell proliferation and DNA replication) [8].
12. As a suggestion, a homogenous sample of the fermentation broth should be removed aseptically at 0, 12, 24, 48, 72, and 96 h for analysis, depending on what analysis is desired. Some examples of techniques that are typically utilized in

characterizing industrial or agro-waste in the microbial lipid production include nuclear magnetic resonance spectroscopy (NMR) [38–41] and Fourier transform infrared spectroscopy (FTIR) analysis [42, 43], which provide structural information on the substrate, gel permeation chromatography (GPC), a size-exclusion chromatography technique separates analytes based on size and provides information about the molecular weight distribution of the material present in the fermentation and gives an idea about the its utilization by the cells during fermentation (preferences for particular molecular weight particles) [43, 44], cell dry weight (CDW) to quantify cell density, gas chromatography mass spectrometry (GC-MS) to identify compounds of interest generated by *Rhodococcus* or *Yarrowia* fermented on the industrial/agro-waste substrates, and serial dilute plating (SDP) and OD_{600}, which provide information on cell viability and cell density over time. Additional description of analytical techniques is provided in Le et al. [45].

Example sample removal schedule:

0 h: Remove enough volume so you have enough solid material for analysis (Klason lignin analysis (300 mg minimum), NMR (40 mg, or 100 mg for solid-state), FTIR (5 mg), GPC (5 mg), CDW (10 mL broth), GC-MS (6–10 mg solids), HPLC (<5 mL broth), etc.) ~45 mL or ~600 mg solids, perform SDP (200 μL), OD_{600} measurement (200 μL), spin down cells for CDW.

12 h: Perform SDP, OD_{600} measurement, collect 10 mL for CDW/spin down cells for CDW, 5 mL for HPLC.

24 h: Perform SDP, OD_{600} measurement, collect 10 mL for CDW/spin down cells for CDW, 5 mL for HPLC.

48 h: Remove enough volume so you have enough solid material for analysis (Klason, NMR, FTIR, GPC, CDW, GC-MS, HPLC, etc. ~45 mL or ~600 mg solids), perform SDP, OD_{600} measurement, spin down cells for CDW.

72 h: Perform SDP, OD_{600} measurement, collect 10 mL for CDW/spin down cells for CDW, 5 mL for HPLC.

96 h: Remove enough volume so that you have enough solid material for analysis (Klason, NMR, FTIR, GPC, CDW, GC-MS, HPLC, etc. ~45 mL or ~600 mg solids), perform SDP, OD_{600} measurement, spin down cells for CDW.

13. SDP, serial dilution and plating, is used to determine cell viability and is particularly useful when OD_{600} measurements are not practical to determine cell density, because of the effluent used, that is, the effluent has particles that interfere with the scattering of the bacterial cells/cannot differentiate between cell scattering and effluent scattering. Dilutions typically 10^{-2}, 10^{-4}, 10^{-6} are appropriate until approximately

48 h, depending on effluent/carbohydrate source used to determine viability and distinct colonies. Dilutions greater than 10^{-6} may be necessary to prevent overgrowth on the agar plates.

14. Colonies may initially take more than 24 h to appear. If they appear and are too small to count, allow an additional 24–48 h incubation before counting. If no colonies appear, the cells are not viable on the substrate tested. If colonies result in a lawn, adjust dilution factor for more dilute plating. Typically, 10^{-4}–10^{-6} is optimal and 10^{-2} dilution results in lawn after 48 h of fermentation, but is substrate dependent.
15. Using a sand bath in a metal pan for transesterification is a safe alternative to a water bath heated to 100 °C, which can evaporate over time and result in an unintended hazard if left unattended. However, a sand bath may have hotspots resulting in uneven heating or overheating. If a water bath is used, replace the evaporated water as required. A "Digi-Block" for culture tubes is the ideal instrument for even heating of samples; use if available.
16. For lipid analysis, internal standards are used to account for volume change in the column or measuring transesterification efficiency. For volume change, 100 μL of hexane-containing 2 mg/mL methyl tridecanoate is added and 2 mg/mL glyceryl triheptadecanoate is typically used for transesterification efficiency.
17. Use caution when adding the sulfuric acid as a violent exothermic reaction should occur.
18. Suggested internal standard: Pentadecanoic acid (15:0, Sigma).
19. Methanolic HCl, 0.5 M HCl in methanol (0.5 N) for GC derivatization can be purchased from Sigma, ampule of 20 × 1 mL.

Acknowledgments

This work was supported by US Department of Energy (award #DE—EE0006112), and we would like to acknowledge our collaborator Joshua S. Yuan at Texas A&M University.

References

1. Miao X, Wu Q (2006) Biodiesel production from heterotrophic microalgal oil. Bioresour Technol 97(6):841–846
2. Deeba FV, Pruthi V, Negi YS (2016) Converting paper mill sludge into neutral lipids by oleaginous yeast *Cryptococcus vishniaccii* for biodiesel production. Bioresour Technol 213:96–102
3. Lamers D, van Biezen N, Martens D, Peters L, Van de Zilver E, Jacobs-van Druemel N, Wijffels RH, Lokman C (2016) Selection of

oleaginous yeasts for fatty acid production. BMC Biotechnol 16(1):45

4. Abghari A, Chen S (2014) *Yarrowia lipolytica* as an oleaginous cell factory platform for production of fatty acid-based biofuel and bioproducts. Front Energy Res 2(21). https://doi.org/10.3389/fenrg.2014.00021
5. Ledesma-Amaro R, Nicaud J-M (2016) Metabolic engineering for expanding the substrate range of *Yarrowia lipolytica*. Trends Biotechnol 34(10):798–809
6. Beopoulos A, Cescut J, Haddouche R, Uribelarrea J-J, Molina-Jouve C, Nicaud J-M (2009) *Yarrowia lipolytica* as a model for bio-oil production. Prog Lipid Res 48(6):375–387
7. Beopoulos A, Nicaud JM, Gailardin C (2011) An overview of lipid metabolism in yeasts and its impact on biotechnological processes. Appl Microbiol Biotechnol 90(4):1193–1206
8. Kosa M, Ragauskas AJ (2011) Lipids from heterotrophic microbes: advances in metabolism research. Trends Biotechnol 29(2):53–61
9. Kosa M, Ragauskas AJ (2012) Bioconversion of lignin model compounds with oleaginous rhodococci. Appl Microbiol Biotechnol 93 (2):891–900
10. Kosa M, Ragauskas AJ (2013) Lignin to lipid bioconversion by oleaginous rhodococci. Green Chem 15(8):2070–2074
11. Wei Z, Zeng G, Huang F, Kosa M, Huang D, Ragauskas AJ (2015) Bioconversion of oxygen-pretreated Kraft lignin to microbial lipid with oleaginous *Rhodococcus opacus* DSM 1069. Green Chem 17(5):2784–2789
12. Wei Z, Zeng G, Huang F, Kosa M, Sun Q, Meng X, Huang D, Ragauskas AJ (2015) Microbial lipid production by oleaginous rhodococci cultured in lignocellulosic autohydrolysates. Appl Microbiol Biotechnol 99 (17):7369–7377
13. Wei Z, Zeng G, Kosa M, Huang D, Ragauskas AJ (2015) Pyrolysis oil-based lipid production as biodiesel feedstock by *Rhodococcus opacus*. Appl Biochem Biotechnol 175(2):1234–1246
14. Le RK, Wells T, Das P, Meng X, Stoklosa RJ, Bhalla A, Hodge DB, Yuan JS, Tagauskas AJ (2017) Conversion of corn stover alkaline pre-treatment waste streams into biodiesel via rhodococci. RSC Adv 7(7):4108–4115
15. Alvarez H, Steinbüchel A (2002) Triacylglycerols in prokaryotic microorganisms. Appl Microbiol Biotechnol 60(4):367–376
16. Alvarez HM, Kalscheuer R, Steinbüchel A (1997) Accumulation of storage lipids in species of Rhodococcus and Nocardia and effect of inhibitors and polyethylene glycol. Eur J Lip Sci Technol 99(7):239–246
17. Gomez JA, Höffner K, Barton PI (2016) From sugars to biodiesel using microalgae and yeast. Green Chem 18(2):461–475
18. Meng X, Yang J, Xu X, Zhang L, Nie Q, Xian M (2009) Biodiesel production from oleaginous microorganisms. Renew Energy 34 (1):1–5
19. Pimentel D, Patzek PW (2005) Ethanol production using corn, switchgrass, and wood; biodiesel production using soybean and sunflower. Nat Resour Res 14(1):65–76
20. Beckham GT, Johnson CW, Karp EM, Salvachua D, Vardon DR (2016) Opportunities and challenges in biological lignin valorization. Curr Opin Biotechnol 42:40–53
21. Alvarez HM, Kalscheuer R, Steinbuchel A (2000) Accumulation and mobilization of storage lipids by *Rhodococcus opacus* PD630 and *Rhodococcus ruber* NCIMB 40126. Appl Microbiol Biotechnol 54:218–223
22. Wells T, Ragauskas AJ (2012) Biotechnological opportunities with the β-ketoadipate pathway. Trends Biotechnol 30:627–637
23. Schlegel H, Kaltwasser H, Gottschalk G (1961) A submersion method for culture of hydrogen-oxidizing bacteria: growth physiological studies. Arch Microbiol 38(3):209–222
24. Yang X, Jin G, Gong Z, Shen H, Song Y, Bai F, Zhao ZK (2014) Simultaneous utilization of glucose and mannose from spent yeast cell mass for lipid production by *Lipomyces starkeyi*. Bioresour Technol 158:383–387
25. Nambou K, Zhao C, Wei L, Chen J, Imanaka T, Hua Q (2014) Designing of a cheap to run fermentation platform for an enhanced production of single cell oil from *Yarrowia lipolytica* DSM3286 as a potential feedstock for biodiesel. Bioresour Technol 173:324–333
26. Daggett P-M, Simione FP (1987) Method of culturing freeze-dried microorganisms. US Patent US4672037A
27. Berny J-F, Hennebert G (1991) Viability and stability of yeast cells and filamentous fungus spores during freeze-drying: effects of protectants and cooling rates. Mycologia 83:805–815
28. Barth G, Gaillardin C (1996) Yarrowia lipolytica. In: Nonconventional yeasts in biotechnology. Springer, New York, pp 313–388
29. Gajdoš P, Nicaud JM, Rossignol T, Čertík M (2015) Single cell oil production on molasses by *Yarrowia lipolytica* strains overexpressing dga2 in multicopy. Appl Microbiol Biotechnol 99(19):8065–8074
30. Blagodatskaj V, Kockova-Kratochvilova K (1973) The heterogeneity of the species Candida lipolytica Candida pseudolipolytica new

species and Candida lipolytica var thermotolerans new variety. Biologia (Bratislava) 28 (9):709–716

31. Barnett JA, Payne RW, Yarrow D (1983) Yeasts: characteristics and identification. Cambridge University Press, Cambridge

32. Qiao K, Wasylenko TM, Zhou K, Xu P, Stephanopoulos G (2017) Lipid production in *Yarrowia lipolytica* is maximized by engineering cytosolic redox metabolism. Nat Biotechnol 35(2):173–177

33. Wei Y, Siewers V, Nielsen J (2017) Cocoa butter-like lipid production ability of non-oleaginous and oleaginous yeasts under nitrogen-limited culture conditions. Appl Microbiol Biotechnol 101(9):3577–3585

34. Xu P, Qiao K, Ahn WS, Stephanopoulos G (2016) Engineering *Yarrowia lipolytica* as a platform for synthesis of drop-in transportation fuels and oleochemicals. Proc Natl Acad Sci U S A 113(39):10848–10853

35. Zhang H, Wu C, Wu Q, Dai J, Song Y (2016) Metabolic flux analysis of lipid biosynthesis in the yeast *Yarrowia lipolytica* using 13C-labeled glucose and gas chromatography-mass spectrometry. PLoS One 11(7):E0159187

36. Friedlander J, Tsakraklides V, Kamineni A, Greenhagen EH, Consiglio AL, MacEwen K, Crabtree DV, Afshar J, Nugent RL, Hamilton MA, Shaw AJ, Suth CR, Stephanopoulos G, Brevnova EE (2016) Engineering of a high lipid producing *Yarrowia lipolytica* strain. Biotechnol Biofuels 9(1):77

37. He Y, Li X, Ben H, Xue X, Yang B (2017) Lipid production from dilute alkali corn stover lignin by Rhodococcus strains. ACS Sustain Chem Eng 5(3):2302–2311

38. Korntner P, Sumerskii I, Bacher M, Rosenau T, Potthast A (2015) Characterization of technical lignins by NMR spectroscopy: optimization of functional group analysis by ^{31}P NMR spectroscopy. Holzforschung:807. https://doi.org/10.1515/hf-2014-0281

39. Ben H, Ragauskas AJ (2011) NMR characterization of pyrolysis oils from Kraft lignin. Energy Fuel 25(5):2322–2332

40. Pu Y, Cao S, Ragauska AJ (2011) Application of quantitative ^{31}P NMR in biomass lignin and biofuel precursors characterization. Energy Environ Sci 4(9):3154–3166

41. Sannigrahi P, Ragauskas AJ (2011) Characterization of fermentation residues from the production of bio-ethanol from lignocellulosic feedstocks. J Biobased Mater Bioenergy 5 (4):514–519

42. Kubo S, Kadla JF (2005) Hydrogen bonding in lignin: a Fourier transform infrared model compound study. Biomacromolecules 6 (5):2815–2821

43. Meng X, Sun Q, Kosa M, Huang F, Pu Y, Ragauskas AJ (2016) Physicochemical structural changes of poplar and switchgrass during biomass pretreatment and enzymatic hydrolysis. ACS Sustain Chem Eng 4(9):4563–4572

44. Tolbert A, Akinosho H, Khunsupat R, Naskar AK, Ragauskas AJ (2014) Characterization and analysis of the molecular weight of lignin for biorefining studies. Biofuels Bioprod Biorefin 8 (6):836–856

45. Le RK, Das P, Mahan KM, Anderson SA, Wells T Jr, Yuan JS, Ragauskas AJ (2017) Utilization of simultaneous saccharification and fermentation residues as feedstock for lipid accumulation in *Rhodococcus opacus*. AMB Express 7 (1):185

Chapter 6

Quantification of Lipid Content in Oleaginous Biomass Using Thermogravimetry

Balakrishna Maddi, Agasteswar Vadlamani, Sridhar Viamajala, and Sasidhar Varanasi

Abstract

Laboratory analytical techniques employed for triglyceride quantification in oleaginous biomass (e.g., microalgae and oilseeds) involve multiple steps and typically require use of volatile organic solvents. Here we describe a single-step approach for measurement of triglycerides using thermogravimetry (TG). We have observed that triglycerides undergo complete volatilization over a narrow temperature interval of 370–450 °C, with negligible solid residue under inert atmosphere, whereas other constituents of oleaginous biomass (such as proteins and carbohydrates) primarily degrade below 350 °C. As a result, triglyceride content of biomass can be estimated using TG by determining the mass loss of the sample in the temperature interval of 370–450 °C.

Key words Microalgae, Thermogravimetry, Lipids, Triglycerides, Oleaginous biomass

1 Introduction

Oleaginous biomass such as oilseeds and microalgae are rich in triglycerides and are feedstocks for biofuels and other oleochemicals [1, 2]. For microalgae, the energy-dense triglycerides accumulate and are reutilized over relatively short periods of time (order of days to weeks) [3]. As a result, microalgal cultures must be continually monitored for lipids to match their harvest with periods of highest lipid content and/or productivity. Lipids are commonly analyzed in the laboratory using one of the following methods. The first method involves the extraction of lipids from the solid biomass into organic solvents, often with prior mechanical disruption (e.g., sonication) of solid samples. Lipid content is then assessed either by gravimetric analysis of the organic solvent or by using chromatography techniques [4–8]. The estimates obtained from this method are frequently variable/inconsistent, especially for microalgae samples [9, 10]. A second method that is convenient is the use of time domain NMR to measure lipid content [11]. This

Venkatesh Balan (ed.), *Microbial Lipid Production: Methods and Protocols*, Methods in Molecular Biology, vol. 1995, https://doi.org/10.1007/978-1-4939-9484-7_6,

Table 1
Comparison of triglyceride content of oleaginous biomass estimated via TG method and in situ transesterification method. Triglyceride content determined using TG method is statistically similar to estimates obtained from in situ transesterification method

Feedstock	Triglyceride content, wt% dry-basis	
	TG method	In situ transesterification method
Soy flour	21.3 ± 0.7	19.5 ± 1.2
Sunflower flour	52.4	49.1 ± 4.3
Chlorella sorokiniana Str. SLA-04	37.2 ± 0.3	37.7 ± 0.1
Schizochytrium sp.	44.5 ± 0.1	43.7 ± 0.2

is a noninvasive method, but requires specialized instrumentation and in particular equipped with a larger magnetic for accurate measurement of smaller amounts of lipids. A third potential method is to estimate lipid content by quantifying fatty acid methyl esters (FAMEs) obtained after in situ transesterification of lipid-containing biomass [12]. In situ transesterification combines extraction and transesterification into a single step during which transesterification is catalyzed generally either by an acid or a base. Transesterification catalyzed by a base is a faster reaction, but is sensitive to moisture content in biomass On the other hand, in the acid catalyzed reactions, all fatty acids can be esterified even in the presence of small amounts of moisture [12]. Laurens et al. [9] have successfully demonstrated the accuracy and reliability of acid catalyzed in situ transesterification. Here, we describe the well-established acid catalyzed in situ transesterification method for FAMEs synthesis of small samples. Table 1 shows the FAMEs content of different oleaginous biomass.

While the NMR and in situ transesterification methods generate consistent and more accurate lipid content data, this method involves multiple steps, substantially dry biomass and usage of organic solvents. Further these methods are tedious, require expensive chromatography/NMR instruments and skilled technicians for accurate results, and require instrument calibration and long wait times between sample collection and analysis. Therefore, we have developed a new method using TG analyzer for quantification of lipids that is much more convenient than conventional methods, and yet accurate.

TG is the measurement of change in sample mass with temperature and/or time. Previously, this analytical technique was used to perform proximate analysis of carbonaceous materials (such as coal, biomass, and fossil fuels) [11, 13–16], evaluate the thermal stability of materials (ASTM E2550-11) [17–19], determine the boiling

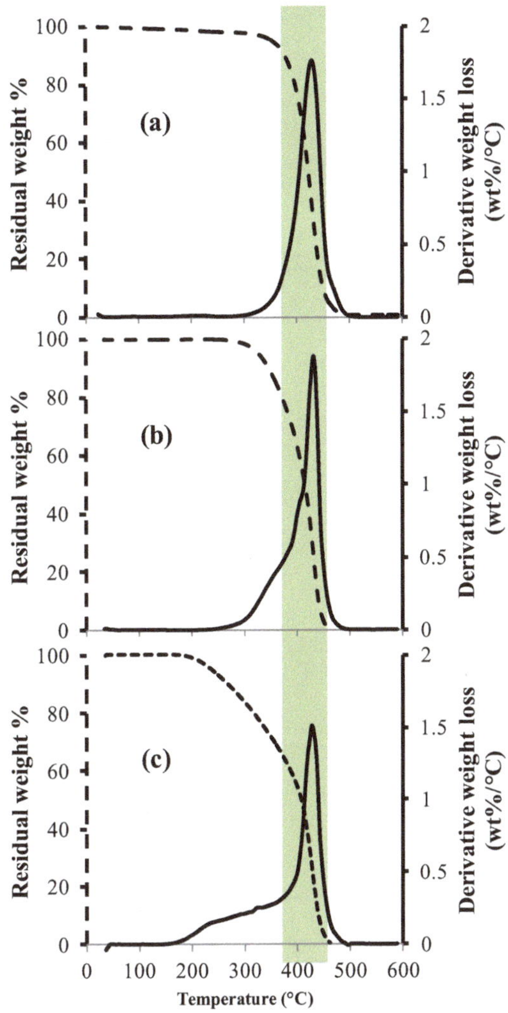

Fig. 1 Derivative weight loss (solid lines) and residual weight (dashed lines) profiles represents the pyrolysis of (**a**) triolein, an example of triglyceride, (**b**) 1,3-diolein, an example of diglycerides, and (**c**) 1-oleoyl-rac-glycerol, an example of monoglycerides

point distribution of liquid fuels [20, 21] and perform kinetic studies of unit operations as well as processes [21–27].

First, we demonstrated the use of TG method to determine the volatilization temperature of triolein—a model triglyceride). It was observed that triolein degraded over a narrow temperature interval of 370–450 °C with a single derivative weight loss peak (Fig. 1a). Additionally, TG analysis showed that 1,3-diolein and 1-oleoyl-rac-glycerol (model compounds for monoglycerides and diglycerides) also volatilized in the same temperature interval of 370–450 °C. These results indicated that volatilization of monoglycerides, diglycerides, and triglycerides all occurred within a narrow temperature window of 370–450 °C.

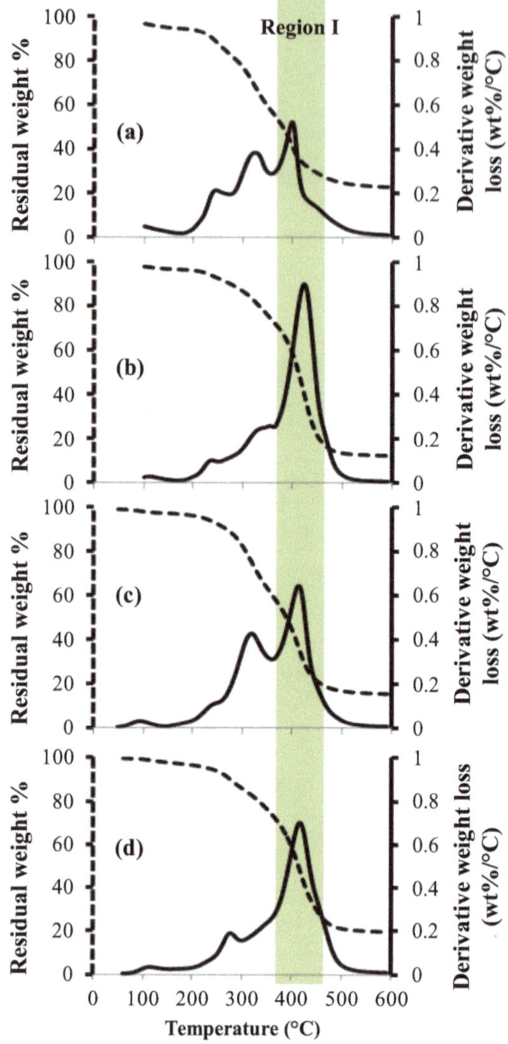

Fig. 2 Derivative weight loss (solid lines) and residual weight (dashed lines) profiles represents the pyrolysis of (**a**) soy flour, (**b**) grounded sunflower seed, (**c**) *Chlorella* sp., and (**d**) *Schizochytrium* sp. Derivative weight loss peak in the Region I indicates volatilization of triglyceride fraction present in these feedstocks. This is the working principle of TG method for triglyceride quantification

Figure 2 depicts the thermal degradation of oleaginous biomass (soy flour, sunflower flour, *Chlorella sorokiniana* str. SLA-04 and *Schizochytrium limacinum* str. SR-21) over the temperature range of 20–600 °C. These data show three distinct peaks. Previous studies have shown that the weight loss peaks at approximately 250 and 320 °C, which correspond to deacetylation and degradation of protein and carbohydrates. Since the thermal degradation temperature interval of triglycerides does not overlap with other

Table 2
Comparison of triglyceride content of wet and dry sample of Chlorella sp. estimated using the TG method. Three samples were analyzed. Average ± standard deviation values are shown

Sample	Triglyceride content estimated by the TG method, wt% dry-basis
Dry *Chlorella* sp.	37.2 ± 0.3
Wet *Chlorella* sp. (moisture content—84.6 ± 0.7)	40 ± 0.7

constituent biopolymers (i.e., protein and carbohydrates), the lipid content in oleaginous biomass can be estimated from weight loss in the temperature interval of 370–450 °C.

The method was demonstrated to quantify triglycerides content in wet samples (Table 2) and could also be extended to verify the presence of triglycerides in biomass as well as the residues obtained after thermochemical or thermocatalytic treatment.

Finally, lipid quantification using TG method was validated by the comparing the results obtained from this TG method with that of the in situ transesterification method (Table 1). More details related to the method theory and validation are described in the recent publication by Maddi et al. [28]

2 Materials

2.1 Instrument

Gas chromatography, Thermogravimetric Analyzer (TGA) with temperature range ≥500 °C.

2.2 Biomass

Oleaginous feedstocks rich in triglycerides—e.g., soy, sunflower, *Chlorella* sp., *Schizochytrium* sp., and other microalgae.

2.3 Chemicals

HPLC-grade chloroform, *n*-hexane, methanol, and FAME neat mix (for calibration curve).

3 Methods

3.1 In Situ Transesterification for FAME Synthesis

3.1.1 Sample Preparation

1. Freeze-dried biomass samples (10–20 mg) is weighed into clean 15 mL serum vials (or equivalent).
2. Then 1 mL of acidified methanol (5% by volume H_2SO_4 in methanol) is added.
3. Following the acidified methanol addition, the vials are crimped with Teflon-lined caps and incubated at 90 °C for 90 min.
4. Thereafter, the vials are cooled to room temperature and 4 mL hexane is added.

5. The mixture is then incubated at 90 °C for 15 min to extract FAMEs into hexane.
6. After cooling down to room temperature, the hexane phase is recovered and analyzed using gas chromatography (GC).

3.1.2 Creating an Instrument Method and Starting an Experiment

1. Carrier gas—Helium.
2. Oven—GC Oven is programmed as follows: initially hold a temperature of 60 °C for 1 min followed by a steady increase to 370 °C at a rate of 10 °C/min. Finally, maintain the oven temperature at 370 °C for 6 min.
3. Column—A 0.32 mm Rtx Biodiesel column approximately 15 m long with a 0.1 μm film thickness (or equivalent) can be used. The carrier gas in the column is kept at a constant linear velocity of 50 cm/s.
4. Detector—Use a flame ionization detector (FID) with detector temperature maintained at 370 °C. Maintain the flow rate for H_2 at 40.0 mL/min, for air at 400.0 mL/min.
5. Injection—Use an autosampler with a 10 μL syringe injection. Perform three preinject and three postinject solvent washes using hexane, methanol and chloroform. Inject 1.0 μL of prepared sample while keeping the injector temperature at 370 °C.

3.1.3 Analyzing the Data

1. Assess FAME concentration in samples by comparison of sample peak areas with FAME calibration standards.
2. After obtaining the data, use following equations to calculate FAMEs content:

$$\text{FAME content } (\%(\text{w/w})) = \frac{\text{FAME conc.}\left(\frac{\text{g}}{\text{L}}\right) \times \text{Vol.of hexane (L)}}{\text{Initial weight (g)}} \times 100$$

3.2 TG Analysis of Dry/Freeze-Dried Biomass

3.2.1 Creating an Instrument Method and Starting an Experiment

1. N_2 is continuously purged (100 mL/min) through the TGA oven to maintain an inert atmosphere during analysis.
2. Furnace is opened and two clean, empty crucibles are placed on the platforms of the sample and reference beams.
3. Following the weight stabilization (i.e., after weight tare), furnace is opened and 10 mg of biomass sample is loaded into the crucible that is placed on sample beam. The crucible placed on reference beam is left empty.
4. Now, the TGA instrument is programmed to heat the sample from room temperature to 600 °C at a constant rate of 20 °C/min.
5. Residual weight of the sample (w) is recorded during thermal treatment and the derivative weight loss $\frac{dw}{dT}$ is calculated. Both

weight loss and derivative weight loss (measured) are plotted versus sample temperature (Figs. 1 and 2).

6. Repeat the same procedure for all samples.

3.2.2 Analyzing the Data

The lipid content can be calculated from the weight loss data using the following equation:

$$\text{Lipid content } (\%(\text{w/w})) = \frac{\text{Weight loss}_{(370^\circ\text{C}-450^\circ\text{C})}}{\text{Initial weight}} \times 100$$

3.3 TG Analysis of Wet Microalgae Biomass

3.3.1 Sample Preparation

1. Microalgae liquid cultures are centrifuged and the supernatant is discarded and the pellet obtained is washed 2–3 times with distilled water in order to remove salts.
2. The washed pellet is then used in the TG analysis.

3.3.2 Creating an Instrument Method and Starting an Experiment

1. Similar to dry biomass analysis (Subheading 3.1.1), N_2 is continuously purged (100 mL/min) through the TGA oven to maintain an inert atmosphere during analysis.
2. Furnace is opened and two clean, empty crucibles are placed on the platforms of the sample and reference beams.
3. Following the weight stabilization, furnace is opened again and 50 mg of wet biomass paste is loaded into the crucible that is placed on sample beam. The alumina crucible placed on reference beam is left empty.
4. The TGA instrument is programmed to heat the sample from room temperature to 105 °C at a constant rate of 20 °C/min. The system is maintained isotheral for 4 h to evaporate water present in the wet sample. The residual weight of the sample obtained at the end of this step is equal to the sample dry weight.
5. The sample is then heated from 105 to 600 °C at a constant rate of 20 °C/min.
6. The residual weight of the sample is recorded over the time by the instrument.

3.3.3 Analyzing the Data

After obtaining the data, following equations are used to calculate dry weight and lipid content of biomass:

$$\text{Dry weight } (\%(\text{w/w})) = \left(1 - \frac{\text{Weight loss}_{(20^\circ\text{C}-105^\circ\text{C})}}{\text{Iinitial weight}}\right) \times 100$$

$$\text{Dry weight (mg)} = \text{Dry weight } (\%(\text{w/w})) \times \text{Initial sample weight}$$

$$\text{Lipid content } (\%(\text{w/w})) = \frac{\text{Weight loss}_{(370^\circ\text{C}-450^\circ\text{C})}}{\text{Dry weight}} \times 100$$

4 Safety Precautions

1. Avoid contact with hot items such as furnaces, serum bottles, and septa nuts during experimental analysis as it can cause burns.
2. Syringe needles should be handled carefully as they can easily puncture skin.
3. Crucibles used should be thoroughly cleaned with methanol to remove all the char formed after the experimental analysis.

Acknowledgments

This work was supported by (1) the US Department of Energy Bioenergy Technologies Office (award# DE-EE0005993) and (2) the National Science Foundation through the Sustainable Energy Pathways Program (award# CHE-1230609)

References

1. Vadlamani A, Viamajala S, Pendyala B, Varanasi S (2017) Cultivation of microalgae at extreme alkaline pH conditions–a novel approach for biofuel production. ACS Sust Chem Eng 5 (8):7284–7294
2. Sanchez DL, Nelson JH, Johnston J, Mileva A, Kammen DM (2015) Biomass enables the transition to a carbon-negative power system across western North America. Nat Clim Chang 5(3):230–234
3. DOE, U.S. Department Of Energy (2016) National algal biofuels technology review. U.S. Department Of Energy, O.O.E.E.A.R.E., Bioenergy Technologies Office, Washington, D.C.
4. Amaro HM, Guedes AC, Malcata FX (2011) Advances perspectives in using microalgae to produce biodiesel. Appl Energy 88 (10):3402–3410
5. Bligh EG, Dyer WJ (1959) A rapid method of total lipid extraction and purification. Can J Biochem Physiol 37(8):911–917
6. Kumari P, Reddy CRK, Jha B (2011) Comparative evaluation and selection of a method for lipid and fatty acid extraction from macroalgae. Anal Biochem 415(2):134–144
7. Lee S, Yoon B-D, Oh H-M (1998) Rapid method for the determination of lipid from the green alga *Botryococcus braunii*. Biotechnol Tech 12(7):553–556
8. Mercer P, Armenta RE (2011) Developments in oil extraction from microalgae. Eur J Lipid Sci Technol 113(5):539–547
9. Laurens LM, Quinn M, Van Wychen S, Templeton DW, Wolfrum EJ (2012) Accurate and reliable quantification of total microalgal fuel potential as fatty acid methyl esters by in situ transesterification. Anal Bioanal Chem 403 (1):167–178
10. Lohman EJ, Gardner RD, Halverson L, Macur RE, Peyton BM, Gerlach R (2013) An efficient and scalable extraction and quantification method for algal derived biofuel. J Microbiol Methods 94(3):159–396
11. Retief L (2018) Analysis of vegetable oils, seeds and beans by TGA and NMR spectroscopy. Department of chemistry and polymer science. University Of Stellenbosch, Stellenbosch, South Africa, p 196
12. Nelson DR, Viamajala S (2016) One-pot synthesis and recovery of fatty acid methyl esters (fames) from microalgae biomass. Catal Today 269:29–39
13. Elder JP (1983) Proximate analysis by automated thermogravimetry. Fuel 62(5):580–584
14. Kök M (2008) Recent developments in the application of thermal analysis techniques in fossil fuels. J Therm Anal Calorim 91 (3):763–773
15. Maddi B, Viamajala S, Varanasi S (2011) Comparative study of pyrolysis of algal biomass from

natural lake blooms with lignocellulosic biomass. Bioresour Technol 102 (23):11018–11026
16. Petrakis L, Grandy DW (1980) Coal analysis, characterization and petrography. J Chem Educ 57(10):689–694
17. Canetti M, Bertini F, Chirico AE, Audisio G (2006) Thermal degradation behaviour of isotactic polypropylene blended with lignin. Polym Degrad Stab 91(3):494–498
18. Yang H, Yan R, Chen H, Lee DH, Zheng C (2007) Characteristics of hemicellulose, cellulose and lignin pyrolysis. Fuel 86 (12–13):1781–1788
19. Yang H, Yan R, Chen H, Zheng C, Lee DH, Liag DT (2006) In-depth investigation of biomass pyrolysis based on three major components: hemicellulose, cellulose and lignin. Energy Fuel 20:388–393
20. Du Z, Li Y, Wang X, Wan Y, Chen Q, Wang C, Lin X, Liu Y, Chen P, Ruan R (2011) Microwave-assisted pyrolysis of microalgae for biofuel production. Bioresour Technol 102 (7):4890–4896
21. Laresgoiti MF, Caballero BM, de Marco I, Torres A, Cabrero M, Chomón MJ (2004) Characterization of the liquid products obtained in Tyre pyrolysis. J Anal Appl Pyrolysis 71(2):917–934
22. Friedman HL (1964) Kinetics of thermal degradation of char-forming plastics from thermogravimetry. Application to a phenolic plastic. J Pol Sci Part C Pol Symp 6(1):183–195
23. Harlick PJE, Sayari A (2006) Applications of pore-expanded mesoporous silicas. 3. Triamine silane grafting for enhanced CO_2 adsorption. Ind Eng Chem Res 45(9):3248–3255
24. Damartzis T, Vamvukab D, Sfakiotakis S, Zabaniotou A (2011) Thermal degradation studies and kinetic modeling of cardoon (*Cynara cardunculus*) pyrolysis using thermogravimetric analysis (TGA). Bioresour Technol 102(10):6230–6238
25. Doyle CD (1961) Kinetic analysis of thermogravimetric data. J Appl Pol Sci 5(15):285–292
26. Kok MV, Pamir MR (1995) Pyrolysis and combustion studies of fossil fuels by thermal analysis methods. J Anal Appl Pyrolysis 35 (2):145–156
27. Murugan P, Mahinpey N, Johnson KE, Wilson M (2008) Kinetics of the pyrolysis of lignin using thermogravimetric and differential scanning calorimetry methods. Energy Fuel 22 (4):2720–2724
28. Maddi B, Vadlamani A, Vaimajala S, Varanasi S (2017) Quantification of triglyceride content in oleaginous materials using thermogravimetry. J Anal Appl Pyrolysis 128:232–237

Chapter 7

Extraction and Characterization of Lipids from Macroalgae

David R. Nobles Jr. and Schonna R. Manning

Abstract

Although most algal biofuel research has focused on microalgae, macroalgae are also potential sources of lipid for the production of biodiesel and other liquid fuels. Reliable, accurate methods for assessing the lipid composition of biomass are essential for the development of macroalgae in this area. The conventional methods most commonly used to evaluate lipid composition, such as those of Bligh and Dyer and Folch, do not provide complete extraction of lipids in photosynthetic cells/tissues and therefore do not provide an accurate accounting of lipid production. Here we present a 2-EE lipid extraction protocol, a method which has been demonstrated to be superior to conventional lipid extraction methods for microalgae, adapted for use with macroalgae.

Key words Algae, Macroalgae, Lipid extraction, 2-Ethoxyethanol, Lipid profiling by thin-layer chromatography, Lipid analysis

1 Introduction

Algal biomass is a promising resource for the production of renewable fuels. Algae offer the advantages of high aerial productivity, growth in saline and brackish water, use of marginal land for cultivation, and the coproduction of high value products [1, 2]. The production of biodiesel from triacylglycerols (TAGs) has been central to the efforts to commercialize algae at a very large scale. Microalgae are known to produce TAGs at up to 50% of their biomass dry weight [3] and thus have been the primary focus of most algal biofuels development and research. Macroalgae have generated much less interest as potential biofuel feedstocks because they have generally been characterized as having low-lipid productivity, often producing less than 5% lipids by dry weight. However, there are multiple examples of macroalgae with lipid production exceeding 10% of the dry biomass. These higher values for observed lipid accumulation, combined with the relative ease of cultivation and harvest of macroalgae as compared to microalgae,

Venkatesh Balan (ed.), *Microbial Lipid Production: Methods and Protocols*, Methods in Molecular Biology, vol. 1995, https://doi.org/10.1007/978-1-4939-9484-7_7,

make them viable candidates for biofuel production and worthy of additional study in this area [4].

Accurate and repeatable methods for the extraction and quantification of lipids are key to the meaningful evaluation of biomass from diverse algae and their suitability as biodiesel feedstocks. To date, the most commonly employed protocols for lipid extraction from microalgal biomass have been the methods of Folch and of Bligh and Dyer [4–7]. However, it has been demonstrated that the use of 2-ethoxyethanol (2-EE) provides superior lipid recovery (150–200%) in microalgae when compared to these methods [8]. Here we describe methods for biomass preparation and related metrics, total lipid extraction using 2-EE, and the identification of lipid classes by thin layer chromatography (TLC). The standardized 2-EE protocol is widely used for microalgae and is adapted here for the extraction of lipids from macroalgae.

2 Materials

Prepare all solutions using ultrapure or double-distilled water and analytical grade reagents. Adhere to all waste disposal regulations when disposing of waste materials.

2.1 Fresh Weight (FW), Dry Weight (DW), and Ash-Free Dry Weight (AFDW) Determination

1. Porcelain crucibles.
2. Scalpel, scissors, or razor blade for harvesting biomass.
3. Flat-tipped forceps.
4. Filter paper or Kimwipes.
5. Weigh boat or weighing paper.
6. Drying oven (80 °C).
7. Crucible tongs.
8. Desiccator with desiccant (e.g., calcium sulfate, *Drierite*) (*see* **Note 1**).
9. Muffle furnace (550 °C).
10. Analytical balance.
11. 15-mL conical centrifuge tubes.
12. 5-mL serological pipettes with pipette pump.
13. 0.5 M NaCl solution (filter sterilized and stored at 4 °C).
14. 0.5 M Ammonium Bicarbonate (filter sterilized and stored at 4 °C).
15. Fresh macroalgal tissue (generally material from blades), 100–200 mg per sample/replicate.

2.2 Lipid Extraction

1. 100-mL round-bottom flask (RBF).
2. Vortex mixer.

3. Water bath (60 °C).
4. 40-mL glass vials with PTFE-lined screw caps.
5. 5-mL glass vials with PTFE-lined screw caps.
6. P1000 micropipettor and 1 mL micropipette tips.
7. 10-mL serological pipettes and pipette bulb.
8. Methanol in squeeze bottle.
9. Acetone in squeeze bottle.
10. Rotary evaporator.
11. Resuspension solvent—toluene–hexane–acetone–methanol 2:2:1:1 (v/v/v/v).
12. 2-Ethoxyethanol (2-EE).
13. Macroalgal tissue, 100–200 mg per sample/replicate.
14. Glass Pasteur pipettes and bulb.
15. Glass fiber membrane, 1.2 μm porosity (*see* **Note 2**).
16. Parafilm (*see* **Note 3**).

2.3 Analysis of Biomass Lipids by Thin-Layer Chromatography (TLC)

1. TLC plates—silica gel 60 aluminum sheets with 254 nm fluorescent indicator, F254 (EMD-Millipore 1055670001).
2. 5-μL semiquantitative capillary micropipettor.
3. Chromatography tank.
4. Iodine crystals (Fisher Scientific 7553-56-2) and silica gel mesh 60–200 (Fisher Scientific 7631-86-9) mixed 1:1, sealed in plastic bin, and stored in a dedicated fume hood.
5. Lipid class reference standard mixture with 2 mg/mL final concentration of each analyte, including TAG, DAG, MAG, FFA, PC, PE, FAME, and BC (*see* **Note 4**).
6. Polar solvent system—chloroform–methanol–acetone–acetic acid–water 65:10:20:10:3 (v/v/v/v/v; *see* **Note 5**).
7. Nonpolar solvent system—hexane–diethyl ether–acetic acid 80:20:1 (v/v/v; *see* **Note 5**).
8. Ultraviolet (UV) light handwand (254 nm and 365 nm).

3 Methods

If dried tissue will be used for the lipid extraction, prepare two sets of tissue samples: one set of samples to be reserved for lipid extraction after drying and one set to determine ash-free dry weight (*see* Fig. 1; *see* **Note 6**).

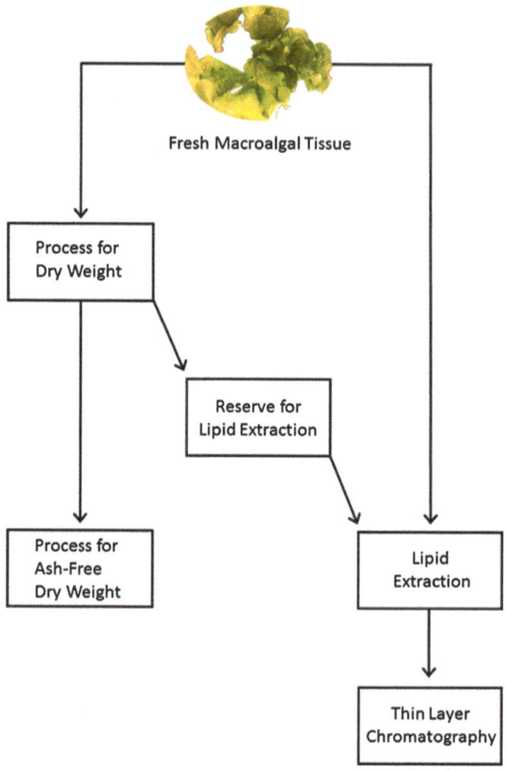

Fig. 1 This flow diagram illustrates the workflow streams for lipid extraction from fresh and dried biomass. Lipid extractions can be performed directly using fresh biomass or alternatively, on dried biomass after determination of dry weight. For the latter option, samples are split with half utilized to determine ash-free dry weight and half utilized for lipid extraction

3.1 Fresh Weight (FW), Dry Weight (DW), and Ash-Free Dry Weight (AFDW) Determination

1. Weigh a labeled crucible for each sample to be processed. Record the values.
2. Using a scalpel, scissors, or razor blade, harvest macroalgal tissue (typically from the blades) and place in a 15-mL conical tube.
3. Use a serological pipette to wash the tissue with 5 mL of sterile 0.5 M NaCl; remove epiphytes and any other foreign material by gently inverting the sealed tube 10–15 times.
4. Decant the NaCl solution. Use a serological pipette to rinse the tissue with 5 mL of 0.5 M ammonium bicarbonate and gently invert 10–15 times.
5. Remove tissue with flat-tipped forceps and gently pat dry with filter paper or Kimwipes.
6. Weigh the washed tissue in a tared weigh boat or on weighing paper to determine the FW. Record the value.

7. Transfer tissue to a preweighed crucible and place in a drying oven set to 80 °C. Check the weight of the crucible every 2 h until a constant weight is obtained.
8. Store the sample in the desiccator overnight to remove the last traces of water.
9. Transfer the crucible to the analytical balance and weigh. Record the value.
10. If using dried tissue for lipid extraction, one set of samples should be returned to the desiccator and reserved for lipid extraction. The other set will proceed to **step 11** for ashing (*see* Fig. 1).
11. Place the crucible in a muffle furnace set to 550 °C.
12. Combust the tissue sample for 2 h then allow the furnace to cool for 1 h.
13. Remove the crucible using crucible tongs and place on an insulated surface—items will be very hot. Allow the crucible and contents to cool to room temperature, place in a desiccator, and weigh once an hour until a constant weight is achieved.
14. Transfer the crucible to the analytical balance and weigh. Record the value. Subtract the weight obtained from **step 1** to determine the weight of the ash.
15. Calculate the tissue DW, AFDW, % ash, and % moisture for sample Duplicate 1 (DWD1, AFDWD1, etc.) as follows:

$$\text{DWD1} = (\text{crucible} + \text{dried tissue}) - \text{crucible}$$
$$\text{AFDWD1} = (\text{crucible} + \text{dried tissue}) - (\text{crucible} + \text{ashed tissue})$$
$$\%\text{Ash} = ((\text{crucible} + \text{ash}) - \text{crucible})/\text{DWD1} \times 100\%$$
$$\%\text{Moisture} = (\text{FWD1} - \text{DWD1})/\text{FWD1} \times 100\%$$

where

FWD1 = fresh weight of duplicate #1 in grams (*see* Subheading 3.1, **step 4**).

DWD1 = dry weight of duplicate #1 in grams.

AFDW1 = ash-free dry weight of duplicate #1 in grams.

Repeat calculations for each sample Duplicate 2 (DWD2, AFDWD2, etc.).

3.2 Lipid Extraction

1. Weigh 25–50 mg of dried tissue reserved in **step 8** from Subheading 3.1 and transfer to a 40-mL glass vial. Alternatively, fresh biomass can be used. In which case, an amount of fresh tissue proportional to that of the dried material, based on the ratio of fresh weight to dry weight can be used (*see* Subheading 3.1, **step 13** for dry weight calculation method).
2. Use a serological pipette to transfer 5 mL of MeOH to the vial containing harvested tissue. Cap the vial tightly and extract the tissue by heating in a 60 °C water bath for 30 min.

3. Use a serological pipette to transfer 5 mL 2-EE to the vial, cap tightly, and extract lipids by heating in a 60 °C water bath for 60 min. Mix by gentle vortexing for a few seconds every 5 min.
4. Transfer the liquid extract to a preweighed 100-mL RBF using a glass Pasteur pipette. To maximize recovery, wash the vial containing the residual solid biomass with acetone and methanol (1:1, v/v), to recover all of the lipids, using chlorophyll as the indicator pigment (*see* **Note 2**). Pool the washes for each sample in the 100-mL RBF containing the lipid extract.
5. Remove the solvents using the rotary evaporator at 60 °C at a moderate speed. Weigh the flask periodically. The flask with residue will ultimately achieve a constant dry weight with no appreciable reduction due to loss of solvent(s). Record the weight of the RBF with lipid residue.
6. Using a micropipettor, resuspend the lipid extract in 2 mL 2:2:1:1 (hexane–toluene–acetone–methanol, v/v/v/v) with gentle swirling and transfer to a 5-mL glass vial; rinse the RBF with another 2 mL of 2:2:1:1 and pool with the previous (*see* **Note 3**).
7. Dry the RBF using the rotary evaporator at 60 °C until a constant weight is achieved. Record the value.
8. Subtract the weight of the RBF minus extract from the weight of the RBF with extract to obtain the grams of TLE. Divide the grams of total lipids extracted by the resuspension volume (4 mL) to obtain the concentration of lipids in grams per liter for subsequent TLC analyses.

   If beginning with dried material, calculate the % biomass extracted as total lipid as follows for sample Duplicate 1 (D1):

$$\%\text{TLED1 of dry weight} = \text{TLED1}/\text{DWD1} \times 100\%$$
$$\%\text{TLED1 of ash-free dry weight} = \text{TLED1}/\text{AFDWD1} \times 100\%$$

   where

   TLED1 = total lipid extract of duplicate #1 in grams.

   DWD1 = dry weight of duplicate #1 in grams.

   AFDW1 = ash-free dry weight of duplicate #1 in grams.

   Repeat calculations with sample Duplicate 2 (D2).

   Perform statistical averages, %RSD, and SD of replicate samples, D1 and D2.
9. If beginning with fresh material, calculate the moisture correction (mc) for duplicate samples D1 and D2 as follows to estimate the sample DW and AFDW:

$$\%\text{Moisture of D1} = (\text{FWD1} - \text{DWD1})/\text{FWD1} \times 100\%$$
$$\%\text{Moisture of D2} = (\text{FWD2} - \text{DWD2})/\text{FWD2} \times 100\%$$
$$\text{Average}\%\text{moisture of D1\&D2} = (\%\text{moisture D1} + \%\text{moisture D2})/2$$
$$\text{DWD1}_{mc} = \text{FWD1} \times (100 - \text{average}\%\text{moisture of D1\&D2})/100$$
$$\text{AFDWD1}_{mc} = \text{DWD1}_{mc} - \text{average ash weight for D1\&D2}$$

Perform statistical averages, %RSD, and SD of replicate samples, D1 and D2.

FWD1 = fresh weight of duplicate #1 in grams.

DWD1_{mc} = moisture-corrected dry weight of duplicate #1 in grams.

DWD2_{mc} = moisture-corrected dry weight of duplicate #2 in grams.

AFDWD1_{mc} = moisture-corrected ash-free dry weight of duplicate #1 in grams.

3.3 Analysis of Biomass Lipids by Thin-Layer Chromatography (TLC)

1. Prepare a 10 cm × 10 cm TLC plate; using a pencil, label the origin, solvent front, mid-way point, and sample spots as shown in Fig. 2 (*see* **Note 7**).
2. Spot 5 and 10 μL of each extract for TLC with capillary micropipettors along with 5 μL of lipid reference standards (*see* Fig. 2; *see* **Note 4**).
3. Add polar solvent system 65:10:20:10:3 chloroform–methanol–acetone–acetic acid–water (v/v/v/v/v) to the chromatography tank and cover (*see* **Notes 5** and **8**).
4. Carefully insert the TLC plate straight down and lean behind the ridge with the face of the plate toward the front, and then cover the tank. Develop the plate to the mid-way point, then remove the plate with flat-tipped forceps, lean against the front of the tank, and allow to dry completely (front and back should be dry).
5. Next, develop the plate with the nonpolar solvent system 80:20:1 hexane–diethyl ether–acetic acid (v/v/v) up to the solvent front, then remove the plate from tank (*see* **Notes 5** and **8**).
6. Permit to dry, then expose the TLC plate to UV light using the handwand at each wavelength to examine absorption and fluorescence properties. Take images of the plate in visible light and with UV.
7. Place the TLC plate in the chamber containing iodine–silica 1:1 (w/w) and seal. Expose for 2–3 min; lipids will stain darker with increased exposure time.
8. Diagram the TLC plate as shown in Fig. 3—photograph and indicate where lipids are observed (plates can be reexposed to iodine if needed).

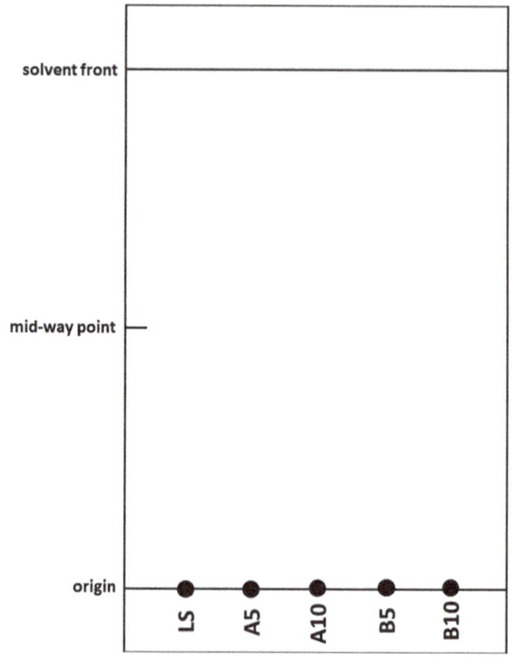

Fig. 2 This diagram illustrates the setup and labeling of a TLC plate for analysis of TLE. LS = Lipid Standards. Lanes A5 and B5 represent 5 μL of TLE from samples A and B, Lanes A10 and B10 represent 10 μL of TLE from samples A and B

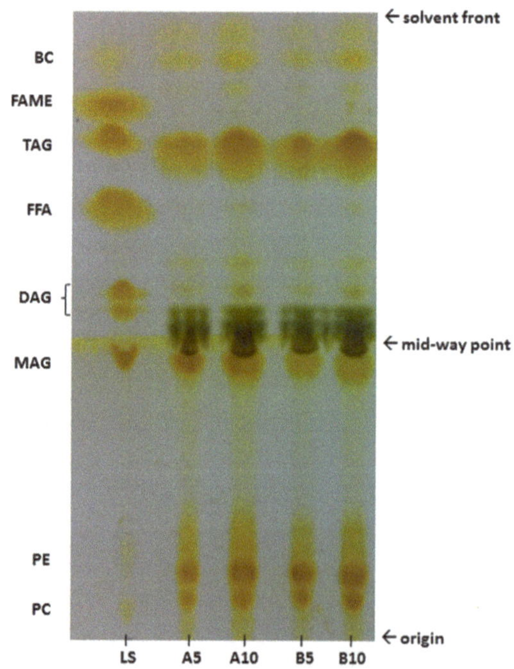

Fig. 3 The TLC shows the development of TLE samples A and B, applied at two different concentrations. Lane 1 contains the lipid standards (LS), lane 2 contains 50 μg of TLE A (A5), lane 3 contains 100 μg of TLE A (A10), lane 4 contains 50 μg of TLE B (B5), and lane 5 contains 100 μg of TLE B (B10)

4 Notes

1. Exercise caution if using color-changing desiccant containing cobalt(II) chloride, a known carcinogen, and follow local guidelines regarding the handling and disposal of hazardous waste.
2. If the tissue is not completely decolorized, rinse the biomass with methanol and acetone and homogenize with 5 mL of 2-EE. Transfer the homogenate to the RBF, add another 5 mL 2-EE and extract for 30 min at 60 °C or until the tissue is decolorized. Filter extracted solids over a 1.2 μm glass-fiber membrane and pool the filtrates.
3. Do not collect any solids. Cap the vial tightly, wrap with Parafilm, and store at 4 °C.
4. TAG = Triacylglycerol (triolein, Sigma T7140-500MG), DAG = Diacylglycerol (1,2- and 1,3-diolein, Sigma D8894-50MG), MAG = monoacylglycerol (monoolein, Sigma M7765-50MG), FFA = free fatty acid (oleic acid, Sigma O1008-5G), PC = phosphatidylcholine (dioleoyl phosphatidylcholine, Sigma P6354-25MG), PE = phosphatidylethanolamine (dioleoyl phosphatidylethanolamine, Sigma P1223-25MG), FAME = fatty acid methyl ester (methyl oleate, Sigma 311111-5G), and BC = beta-carotene (Sigma C4582-10MG). Resuspend standards in toluene–hexane–acetone–-methanol 2:2:1:1 (v/v/v/v) to a prepare 20 mg/mL stock solutions; prepare in glass vials with screw caps, wrap in Parafilm, and store at −20 °C. Combine 100 μL of each stock solution plus 200 μL of the 2:2:1:1 solvent in a screw-cap glass vial to prepare 1 mL of the Lipid Standard (LS) mixture containing 2 mg/mL of each analyte (1,2- and 1,3-DAG are counted together as total DAG). Wrap in Parafilm and store at −20 °C.
5. Use glacial acetic acid.
6. Two replicates, at a minimum, are recommended for each measurement.
7. Use a dull #2 pencil to *lightly* draw a line at the origin, a line at the solvent front, a dash at the mid-way point, small dots where spots are to be applied, and sample labels. Take care not to break the surface of the silica gel. A 1-cm rule should be applied—the origin and solvent front are 1-cm from the bottom and top of the plate, respectively, and sample spots should be 1 cm from the edges and 1 cm between samples.
8. Do not exceed 0.8 cm in depth (usually just below the ridge in standard tanks) to avoid loss of sample. Equilibrate the chromatography tank with developing solvent for at least 15 min prior to developing the plate.

Acknowledgments

This work was supported by NSF grant DBI-1560772 and DOE grant DE-1162-1511.

References

1. Hu Q, Sommerfeld M, Jarvis E, Ghirardi M, Posewitz M, Seibert M, Darzins A (2008) Microalgal triacylglycerols as feedstocks for biofuel production: perspectives and advances. Plant J 54:621–639
2. Quinn JC, Davis R (2015) The potentials and challenges of algae based biofuels: a review of the techno-economic, life cycle, and resource assessment modeling. Bioresour Technol 184:444–452
3. Scott SA, Davey MP, Dennis JS, Horst I, Howe CJ, Lea-Smith DJ, Smith AG (2010) Biodiesel from algae: challenges and prospects. Curr Opin Biotechnol 21(3):277–286
4. Magnusson M, Mata L, de Nys R, Paul NA (2014) Biomass, lipid and fatty acid production in large-scale cultures of the marine macroalga *Derbesia tenuissima* (Chlorophyta). Mar Biotechnol 16:456–464
5. Xu X, Kim JY, Oh YR, Park JM (2014) Production of biodiesel from carbon sources of macroalgae, Laminaria japonica. Bioresour Technol 169:455–461
6. Folch J, Lees M, Stanley GHS (1957) A simple method for the isolation and purification of total lipids from animal tissues. J Biol Chem 226:497–509
7. Bligh E, Dyer WJ (1959) A rapid method of total lipid extraction and purification. Can J Biochem Physiol 37:911–917
8. Jones J, Manning S, Montoya M, Keller K, Poenie M (2012) Extraction of algal lipids and their analysis by HPLC and mass spectrometry. J Am Oil Chem Soc 89:1371–1381

Chapter 8

Genetic Engineering Approaches Used to Increase Lipid Production and Alter Lipid Profile in Microbes

Xiao-Ling Tang and Ya-Ping Xue

Abstract

Microbial production of lipids provides important alternative sources for a variety of fine chemicals and fuels. With the development of biotechnology, genetic engineering approaches are widely used to increase lipid production in microbes, as well as to alter the lipid profile with unique physicochemical properties. In this chapter, based on the well-known information of de novo lipid accumulation mechanisms in microbes, genetic engineering strategies at the direction of increased supply of substrates, regulation of lipid synthesis pathway, regulation of lipid catabolic pathway, and regulation of lipid profiles are described. These methods provide promising insights to promote the optimization of lipid accumulation and properties.

Key words Microbial lipids, Genetic engineering, Lipid accumulation, Lipid profile alternation

1 Introduction

Lipids are the essential components of microbial cells for the main in vivo biological functions include energy storage, signal recognition, and structural components. Lipids produced from microbes including bacteria, yeasts and microalgae offer promising sources of valuable chemicals that can replace petroleum-derived compounds. In the last decade, a great deal of attention has been focused on lipid accumulation in microorganisms, as well as their conversion to chemicals such as fatty aldehydes, fatty alcohols, and alkanes/alkenes. However, the microbial processes for production of lipids and their derivatives always suffer from low yield and non-ideal properties. Understanding of de novo lipid synthesis is of great significance to modify the relevant pathways to improve microbial lipid production and alter the lipid profile. The application of genetic engineering significantly promotes microbial lipid accumulation and synthesis of lipids with desired properties.

Venkatesh Balan (ed.), *Microbial Lipid Production: Methods and Protocols*, Methods in Molecular Biology, vol. 1995, https://doi.org/10.1007/978-1-4939-9484-7_8,

2 Overview of Lipid Metabolism in Microbes

De novo lipid accumulation is an anabolic biochemical process and most lipids are stored as cellular components such as triacylglycerols (TAGs). Biosynthesis of lipids is ubiquitous in microorganisms although production levels and composition of lipids are microbe-specific. For some gram-positive actinomycetes including *Cryptococcus* sp., *Mycobacterium* sp., *Nocardia* sp., *Rhodococcus* sp., *Micromonospora* sp., the cellular lipid accumulation could reach to a level above 20% of their cell mass on average and in some conditions, the maximum lipids content exceeds 70% [1, 2]. As such, these microbes are defined as oleaginous organisms. Eighty to ninety percent of the lipids are stored as TAGs, whereas a small fraction of them are found in the form of steryl esters, which are called neutral lipids [3]. For some gram-negative proteobacteria including *Alcanivorax* sp., *Marinobacter* sp., or *Acinetobacter* sp., TAGs are also reported as the storage form of lipids although the amount of accumulated lipids is much smaller than oleaginous microbes.

Except for fatty acid generation from an external feed and lipid turnover within microbe cells, the de novo biosynthesis plays an important role in microbial lipid accumulation. Despite the early development of fatty acid biosynthesis during the evolution of life, the biosynthesis pathway is highly conserved among the kingdoms of life. A series of enzymes including acetyl-CoA carboxylase, fatty acid synthetase complex, fatty acid desaturase and fatty acyl elongase are responsible for the de novo lipid synthesis in microbes (Fig. 1). Generally, the biosynthesis of lipids starts from acetyl-CoA, which is carboxylated to malonyl-CoA by the enzyme of

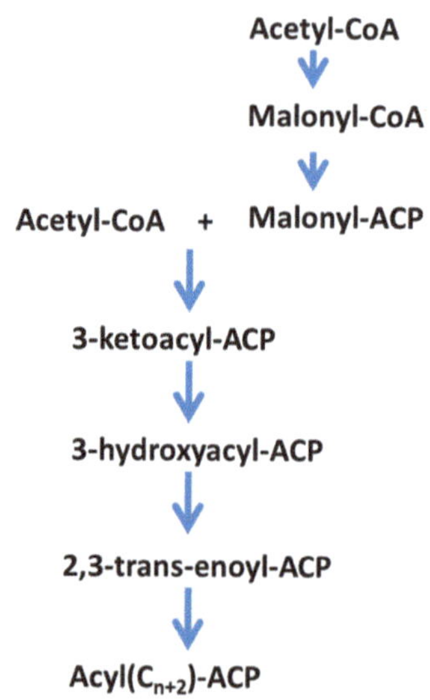

Fig. 1 Schematic diagram of fatty acid synthesis and elongation

acetyl-CoA carboxylase (ACC), followed by its exchange with the acyl carrier protein (ACP) to form malonyl-ACP. The acetyl-CoA, together with the malonyl-ACP, is then destined to feed into the fatty acid synthesis pathway. In the first turn, the condensation of acetyl-CoA and malonyl-ACP results in acetoacetyl-ACP. Acetoacetyl-ACP is then reduced to 3-hydroxybutyryl-ACP, which is further dehydrated to 2-butenoyl-ACP, followed by further reduction to butyryl-ACP. The butyryl-ACP enters the second round of the fatty acid synthesis by condensation with malonyl-ACP again (Fig. 1). The end fatty acid derivative products vary between different species. The further elongation and desaturation are carried out by the specific elongase and desaturase of microbe species. After cyclic series of elongation reactions, different carbon length contained fatty acyl-ACPs of fatty acyl-CoAs are formed and released then.

The fatty acyl-CoA precursors are then destined to the formation of lipids, such as TAGs and steryl esters. The lipid storage pathway starts from the generation of glycerol-3-phosphate (G-3-P), which is synthesized by glycerol-3-phosphate dehydrogenase (GPD1) from dihydroxyacetone phosphate (DHAP) or by acylation of glycerol via glycerol kinase (GUT1). The lysophosphatidic acid (LPA) is subsequently formed from the condensation of G-3-P and fatty acyl-CoA via the enzyme of G-3-P acyltransferase (GPAT). The acylation of LPA results in the formation of phosphatidic acid (PA) via the enzyme LPA acyltransferase (LPAT). Finally, after dephosphorylation of PA, the diacylglycerol (DAG) is formed, and it is catalyzed by DGA acyltransferase (DGAT) to generate triacylglycerol (TAG) (Fig. 2). The lipids are stored in the microbial cells as lipid particles or lipid bodies, which contribute more than 70% of the total lipid content. The stored lipids later serve as energy reserves if necessary.

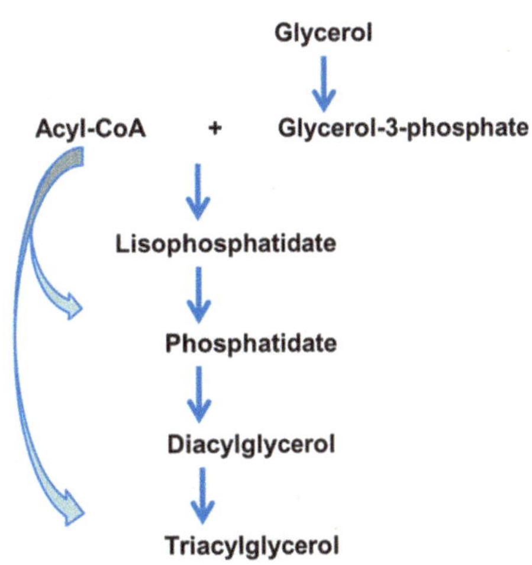

Fig. 2 Schematic diagram of TAG synthesis

Despite the highly conserved pathway for lipid synthesis, different enzymes and genetic organizations have evolved in different microbes. Investigation and regulation of the involved enzymes together with the related transcription factors provides important strategies for the modification of lipid components and synthesis levels.

The catabolism of lipids is initiated by the activation of fatty acids to fatty acyl-CoAs, which is consumed by the β-oxidation pathway. The fatty acyl-CoAs with different carbon chain length undergo different process of degradation and generally, after each degradation cycle, a two carbon shortened fatty acyl-CoAs are generated and then fed into the second degradation cycle again. The acyl-CoA oxidases in peroxisomes are mainly responsible for the β-oxidation and the finally obtained acetyl-CoAs will further enter the citric acid cycle transported by the carnitineacetyl-transferases, providing necessary energy required by the microbes.

3 Genetic Engineering for Increased Lipid Production

Based on the well-known metabolism mechanism of lipids in microbes, the regulation of the involved metabolic pathways could either increase the lipid production or alter the lipid profile in microbes.

3.1 Genetic Engineering for Increased Supply of Precursors and Coenzymes

3.1.1 Background

The de novo synthesis of lipids requires substrates and cofactors. The acetyl-CoA is the most important precursor. The lipid synthesis is initiated by loading the acetyl-CoA onto the FAS complex, and during the elongation process, the acetyl-CoA needs to be constantly supplied. The NADPH is the most important cofactor for the lipid synthesis, and in each step of the chain elongation, two molecules of NADPH are required. Besides, the lipid biosynthesis is a process depending on the condition of high energy loading. The ATP is the main energy form and during the lipid synthesis, the ATP–AMP ratio has been found to be increased in the microbial cells. As such, one efficient strategy to increase lipid production in microbes is to engineer the metabolic pathways for increased supply of precursors and coenzymes.

3.1.2 Genetic Engineering of Pyruvate–-Acetaldehyde–Acetate Pathway

One pathway for acetyl-CoA generation ubiquitous in microbes is the pyruvate–acetaldehyde–acetate pathway (Fig. 3). The acetyl-CoA is produced from the glycolysis pathway. The sugars first undergo fermentation process in microbial cells to generate pyruvate. By the enzyme of pyruvate decarboxylase, the pyruvate is converted to acetaldehyde, which is further converted to acetate by the enzymes of aldehyde dehydrogenase and acetyl-CoA synthetase. Acetate is finally catalyzed to generate cytosol acetyl-CoA. In

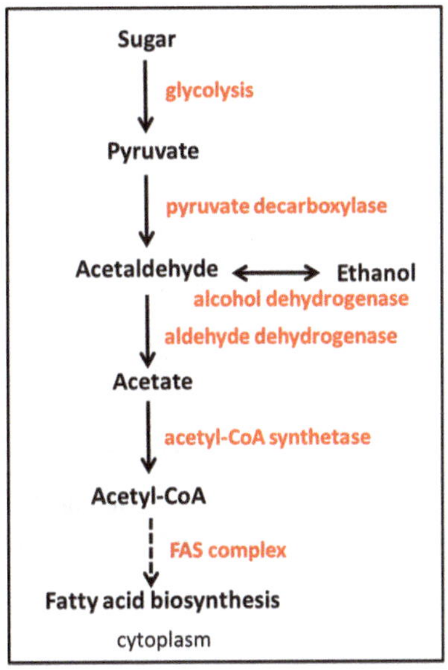

Fig. 3 The pathway for cytosol acetyl-CoA generation

addition, the intermediate acetaldehyde could be partially reversible converted to ethanol by the enzyme alcohol dehydrogenase.

The genetic strategy aims to reengineer the above-described pathway and redirect the carbon flux towards the acetyl-CoA production. The acetyl-CoA synthetase is demonstrated to be the rate-limiting step in microbes and the overexpression of genes encoding acetyl-CoA synthetase is reported to significantly increase the intracellular acetyl-CoA levels, resulting in increased production of lipids or fatty acid derivatives [4]. From another aspect, the competing pathway from acetaldehyde to ethanol can be blocked and the involved genes encoding the isoenzymes of alcohol dehydrogenases knocked out to decrease the conversion of acetaldehyde to ethanol. Through this strategy, both acetaldehyde and acetate are reported to be accumulated to a much higher level and they are conducive to the supply of acetyl-CoA for lipid synthesis [5].

3.1.3 Genetic Engineering of Pyruvate–Citrate–Acetyl-CoA Pathway

In some oleaginous microorganisms, an additional cytosol acetyl-CoA generation pathway exists. The citrate generated from the citric acid cycle could be transported out of mitochondria to cytosol by the citrate/malate shuttle and then cleaved by the enzyme of ATP-citrate lyase, which exists widely in the oleaginous microbes, but is absent in the nonoleaginous microbes. The overexpression of heterologous ATP-citrate lyase in some microorganisms, together with the increased supply of citric acid could successfully increase the production level of acetyl-CoA [6]. On the other hand, the citric acid could be accumulated under nitrogen limited culture condition for some microbes due to the decrease of cellular AMP concentration. This results in the depression of enzyme activity for

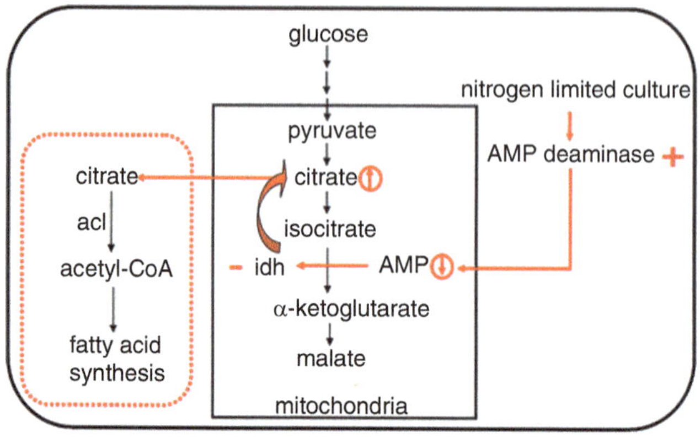

Fig. 4 Schematic diagram of lipid accumulation in microbes with ATP-citrate lyase (acl) under nitrogen limited culture conditions

conversion of citric acid to β-ketoglutarate in the citric acid cycle (Fig. 4). As such, the microbes with original or heterologous expressed ATP-citrate lyase could successfully convert the accumulated citric acid to cytosol acetyl-CoA for lipid synthesis.

3.1.4 Genetic Engineering of NADPH Supply

NADPH is an essential cofactor for lipid synthesis in microbes. NADPH is mainly generated by the pentose phosphate pathway in cytosol, which includes the enzyme activity of glucose-6-dehydrogenase, 6-phosphogluconate dehydrogenase, and NADPH-dependent isocitrate dehydrogenase. In some oleaginous microbes, the main parts of intracellular NADPH are generated by malic enzyme, which is responsible for the conversion of malate into pyruvate and meanwhile regenerating NADPH from $NADP^+$, to provide the vital cofactor for lipid synthesis. Some efforts have been made on the increase of NADPH supply through regulation of enzyme activity involved in the NADPH metabolism pathway or introduction or overexpression of malic enzymes in certain microbes, resulting in an increased production of lipids [7, 8].

3.2 Genetic Engineering for Regulation of Lipid Synthesis Pathway

3.2.1 Background

As described above, the lipid synthesis pathway is well-known in most microbes. A promising strategy to increase lipid production is to alter the regulation or pathway direction.

3.2.2 Genetic Engineering of Malonyl-CoA Synthesis

The carboxylation of acetyl-CoA to malonyl-CoA initiates the lipid biosynthesis. The acetyl-CoA carboxylase is responsible for generating malonyl-CoA by condensation of the acetyl-CoA unit with the bicarbonate anion. It is regarded as the crucial enzyme for de novo lipid synthesis. The overexpression of acetyl-CoA carboxylase in microbes has been shown to result in significant increase of lipid production [9]. In addition, the acetyl-CoA carboxylase is transcriptionally regulated and abolishing the transcription factor through enzyme mutagenesis or gene knock out strategy resulted in further accumulation of lipids [10]. In some microbes such as bacteria and plant cells, the existence of malonyl-CoA synthetase is responsible for the generation of malonyl-CoA by condensation of malonic acid and CoA. As such, overexpression of malonyl-CoA synthetase in microbes provides an alternative source for lipid precursor production [11].

3.2.3 Genetic Engineering of Fatty Acid Elongation

With acetyl-CoA and malonyl-CoA as precursors, the fatty acid elongation is carried out by the fatty acid synthase complex. Different microbes contain different types of fatty acid synthase system and regulation of the enzyme complex not only changes the lipid production level but also alters the lipid profile. Similar to the function of acetyl-CoA carboxylase, the fatty acid synthesis is also controlled at the transcriptional level. Therefore, modification of the transcription factors is also of great significance for lipid synthesis. In some condition, the multiple regulation of acetyl-CoA carboxylase and fatty acid synthase results in accumulation of lipids [12].Since both the acetyl-CoA carboxylase and fatty acid synthase have a large nucleotide size, the direct expression with plasmid-based method becomes difficult. Thus, the strategies of replacement of the native promoters of the involved enzymes with more appropriate promoters, as well as the genetic engineering approaches including circular polymerase extension and in-fusion ligation are always applied to increase the expression and activities of the enzymes.

3.2.4 Genetic Engineering of TAG Synthesis

In microbial cells, lipids are stored in the forms of cellular lipid components such as triacylglycerols and sterol esters, while the free fatty acyl-CoAs or fatty acyl-ACPs always generate feedback inhibition to the cells. Therefore, another efficient strategy to accumulate lipid is to strengthen the TAGs synthesis pathway and reduce the cellular concentration of acyl-CoAs or ACPs. Overexpression of enzymes responsible for the rate-limiting steps of the TAG synthesis pathway is demonstrated to successfully increase of lipid production in certain microbes [13, 14].

3.3 Genetic Engineering for Regulation of Lipid Catabolic Pathway

3.3.1 Background

The decrease or inactivation of lipid degradation is considered as another strategy to increase the intracellular lipid content. To initiate the lipid degradation, the fatty acids should be activated to acyl-CoA esters and the formed acyl-CoAs are consumed through the β-oxidation pathway. A series of enzyme activities provide the β-oxidation machinery with specific precursors and regulation of the involved enzymes could efficiently regulate the lipid degradation.

3.3.2 Genetic Engineering of Fatty Acid Degradation

Taking yeast as an example, the activated long-chain fatty acids are transported into peroxisome by the transporters Pxa1p and Pxa2p. Disruption of either the Pxa1p or Pxa2p resulted in growth deficiency in medium with long-chain fatty acids as the sole carbon source. The acy-CoA oxidase (POX) in peroxisome is responsible for oxidation of the fatty acyl-CoAs to *trans*-2-enoyl-CoA, which is the core step involved in the β-oxidation degradation process. It has been reported that the POX genes mutants could alter the lipid profiles in *Y. lipolytica*. In the POX2 gene disrupted strain, the utilization of long-chain fatty acids was blocked, whereas, in the POX3 gene disrupted strain, the degradation of short-chain fatty acids decreased. In the POX2, POX3, and POX5 gene disrupted strain, a "slim" phenotype was observed because of the feedback regulation of oxidation. On the other hand, overexpression of the POX2 gene resulted in an obese phenotype [15]. Although such kind of manipulation is not as applicable as in nonoleaginous microbes such as *S. cerevisiae,* it provides a good idea for the modification of microbial cells with enhanced lipid accumulation.

4 Genetic Engineering to Alter Lipid Profile in Microbes

4.1 Background

Lipid profiles differ among microbes. Based on the specific regulation mechanisms, lipids containing different length of carbon chains with saturated or unsaturated fatty acids are formed as final products. The various unsaturated fatty acids are formed by the respective desaturases, which introduces the double bond into the saturated fatty acyl-CoA/ACP derivatives.

4.2 Genetic Engineering of Elongation and Desaturation for Fatty Acid Synthesis

The most common unsaturated fatty acyl-CoAs in yeast as an example are C16:1, C18:1, C18:2, and C18:3, which are regulated by different functional desaturases. The Δ-9 fatty acid desaturase works on the formation of monounsaturated ones C16:1 and C18:1, whereas the Δ12 andΔ15 desaturases are responsible for introducing the additional double bonds into the location of Z12 and Z15 of the monounsaturated acyl chains, to form C18:2 and C18:3 fatty acyl-CoAs. However, the fatty acid profile could be modified if the elongase and desaturase are regulated, resulting in the alternation of whole lipid profile.

In *S. cerevisiae,* since the Δ9 fatty acid desaturase exists as the sole desaturase, the fatty acids produced are primarily saturated (C16:0, C18:0) and monounsaturated (C16:1, C18:1). In some species, other desaturases such as Δ12 and ω3 fatty acid desaturases exist and they can convert the Δ9-monounsaturated fatty acids (C18:1) into linoleic acid (18:2) and α-linolenic acid (C18:3). Based on these genetic information, the successful example is that the over expression of Δ12 and ω3 fatty acid desaturases from certain species successfully altered the fatty acid profile in *S. cerevisiae*, which could produce linoleic and α-linolenic acid [16].

4.3 Genetic Engineering of Fatty Acid Degradation Process

From another aspect, some types of fatty acids with ideal properties included in lipids are regarded as important precursors for fine chemicals and biofuels. Various strategies to alter the lipid profile for the increased synthesis of desired fatty acids have been tried on different microbes in recent years. The regulation of fatty acid synthesis process, as well as its degradation could be an efficient strategy applied. In *S. cerevisiae*, the β-oxidation pathway is modified to keep the long chain fatty acid oxidation, but prevent the medium chain fatty acid oxidation. Through replacement of gene encoding acyl-CoA oxidase solely involved in β-oxidation in *S. cerevisiae* with one that prefers long chain fatty acyl-CoA substrates, together with expression of enzyme responsible for the transfer the accumulated medium chain fatty acyl-CoAs out of peroxisomes, the medium chain fatty acids are successfully accumulated and the lipid profile is altered [17].

5 Conclusion

With the development of molecular system, synthetic biology, and genetic technologies, the biosynthetic pathways of microorganisms could be easily modified. The genetic approaches described in this chapter are all demonstrated as powerful tools for the optimization of lipid production and desired fatty acid profile.

References

1. Ratledge C (1994) Yeasts, molds, algae and bacteria as sources of lipids. In: Kamel BS, Kakuda Y (eds) Technological advances in improved and alternative sources of lipids. Blackie Academic and Professional, London, pp 235–291
2. Fujimoto T, Ohsaki Y, Cheng J, Suzuki M, Shinohara Y (2008) Lipid droplets: a classic organelle with new outfits. Histochem Cell Biol 130:263–279
3. Janßen HJ, Steinbüchel A (2014) Fatty acid synthesis in *Escherichia coli* and its applications towards the production of fatty acid based biofuels. Biotechnol Biofuels 7:7
4. Chen Y, Daviet L, Schalk M, Siewers V, Nielsen J (2013) Establishing a platform cell factory through engineering of yeast acetyl-CoA metabolism. Metab Eng 15:48–54
5. Lian J, Si T, Nair NU, Zhao H (2014) Design and construction of acetyl-CoA overproducing *Saccharomyces cerevisiae* strains. Metab Eng 24:139–149
6. Tang X, Chen WN (2014) Investigation of fatty acid accumulation in the engineered *Saccharomyces cerevisiae* under nitrogen limited culture condition. Bioresour Technol 162:200–206
7. de Jong BW, Shi S, Siewers V, Nielsen J (2014) Improved production of fatty acid ethyl esters in *Saccharomyces cerevisiae* through up-regulation of the ethanol degradation pathway and expression of the heterologous phosphoketolase pathway. Microb Cell Factories 13:39
8. Zhang Y, Adams IP, Ratledge C (2007) Malic enzyme: the controlling activity for lipid production? Overexpression of malic enzyme in *Mucor circinelloides* leads to a 2.5-fold increase in lipid accumulation. Microbiology 153:2013–2025

9. Runguphan W, Keasling JD (2014) Metabolic engineering of *Saccharomyces cerevisiae* for production of fatty acid-derived biofuels and chemicals. Metab Eng 21:103–113
10. Shi S, Chen Y, Siewers V, Nielsen J (2014) Improving production of malonyl coenzyme A- derived metabolites by abolishing Snf1-dependent regulation of Acc1. MBio 5: e01130–e01114
11. Chen WN, Tan KY (2013) Malonate uptake and metabolism in *Saccharomyces cerevisiae*. Appl Biochem Biotechnol 171:44–62
12. Nevoigt E, Kohnke J, Fischer CR, Alper H, Stahl U, Stephanopoulos G (2006) Engineering of promoter replacement cassettes for fine-tuning of gene expression in *Saccharomyces cerevisiae*. Appl Environ Microbiol 72:5266–5273
13. Mora G, Scharnewski M, Fulda M (2012) Neutral lipid metabolism influences phospholipid synthesis and deacylation in *Saccharomyces cerevisiae*. PLoS One 7:e49269
14. Kamisaka Y, Kimura K, Uemura H, Yamaoka M (2013) Overexpression of the active diacylglycerolacyltransferase variant transforms *Saccharomyces cerevisiae* into an oleaginous yeast. Appl Microbiol Biotechnol 97:7345–7355
15. Mličková K, Luo Y, D'Andrea S, Peč P, Chardot T, Nicaud J (2004) Acyl-CoA oxidase, a key step for lipid accumulation in the yeast *Yarrowia lipolytica*. J Mol Catal B Enzym 28:81–85
16. Yazawa H, Iwahashi H, Kamisaka Y, Kimura K, Uemura H (2009) Production of polyunsaturated fatty acids in yeast *Saccharomyces cerevisiae* and its relation to alkaline pH tolerance. Yeast 26:167–184
17. Chen L, Zhang J, Chen WN (2014) Engineering of *Saccharomyces cerevisiae* β-oxidation pathway to increase medium chain fatty acid production as potential biofuel. PLoS One 9: e84853

Chapter 9

Extraction Methods Used to Separate Lipids from Microbes

Balakrishna Maddi

Abstract

Lipids accumulated inside microbial cells (such as fungi, bacteria, yeast, microalgae, and macroalgae) have to be extracted for production of biodiesel and/or oleochemicals via thermochemical or thermocatalytic processes. Extraction methods for separation as well as quantification of lipids from microbes require organic solvents (such as hexane, methanol, and tetrahydrofuran) and involve multiple unit operations/steps such as thermal treatment (supercritical CO_2 extraction), chemical treatment (acid or base, enzymes), mechanical operations (such as sonication and ball beating), liquid–solid separation, liquid–liquid extraction, and gravimetric or chromatographic analysis.

Key words Solvent extraction methods, Lipids, Triglycerides, Free fatty acids, Algae, Fungi

1 Introduction

Lipids in microbes are classified into storage lipids and membrane lipids [1]. Storage lipids are triglycerides, diglycerides, monoglycerides, and free fatty acids and have wide variety of application especially for production of heavy-duty vehicle fuels (e.g., diesel) [2] as well as high-value oleochemicals [3–7]. Therefore, we discussed analytical methods for extraction and further quantification of storage lipids from microbes in this chapter.

Recovery of lipids from microbes involves extraction into one or more organic solvents such as hexane, methanol, dichloromethane, and tetrahydrofuran followed by solvent recovery (e.g., distillation) operations [8–10]. Cell-disruption operation (e.g., ultrasonication or bead beating) is required for complete lipid extraction from microbes [11–13]. The aim of these cell-disruption operations is to lyse or break the microbial cells for diffusion of organic solvents into microbe cells to dissolve lipid molecules. Lipids extracted into organic solvents can be separated from cell residue and quantified using conventional gravimetric measure-

Venkatesh Balan (ed.), *Microbial Lipid Production: Methods and Protocols*, Methods in Molecular Biology, vol. 1995,
https://doi.org/10.1007/978-1-4939-9484-7_9,

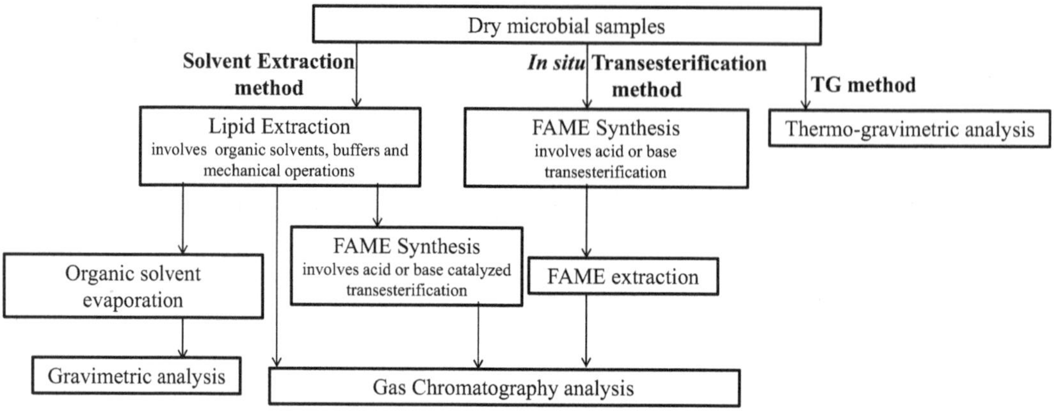

Fig. 1 Operations or steps involved in different analytical methods for quantifying triglyceride content in oleaginous feedstocks. Abbreviation of FAME is fatty acid methyl esters. In this chapter, solvent extraction methods to quantify lipids in microbial samples are extensively discussed. Other methods—transesterification and thermogravimetry (TG) analytical methods—are illustrated in the next chapter

ments or gas chromatography instrument as shown in Fig. 1. Thermal chemical (supercritical CO_2 and enzymes) pretreatments were also employed to lyse the microbial cell walls to replace mechanical operations.

2 Materials

1. Tristerarte, tripalmitate, triolein, hexane, methanol, tetrahydrofuran, chloroform, palmitic acid, stearic acid, oleic acid, 1-oleoyl-rac-glycerol, 1,3-Diolein, triolein, tristearate, glyceryl tripalmitate, 20 mL scintillation glass vials were purchased from Sigma-Aldrich (St. Louis, MO) (*see* **Note 1**).
2. Stainless steel bead beating vials with polypropylene plug caps and three types of beads (0.6 g of 0.1 mm zirconium/silica beads, 0.4 g of 1.0 mm glass beads, two 2.7 mm glass beads) are purchased from BioSpec Products (Bartlesville, OK) (*see* **Note 2**).
3. Analytical balance, a Mini-Beadbeater-1 (BioSpec Products, Bartlesville, OK), Supercritical fluid extraction apparatus (ISCO, SFX 220, USA), desiccator, centrifuge, and a Shimadzu 2010 GC equipped with a Restek (Bellefonte, PA) Rtx Biodiesel column (15 m Length, 0.32 mm ID and 0.1 μm film thickness), and flame ionization detector (FID) are used (*see* **Note 2**).

3 Methods

3.1 Bligh and Dryer Method—Wet Method

1. Estimation of moisture content of wet algal paste—Use aluminum weight boats (*see* **Note 3**) for estimation of moisture content.
2. Weigh 0.5–2.0 g of wet algal into the aluminum weight boat (*see* **Note 3**). Record the wet sample weight.
3. Place the sample with aluminum boat in an air drying oven which has been maintained at 105 ± 5 °C for 12 h [14] and/or dry to obtain constant weight. Constant weight is defined as less than ±0.5 mg weight change upon 1 h in the oven.
4. Carefully remove the crucible from the oven and place directly into a desiccator. Record the weight of sample after drying to the nearest 0.1 mg.
5. Place the sample back into the oven at 105 ± 5 °C and record the weight after 1 h.
6. Carefully remove the crucible from the oven and place directly into a desiccator. Record the weight of sample after drying to the nearest 0.1 mg. Store the dried sample in a desiccator for further analysis if the weight change is below ±0.5 mg.
7. Calculate moisture content as shown below:

$$\%\ \text{Moisture content} = \frac{(\text{Weight of wet sample} - \text{Weight of dried sample})}{\text{Weight of wet sample}} \times 100$$

8. Dry the 20 mL vials following the procedure described in Subheading 3.1, **step 5**.
9. Weigh 0.5–2.0 g of wet algal into 20 mL vials (*see* **Note 3**). Record the wet sample weight to the nearest 0.1 mg.
10. Add 4 mL of methanol and 2 mL of chloroform to the vial. Place in ice bath.
11. Agitate vial containing microbial sample, methanol and chloroform for 1 min using Fisher Scientific Model 120 sonic dismembrator (Pittsburgh, PA). During sonication operation, sample vials should be placed in ice bath to prevent evaporation of organic solvents. Evaporation of organic solvents may result in incomplete extraction of lipids.
12. Agitate sample vial for 3 min (*see* **Note 4**). This operation is performed to break cell walls and extract the lipids into the organic solvent mixture. Optimum agitation time for complete extraction of lipid is not standardized in the literature and depends upon organic solvents used, microbial species, and size of the sample. During sonication, lipids are extracted into chloroform layer.

13. Separate chloroform phase using separating funnel and transfer into another vial. Record the weight of vial containing chloroform phase.
14. Evaporate organic solvents using continuous-flow of nitrogen for 12 h. After 12 h, record the weight of vial after evaporation of chloroform.
15. Calculate % lipid content using the following formula:

$$\% \text{ Lipids} = \frac{\begin{pmatrix}\text{Weight of vial before chloroform evaporation} \\ - \text{Weight of vial after chloroform evaporation}\end{pmatrix}}{\text{Weight of wet sample} \times (100 - \%\text{Moisture content})} \times 100$$

16. Inject the chloroform phase into gas chromatography (GC) to quantify triglycerides, diglycerides, monoglycerides, and free fatty acids.

3.2 Solvent Extraction Using Sonication—Dry Method

1. Weigh 50–100 mg of wet algal into 20 mL vials. Record the wet sample weight to the nearest 0.1 mg.
2. Add 5 mL of equimolar solution of hexane, chloroform, and tetrahydrofuran to the vial. Place in ice bath. Selection of organic solvent depends upon microbial species.
3. Agitate vial containing microbial sample, and mixture of organic solvents for 1 min using Fisher Scientific Model 120 sonic dismembrator (Pittsburgh, PA). During sonication operation, sample vials should be placed in ice bath to prevent evaporation of organic solvents. Evaporation of organic solvents may result in incomplete extraction of lipids.
4. Agitate sample vial for 3 min (*see* **Note 4**). This operation is performed to break cell walls and extract the lipids into organic solvent mixture. Optimum agitation time for complete extraction of lipid is not standardized in the literature and depends upon organic solvents used, microbial species, and size of the sample. During sonication, lipids are extracted into organic layer.
5. Centrifuge and separate cell residue and transfer liquid into another vial. Record the weight of vial containing liquid phase.
6. Evaporate organic solvents using continuous-flow of nitrogen for 12 h. After 12 h, record the weight of vial after evaporation of chloroform.
7. Calculate % lipid content using the following formula:

$$\% \text{ Lipids} = \frac{\begin{pmatrix}\text{Weight of vial before organic solvent evaporation} \\ - \text{Weight of vial after organic solvent evaporation}\end{pmatrix}}{\text{Weight of wet sample} \times (100 - \%\text{Moisture content})} \times 100$$

8. Alternatively, analyze the organic liquid extract using gas chromatography (GC) to quantify lipids.

3.3 Solvent Extraction Using Bead Beating—Dry Method

1. Weigh 15 mg of dry microbe sample and transfer into 2 mL stainless steel bead beating vials with polypropylene plug cap that contain 1 mL of chloroform. Beads used in this analytical method were described in the literature [15]. Three types of beads (0.6 g of 0.1 mm zirconium/silica beads, 0.4 g of 1.0 mm glass beads, and two 2.7 mm glass beads) are added to each vial before capping.
2. Agitate vial for 30 s at 2500 oscillations per minute using a Mini-Beadbeater-1 (BioSpec Products, Bartlesville, OK) and then cooled in an ice bath for 1 min. Repeat this step for 2–45 times. The number of times depends upon the microbe strain (*see* **Note 4**).
3. Centrifuge these vials to separate organic solvent mixture from the cell residue.
4. Transfer the organic phase in stainless steel vials into 5 mL glass vial.
5. Rinse stainless steel vials with 1 mL of chloroform and add to the organic phase collected.
6. Filter and transfer this organic phase into GC vials for quantification of triglycerides, monoglycerides, diglycerides, and fatty acids using gas chromatography equipped with flame ionization detector (GC-FID).
7. Alternatively, evaporate organic solvents using continuous-flow of nitrogen for 12 h. After 12 h, record the weight of vial after evaporation of chloroform.
8. Calculate % lipid content using the following formula:

$$\%\,\text{Lipids} = \frac{\begin{pmatrix}\text{Weight of vial before chloroform evaporation}\\ -\ \text{Weight of vial after chloroform evaporation}\end{pmatrix}}{\text{Weight of wet sample} \times (100 - \%\text{Moisture content})} \times 100$$

3.4 Supercrictical CO_2 Extraction—Dry Method

1. Weigh 3 ± 0.1 g of dry microbe sample and transfer into extraction vessel of Supercritical fluid extraction apparatus. Description of this apparatus is discussed in the literature [16].
2. Place the extraction vessel in heating chamber and maintain this chamber at 55 °C, 500 bar and purge the surge the sample with continuous flow of supercritical CO_2 at 1.9 g/min.
3. Allow the system to equilibrate for 15 min before collecting lipid extracts.
4. Collect the extract for every 5 cm^3 of supercritical CO_2 passed over the sample.

5. Weigh the extract and calculate lipid content using the following formula:

$$\% \text{ Lipids} = \frac{(\text{Lipid extract})}{(\text{Initial weight of sample})} \times 100$$

6. Alternatively, transfer this lipid extract into GC vials and dilute with chloroform for quantification of triglycerides, monoglycerides, diglycerides, and fatty acids using gas chromatography equipped with flame ionization detector (GC-FID).

3.5 Enzymatic Hydrolysis Followed by Solvent Extraction—Wet Method [17]

1. Determine the moisture content of microbial culture as described in Subheading 3.1, **steps 1–7**.
2. Record the weight of empty 20 mL glass vial.
3. Weigh 2.0 ± 0.1 g of wet microbe culture and add 0.04 mL of cellulases and 0.02 mL of β-glucosidases to 20 mL scintillation glass vial. Add citrate buffer solution to maintain pH of 4.8.
4. Place the vial in oven and maintained at 50 °C for 72 h.
5. After hydrolysis of cell walls, add mixture of organic solvent containing hexane, chloroform, and tetrahydrofuran to the vial for extraction of lipids into organic phase.
6. After mixing the organic phase for 6 h, evaporate organic solvents using continuous-flow of nitrogen for 12 h. After 12 h, record the weight of the vial after evaporation of chloroform.
7. Calculate % lipid content using the following formula:

$$\% \text{ Lipids} = \frac{\begin{pmatrix} \text{Weight of vial before evaporation of organic solvents} \\ - \text{Weight of vial after evaporation of organic solvents} \end{pmatrix}}{\text{Weight of wet sample} \times (100 - \%\text{Moisture content})} \times 100$$

8. Alternatively, transfer this organic phase into GC vials for quantification of triglycerides, monoglycerides, diglycerides, and fatty acids using gas chromatography equipped with flame ionization detector (GC-FID).

3.6 Gas Chromatography Method [4, 15]

1. Use a gas chromatography (GC) equipped with a Restek (Bellefonte, PA) Rtx Bio-diesel column—15 m × 0.32 mm ID × 0.1 μm and flame ionization detector (FID) to quantify triglycerides, diglycerides, monoglycerides, free fatty acids, and fatty acid methyl esters. One microliter of sample is injected into GC instrument for analysis.
2. Use helium as a carrier gas with a constant linear velocity of 50 cm/s in the column.
3. Maintain GC injector and FID temperatures at 370 °C.

4. Program the GC oven initially hold a temperature of 60 °C for 1 min followed by a steady increase to 370 °C at a ramp rate of 10 °C/min. Finally, the oven temperature is maintained at 370 °C for 6 min.
5. Prepare a series of calibration standards containing palmitic acid, stearic acid, oleic acid, 1-oleoyl-rac-glycerol, 1,3-diolein, triolein, tristearate, and glyceryl tripalmitate dissolved in equimolar mixture of hexane, chloroform, and tetrahydrofuran. Prepare standards in concentration range of 10–0.5 mg/mL. Use a five or six point calibration. If standards are prepared outside of the suggested ranges, the new range for these calibration curves must be validated.
6. Analyze sample and calibration standards to quantify triglyceride, diglyceride, monoglyceride, and free fatty acid content in microbial samples (*see* **Note 5**).
7. GC retention time of free fatty acids, monoglycerides, diglycerides, and triglycerides is shown in Fig. 1 [15].

4 Notes

1. Follow standard chemical handling procedures.
2. Operate all equipment according to procedures described in the instrument manual.
3. Ensure that a representative microbial sample is taken for estimation of lipid content.
4. Estimates obtained from these solvent extraction techniques for microalgae feedstocks have been reported to be inconsistent and potentially inaccurate [4, 9, 15]. Agitation time for sonication and bead beating of microbe-solvent mixture is not constant and depends upon the type of microbial species. Hence, it is recommended to estimate the optimum agitation time for individual microbes.
5. Run all samples and GC standards in duplicate, at minimum.

5 Subcritical Water Hydrolysis

Subcritical water also known as pressurized hot water, is liquid water under pressure at temperatures between the atmospheric boiling point 100 °C and the critical water temperature 374 °C. In subcritical state the water is maintained in the liquid form by applying pressure and this way the liquid water is in equilibrium with vapor at the saturated vapor pressure. Under these conditions the water properties change and an increase of solvating power, diffusivity, and self-ionization of water (acidification) could be observed. On the other hand, a decrease in viscosity and surface tension as well as of its polarity could be observed. Water is non-toxic; therefore, it is considered a "green" solvent. Recently, subcritical water technology has been chosen for the extraction of various active compounds from biomass, to remove non-oil biomass leaving an oil-rich biomass for biodiesel production [18] or to release nitrogen containing compounds (amino acids and peptides) to the water phase producing an oil-rich biomass residue [19]. Subcritical water extraction was also applied to microalgae and yeasts biomass, and the aim was to recover polysaccharides and protein-derived products [20, 21], or potentially can be used to produce biocrude which can be subsequently be upgraded to liquid hydrocarbons [21]. Using flash hydrolysis of microalgae under subcritical water conditions, the proteins are solubilized as water-soluble peptides and amino acids, and the remaining material which comprises lipids and algal cell can be filtered and used in biofuel production [22].

References

1. Harwood J (2012) Lipids in plants and microbes. Springer Science & Business Media, Berlin
2. Chu S, Majumdar A (2012) Opportunities and challenges for a sustainable energy future. Nature 488(7411):294
3. Maddi B, Vadlamani A, Viamajala S, Varanasi S (2017) Quantification of triglyceride content in oleaginous materials using thermogravimetry. J Anal Appl Pyrolysis 128:232–237
4. Maddi B, Viamajala S, Varanasi S (2018) Pyrolytic fractionation: a promising thermochemical technique for processing oleaginous (algal) biomass. ACS Sustain Chem Eng 6 (1):237–247
5. Shirazi Y, Tafazolian H, Viamajala S, Varanasi S, Song Z, Heben MJ (2017) High-yield production of fatty nitriles by one-step vapor-phase thermocatalysis of triglycerides. ACS Omega 2(12):9013–9020
6. Thakur DS, Kundu A (2016) Catalysts for fatty alcohol production from renewable resources. J Am Oil Chem Soc 93(12):1575–1593
7. Gnansounou E, Raman KJ (2016) Life cycle assessment of algae biodiesel and its co-products. Appl Energy 161:300–308
8. Bligh EG, Dyer WJ (1959) A rapid method of total lipid extraction and purification. Can J Biochem Physiol 37(8):911–917
9. Laurens LML, Quinn M, Van Wychen S, Templeton DW, Wolfrum EJ (2012) Accurate and reliable quantification of total microalgal fuel potential as fatty acid methyl esters by in situ transesterification. Anal Bioanal Chem 403 (1):167–178
10. Roux J-M, Lamotte H, Achard J-L (2017) An overview of microalgae lipid extraction in a biorefinery framework. Energy Procedia 112:680–688
11. Amaro HM, Guedes AC, Malcata FX (2011) Advances and perspectives in using microalgae to produce biodiesel. Appl Energy 88 (10):3402–3410
12. Lee SJ, Yoon B-D, Oh H-M (1998) Rapid method for the determination of lipid from the green alga *Botryococcus braunii*. Biotechnol Tech 12(7):553–556
13. Mercer P, Armenta RE (2011) Developments in oil extraction from microalgae. Eur J Lipid Sci Technol 113(5):539–547
14. Maddi B, Viamajala S, Varanasi S (2011) Comparative study of pyrolysis of algal biomass from natural lake blooms with lignocellulosic biomass. Bioresour Technol 102 (23):11018–11026
15. Lohman EJ, Gardner RD, Halverson L, Macur RE, Peyton BM, Gerlach R (2013) An efficient and scalable extraction and quantification method for algal derived biofuel. J Microbiol Methods 94(3):235–244
16. Taher H, Al-Zuhair S, Al-Marzouqi AH, Haik Y, Farid M, Tariq S (2014) Supercritical carbon dioxide extraction of microalgae lipid: Process optimization and laboratory scale-up. J Supercrit Fluids 86:57–66

17. Cho H-S, Oh Y-K, Park S-C, Lee J-W, Park J-Y (2013) Effects of enzymatic hydrolysis on lipid extraction from *Chlorella vulgaris*. Renew Energy 54:156–160
18. Levine RB, Pinnarat T, Savage PE (2010) Biodiesel production from wet algal biomass through in situ lipid hydrolysis and supercritical transesterification. Energy Fuel 24:5235–5243
19. Moscoso-Garcia JL, Obeid W, Kumar S et al (2013) Flash hydrolysis of microalgae (*Scenedesmus sp.*) for protein extraction and production of biofuels intermediates. J Supercrit Fluids 82:183–190
20. Chakraborty M, Miao C, McDonald A et al (2012) Concomitant extraction of bio-oil and value added polysaccharides from *Chlorella sorokiniana* using a unique sequential hydrothermal technology. Fuel 95:63–70
21. Moscoso-Garcia JL, Teymouri A, Kumar S (2015) Kinetics of peptides and arginine production from microalgae (*Scenedesmus sp.*) by flash hydrolysis. Ind Eng Chem Res 54 (7):2048–2058
22. Kumar S, Hablot E, Moscoso-Garcia JL et al (2014) Polyurethanes preparation using proteins obtained from microalgae. J Mater Sci 49(22):7824–7783

Chapter 10

Novel Microbial Modification Tools to Convert Lipids into Other Value-Added Products

Priya Kumari, Farnaz Yusuf, and Naseem A. Gaur

Abstract

CRISPR-Cas9 Clustered Regularly Interspaced Short Palindromic Repeats (CRISPR) and CRISPR associated (Cas) is a microbial adaptive immune system that has revolutionized the field of molecular biology and genome engineering. The Type II CRISPR system consists of Cas9 nuclease of *Streptococcus pyogenes* and the RNA complex that guides Cas9 nuclease to a specific sequence of DNA in the genome. The CRISPR-Cas9 technology has reformed our ability to edit DNA and to regulate expression levels of genes of interest to high precision and accuracy. It is a powerful technology, which is used for genome engineering of a wide range of organisms for various applications. Here, we describe a method involving CRISPR-Cas9-mediated genome editing via nonhomologous end joining (NHEJ) or homology-directed repair (HDR) mechanisms for biotechnological applications in yeast. The complete procedure of genome editing including target sequence selection, cloning gRNA with a target sequence, transformation, and verification of the desired mutation/deletion or insertion can be achieved within 2–3 weeks in yeast.

Key words CRISPR-Cas9, gRNA, CRISPR RNA, Trans-activating CRISPR RNA, Genome editing, Homology-directed repair, Nonhomologous end joining

1 Introduction

Lipids comprise a vast and diverse group of organic molecules such as fat, oil, wax, certain vitamins, hormones, monoglycerides, diglycerides, triglycerides, and phospholipids. Lipids play vital roles in many biological processes including energy storage, signaling and structural components of the cell membrane. They have wide industrial applications as a raw material for the production of surfactants, detergents, soaps, lubricants, solvents, paints, cosmetics, biofuels, and other biochemicals. Lipids from animal and plant sources are often insufficient for industrial production of lipid-derived chemicals. Moreover, the efficiency and yield of lipid derivatives produced by chemical methods is poor and require harsh reaction conditions [1].

Venkatesh Balan (ed.), *Microbial Lipid Production: Methods and Protocols*, Methods in Molecular Biology, vol. 1995, https://doi.org/10.1007/978-1-4939-9484-7_10,

Microbial metabolic engineering for production of lipid-derived chemicals is an attractive strategy to overcome the problems associated with chemical methods [2–5]. Drug or nutritional selection based traditional metabolic engineering strategies frequently fail to produce the desired strains, because multiple genetic modifications in a single strain limit their use. Additionally, recycling of markers by recombination based strategies leave a copy of repeat sequence in the genome, which leads to genome instability. Recently, several genome editing technologies, such as transcription activator-like effector nucleases (TALENs) and zinc-finger nucleases (ZFNs) have emerged, but they are tedious, time-consuming, and expensive.

CRISPR-Cas9 is simple, efficient, cost-effective, highly precise, scarless, and marker-free genome editing tool [6–10]. CRISPR-Cas9 consists of a Cas9 nuclease and the RNA complex known as guide RNA (gRNA), which directs Cas9 to a specific DNA sequence in the genome [11]. The gRNA consists of two RNA molecules the CRISPR RNA (crRNA) and the trans-activating CRISPR RNA (tracrRNA). The crRNA includes a 20-nucleotide target sequence, which bind to the complementary strand in the genome along with the tracrRNA forming a DNA-RNA duplex recognized by the Cas9 nuclease. The target sequence must be followed by a PAM (protospacer adjacent motif) sequence (which is –NGG in case of *S. pyogenes* Cas9). The binding of Cas9 generates double strand break (DSB) three base pairs upstream of the PAM sequence in the genome (Fig. 1).

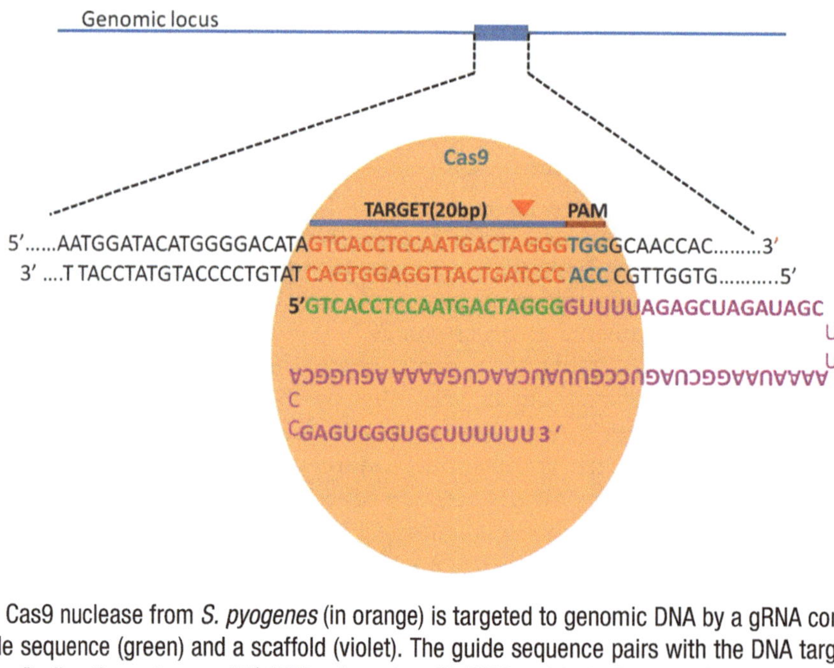

Fig. 1 The Cas9 nuclease from *S. pyogenes* (in orange) is targeted to genomic DNA by a gRNA consisting of a 20-nt guide sequence (green) and a scaffold (violet). The guide sequence pairs with the DNA target (blue bar on top strand), directly upstream of 5′-NGG adjacent motif (PAM; red bar). Cas9 creates a DSB 3 bp upstream of the PAM site (red triangle)

Yeast has a versatile lipid metabolic pathway and serves as a promising organism for industrial lipid derived chemicals production. Metabolic engineering strategies can be applied to develop yeast cells factories for desired lipid-derived chemicals such as alkane, alkene, fatty ketone, fatty alcohols, and fatty esters (Fig. 2). Engineering of yeast for production of fatty acid-ethyl ester (FAEEs) requires heterologous expression of wax ester synthase which catalyzes the esterification of acyl-CoA and alcohols. For increased production of FAEEs, yeast metabolic pathways need to be optimized to inhibit lipid storage and block degradation of fatty acids by beta-oxidation.

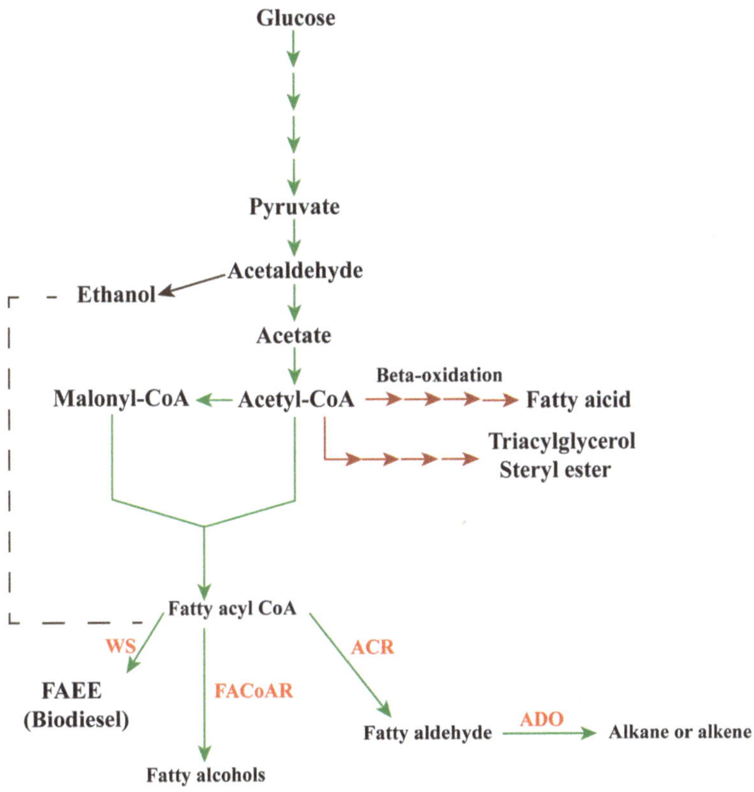

Fig. 2 Metabolic pathways for lipid-derived chemical. *WS* wax ester synthase, *FaCoAR* fatty acyl-CoA reductase, *ACR* fatty acyl-CoA reductase, *ADO* aldehyde-deformylating oxygenase, *ADH* alcohol dehydrogenase. The enzymes in red color catalyze the transformation of lipid intermediates into different lipid-derived valuable products. Arrows in maroon shows the pathways required to be deleted while green color arrows indicate the pathways to be overexpressed in order to produce high amount of lipid-derived value added products

2 Materials

2.1 Reagents

1 M sorbitol, 10× TAE buffer, 1 M DTT, 6× DNA Loading dye, dNTPs, Agarose, DNA ladder, Plasmid MINI PREP KIT, TOPO cloning kit, PCR cleanup kit, DNase/RNase-free ultrapure water.

2.2 Strains

Yeast strain (BY4741; *MATa his3Δ1 leu2Δ0 met15Δ0 ura3Δ0*), chemically competent *E. coli* cells (*DH5α*).

2.3 Enzymes

High-fidelity Phusion DNA polymerase, *Dpn*I, T4 DNA ligase, T4 polynucleotide kinase.

2.4 Media

LB broth and agar, YPD (Yeast extract, Peptone, Dextrose), Synthetic defined media with amino acid dropout (−LEU) broth and agar.

2.5 Plasmids

Cas9-NAT plasmid (Catalog No. 64329), pRPR1p_gRNA_handle_RPR1t plasmid (Catalog No. 49014).

2.6 Antibiotics

Ampicillin, nourseothricin (NAT).

2.7 Instruments Required

Gel imager, heat block or water bath (42 °C), incubator shakers (37 and 30 °C), NanoDrop, spectrophotometer, tabletop refrigerated centrifuge, thermocycler, freezers (−20 °C, −80 °C), Gene Pulser Xcell (Bio-Rad), gel electrophoresis unit.

2.8 Other Requirements

Microcentrifuge tubes, PCR tubes, 250 ml shake flasks, 2 mm cuvettes for yeast electroporation, Falcon tubes (15 ml, 50 ml), spreader, petri plates.

3 Experimental Design

3.1 Target Sequence Selection

The most critical step of CRISPR-Cas9 mediated genome editing is the selection of an appropriate target sequence in the genome. The guidelines for target sequence selection are as follows: (1) the target sequence must be unique in the genome to minimize the off-target effects (the risk of Cas9-mediated cleavage at nonspecific locations in the genome) and maximize on-target activity (cutting at the specific sites); (2) choose proximal region of the ORF to generate premature stop codons; (3) begin with 5′G (if it does not begin with G in the genome add one to increase on-target efficiency); (4) the GC content of target sequence should be between 40% and 80% (higher GC content stabilizes RNA–DNA duplex while destabilizing off-target hybridization). Target sequences can be selected either manually or by using online tools (e.g., CRISPRdirect, CHOPCHOP, CRISPy) (*see* **Note 1**).

3.2 Strategies for gRNA and Cas9 Expression

Genome editing using CRISPR-Cas9 can be achieved by using either single plasmid or two plasmids systems. In single plasmid system both Cas9 and gRNA are expressed by same plasmid, while Cas9 and gRNA are expressed by two independent plasmids in two plasmids system. Overexpression of Cas9 leads to cell toxicity; hence, gRNA and Cas9 levels should be optimized for efficient genome editing. Therefore, it appears reasonable to maintain Cas9 expression relatively low by using low copy number vectors and the expression of gRNA high to ensure high editing efficiency.

4 Methods

4.1 Cloning of Target Sequence in gRNA Expression Vector

Clone target sequence in gRNA vector (pRPR1p_gRNA_handle_RPR1t plasmid) as below

1. To clone the desired target sequence in pRPR1p_gRNA_handle_RPR1t vector, amplify the entire vector with forward primer (harboring the target sequence at the 5′ end) and reverse primer (Fig. 3) by using following PCR reaction. (Set the reaction on ice).

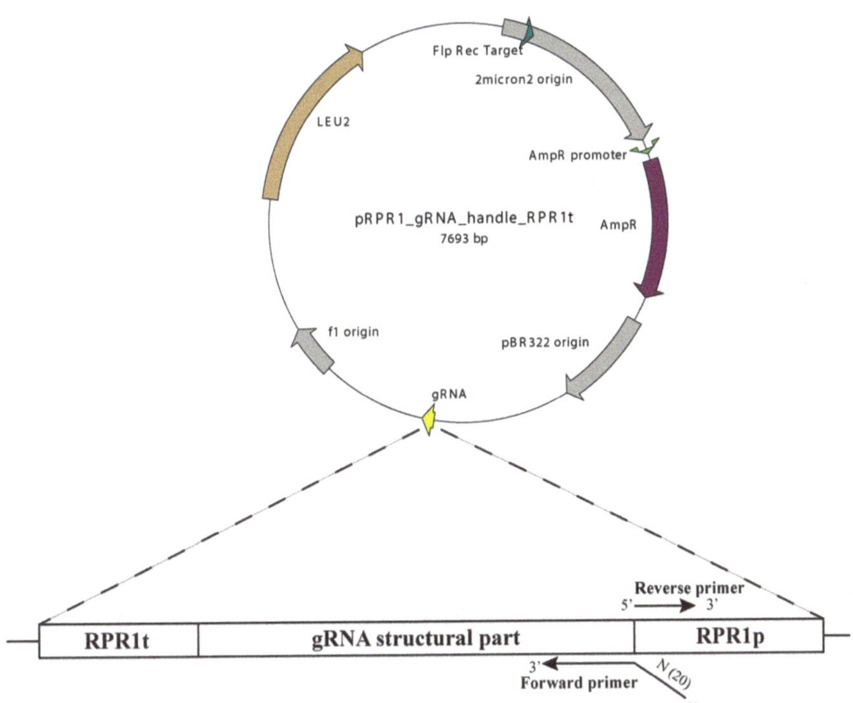

Fig. 3 Pictorial representation of pRPR1_gRNA_handle_RPR1t vector construction strategy with detailed illustration of gRNA cassette and primers for insertion of the target sequences. Forward primer contains overhang of 20 bp target sequence at 5′ end

Table 1
PCR cyclic reaction mixture

Components	Volume (μl)	Final conc.
5× Phusion buffer	5.0 μl	1×
10 mM dNTPs	1.0 μl	200 μM each
10 μM Forward primer with gRNA target sequence	2.5 μl	0.5 μM
10 μM Reverse Primer	2.5 μl	0.5 μM
(pRPR1p_gRNA_handle_RPR1t) plasmid	*X* μl	1–5 ng
Phusion DNA polymerase	0.5 μl	0.02 U/μl
H_2O	Add to 50 μl	

2. Perform the PCR with the following cyclic parameters:
   Initial template denaturation at 98 °C for 2 min then 30 cycles with following parameters 98 °C for 30 s; annealing (50–60 °C) for 30 s; extension (72 °C) for 4 min followed by final extension at 98 °C for 10 min and hold at 4 °C (Table 1) (*see* **Note 2**).
3. Check the amplified product by loading 5 μl of the PCR product on an agarose gel (1%) to verify successful amplification.
4. Treat PCR product by T4 polynucleotide kinase for 5′ phosphorylation (1–2 h at 37 °C). For recircularization of vector add 1 μl of T4 DNA ligase and incubate at 16 °C for overnight.
5. Add 1 μl of *Dpn*I to the ligation reaction. Mix and incubate at 37 °C for 6 h (You can also give the *Dpn*I treatment overnight to ensure complete template DNA removal).
6. Transform the *DpnI* treated ligated product into DH5α competent cells.
   (a) Add 5 μl of ligation mix to 50 μl of DH5α competent cells.
   (b) Mix by flicking the tube and incubate on ice for 30 min.
   (c) Give a heat shock: 42 °C for exactly 90 s and place immediately on ice for 5 min.
   (d) Add 950 μl LB broth without antibiotic and incubate at 37 °C for 1 h with shaking (200 rpm).
   (e) Spin at 2600 × *g* for 60 s, discard the supernatant and resuspend the pellet in 50–100 μl of LB broth.
   (f) Spread on LB agar plate with ampicillin (100 μg/ml) and incubate the plate at 37 °C for 16 h.
7. Pick two to three transformants from the above incubated plate and inoculate in to 3 ml LB broth with ampicillin (with 100 μg/ml), incubate at 37 °C for 16 h with shaking.

8. Isolate plasmid DNA using Miniprep Kit and quantify by using NanoDrop spectrophotometer.
9. Confirm the insertion of target sequence by sequencing and PCR.

4.2 Design of Repair DNA or Donor DNA Cassettes

Yeast transformants with Cas9 and gRNA expressing vectors produce double stranded breaks (DSB) in the genome. DSB is repaired either by NHEJ, which will cause random insertion, deletion or a frameshift mutation (indel) or by HDR in the presence of repair DNA cassette (with homologous arm) (Fig. 4). For gene disruption a single gRNA and Cas9 vector can be transformed to create indel mutation in the desired sequence. However, to delete a specific gene sequence in the genome, two separate gRNAs with distinct target sequences are required along with the Cas9 expressing plasmid (Fig. 5). If knockout of a full gene is not possible due to the presence of overlapping coding sequences in genome, supplying of a donor DNA cassette with premature stop codon along with Cas9 and a gRNA expressing vector will lead to truncated protein and ultimately create gene deletion effect. In case of simultaneous gene knock out and knock in, generate a DNA cassette harboring the desired sequence with the homology arms for genomic integration. Cotransform the repair or donor DNA cassette along with Cas9-expressing plasmid and gRNA vector with two separate target sequences specific to DNA locus where knock in is required (*see* **Notes 3**, **4** and **5**).

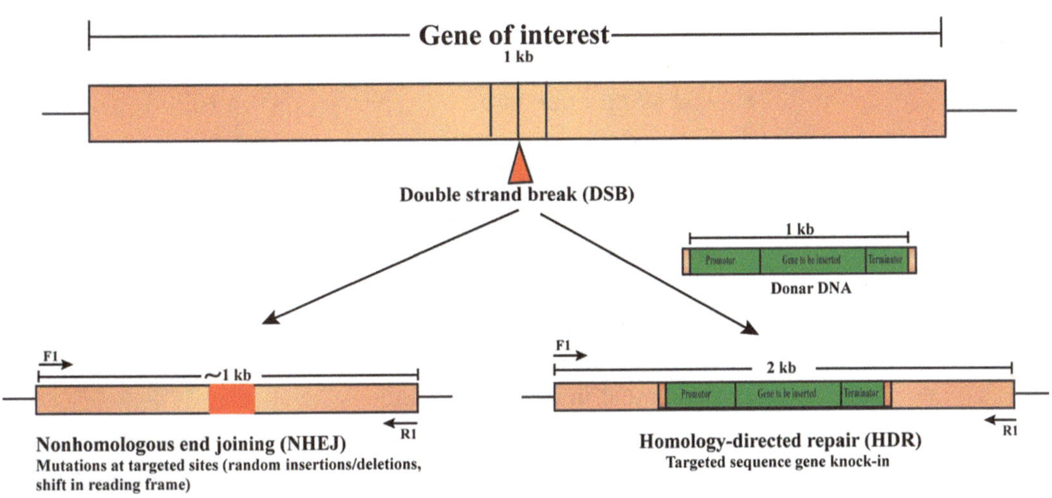

Fig. 4 Schematic representation of DSB repairs. Double-stranded breaks generated by CRISPR-Cas9 can be repaired by two different mechanisms, In NHEJ, ends of DSB rejoined by endogenous DNA repair machinery resulting in random insertion, deletion or frameshift mutations. Alternatively, presence of a repair template favors the HDR pathway, which allows for precise editing of the genome

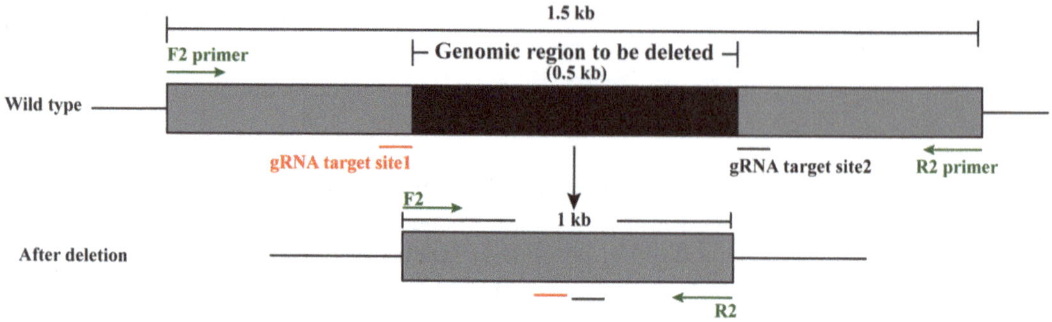

Fig. 5 Graphical representation of targeted gene deletion with CRISPR/Cas9 system. A desired genomic region is deleted by using two separate gRNA target sequence and the ends of break rejoin by NHEJ. The deletion can be confirmed by PCR analyses using primers flanking the targeted regions. Targeted deletions can be further verified by sequencing

4.3 Preparation of Yeast Electrocompetent Cells and Transformation

1. Inoculate BY4741 cells in 5 ml of YPD broth and incubate overnight at 30 °C with constant shaking at 200 rpm.
2. Dilute the overnight culture to an OD_{600} of 0.2 in 100 ml of YPD in a shake flask.
3. Grow culture at 30 °C with shaking at 200 rpm till mid log phase (OD_{600} 0.6–0.8).
4. Harvest the cells in 50 ml falcon tubes by centrifugation at 1000 × *g* for 5 min.
5. Remove the supernatant and wash cells with 20–30 ml water.
6. Resuspend the pellet in 10 ml buffer (1XTE, 100 mM LiOAc) and incubate at 30 °C for 30 min with shaking (130 rpm).
7. Add 250 μl of 1 M DTT and incubate at 30 °C for 60 min with shaking at 130 rpm.
8. Add 40 ml chilled water and harvest cells at 1000 × *g* for 5 min at 4 °C.
9. Add 5 ml of chilled 1 M sorbitol after removing supernatant.
10. Spin at 1000 × *g* for 5 min at 4 °C, decant the supernatant, resuspend in 500 μl of 1 M sorbitol, and keep on the ice.
11. Add 40–45 μl of cells in a microcentrifuge tube with 1–3 μg of each gRNA and Cas9 harboring plasmids and mix gently (for an HDR add donor or repair DNA in the above mix) and incubate it in ice for 10 min.
12. Transfer the mix in to 2 mm cuvette to proceed for electric pulse at 2.5 kV, 200 Ω, and 25 μF for 5 m s by using Gene Pulser Xcell (Bio-Rad).
13. Resuspend the cells in 950 μl YPD broth, incubate at 30 °C for 2–3 h with shaking and spread on synthetic complete medium without leucine, with NAT (200 μl/ml).

4.4 Confirmation of Gene Disruption/Deletion/Insertion/Replacement

Deletion can be verified by PCR based methods.

1. Isolate genomic DNA from the transformants.
2. Assemble the PCR reaction as given in Table 2.
3. Run samples in a thermocycler using the following parameters:

   98 °C for 2 min followed by 35 cycles of (98 °C for 30 s, annealing temp for 30 s, 72 °C for 1 min), and 72 °C for 10 min. Optimize PCR conditions for each primer pair.
4. Analyze the PCR product on 1% agarose gel.
5. Examine samples for the presence/absence of expected DNA fragments.
6. PCR amplification using F2 and R2 primers (Fig. 5) leads to full gene amplification; however, in the case of deletion size of PCR amplicon will be smaller (Fig. 6).

For the confirmation of frame shift mutation or insertion/deletion of very small fragment (indel mutations) sequencing of

Table 2
PCR reaction mixture

Components	Volume (μl)	Final conc.
5× Phusion buffer	5.0 μl	1×
10 mM dNTPs	1.0 μl	200 μM each
10 μM Forward primer (F2)	2.5 μl	0.5 μM
10 μM Reverse Primer (R2)	2.5 μl	0.5 μM
Template genomic DNA	*X* μl	50–100 ng
Phusion DNA polymerase	0.5 μl	0.02 U/μl
H_2O	Add to 50 μl	

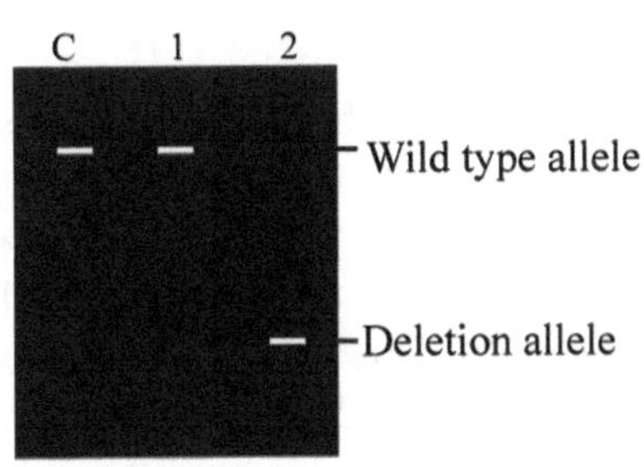

Fig. 6 Pictorial representation of PCR amplification for screening for CRISPR/Cas9-mediated gene deletion. For screening of transformants with deletion, PCR reaction was assembled using primers (F2 and R2) flanking the gRNA target sequence. Lane C: control; Lane 1 and 2: different clones for the screening of deletion

Table 3
Verification of gene editing

Type of editing	Primer pairs	Amplicon size (in kb)
Insertion	F1 and R1	2 kb
Indel	F1 and R1	~1 kb
Disruption	F1 and R1	2 kb
Deletion	F2 and R2	1 kb

amplified PCR product is required. Amplify the targeted region by using primers (F1 and R1) (Fig. 4) outside the gRNA target sequence and send for sequencing after cloning in TOPO blunt vector.

For screening of simultaneous insertion and disruption, do the PCR to amplify using primer outside the cleavage sites (F1 and R1) (Fig. 4). Run the amplified product on agarose gel (1%) and sequence it after cloning in TOPO blunt end vector. The gene manipulation can be verified as given in Table 3.

4.5 Validation of Deletion/Disruption/Insertion in the Genome

Transformants of the desired phenotypes can be evaluated by analyzing the RNA and protein levels.

1. To evaluate the RNA expression level, perform RT-qPCR by using one primer outside DSB while other within the DSB.
2. To assess the protein, perform an immunoblot using an antibody against the relevant protein. This can be done by dot blot. Blot the cell lysate from different transformants on nitrocellulose membrane along with control protein and probe it with an antibody against the protein of interest.

5 Notes

1. Web-based tools for selecting target sequences leads to multiple target sequence options. Since different tools run with an independent algorithm, it is recommended to use more than one tool during selection of the target sequences.
2. It is recommended to use a high fidelity DNA polymerase for all PCR reactions.
3. For repeated rounds of genome editing, plasmids harboring gRNA should be looped out of the cells. This could be achieved by growing multiple generations of the transformants without selection pressure. Then a new gRNA vector can be transformed to introduce the next round of modification in the

genome. After the final modification, both vectors can be removed in the absence of selection pressure to generate a strain free of plasmids.

4. It is recommended to remove PAM site from the donor sequence to prevent nonspecific Cas9-mediated cleavages.
5. The repair template for HDR can also deliver as a part of an expression vector.

References

1. Song J-W, Jeon E-Y, Song D-H, Jang H-Y, Bornscheuer UT, Oh D-K, Park J-B (2013) Multistep enzymatic synthesis of long-chain α,-ω-dicarboxylic and ω-hydroxycarboxylic acids from renewable fatty acids and plant oils. Angew Chem Int Ed 52(9):2534–2537
2. Keasling JD, Chou H (2008) Metabolic engineering delivers next-generation biofuels. Nat Biotech 26(3):298–299
3. Schirmer A, Rude MA, Li X, Popova E, del Cardayre SB (2010) Microbial biosynthesis of alkanes. Science 329(5991):559
4. Lee JW, Na D, Park JM, Lee J, Choi S, Lee SY (2012) Systems metabolic engineering of microorganisms for natural and non-natural chemicals. Nat Chem Biol 8(6):536–546
5. Ostergaard S, Olsson L, Nielsen J (2000) Metabolic engineering of *Saccharomyces cerevisiae*. Microbiol Mol Biol Rev 64(1):34–50
6. Lartigue C, Vashee S, Algire MA, Chuang RY, Benders GA, Ma L, Noskov VN, Denisova EA, Gibson DG, Assad-Garcia N, Alperovich N, Thomas DW, Merryman C, Hutchison CA 3rd, Smith HO, Venter JC, Glass JI (2009) Creating bacterial strains from genomes that have been cloned and engineered in yeast. Science 325(5948):1693–1696
7. Gibson DG, Young L, Chuang RY, Venter JC, Hutchison CA 3rd, Smith HO (2009) Enzymatic assembly of DNA molecules up to several hundred kilobases. Nat Methods 6 (5):343–345
8. Zhu B, Cai G, Hall EO, Freeman GJ (2007) In-fusion assembly: seamless engineering of multidomain fusion proteins, modular vectors, and mutations. BioTechniques 43(3):354–359
9. Mali P, Yang L, Esvelt KM, Aach J, Guell M, DiCarlo JE, Norville JE, Church GM (2013) RNA-guided human genome engineering via Cas9. Science 339(6121):823–826
10. Sander JD, Joung JK (2014) CRISPR-Cas systems for editing, regulating and targeting genomes. Nat Biotech 32(4):347–355
11. La Russa MF, Qi LS (2015) The new state of the art: Cas9 for gene activation and repression. Mol Cell Biol 35(22):3800–3809

Chapter 11

Alkaline and Alkaline-Oxidative Pretreatment and Hydrolysis of Herbaceous Biomass for Growth of Oleaginous Microbes

Jacob D. Crowe, Muyang Li, Daniel L. Williams, Alex D. Smith, Tongjun Liu, and David B. Hodge

Abstract

This chapter describes methods for generation of hydrolysates amenable to conversion to microbial lipids from herbaceous lignocellulosic biomass utilizing either mild alkali pretreatment with NaOH or alkaline hydrogen peroxide pretreatment with NaOH and H_2O_2. This pretreatment is followed by enzymatic hydrolysis of the plant cell wall polysaccharides to yield hydrolysates. These hydrolysates are composed primarily of the monosaccharides glucose and xylose as well as acetate and phenolic monomers that may all serve as a source of renewable carbon to produce microbial lipids. Application of these mild pretreatment conditions minimizes the generation of inhibitors, enabling microbial cultivations to often be performed without the need for detoxification.

Key words Alkaline pretreatment, Plant cell wall deconstruction, Cellulosic sugars, Alkaline hydrogen peroxide pretreatment

1 Introduction

The carbon contained in the cell wall polymers of plants (i.e., lignocellulosic biomass) is a potential as a renewable feedstock for the production of fuels and chemicals such as microbial lipids. Due to the "recalcitrance" of plant cell walls set by their higher order structure, lignocellulosic biomass must be subjected to chemical, physical, thermal, and/or biological treatments to yield intermediates that are amenable to biocatalytic conversion routes. A chemical pretreatment followed by enzymatic hydrolysis of the cell wall polysaccharides is one well-recognized route to "deconstructing" the cell wall to yield these intermediates, with a diverse range of pretreatment chemistries available [1].

Venkatesh Balan (ed.), *Microbial Lipid Production: Methods and Protocols*, Methods in Molecular Biology, vol. 1995, https://doi.org/10.1007/978-1-4939-9484-7_11,

Alkaline delignification of woody biomass (i.e., the Kraft process) is one of the primary industrial processes for producing cellulose-rich pulps for papermaking. While effective alkaline delignification of woody biomass requires elevated temperature (i.e., >150 °C), pressure, and alkali loadings, when applied to nonwoody feedstocks such as grasses (i.e., the Poaceae or Gramineae), substantially milder conditions can be employed [2, 3]. Additionally, alkali-labile ferulate cross-links within and between grass lignins and hemicelluloses can be broken by alkaline pretreatments under mild conditions resulting in significant solubilization of xylan and lignin relative to dicots [3]. Consequently, when applied as a pretreatment for the production of cellulosic sugars, lower alkali loadings and temperatures can be employed for herbaceous grasses and mild alkali treatments have been proposed as a standalone pretreatment for grasses including corn stover [4, 5], wheat straw [6], sweet sorghum [7], and switchgrass [8] as well as a "deacetylation" step prior to mechanical refining for corn stover as developed at NREL either coupled to dilute acid post-treatment or to mechanical disc refining [9].

Addition of H_2O_2 during NaOH pretreatment at mild temperatures, while maintaining the pH at the pK_a (i.e., 11.5) of H_2O_2 to maximize peroxide dissociation can yield effective lignin solubilization and mild lignin oxidation; presumably due to oxidation by reactive oxygen species. This alkaline hydrogen peroxide (AHP) pretreatment has been explored for a range of grasses [10, 11]. Notable features of AHP pretreatment of grasses include generation of relatively high biomass digestibilities when oxidant loading is not limiting as demonstrated for a range of feedstocks (Fig. 1).

These alkaline and alkaline-oxidative pretreatments offer the potential for the generation of sugar- and aromatic monomer-containing hydrolysates. One important feature of these mild alkaline and alkaline-oxidative pretreatments is that, unlike many other pretreatment chemistries, minimal inhibitors are generated during the pretreatment stage other than extractives and compounds saponified from cell wall biopolymers (i.e., acetate, hydroxycinnamic acids). Consequently, concentrated hydrolysates derived from alkaline and alkaline-oxidative pretreatments have been demonstrated to be capable of biological conversion to biofuels at high titers without any detoxification [12–14]. As described previously, alkaline pretreatments result in the solubilization of ester-linked phenolic compounds that are abundant in grasses including the hydroxycinnamic acids *p*-coumaric acid and ferulic acid (Fig. 2). While inhibitory at high enough concentrations, these compounds in alkaline and AHP pretreatment liquors have also been shown to serve as suitable substrates for biological conversion [15].

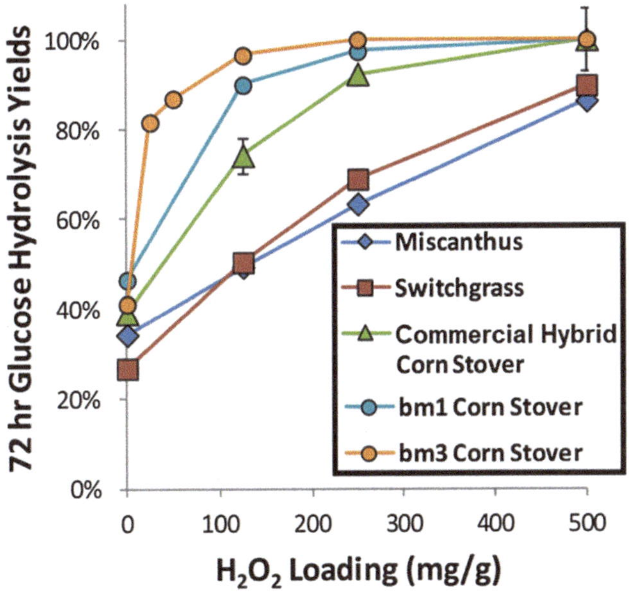

Fig. 1 Example hydrolysis yields for diverse bioenergy grasses subjected to AHP pretreatment at various H_2O_2 loadings. Biomass was treated at 2% (w/v) solid loading, at 30 °C and 180 rpm orbital shaking for 24 h at a pH of 11.5. Washed, pretreated biomass was subjected to enzymatic hydrolysis at pH 4.8, 50 °C with 30 mg protein/g biomass protein loading of CTec2 with orbital shaking at 180 rpm for 72 h

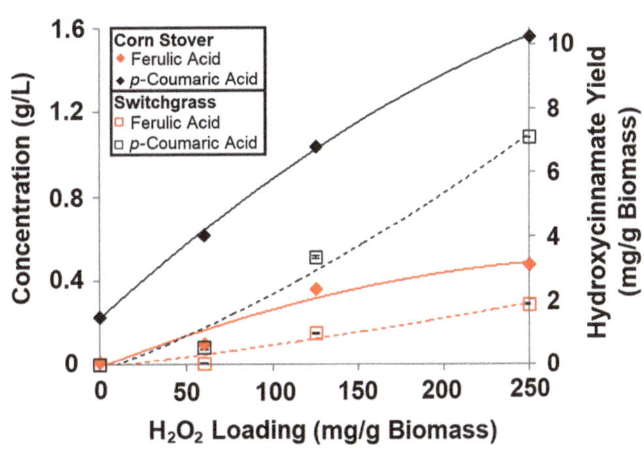

Fig. 2 Hydroxycinnamate concentrations and yields in AHP pretreatment liquors from corn stover and switchgrass as a function of H_2O_2 loading

2 Materials

2.1 Feedstock Preparation

1. Herbaceous lignocellulosic biomass feedstock.
2. Temperature controlled isothermal oven capable of holding biomass at 105 °C.
3. US Standard Sieve series with 400 mesh screen.
4. Desiccator for biomass storage and cooling after oven drying.
5. Willey mill or hammer mill for particle size reduction.

2.2 Alkaline Hydrogen Peroxide Pretreatment

1. 250-mL Erlenmeyer flasks as pretreatment vessels.
2. NaOH solution: Prepare a 500 mL 5 M NaOH solution by adding 100 g of NaOH in a 500-mL volumetric flask. Place volumetric flask in ice bath and add deionized water to volumetric flask slowly, with intermediate mixing to limit heat released from exothermic mixing.
3. Hydrogen peroxide: 30% (w/w) H_2O_2 in aqueous solution stock reagent.
4. Sulfuric acid: 72% (w/w) in aqueous solution stock reagent.
5. Parafilm wax and aluminum foil.
6. Orbital shaker incubator with temperature control from 25 to 50 °C. Racks for holding 250-mL Erlenmeyer flasks.

2.3 Alkaline Pretreatment

1. NaOH solution: Prepare a 1 L 0.25 M NaOH solution by adding 10 g of NaOH in a 1-L volumetric flask. Place volumetric flask in ice bath and add deionized water to volumetric flask slowly, with intermediate mixing to limit heat released from exothermic mixing.
2. Water bath capable of holding 250-mL Erlenmeyer flasks, with temperature control to 80 °C.

2.4 Enzymatic Hydrolysis

1. Citric acid buffers: Prepare a 500-mL stock solution of 1 M citric acid by adding 90.06 g reagent grade anhydrous citric acid to a 500-mL volumetric flask along with deionized water. 1 M sodium citrate conjugate base is prepared in a similar manner by addition of 107.06 g of anhydrous sodium citrate.
2. Commercial enzyme cocktails of either cellulase (e.g., Cellic CTec2®, Novozymes, Inc. or Accellerase TRIO®, DuPont Industrial Biosciences) xylanase (e.g., Cellic HTec2®, Novozymes, Inc. or Accellerase XY®), and pectinase (e.g., Pectinex® Ultra SPL, Novozymes, Inc.).
3. 50 mL centrifuge tubes for solid–liquid separation after enzymatic hydrolysis.
4. 0.22 μm Stericup sterile filter and either an aspirator system or vacuum pump for sterilization of hydrolysate.

3 Methods

3.1 Feedstock Preparation

1. Determine moisture content by gravimetric difference after oven drying a known mass of feedstock overnight in a 105 °C temperature controlled oven, followed by cooling in a desiccator.
2. Refrain from using high-ash content material if possible; sieve material using a 400-mesh screen to remove ash prior to milling.
3. Mill material to reduce particle size if using whole biomass as a starting feedstock. Multiple particle sizes are acceptable for generating a substrate fraction as long as particle sizes are small enough to allow adequate mixing.

3.2 Alkaline Hydrogen Peroxide Pretreatment

1. Depending upon the amount of substrate needed, prepare multiple flasks using the following procedure for bench scale pretreatment.
2. Add 10 g of milled feedstock to a 250 mL-Erlenmeyer flask.
3. Prepare 1 L of deionized water with pH adjusted to 12 through addition of 5 M NaOH (*see* **Notes 1** and **2**).
4. A total solids content of 15% (mass biomass solids per liquid volume) will be used for pretreatment, with 10 g of biomass with 56.7 mL liquid yielding a total solution mass of 66.7 g.
5. Add 50 mL of pH 12 adjusted water to flask using either a disposable serological pipette or a graduated cylinder. Using a metal spatula, mix biomass until all material has been presoaked with solution (*see* **Note 3**). Allow presoaked solution to stand for at least 5 min.
6. Measure the pH of the presoaked mixture using a general electrode probe (*see* **Notes 2** and **4**). Adjust pH to around pH 11 through addition of small aliquots of 5 M NaOH (*see* **Note 5**). Record the total volume of NaOH added.
7. A H_2O_2 loading of 0.125 g H_2O_2/g biomass is desired for effective delignification of biomass. Using a 30% (w/w solution) stock, add 3.754 mL of H_2O_2 to biomass multiple aliquots (*see* **Note 6**). Mix after each addition.
8. After addition of peroxide, measure pH. Adjust to pH 11.5 using 5 M NaOH. Record volume added after desired pH is achieved.
9. If total volume is less than 66.7 mL, add deionized water until volume is achieved. If volume is over desired amount, then note changes to actual solids loading of pretreatment (*see* **Notes 5** and 7).

10. Use Parafilm to seal the top of each flask and prevent evaporation. Parafilm also allows for volume expansion due to oxygen evolution during pretreatment without altering pressure of pretreatment conditions.
11. Add flasks to an incubator with orbital shaking set to 30 °C, with mixing at 180 rpm. The pretreatment will be performed at these incubator conditions.
12. Allow pretreatment to occur. Visible bleaching should be observed, as well as oxygen evolution manifested as bubbles. If observations are not made during initial 30 min of pretreatment, remove flasks and measure pH. Adjust to 11.5 and note volume additions.
13. After 3-h total mixing time, remove flasks and measure pH (*see* **Note 8**). Adjust pH with 5 M NaOH and return to mixing. Repeat for 6-h, 9-h, and 12-h pretreatment times.
14. After 24 h total pretreatment time, remove flasks from incubator (*see* **Note 9**). Adjust pH to a range of 6–7 using 72% (w/w) H_2SO_4 and dilute pretreated slurry with an appropriate volume of deionized water in preparation for enzymatic hydrolysis at 10% solids loading (mass original dry biomass per total liquid volume; *see* Subheading 3.4).

3.3 Alkaline Pretreatment

1. Prepare 10 g of milled feedstock and add to a 250-mL Erlenmeyer flask.
2. Using a pipette or graduated cylinder, add 100 mL of 0.25 M NaOH solution to the flask, yielding a total solids content of 10% (mass biomass solids per liquid volume) for pretreatment with an NaOH loading of 100 mg NaOH per g biomass. Mix solution with spatula or gentle swirling of flask.
3. Seal flasks with Parafilm followed by aluminum foil to prevent evaporation.
4. Immerse flasks in water bath set at 80 °C. Ensure water bath water level is sufficient to keep most of the flasks immersed (*see* **Note 10**).
5. Total pretreatment time will be 1 h, with intermittent mixing every 10 min to ensure even pretreatment.
6. After 1 h pretreatment time, remove flasks and allow to cool for 10 min (*see* **Note 11**).
7. Adjust pH to a range of 6–7 using 72% (w/w) H_2SO_4 in preparation for enzymatic hydrolysis (*see* Subheading 3.4).

3.4 Enzymatic Hydrolysis

1. Add deionized water following a total volume addition of 10 mL liquid per 1 g of original dry biomass to prepare for enzymatic hydrolysis at 10% original mass to total volume) solids loading.

2. A 50 mM Na-citrate buffer (pH 5.2) will be used to maintain pH during duration of hydrolysis. Use 1 M citric acid and 1 M sodium citrate conjugate base stocks added to pretreatment slurry for buffering (*see* **Note 12**).
3. Adjust pH to 5.2 using 72% (w/w) H_2SO_4 and record the final hydrolysis liquid volume.
4. Multiple enzyme cocktails are acceptable for effective enzymatic hydrolysis of the pretreated slurry at pH 5.2 (*see* **Note 13**). Determine protein content of enzyme cocktail using an appropriate quantification method, and load enzyme to a total 30 mg/g glucan of material based on the original composition (*see* **Note 14**). If initial composition is not known, an assumption for 0.4 g glucan/g biomass may suffice, resulting in a loading of 12 mg/g biomass.
5. After addition of enzyme, seal flasks with Parafilm and cover Parafilm with a layer of aluminum foil. Set incubator temperature to 50 °C and allow enzymatic hydrolysis to proceed for at least 48 h. Ensure no pH drift during enzymatic hydrolysis (*see* **Note 15**).
6. After desired hydrolysis time has elapsed, remove flasks from incubator and transfer hydrolysate to centrifuge-capable tubes. Centrifuge solution at 8000 × *g* for 10 min for solid–liquid separation. Other techniques for solid–liquid separation are applicable here depending upon available equipment.
7. Transfer liquid fraction to 0.22 μm Stericup and vacuum filter to sterilize solution.
8. Hydrolysate liquor can now be utilized as a feedstock for oleaginous microbe production. An additional nitrogen source may be required as a supplement. Quantification of sugars and determination of sugar yields can be performed as reported in our previous work [16].

4 Notes

1. Deionized water is adjusted to pH 12.0 prior to addition of biomass to promote available liquid fraction for analysis of pH. Deionized water at pH 7.0 is absorbed into the biomass and results in no available liquid fraction for pH measurement. Refer to [17, 18] for clarification on biomass swelling.
2. A standard pH meter may be used for all pH adjustments. A four-point calibration with an expanded linear range in the high pH values was used, with pH calibration buffers of 4.01, 7.00, 10.00, and 12.45; however, a three-point calibration may be sufficient.

3. Care should be taken to ensure minimal biomass is lost during mixing, as biomass will stick the spatula or stir rod used. Gently scrape the spatula edges along the flask side to retain biomass stuck on spatula.
4. A general all-purpose pH probe is sufficient for determination of pH in biomass slurry; however, we found a spear-tip probe to be easier for determining a stable pH value of the solution.
5. Make sure to mix thoroughly and wait a short time after each addition of 5 M NaOH to not overshoot desired pH. If desired pH range is exceeded, use 72% (w/w) H_2SO_4 to adjust pH. The amount of 5 M NaOH to reach pH 11 will vary depending upon feedstock.
6. H_2O_2 loading was performed using the following calculations and assuming a density of 1.11 g/mL at 20 °C

$$10\ \text{g biomass} \times \frac{0.125\ \text{g}\ H_2O_2}{\text{g biomass}} \times \frac{\text{mL stock}}{0.3\ \text{mL}\ H_2O_2} \times \frac{\text{mL}\ H_2O_2}{1.11\ \text{g}\ H_2O_2}$$

7. If solids loading is significantly off due to excessive addition of acid or base, repeat procedure using a lower starting volume of pH 12 adjusted deionized water as outlined in **step 5**.
8. pH drift will occur during pretreatment due to the generation of acid groups in solution during oxidation. pH drift will be most significant in the initial 3-h pretreatment time.
9. Pretreatment for 24 h should be sufficient to degrade all H_2O_2 (Fig. 3) at low oxidant loadings. If H_2O_2 is still present (e.g., at higher oxidant loadings), incubation with a catalase can be performed prior to enzymatic hydrolysis as in our previous work [12].
10. Complete immersion of the flask should be avoided; however, the water level of water bath should be higher than the volume of pretreated slurry in the flask. In addition, evaporation will occur at 80 °C in the water bath and should be anticipated when heating and performing the alkaline pretreatment.
11. When removing flask from the water bath, take care and use a heat resistant oven glove. Avoid placing hot flasks directly on lab bench, and use either a rack or cardboard to insulate the glassware while cooling.
12. Acid and base additions were determined using the Henderson–Hasselbalch equation.
13. Multiple enzyme cocktails have been utilized successfully for effective hydrolysis of alkaline hydrogen peroxide pretreated biomass. Prior optimized cocktails using Accelerase 1000, xylanase, and Multifect Pectinase [19] have been determined. Newer Accelerase 1500 cocktails or Novozymes Cellic CTec2 and CTec3 may also be used instead. CTec cocktails should be

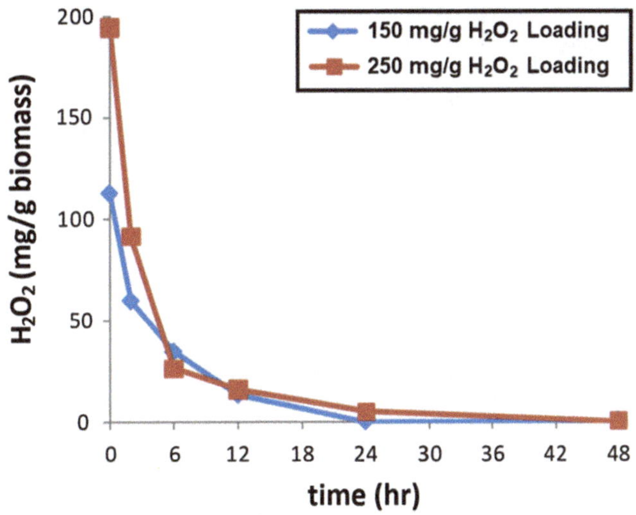

Fig. 3 Hydrogen peroxide decomposition during AHP pretreatment of corn stover

supplemented with HTec hemicellulose cocktails to facilitate hemicellulose depolymerization and improve glucose yields [20].

14. Composition analysis of the original biomass can be performed using the NREL two-stage acid hydrolysis outlined in [21].

15. Measure pH after 12-h hydrolysis time to ensure no pH drift. Adjust pH using 72% (w/w) H_2SO_4 or 5 M NaOH if necessary.

References

1. Ong RG, Chundawat SPS, Hodge DB, Keskar S, Dale BE (2014) Linking plant biology and pretreatment: understanding the structure and organization of the plant cell wall and interactions with cellulosic biofuel production. In: McCann MC, Buckeridge MS, Carpita NC (eds) Plants and bioenergy. Springer, New York, pp 231–253
2. Stoklosa RJ, Hodge DB (2015) Fractionation and improved enzymatic deconstruction of hardwoods with alkaline delignification. Bioenergy Res 8(3):1224–1234
3. Stoklosa RJ, Hodge DB (2012) Extraction, recovery, and characterization of hardwood and grass hemicelluloses for integration into biorefining processes. Ind Eng Chem Res 51 (34):11045–11053
4. Karp EM, Donohoe BS, O'Brien MH, Ciesielski PN, Mittal A, Biddy MJ, Beckham GT (2014) Alkaline pretreatment of corn stover: bench-scale fractionation and stream characterization. ACS Sustain Chem Eng 2 (6):1481–1491
5. Li M, Heckwolf M, Crowe JD, Williams DL, Magee TD, Kaeppler SM, de Leon N, Hodge DB (2015) Cell-wall properties contributing to improved deconstruction by alkaline pre-treatment and enzymatic hydrolysis in diverse maize (*Zea mays* L.) lines. J Exp Bot 66(14):4305–4315
6. Pavlostathis SG, Gossett JM (1985) Alkaline treatment of wheat straw for increasing anaerobic biodegradability. Biotechnol Bioeng 27 (3):334–344
7. Wu L, Arakane M, Ike M, Wada M, Takai T, Gau M, Tokuyasu K (2011) Low temperature alkali pretreatment for improving enzymatic digestibility of sweet sorghum bagasse for ethanol production. Bioresour Technol 102 (7):4793–4799
8. Karp EM, Resch MG, Donohoe BS, Ciesielski PN, O'Brien MH, Nill JE, Mittal A, Biddy MJ,

Beckham GT (2015) Alkaline pretreatment of switchgrass. ACS Sustain Chem Eng 3 (7):1479–1491

9. Chen X, Kuhn E, Jennings EW, Nelson R, Tao L, Zhang M, Tucker MP (2016) DMR (deacetylation and mechanical refining) processing of corn stover achieves high monomeric sugar concentrations (230 g L^{-1}) during enzymatic hydrolysis and high ethanol concentrations (>10% v/v) during fermentation without hydrolysate purification or concentration. Energ Env Sci 9(4):1237–1245
10. Li M, Foster C, Kelkar S, Pu Y, Holmes D, Ragauskas A, Saffron C, Hodge DB (2012) Structural characterization of alkaline hydrogen peroxide pretreated grasses exhibiting diverse lignin phenotypes. Biotechnol Biofuels 5(1):38
11. Gould JM (1985) Studies of the mechanism of alkaline peroxide delignification of agricultural residues. Biotechnol Bioeng 27(3):225–231
12. Banerjee G, Car S, Williams DL, López Meza S, Walton J, Hodge DB (2012) Scale-up and integration of alkaline hydrogen peroxide pretreatment, enzymatic hydrolysis, and ethanolic fermentation. Biotechnol Bioeng 109(4):922–931
13. Sato TK, Liu T, Parreiras LS, Williams DL, Wohlbach DJ, Bice BD, Breuer RJ, Ong IS, Qin L, Bulsalacchi D, Deshpande S, Daum C, Gasch AP, Hodge DB (2014) Harnessing genetic diversity in *Saccharomyces cerevisiae* for fermentation of xylose in hydrolysates of alkaline hydrogen peroxide-pretreated biomass. Appl Environ Microb 80(2):540–554
14. Liu T, Williams DL, Pattathil S, Li M, Hahn MG, Hodge DB (2014) Coupling alkaline pre-extraction with alkaline-oxidative post-treatment of corn stover to enhance enzymatic hydrolysis and fermentability. Biotechnol Biofuels 7(1):48
15. Le RK, Wells T, Das P, Meng X, Stoklosa RJ, Bhalla A, Hodge DB, Yuan JS, Ragauskas AJ (2017) Conversion of corn stover alkaline pre-treatment waste streams into biodiesel via Rhodococci. RSC Adv 7(7):4108–4115
16. Hodge DB, Karim MN, Schell DJ, McMillan JD (2008) Soluble and insoluble solids contributions to high-solids enzymatic hydrolysis of lignocellulose. Bioresour Technol 99 (18):8940–8948
17. Williams DL, Hodge DB (2014) Impacts of delignification and hot water pretreatment on the water induced cell wall swelling behavior of grasses and its relation to cellulolytic enzyme hydrolysis and binding. Cellulose 21 (1):221–235
18. Williams DL, Crowe JD, Ong RG, Hodge DB (2017) Water sorption in AFEX- and AHP-pretreated grasses as a predictor of enzymatic hydrolysis yields. Bioresour Technol 245:242–249
19. Banerjee G, Car S, Scott-Craig JS, Hodge DB, Walton JD (2011) Alkaline peroxide pretreatment of corn stover: effects of biomass, peroxide, and enzyme loading and composition on yields of glucose and xylose. Biotechnol Biofuels 4(1):16
20. Crowe JD, Zarger RA, Hodge DB (2017) Relating nanoscale accessibility within plant cell walls to improved enzyme hydrolysis yields in corn stover subjected to diverse pretreatments. J Agr Food Chem 65(39):8652–8662
21. Sluiter A, Hames B, Ruiz R, Scarlata C, Sluiter J, Templeton D, Crocker D (2012) Determination of structural carbohydrates and lignin in biomass. NREL/TP-510-42618

Chapter 12

Laboratory Conversion of Cultivated Oleaginous Organisms into Biocrude for Biofuel Applications

Eboibi Blessing, Umakanta Jena, and Senthil Chinnasamy

Abstract

Hydrothermal liquefaction (HTL) is a thermochemical process for the wet conversion of oleaginous microorganisms and other biomass and carbon rich feedstocks to biofuels under subcritical conditions. It is a novel green process that produces biocrude as a primary product along with other by-products which include gases, aqueous phase coproduct (ACP) and solid residues. Here we describe in detail the protocols for the conversion of biomass to biocrude through HTL and separation, quantification and analyses of HTL products.

Key words Aqueous phase, Biocrude, Biofuel, Biomass, Dichloromethane, Energy recovery, Hydrothermal liquefaction, Microalgae, Oleaginous, Solid residue, Solvents

1 Introduction

Hydrothermal liquefaction (HTL) is an advanced thermochemical conversion process of wet biomass obtained from oleaginous algae, yeast, bacteria and fungi and other carbon rich feedstocks with 80–90% moisture content into biocrude using subcritical water. Conventional processes convert biomass derived oils/lipids to biodiesel and carbohydrates to bioethanol and the yield and recovery of biodiesel and bioethanol are dependent on the lipid or carbohydrate content of the biomass. In comparison to the conventional transesterification process used for conversion of lipids to biodiesel and the fermentation process used for the conversion of carbohydrates to bioethanol, the HTL process converts the entire whole-cell biomass to hydrocarbons where the carbon and hydrogen are derived from lipids, carbohydrates and proteins [1]. Importantly, it is an energy saving process compared to other thermal conversion processes such as pyrolysis, as it avoids energy-intensive steps involved in the drying of biomass feedstocks for the conversion into fuel products. Unlike biodiesel produced from the transesterification process and bioethanol produced through fermentation

Venkatesh Balan (ed.), *Microbial Lipid Production: Methods and Protocols*, Methods in Molecular Biology, vol. 1995, https://doi.org/10.1007/978-1-4939-9484-7_12,

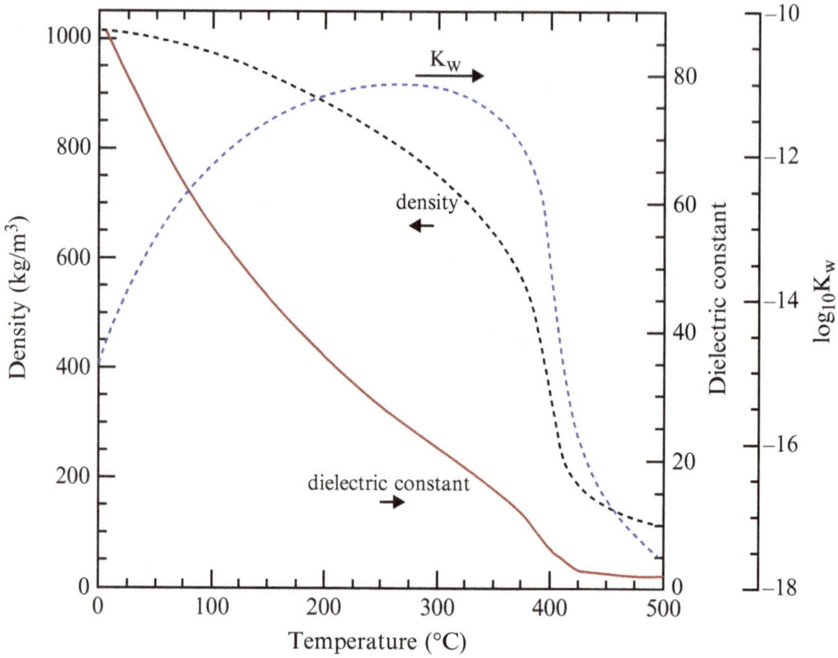

Fig. 1 Properties of water as a function of temperature (Adapted from Kruse and Dahmen [5])

processes, HTL biocrude can be used along with petrocrude in the existing oil refineries to produce various fuel fractions such as green petrol/gasoline, green diesel, and green jet fuels. HTL is a high pressure thermal conversion process carried out at the following reaction conditions: temperature 100 °C to 374 °C, pressure 5–25 MPa (50–250 bar), reaction time 5–60 min, solids loading rate 10–20% (w/w) with or without catalysts [2–4]. Hot compressed water under subcritical condition has significant changes in properties which include decrease in dielectric constant and increase in ionic properties (Fig. 1). As temperature increases, the density decreases leading to weaker hydrogen bonds, and the density decrease is severe as water approaches the critical temperature [5]. Decrease in dielectric constant in the subcritical regime leads to change in polarity of water. Subcritical water can dissolve numerous organic compounds that are not readily dissolved in water at room temperature. Subcritical water shows both acidic and basic properties simultaneously due to significant increase in the ionic product (K_w). Breakdown of macromolecules becomes easier in the highly reactive subcritical water under HTL conditions compared to ambient water, because it enhances the hydrolysis, decarboxylation, depolymerization, and repolymerization/condensation reactions to produce biocrude (also called "bio-oil") along with gases, water solubles and char as coproducts.

HTL is a novel technology to produce renewable liquid transportation fuels in the future. This chapter describes the procedures and protocols involved in the laboratory conversion of oleaginous biomass using HTL technology and separation, quantification and analyses of HTL products.

2 Materials

Store the dried biomass feedstock samples in the form of powder in refrigerator until further use. Prepare the feedstock with distilled water. Strictly follow safe disposal of wastes.

Feed stocks:	Oleaginous biomass (e.g., microalgae, yeast, bacteria, fungi)
Solvents:	Dichloromethane or Acetone
Homogeneous catalysts:	Na_2CO_3, KOH, K_2CO_3, NaOH, formic acid, acetic acid
Heterogeneous catalysts:	HZSM-5, Pd/C, Pt/C, Ru/C, Ni/SiO_2-Al_2O_3, CoMo/γ-Al_2O_3 (sulfided), Pd/γ-Al_2O_3, Pt/γ-Al_2O_3, Mo_2C, and MoS_2
Reactor:	A custom made high pressure hydrothermal liquefaction batch reactor manufactured by Parr Instruments Co. or any other suppliers with all safety arrangements to treat biomass at subcritical reaction conditions (100–374 °C and 50–250 bar pressure)
Analytical equipment:	Bomb calorimeter, elemental CHNS analyzer (for measuring carbon, hydrogen, nitrogen, sulfur), gas chromatography–mass spectrometry (GC-MS) for biocrude analysis, gas chromatography for analysis of gas phase

3 Methods

3.1 Hydrothermal Liquefaction

3.1.1 Solids Concentration in the Feedstock

1. Mix 10–30 *g* of dry biomass of oleaginous algae/yeast/bacteria/fungi in 100 mL of deionized water in a beaker to prepare a homogeneous suspension of feedstock with 10–30% w/w solids.
2. Loading of 10–30% w/w of solids has been found to be optimum for the hydrothermal conversion process (*see* **Note 1**).
3. Load the feedstock slurry into the HTL reactor.

3.1.2 Volume of Feedstock in the Reactor

1. Ensure that the volume of feedstock slurry added into the reactor would expand 95% of the rector volume at the reaction temperature.

3.1.3 Catalyst

1. Carry out the HTL runs with or without catalysts (*see* **Note 2**).
2. Add catalysts (homogeneous or heterogeneous) at the rate of 1–20% w/w (5% w/w is considered ideal) of the solids loaded in the reactor [3, 6–8].

3.1.4 Initial Pressure

1. Seal the reactor after adding the feedstock and catalyst.
2. Purge and pressurize the reactor with nitrogen gas to an initial pressure of 2 MPa or 20 bar to prevent vaporization of water during the reaction (*see* **Note 3**).

3.1.5 Reactor Operation

1. Heat the reactor to the predefined reaction temperature using an electrical heating jacket (*see* **Note 4**).
2. Stir the reactor continuously at 300 rpm to maintain homogeneity of reactants.
3. Maintain the reactor at preset temperature (200 to 350 °C ± 2 °C) for a predefined reaction time (5–60 min).
4. Switch off the reactor and allow it to cool to ambient temperature by circulating tap water through the cooling coils, and in some cases by submerging (for microreactors) the reactor in to a water-bath (*see* **Note 5**).

3.1.6 Product Recovery from Reaction Mixture

1. Collect the gas phase in Tedlar air sampling bags through the reactor gas vent after cooling the reactor.
2. Open the reactor and transfer the reaction product mixture (*see* **Note 6**) to a separating funnel.
3. Wash the reactor walls and parts (magnetic stirrer) with deionized water and solvent such as dichloromethane (DCM) or acetone in order to recover residual product (*see* **Note 7**).
4. Add the washed fractions subsequently to the separating funnel containing the rest of the reaction mixture.
5. Agitate the separating funnel for at least 2 min and leave it to stand for 12 h after adding required quantity of solvent (at least in ration 1:1), in order to allow phase separation (*see* **Note 8**).
6. After phase separation, remove various product phases by following a series of filtration and extraction protocols (*see* **Note 9**).
7. Decant the top phase consisting of light fraction biocrude first.
8. Filter the remaining two phases under vacuum to obtain water soluble components (aqueous phase coproduct), and water insoluble components (solid residues and heavy fraction biocrude).
9. The water soluble components obtained as filtrate is referred as aqueous phase coproduct (ACP).

10. Wash all the water insoluble components retained on the filter and wall of the reactor thoroughly with suitable solvent (Dichloromethane/Acetone) to extract the hydrocarbons.
11. Vacuum filter the solvent containing hydrocarbons to remove the solid particles.
12. Oven dry the solids [9, 10] recovered after the solvent extraction at 105 °C for 24 h to obtain solid residue (SR).
13. Vacuum distill the solvent fraction with extracted hydrocarbon at a temperature above the boiling point of the solvent to separate the heavy fraction biocrude.
14. Separate light fraction biocrude and heavy fraction biocrude.
15. Continuously flush the oil sample with nitrogen for 24 h to remove the residual solvents.
16. Weigh the biocrude to quantify biocrude yield and recovery, after the complete removal of the solvent (*see* **Note 10**).
17. Quantify the water soluble organics [9, 10] by evaporating a known quantity of aqueous phase coproduct (ACP) at 70 °C for 24 h.
18. Record the weight of organic and inorganic fractions in the aqueous phase coproduct (ACP) treated at 70 °C.
19. Treat this mixture with organic and inorganic compounds at 550 °C for 4 h to quantify ash content.
20. Subtract the ash content from the previous fraction containing both organics and ash to quantify the organics present in the aqueous phase coproduct (ACP).
21. A general product separation protocol performed in the laboratory set up is shown in Fig. 2.

3.1.7 Product Yield

Gas Phase Yield

1. Estimate the weight of the Tedlar bag containing the gas phase by calculating the difference in the final and initial weight of Tedlar bag.
2. Express the gas phase yield as percentage weight relative to initial algae biomass charged to reactor.
3. Alternatively, calculate the gas yield by noting the difference in the final and initial weight of reactor after the treatment (*see* **Note 11**).

Estimation of the Yield of Biocrude, Solid Residue and Aqueous Phase Coproduct (ACP)

1. The gravimetric yields of biocrude, solid residue and aqueous phase coproduct (ACP) are determined using Eq. 1.

$$\Upsilon_{\mathrm{i}} = \frac{M_{\mathrm{i}}}{M_{\mathrm{f}}} \times 100\% \tag{1}$$

where Υ is the yield; i represents the biocrude, solid residue, or aqueous phase coproduct (ACP); f is the feedstock; and M represents mass (grams).

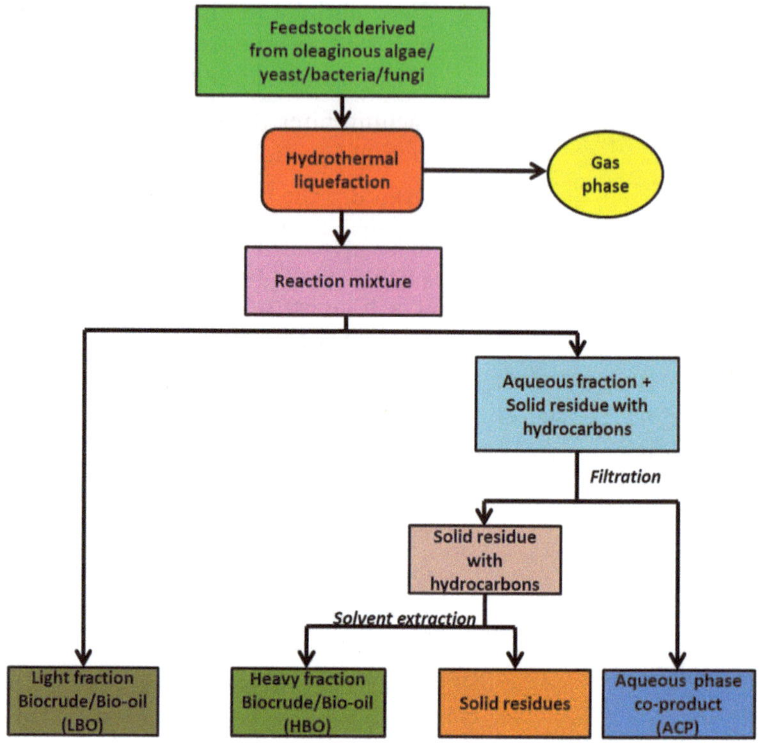

Fig. 2 HTL process and its various products

3.1.8 *Analyses*

Proximate and Ultimate Analysis

1. Determine the composition of elemental carbon (C), hydrogen (H), nitrogen (N), sulfur (S) of the biocrude, solid residues and aqueous phase components using an elemental analyzer in accordance with ASTM D-5291 and D-3176 methods.
2. Calibrate the elemental analyzer using known standards like BBOT standard (2,5-Bis(5-*tert*-butyl-benzoxazol-2yl)thiophene) or sulfanilamide.
3. Weigh known quantities of samples and standards in triplicate in tin capsules.
4. Combust the samples in excess oxygen in the elemental analyzer.
5. Calculate the elemental composition of samples by analysing the concentrations in the product gas.
6. Calculate the atomic ratios such as H/C, O/C and N/C using the elemental composition.
7. The oxygen content could be determined by difference, using Eq. 2.

$$\text{Oxygen content}\left(\frac{w}{w}\%\right) = 100 - (\mathrm{C} + \mathrm{H} + \mathrm{N} + \mathrm{S}) \quad (2)$$

Higher Heating Value

1. Measure higher heating value (HHV) by using an isoperibol bomb calorimeter following ASTM D-5865 and D-4809 methods.
2. Alternatively, calculate HHV using the data obtained from the elemental analysis in to a unified correlation equation proposed by Channiwala and Parikh [11] (Eq. 3a), Garcia Alba et al. [2]—Boie's formula (Eq. 3b); and Valdez et al. [12]—Dulong's formula (Eq. 3c).

$$\begin{aligned}\text{HHV (MJ/kg)} &= 0.3491\text{C} + 1.1783\text{H} + 0.1005\text{S} \\ &\quad - 0.1034\text{O} - 0.0151\text{N} - 0.0211\text{A}\end{aligned} \tag{3a}$$

$$\begin{aligned}\text{HHV (MJ/kg)} &= 0.516\text{C} + 1.16225\text{H} - 0.1109\text{O} \\ &\quad + 0.0628\text{N}\end{aligned} \tag{3b}$$

$$\text{HHV (MJ/kg)} = 0.0338\text{C} + 1.428\left(\text{H} - \frac{\text{O}}{8}\right) + 0.095\text{S} \tag{3c}$$

where C, H, N, S, O, and A are the mass percentages of carbon, hydrogen, nitrogen, sulfur, oxygen and ash, respectively, on a dry weight basis.

Atomic Ratios

1. Calculate the atomic molar ratios using the weight percentages of C, H, N, S and O data and the molecular weight of a specific element, shown in Eqs. 4a, 4b, and 4c.

$$\text{H/C} = \frac{\text{H wt\%} \times \text{MW}_\text{C}}{\text{C wt\%} \times \text{MW}_\text{H}} \tag{4a}$$

$$\text{O/C} = \frac{\text{O wt\%} \times \text{MW}_\text{C}}{\text{C wt\%} \times \text{MW}_\text{O}} \tag{4b}$$

$$\text{N/C} = \frac{\text{N wt\%} \times \text{MW}_\text{C}}{\text{C wt\%} \times \text{MW}_\text{N}} \tag{4c}$$

Estimation of Energy Input and Energy Recovery

1. Knowing the amount of energy input to HTL is necessary to assess if the process is a net energy producer.
2. Estimate the energy input required to produce a unit of biocrude from microalgae biomass by using the enthalpies of saturated liquids (h_f), shown in Eq. 5.

$$E_{\text{HTL}} = \frac{\left[\left(\left(h_f(T_2) - h_f(T_1)\right) \times M_{\text{water}}\right] + 0.5 \times \left[\left(\left(h_f(T_2) - h_f(T_1)\right) \times M_{\text{algae}}\right]}{M_{\text{biocrude}}} \tag{5}$$

where E_{HTL} represents the HTL energy input, h_f represents the enthalpies of water at base temperature assumed to be room temperature (T_1), and T_2 is the predefined reaction temperature. M_{algae} and M_{biocrude} are the masses (kg) of initial algae and biocrude, respectively.

3. The energy recovery (ER) and the energy consumption ratio (ECR) are terms used to estimate the HTL process energy balance [13].
4. Calculate the ER using (Eq. 6) from the yield and HHVs data, and expressed as percentage of a specific product.

$$\mathrm{ER} = \frac{\text{Higher heating value of product (MJ/kg)} \times \text{Mass of product (kg)}}{\text{Higher heating value of feedstock (MJ/kg)} \times \text{Mass of feedstock (kg)}} \times 100\% \quad (6)$$

5. The energy consumption ratio (ECR) is normally estimated (using Eq. 7) based on the required energy input for liquefaction (E_{HTL}) and the available energy of biocrude produced. The available energy is the HHV (MJ/kg) of biocrude, which is either estimated using a bomb calorimeter or the CHNSO data. An ECR value >1 means that the process energy balance is negative, below 1 means that there is a net energy gain.

$$\mathrm{ECR} = \frac{\text{Amount of energy required for liquefaction (MJ/kg)}}{\text{Availbale energy of biocrude produced (MJ/kg)}} \quad (7)$$

Estimation of Carbon Recovery

1. Calculate carbon recovery or CR (%), in the product fractions (i.e., biocrude and solid residues) as per the equation given below where C indicates carbon (Eq. 8):

$$\mathrm{CR} = \frac{\text{C\%of the product} \times \text{Mass of product (kg)}}{\text{C\%of the feedstock} \times \text{Mass of feedstock (kg)}} \times 100\% \quad (8)$$

3.1.9 *Identification of Chemical Compounds*

GC-MS Analyses: Biocrude and Aqueous Phase

1. Prepare 2.5% v/v solution of biocrude sample diluted with a solvent (e.g., dichloromethane).
2. For aqueous phase coproduct (ACP) analyses, add 10% v/v solution of the resultant aqueous phase to a mixture of acetone and methanol at a ratio of 1:1(v/v).
3. Carry out the analyses of HTL product samples using a gas chromatography–mass spectrometer (GC-MS) with a selective detector, and a HP-5 MS column of 30 m length × 0.25 mm i.d.
4. Maintain the inlet temperature at 230 °C and detector temperature at 280 °C.
5. Inject a sample size of 1 μL and maintain the flow rate of helium at 1 mL/min.

6. Maintain the oven temperature at 40 °C for 2.5 min which is then followed with a ramp at 8 °C/min to 250 °C and hold it for 5 min.
7. Use suitable mass spectrometer scan range of 15–500 mass units, and identify compounds using a mass spectral library (NIST 98).

Gaseous Products

1. Analyse HTL gaseous products using a gas chromatography equipped with a thermal conductivity detector and a 15 ft. × 1/8 in. i.d. stainless steel column using argon as carrier gas with a flow rate 15 mL/min.
2. Connect the reactor gas vent to the GC gas sampling valve.
3. Allow a fraction of gas in the reactor to flow into the sample loop.
4. Open the reactor valve slightly to allow about 1 mL sample.
5. Close the GC valve followed by that of the reactor.
6. Maintain the temperature of the column at 35 °C for 5 min initially, and then ramp to 225 °C at a rate of 20 °C/min and hold it for 15 min.
7. Generate calibration curves from the analysis of analytical standard with known composition, for quantifying mole fractions of specific gaseous components.
8. Use the quantity of helium added to the reactor as an internal standard in order to calculate the molar amount of each constituent.
9. Estimate the yield of each gas component by dividing its molar amount by the mass of dry algae loaded in to the reactor [7].

4 Notes

1. Based on the nature of the feedstock slurry, the solids concentration can be varied from 10–30 wt.% (w/w) for batch studies in HTL reactor. In laboratory-scale continuous HTL operation the solids loading is maintained significantly lower and is approximately 1% in order to minimize pumping issues and avoid clogging the pipelines and valves.
2. Catalyst enhances HTL reaction by lowering the activation energy and subsequently improves biocrude yield and quality by reducing the level of heteroatoms such as nitrogen, sulfur, and oxygen. Often, a hydrogen donor cosolvent (e.g., isopropanol) is used for lowering the HTL operational temperature and pressure [14].

3. The main purpose of reaction pressure is to maintain the water in liquid state to avoid vaporization and energy loss.
4. In some laboratory HTL experiments, a fluidized bath reactor is used for heating up a microreactor.
5. Cooling of reactor at the end of experimental run is also achieved by air-cooling using an electric fan.
6. Liquefaction produces a reaction mixture, which can be separated into four products. The primary product of HTL is biocrude or green crude or renewable crude. Other HTL by-products include solid residues, aqueous phase coproduct (ACP) and gas phase (Fig. 2). Beside the gas phase (which is collected from headspace), biocrude, solid residues and aqueous phase coproduct (ACP) are extracted from the liquid reaction mixture. It is important that these products are quantified and analyzed following liquefaction.
7. Dichloromethane or methylene chloride is considered to be very efficient solvent for HTL biocrude extraction as its boiling point is very low 39.6 °C and it can dissolve a wide range of organic compounds. However, acetone is often used as a solvent for extracting organics from the reaction mixture due to its less toxicity than DCM. In laboratory methods, selection of a solvent is performed based on its polarity. Adding equal volume of solvents such as dichloromethane or acetone to reaction mixture in the separating funnel is enough for phase separation.
8. The bottom layer is a combination of char residue and heavy oil. The aqueous phase coproduct (ACP) is rich in water-soluble organics and nutrients (such as Nitrogen and Phosphorus).
9. Gravity separation and other nonsolvent methods will be ideal and economical for large-scale HTL operations.
10. Bottom heavy oil obtained from vacuum distillation is referred to as heavy biocrude oil (HBO). Biocrude oil obtained from top layer is referred as light biocrude oil (LBO).
11. A more accurate method of estimation practiced in laboratory studies involves the measurement of headspace gas composition and measurement of final reactor pressure and temperature.

Acknowledgments

The authors thank the management of Aban Group, Chennai, Tamil Nadu, India for their support to publish this book chapter.

References

1. Barreiro DL, Prins W, Ronsse F, Brilman W (2013) Hydrothermal liquefaction (HTL) of microalgae for biofuel production: state of the art review and future prospects. Biomass Bioenergy 53:113–127
2. Garcia Alba L, Torri C, Samorì C, van der Spek J, Fabbri D, Kersten SR, Brilman DW (2011) Hydrothermal treatment (HTT) of microalgae: evaluation of the process as conversion method in an algae biorefinery concept. Energy Fuel 26(1):642–657
3. Jena U, Das KC, Kastner JR (2011) Effect of operating conditions of thermochemical liquefaction on biocrude production from Spirulinaplatensis. Bioresour Technol 102 (10):6221–6229
4. Peterson AA, Vogel F, Lachance RP, Fröling M, Antal MJ Jr, Tester JW (2008) Thermochemical biofuel production in hydrothermal media: a review of sub-and supercritical water technologies. Energy Environ Sci 1 (1):32–65
5. Kruse A, Dahmen N (2015) Water—a magic solvent for biomass conversion. J Supercrit Fluids 96:36–45
6. Biller P, Riley R, Ross AB (2011) Catalytic hydrothermal processing of microalgae: decomposition and upgrading of lipids. Biores Technol 102(7):4841–4848
7. Duan P, Savage PE (2010) Hydrothermal liquefaction of a microalga with heterogeneous catalysts. Ind Eng Chem Res 50(1):52–61
8. Ross AB, Biller P, Kubacki ML, Li H, Lea-Langton A, Jones JM (2010) Hydrothermal processing of microalgae using alkali and organic acids. Fuel 89(9):2234–2243
9. Eboibi B, Lewis D, Ashman P, Chinnasamy S (2014) Effect of operating conditions on yield and quality of biocrude during hydrothermal liquefaction of halophytic microalga *Tetraselmis* sp. Bioresour Technol 170:20–29
10. Theegala CS, Midgett JS (2012) Hydrothermal liquefaction of separated dairy manure for production of bio-oils with simultaneous waste treatment. Bioresour Technol 107:456–463
11. Channiwala SA, Parikh PP (2002) A unified correlation for estimating HHV of solid, liquid and gaseous fuels. Fuel 81(8):1051–1063
12. Valdez PJ, Nelson MC, Wang HY, Lin XN, Savage PE (2012) Hydrothermal liquefaction of Nannochloropsis sp.: systematic study of process variables and analysis of the product fractions. Biomass Bioenergy 46:317–331
13. Biller P, Ross AB (2011) Potential yields and properties of oil from the hydrothermal liquefaction of microalgae with different biochemical content. Bioresour Technol 102(1):215–225
14. Jena U, McCurdy AT, Warren A et al (2015) Oleaginous yeast platform for producing biofuels via co-solvent hydrothermal liquefaction. Biotechnol Biofuels 8(1):167

Chapter 13

Life Cycle Analysis of Producing Microbial Lipids and Biodiesel: Comparison with Plant Lipids

Tom Bradley and Daniel Maga

Abstract

Life Cycle Assessment is the *"compilation and evaluation of the inputs, outputs and the potential environmental impacts of a product system throughout its life cycle"* (ISO 14040). Examples of environmental impacts include climate change, ozone depletion, toxicity, eutrophication, particulate matter, radiation, and more. In this chapter we describe the process of undertaking an LCA for algal products, considering the ISO 14040 and ISO 14044, standards, as well as information from the European Renewable Energy Directive. We describe popular software packages, and the approach of purely using a spreadsheet with an example of algae- and soy-based biodiesel.

Key words Life Cycle Assessment, LCA, Carbon footprint, Environmental impacts, Energy balance, Sustainability

1 Introduction

Life Cycle Assessment was first used by the Coca Cola Company in 1969 to understand the environmental impact of the production of their bottles [1]. Since then it has grown into a widely used methodology to understand a range of environmental impacts for the full life cycle of a product or service. Within biofuels, there is widespread debate over the true sustainability of various fuel types. Therefore, LCA is a very important tool to gain some understanding of how biofuels compare from an environmental perspective with other fuel types, and where issues exist how to rectify them.

The environmental impacts considered within LCA address environmental themes such as on global warming (mid-points) or their eventual damage on human health, on the environment and on resources (end-points). For example, a mid-point of *Toxicity—Terrestrial ecosystems* would be measured in kg 1.4-DB eq, whereas the end-point is measured in species extinct/year. The end point of *Photochemical ozone formation—Human health* is in Disability Adjusted Life Years (DALY), which is a measure of lost years of

Venkatesh Balan (ed.), *Microbial Lipid Production: Methods and Protocols*, Methods in Molecular Biology, vol. 1995, https://doi.org/10.1007/978-1-4939-9484-7_13,

"healthy" life. Impact categories include climate change, ozone depletion, terrestrial acidification, eutrophication, toxicity, ecotoxicity, photochemical oxidant formation, particulate matter formation, fossil depletion, and a range of others [2, 3].

The outline flow of work for an LCA is controlled by the standards ISO 14040 [4] and ISO 14044 [5], further guidance is given by the Joint Research Centre "*International Reference Life Cycle Data System (ILCD) Handbook*" [6] and the United Nations Environmental Program "*Global Guidance Principles for Life Cycle Assessment Databases*" [7]. Specific methodologies for the LCA of biofuels, purely with regard to climate change impacts, are given by the European Commission's "*Renewable Energy Directive*" (*RED*) [8]. Despite this, LCA is an extremely subjective science, with the standards allowing for a wide range of choices on the boundary conditions, functional units, impact categories, and other methodological factors [9]. This does unfortunately mean that despite using very sensible assumptions, many LCA studies are not comparable. There are attempts to unify methodologies within various industries, one example relevant to the authors of this chapter, who work within the field of algae research, is the method within the Algae Cluster [10]. Within this chapter we guide the reader through the different value choices, and how to undertake LCA.

Although this chapter is aimed at microbial and plant-based lipid biofuel LCA, these methods can adapted for any LCA. The chapter is designed in a similar format to the ISO standards. The choices made within this chapter are subjective, as it is in all LCA. However, following this methodology will allow the user to understand LCA and then understand how to make different methodological decisions if considered relevant.

The outline flow of work for an LCA as dictated by ISO 14040 and ISO 14044 requires the LCA researcher to follow four main steps:

- *Goal and scope definition*
- *Life cycle inventory analysis* (*LCI*)
- *Life cycle impact assessment* (*LCIA*)
- *Life cycle interpretation*

This chapter will provide the reader with a guide for LCA, unlike lab-based processes, the subjectivity inherent in LCA means that this cannot be a simple rigid set of instructions, but more a guide for the understanding of how to undertake a basic LCA for biofuels. This does mean that this chapter is more discursive than a standard research methods chapter; however, that is unavoidable for this topic.

2 Materials

1. LCA software package, this can take the form of a spreadsheet, GaBi, SimaPro, Umberto, OpenLCA, or others.
2. LCA database—there are a range of free or subscription-based databases, some examples include Agribalyse, BioEnergieDat, Ecoinvent, ELCD, ESU World Food, LCAFood, LC-Inventories, NEEDS, NREL US Life Cycle Inventory Database, OEKOpro, Ökobau.dat, ProBas, and thinkstep Professional. Generally within the literature EcoInvent is used, but the most important element is to ensure where possible only one database is used. An LCA dataset search of various LCA databases can be conducted on the Nexus website at: https://nexus.openlca.org/search or the AMEE website at: https://discover.amee.com.

3 Methods

3.1 Goal and Scope Definition

The Goal and Scope is essentially the decision making process for what will be studied, and how. The following decisions must be made.

Functional unit. This is the unit which the whole process will measure the impacts against. To remain consistent with other researchers, we recommend using 1 MJ of fuel burnt within a car engine (other options are 1 kg of fuel, 1 kg of lipid oil, 1 MJ of lipid oil, 1 km traveled by a particular vehicle, etc.)

Boundary conditions. The boundary conditions are where the analysis will start and end. Will it include the construction of the facility? Will it include the use of the product? We recommend including the following stages. (1) Cultivation, (2) Harvesting, (3) Extraction, (4) Biodiesel conversion, (5) transport and distribution, and (6) use of biodiesel. It is important to also consider (7) build of facility, (8) land use change where possible, and (9) the residuals and coproducts, as detailed below.

Multifunctionality and allocation. An entire chapter could easily be devoted to this area. A process may produce two products, or waste to be used for another product. In such a case it is important to fairly allocate the impacts to each product. Five options are detailed within [11].

1. *Mass-based method*: This splits the impacts between the product and coproducts based on the ratio of masses.
2. *Energy content-based method*: This splits the impacts between products and coproducts based on the ratio of energy contents. This is the methodology favored by the RED.

3. *Market-based method*: With this the ratio of impacts is based purely on the commercial value of the product and coproducts.
4. *Process purpose-based method*: This is based on understanding if individual processes are to be allocated toward the production of the product or the coproduct. However, as various processes within a facility may be involved in the production of multiple products, these processes need to use the above methods to allocate impacts.
5. *System expansion*: This method, as used by ISO 14040/44, involves the expansion to the modelling of the coproducts, and the calculating the impacts of the coproducts relative to the production of this same coproduct by the status quo. The difference between these is then a benefit or addition to the main product's impact.

These five different methods are used throughout algae LCA, although due to the preference for following either the RED or ISO standards, methods two or five are usually used. However, it is very important to be aware that as shown by [11], the choice of coproduct method can possibly lead to significant differences between impacts. We show this issue within the examples at the end of this chapter.

Recycling. In terms of Ecoinvent (version 3.3) there are three options, "*Cut-Off System Model*," "*Allocation at the Point Of Substitution (APOS) System Model*," and "*Consequential System Model*" [12]. These all treat recycling in different ways. There is no "right" choice; however, when using some software, such as thinkstep GaBi the only choice is the "*Cut-Off System Model*." This assumes a producer is fully responsible for the disposal of wastes, and that they do not receive any credit for the provision of any recyclable materials.

Attributional or consequential. Attributional models present the impacts, within firm system boundaries, of the LCA study. This is the more "traditional" form of LCA. However, Consequential models seek to identify the consequences of a decision, including how it will impact on the supply chain, and the consumption of other products [7].

Impact categories. The impact categories are how the Life Cycle Inventory results are assigned to different impacts (e.g., climate change or ozone depletion). For this chapter, we recommend the use of ReCiPe as it can consider both mid-points and end-points. In term of Climate Change we strongly recommend that you ensure it uses the latest climate change global warming potentials from the IPCC AR5 (2013) report [13]. The 2016 version of ReCiPe uses these up to date figs [3], but some others impact category methods still use data from the IPCC AR4 (2007) report [14], which varies on gases such as methane, importantly there are subtle differences between applications of AR5, so consistency on method used is always important. We recommend using both 20- and 100-year time horizon climate change impacts, to understand the short and longer term impacts on climate change from a process.

Having made these decisions, then you are ready to begin the construction of an LCA model. Generally an LCA process is iterative, and the Goal and Scope may alter slightly due to issues during the analysis.

3.2 Life Cycle Inventory Analysis

Throughout the process under study, detailed data must be collated. For this case, data should include the following:

1. Energy (in its different forms) used per phase (**Note 2**).
2. Input chemicals per phase.
    (a) Quantity (mass or volume).
    (b) Manufacturer.
    (c) Origin.
    (d) Method of transportation to site.
3. Disposables used (including their construction materials).
    (a) Breakdown of materials and masses.
    (b) Manufacturer.
    (c) Origin.
    (d) Method of transportation to site.
4. Treatment of waste (recycled? Landfill? Burnt?).
5. Amount of biodiesel produced over measurement period.
6. Lipid content.
7. Percentage of lipid fraction that can be converted to biodiesel.
8. Energy content of biodiesel.
9. Level of uncertainty of data.

In terms of construction data.

1. Quantity (mass or volume).
2. Manufacturer.
3. Origin.
4. Method of transportation to site.
5. Energy use (and type) by machinery on site.
6. Land type before construction (grassland, forest, brownfield, etc.).
7. Method of land conversion (slash and burn?).
8. Level of certainty of data.

This is divided into several phases, in terms of the Algae Cluster, the four phases were as follows:

1. Growth.
2. Harvesting and extraction.
3. Biodiesel production.
4. Use.

To collect all this data to a high precision is quite challenging, but worth spending as much time as possible. As with any modeling, the higher the quality of the input data, the higher the quality of the final results.

3.3 Life Cycle Impact Assessment

This is the core of the LCA, and is the process of analysing data. This requires complex tools, so this section will provide a short introduction to four different software tools which are used for carrying out a Life Cycle Impact Assessment. We strongly encourage researchers to grow comfortable with the software and undertake training prior to producing an LCA intended for a publication (**Note 1**).

Thinkstep GaBi. GaBi is a software package developed by thinkstep AG, previously known as PE International. This provides a simple flowchart-based visualization of LCA studies, it is a very intuitive software package, and is the most intuitive of the four discussed. However, it does offer limited uncertainty analysis options and does not allow all allocation procedures principally possible for Ecoinvent data.

SimaPro. SimaPro is currently the global leader in terms of LCA. While if offers a less intuitive interface it does allow for further analysis of uncertainty.

Open LCA. This is an open source software package which allows users greater flexibility, it also has substantial uncertainty analysis capabilities.

Spreadsheet (Excel, OpenOffice, etc.). Data can be collated from databases such as Ecoinvent on each individual impact category for products feeding into a process, and their uncertainty, and summed together for a whole process.

To explain how to use these software packages is outside the scope of a single chapter, but the essential workflow is as follows.

1. Ensure that the correct version of the database to be used is installed within the software, and activated. For consistency with other work within academia, we would generally recommend Ecoinvent; however, others do exist. This comes as standard within SimaPro, or can be purchased to add into GaBi or OpenLCA. Although it is possible to mix data from databases, it is certainly not recommended, as they do use slightly different methodologies.

2. Identify the processes within your chosen database to fit with the quantities identified within the Life Cycle Inventory. For example, for the electricity is used from the grid in the USA, if using Ecoinvent 3.3 then the process "*(WECC, US) market for electricity, low voltage*" may be appropriate. It is important to make consistent decisions on geographical locations. Ecoinvent contains "*production*" LCA models, but also "*market*." These include the average travel distances and modes of travel. If the original origin and method of travel used for a product are not known, then the "market" models offer a reasonable approximation. In some cases you may need to search a process database using multiple names, for example "*sodium oxidanide*" will not find a process within Ecoinvent 3.3, whereas it can be found using the alternative name of "*sodium hydroxide*." Some packages include the CAS numbers on flows and processes (**Notes 3** and **4**).
3. Create models for the processes that are not known. In some cases, these will not exist within the particular database you are using. You will have several options, create the process from primary/literature data, import from another database (not recommended but sometimes necessary) or purchase further databases (e.g., thinkstep AG offers "*Data on Demand*") however this can quickly become very expensive. It is best to create a new model based on primary data that you have collected. For example, there is no centrifuge model within Ecoinvent, so for the process of algae harvesting this will need to be created. This dataset would have input flows of "electricity," "water," and algal biomass. The outputs would be water and dried algal biomass. If including the construction, there would also be the inputs of the energy and materials for the centrifuge's manufacture.
4. When you have identified all of the relevant processes to use for the inputs and outputs from the LCI stage, these can be linked together. In the spreadsheet example this is a simple process of meticulously summing the individual impacts. Within GaBi this involves the creation of a flow diagram, while within OpenLCA this would involve the creation of several linked processes through specifying the source process for each individual input flow. Ensure that all impacts are scaled against the "functional unit." Within a software package, having created the model, the final stage is to undertake an analysis to provide a list of all impacts, within GaBi this is through running a "balance calculation," or within OpenLCA, creating a product system and then calculating the impacts.

3.4 Life Cycle Interpretation

Within this part of an LCA, there are several areas to be considered, the impact categories, scenario analysis and uncertainty analysis.

In terms of uncertainty, Ecoinvent data comes with data on the uncertainty of individual processes, provided as lognormal distributions via the semiquantitative pedigree matrix approach. By using a log-normal distribution, this prevents negative results, and is generally more representative of natural systems. This methodology is described in detail within [15]. The semiquantitative nature of the pedigree matrix approach does mean that the uncertainty values provided within Ecoinvent do have a subjective element. Using this uncertainty data, it is possible to calculate the uncertainties between processes utilizing a Monte-Carlo analysis. However, depending on software package this may be a complex procedure. Within SimaPro and OpenLCA this is a reasonably simple analysis to undertake, whereas with GaBi does not support log-normal distributions, only normal and uniform distributions. Good general descriptions of approaches to log-normal distributions can be found within financial textbooks such as [16]. To undertake more detailed analysis, the results from the Monte-Carlo analysis should be exported into a specialized statistics package, such as Minitab.

If using primary data with known uncertainties, such as the range of volumes of input chemicals and energy, then this can be summed through standard statistical methods. This, however, does not include the known uncertainties within the impact categories, which should also be included. For example, the uncertainties of the climate change environmental impacts can be found from the corresponding IPCC report from which they originate.

The preferred method within LCA for statistical analysis is via "sensitivity analysis". Through this, the levels of contribution created by particular processes can be highlighted, which can also inform how to reduce environmental impacts. This is undertaken through varying the quantities of various processes, and recording the impact on the final results.

Various scenarios can be created, for example, changing the electricity source, or using similar products but from different geographical locations.

One important result from LCA which will be noticed through creating different scenarios is the observation of the sources of individual impacts. You will generally observe (if using a wide enough range of environmental impact categories) that reducing the impacts from one category may increase those from another impact. For example, increasing the use of photovoltaics may decrease the climate change impacts but increase the toxicity impacts. This is where choices must be made on which environmental impacts are most important. The LCA should inform you of how to reduce the impacts of your process, by identifying the main sources. Additionally, if you are modelling two or more processes, you can make value choices between fuel types.

Ultimately, it is important to understand what the LCA means, and also what it does not mean. Due to variations in LCA methodologies, it is often not informative to compare with work by other researchers. It is very important that uncertainties are considered, either using methods described here or other methods within the literature.

4 Worked Examples

Using the approaches described above, we shall work through two examples, to make LCA more clear to the reader. It is important to note that these are only examples to improve your understanding of LCA, and should not be taken as definitive examples on the environmental impacts of either algae or plant-based biodiesel. For both examples, we shall use the same Goal and Scope. The data used from the literature within the examples has been simplified, and analyzed using different databases, hence the results are different to the results within the source publications. It must be stated that it is not advisable to use these numbers for your own research, and instead refer back to the base data these examples are based on, specifically, [17–19].

4.1 Goal and Scope Definition

1. Goal: To compare algae- and soy-based biodiesel.
2. Audience: Readers of this book.
3. Functional Unit: 1 MJ of fuel burnt in an engine.
4. Boundary Conditions: Well to Wheel.
5. Database: Ecoinvent 3.3 (this can be sourced from http://www.ecoinvent.org/ although does require a license which costs in the region of EUR 2000).
6. Allocation Method: Energy allocation (to keep the example simple), and also Mass allocation, to show how these can give quite different answers.
7. Geographical Location: The USA.

4.2 Life Cycle Inventory Analysis (Algae)

The data for algae for the growth, harvesting and extraction has been taken from [17], which is one of the few sources of industrial data within the literature. The transesterification process has been adapted from the NREL US Life Cycle Inventory (LCI) database [18] for soy-based biodiesel. The estimates of CO_2 emissions from a car are taken from information within the Ecoinvent 3.3 database.[1] The combined data is given within Table 1.

4.3 Life Cycle Inventory Analysis (Soy-Based Biodiesel)

In terms of the plant-based lipids, we have used data from the US LCI database for the whole process. This has been simplified to allow this example to purely use Ecoinvent data, so some similar inputs have been combined. We also have not included the

[1] Note that tailpipe emissions from vehicles contain far more than CO_2, which would be relevant should toxicity or particulate matter impact categories be considered. However, this is only a simplified example.

Table 1
Example Life Cycle Inventory (LCI) data (algae)

	Process		
Input/output	**Name**	**Quantity**	**Unit**
Growth			
Input	Nitrogen fertilizer	6.842E-03	kg
Input	Phosphorus fertilizer	1.184E-03	kg
Input	Freshwater	4.395E+01	kg
Input	Electricity	9.737E-01	kWh
Input	Flue gas pumped in	4.763E+00	kg
Output	Flue gas emitted	4.711E+00	kg
Harvesting			
Input	Electricity	4.737E-01	kWh
Extraction			
Input	Electricity	5.816E-02	kWh
Input	Heat	4.163E-02	kWh
Input	Hexane	8.684E-03	kg
Transesterification			
Input	Citric acid	1.942E-05	kg
Input	Hydrochloric acid	1.157E-03	kg
Input	Phosphoric acid	1.688E-05	kg
Input	Sodium methylate	6.159E-04	kg
Input	Electricity	9.512E-04	kWh
Input	Methanol, at plant	2.418E-03	kg
Input	Heat	6.743E-03	kWh
Input	Sodium hydroxide, production mix, at plant	2.592E-05	kg
Input	Freshwater	9.036E-03	kg
Use			
Output	CO_2 from tailpipe	0.082631579	kg

transport, as this was not known for the algae so it would introduce an inconsistency within the example. The models adapted from the US database are:

1. Soybean grains, at field.
2. Soybean oil, crude, degummed, at plant.
3. Soy biodiesel, production, at plant.

Within the Soy models above, harvesting is included within (1. Soybean grains, at field). All figures have been scaled in order to provide a model for 1 MJ of fuel. The combined data is given within Table 2.

4.4 Life Cycle Impact Assessment (Algae)

The next stage is to create a model. This can be undertaken in a range of software packages as described before (GaBi, Umberto, OpenLCA, SimaPro, etc.); however, in order to provide a simple example, the LCA will be undertaken with a spreadsheet.

Table 2
Example Life Cycle Inventory (LCI) data (soy)

Input/output	Process Name	Quantity	Unit
Growth			
Input	Agrochemicals	7.085E-05	kg
Input	Phosphorous fertilizer (TSP as P_2O_5)	6.813E-04	kg
Input	Potash fertilizer (K_2O)	1.267E-03	kg
Input	Diesel, combusted in industrial equipment	9.411E-02	MJ
Input	Electricity	3.406E-03	kWh
Input	Natural gas, combusted in industrial boiler	1.917E-03	kWh
Input	Nitrogen fertilizer, production mix, at plant	2.180E-04	kg
Input	Quicklime, at plant	1.285E-02	kg
Input	Water, river	2.166E+00	kg
Input	Water, well, in ground	4.728E+00	kg
Oil extraction			
Input	Bituminous coal, combusted in industrial boiler	6.125E-02	MJ
Input	Heat, from biomass	1.641E-03	MJ
Input	Hexane, at plant	7.697E-05	kg
Input	Diesel, combusted in industrial boiler	7.605E-04	MJ
Input	Electricity	7.515E-03	kWh
Input	Natural gas, combusted in industrial boiler	3.129E-02	kWh
Input	Residual fuel oil, combusted in industrial boiler	1.540E-03	MJ
Input	Water	6.631E-02	kg
Transesterification			
Input	Citric acid	1.919E-05	kg
Input	Hydrochloric acid	1.143E-03	kg
Input	Phosphoric acid	1.668E-05	kg
Input	Sodium methylate	6.086E-04	kg
Input	Electricity	9.398E-04	kWh
Input	Methanol, at plant	2.389E-03	kg
Input	Heat	6.663E-03	kWh
Input	Sodium hydroxide, production mix, at plant	2.561E-05	kg
Input	Freshwater	8.929E-03	kg
Use			
Output	CO_2 per kg diesel	8.263E-02	kg

It is important to stress that under ISO 14040, all possible impacts should be considered. To purely look at Greenhouse Gas emissions *is not a true LCA*. Despite this, within this example we will purely look at the GHG emissions, but the same approach should be taken for all relevant impacts to a process.

The data has been taken from Ecoinvent 3.3, from the Ecoinvent Institute [19]. For this example, we have used the IPCC AR5 climate change impacts for 100-year and 20-year time horizons based on data from [13]. The Ecoinvent impacts have been multiplied by the total mass or energy of each process, to show the total impacts. This can be seen within Table 3.

Table 3
Example algae data with Ecoinvent models and IPCC AR5 impacts

Input/ output	Process			Ecoinvent model	IPCCAR5 w/out biogenic impact		Impact	
	Name	Quantity	Unit		GWP100/unit [$kgCO_2eq$]	GWP20/unit [$kgCO_2eq$]	GWP100 [$kgCO_2eq$]	GWP20 [$kgCO_2eq$]
Growth								
Input	Nitrogen fertilizer	6.842E-03	kg	(GLO) market for nitrogen fertiliser, as N	1.003E+01	1.054E+01	6.865E-02	7.214E-02
Input	Phosphorus fertilizer	1.184E-03	kg	(GLO) market for phosphate fertiliser, as P205	1.923E+00	2.239E+00	2.277E-03	2.651E-03
Input	Freshwater	4.395E+01	kg	(GLO) market group for tap water	5.820E-04	6.930E-04	2.558E-02	3.046E-02
Input	Electricity	9.737E-01	kWh	(WECC, US) market for electricity, low voltage	5.087E-01	5.529E-01	4.954E-01	5.383E-01
Input	Flue gas pumped in	4.763E+00	kg	n/a	−1.000E+00	−1.000E+00	−4.763E+00	−4.763E+00
Output	Flue gas emitted	4.711E+00	kg	n/a	1.000E+00	1.000E+00	4.711E+00	4.711E+00
Harvesting								
Input	Electricity	4.737E-01	kWh	(WECC, US) market for electricity, low voltage	5.087E-01	5.529E-01	2.410E-01	2.619E-01
Extraction								
Input	Electricity	5.816E-02	kWh	(WECC, US) market for electricity, low voltage	5.087E-01	5.529E-01	2.959E-02	3.215E-02
Input	Heat	4.163E-02	kWh	(RoW) steam production in chemical industry	1.037E-01	1.178E-01	4.316E-03	4.904E-03
Input	Hexane	8.684E-03	kg	(GLO) market for hexane	6.415E-01	7.691E-01	5.571E-03	6.679E-03

Transesterification								
Input	Citric acid	1.942E-05	kg	(RNA) citric acid production	0.000E+00	0.000E+00	0.000E+00	0.000E+00
Input	Hydrochloric acid	1.157E-03	kg	(CA-QC) hydrochloric acid production, from the reaction of hydrogen with chlorine	0.000E+00	0.000E+00	0.000E+00	0.000E+00
Input	Phosphoric acid	1.688E-05	kg	(GLO) market for phosphoric acid, industrial grade, without water, in 85% solution state	0.000E+00	0.000E+00	0.000E+00	0.000E+00
Input	Sodium methyl ate	6.159E-04	kg	(GLO) market for sodium methoxide	0.000E+00	0.000E+00	0.000E+00	0.000E+00
Input	Electricity	9.512E-04	kWh	(WECC, US) market for electricity, low voltage	5.087E-01	5.529E-01	4.839E-04	5.259E-04
Input	Methanol.at plant	2.418E-03	kg	(GLO) market for methanol	0.000E+00	0.000E+00	0.000E+00	0.000E+00
Input	Heat	6.743E-03	kWh	(RoW) steam production in chemical industry	0.000E+00	0.000E+00	0.000E+00	0.000E+00
Input	Sodium hydroxide, production mix, at plant	2.592E-05	kg	(GLO) sodium hydroxide to generic market for neutralising agent	0.000E+00	0.000E+00	0.000E+00	0.000E+00
Input	Freshwater	9.036E-03	kg	(GLO) market group for tap water	5.820E-04	6.930E-04	5.259E-06	6.262E-06
Use								
Output	CO_2 from tailpipe	8.263E-02	kg	n/a	1.000E+00	1.000E+00	8.263E-02	8.263E-02
						Total	8.197E-01	9.797E-01

4.5 Life Cycle Impact Assessment (Soy)

As with the algae, a similar process has been undertaken with the Soy data as shown in Table 4.

4.6 Life Cycle Interpretation

Allocation is an important issue. In terms of algae, the mass and energy balances adapted from [17] and [18] are given in Table 5:

Therefore, if we take a mass based allocation, then 27.53% of the impacts are to be applied to the biodiesel, whereas if we take an energy-based allocation (such as within the Renewable Energy Directive) then we have 61.62% of the impacts applied to the algae-based biodiesel. In the case of the Soy biodiesel, then the results are as provided in Table 6.

These results, like the algae, show the allocation method changes the results. In the case of mass based allocation, 6.22% of the impacts are attributed to the soy biodiesel, whereas an energy based allocation will attribute 16.29% of the impacts to the soy biodiesel. This then gives the different environmental impacts from these examples as in Table 7.

As can be seen, this leads to two very different answers. The actual choice of the approach used is highly subjective, the most important thing is that whatever approach is used, it is used consistently. The above figures may change significantly where the impacts of land use change are considered, as soy biodiesel requires higher quality land (and more of it) than a photobioreactor based algae facility.

From the above data, the environmental hotspots in terms of climate change can be identified. In terms of algae for example the electricity use is the source of around 80% of the Climate Change impacts. A sensitivity analysis could be undertaken through varying the sources of the electricity and feedstocks. In terms of detailed analysis of the errors, Monte-Carlo analysis within some LCA packages would allow the lognormal distributions of each process (not included in the tables in this chapter) to be used to produce a range of values.

5 Summary

The purpose of this chapter is to explain the basics of LCA, to allow a researcher to undertake a basic LCA, and use it to identify areas for possible improvement within their biofuel system (be it microbial or plant based). The examples given are simplified from literature data but should not be taken as definitive; they are purely for a learning exercise. There are far more impacts than just Climate Change, and it is important to consider all of them. In addition, other environmental aspects such as Land Use Impacts should be considered.

Table 4
Example soy data with Ecoinvent models and IPCC AR5 impacts

Input/output	Process: Name	Process: Quantity	Process: Unit	Process: Ecoinvent model	IPCCAR5 w/out biogenic impact: GWP100/unit [$kgCO_2eq$]	IPCCAR5 w/out biogenic impact: GWP20/unit [$kgCO_2eq$]	Impact: GWP100 [$kgCO_2eq$]	Impact: GWP20 [$kgCO_2eq$]
Growth								
Input	Agro chemicals	7.085E-05	kg	(GLO) market for pesticide, unspecified	1.051E+01	1.279E+01	7.448E-04	9.064E-04
Input	Phosphorous fertilizer (TSP as P205)	6.813E-04	kg	(GLO) market for phosphate fertiliser, as P205	1.923E+00	2.385E-01	1.310E-03	1.625E-04
Input	Potash fertilizer (K20)	1.267E-03	kg	(GLO) market for potassium fertiliser, as K20	3.553E-01	3.957E-01	4.502E-04	5.014E-04
Input	Diesel, combusted in industrial equipment	9.411E-02	MJ	(GLO) diesel, burned in agricultural machinery	2.098E-01	2.432E-01	1.974E-02	2.289E-02
Input	Electricity	3.406E-03	kWh	(WECC, US) market for electricity, low voltage	5.052E-01	5.478E-01	1.721E-03	1.866E-03
Input	Natural gas, combusted in industrial boiler	1.917E-03	kWh	(RoW) steam production in chemical industry	1.174E-01	1.334E-01	2.250E-04	2.557E-04
Input	Nitrogen fertilizer, production mix, at plant	2.180E-04	kg	(GLO) market for nitrogen fertiliser, as N	1.003E+01	1.054E+01	2.187E-03	2.299E-03
Input	Quicklime, at plant	1.285E-02	kg	(RoW) market for quicklime, milled, loose	1.207E+00	1.258E+00	1.551E-02	1.616E-02
Input	Water, river	2.166E+00	kg	No Climate Change Impact	0.000E+00	0.000E+00	0.000E+00	0.000E+00
Input	Water, well, in ground	4.728E+00	kg	No Climate Change Impact	0.000E+00	0.000E+00	0.000E+00	0.000E+00
Oil Extraction								
Input	Bituminous coal, combusted in industrial boiler	6.125E-02	MJ	(RoW) heat production, at hard coal industrial furnace 1-10 MW	1.493E-01	1.810E-01	9.144E-03	1.109E-02
Input	Heat, from biomass	1.641E-03	MJ	(RoW) heat production, hardwood chips from forest, at furnace 5000 k	9.849E-03	1.257E-02	1.616E-05	2.063E-05

(continued)

Table 4
(continued)

	Process				IPCCAR5 w/out biogenic impact		Impact	
Input/ output	Name	Quantity	Unit	Ecoinvent model	GWP100/ unit [$kgCO_2eq$]	GWP20/ unit [$kgCO_2eq$]	GWP100 [$kgCO_2eq$]	GWP20 [$kgCO_2eq$]
Input	Hexane, at plant	7.697E-05	kg	(GLO) market for hexane	6.415E-01	7.691E-01	4.937E-05	5.919E-05
Input	Diesel, combusted in industrial boiler	7.605E-04	MJ	(GLO) diesel, burned in agricultural machinery	2.098E-01	2.432E-01	1.595E-04	1.850E-04
Input	Electricity	7.515E-03	kWh	(WECC, US) market for electricity, low voltage	5.052E-01	5.478E-01	3.796E-03	4.117E-03
Input	Natural gas,combusted in industrial boiler	3.129E-02	kWh	(RoW) steam production in chemical industry	1.174E-01	1.334E-01	3.672E-03	4.175E-03
Input	Residual fuel oil, combusted in industrial boiler	1.540E-03	MJ	(RoW) heat production, heavy fuel oil, at industrial furnace 1 MW	9.755E-02	1.010E-01	1.502E-04	1.555E-04
Input	Water	6.631E-02	kg	(GLO) market group for tap water	5.557E-04	6.610E-04	3.685E-05	4.383E-05
Transesterification								
Input	Citric acid	1.919E-05	kg	(RNA) citric acid production	2.812E+00	3.228E+00	5.395E-05	6.195E-05
Input	Hydrochloric acid	1.143E-03	kg	(CA-QC) hydrochloric acid production, from the reaction of hydrogen with chlorine	4.756E-01	5.532E-01	5.438E-04	6.326E-04
Input	Phosphoric acid	1.668E-05	kg	(GLO) market for phosphoric acid, industrial grade, without water, in 85% solution state	1.529E+00	1.725E+00	2.550E-05	2.878E-05
Input	Sodium M ethyl ate	6.086E-04	kg	(GLO) market for sodium methoxide	1.653E+00	2.030E+00	1.006E-03	1.235E-03
Input	Electricity	9.398E-04	kWh	(WECC, US) market for electricity, low voltage	5.052E-01	5.478E-01	4.748E-04	5.149E-04
Input	Methanol, at plant	2.389E-03	kg	(GLO) market for methanol	6.036E-01	9.060E-01	1.442E-03	2.164E-03
Input	Heat	6.663E-03	kWh	(RoW) steam production in chemical industry	1.174E-01	1.334E-01	7.820E-04	8.890E-04

Input	Sodium hydroxide, production mix, at plant	2.561E-05	kg	(GLO) sodium hydroxide to generic market for neutralising agent	1.348E+00	1.580E+00	3.452E-05	4.047E-05
Input	Freshwater	8.929E-03	kg	(GLO) market group for tap water	5.557E-04	6.610E-04	4.961E-06	5.901E-06
Use								
Output	CO_2 per kg diesel	8.263E-02	kg	(RER) transport, passenger car, medium size, diesel, EURO 5	1.000E+00	1.000E+00	8.263E-02	8.263E-02
						Total	6.328F-02	7.046E-02

Table 5
Outputs from algae biodiesel production, in terms of mass and energy, in order to allocate impacts

Name	Quantity [kg]	Energy content [MJ/kg]	kg allocation	Energy content allocation
Algae residue (oilcake)	1.848	6.107	50.88%	18.30%
Low value lipids	0.662	16	18.23%	17.18%
Fatty acids	0.002	38	0.06%	0.13%
Glycerin, at biodiesel plant	0.120	14.3	3.30%	2.78%
Algae biodiesel, production, at plant	1.000	38	27.53%	61.62%

Table 6
Outputs from soy biodiesel production, in terms of mass and energy, in order to allocate impacts

Name	Quantity [kg]	Energy content [MJ/kg]	kg allocation	Energy content allocation
Fatty acids	0.002	38	0.01%	0.03%
Glycerin, at biodiesel plant	0.120	14.3	0.75%	0.74%
Soy biodiesel, production, at plant	1.000	38	6.22%	16.29%
Oils, unspecified	0.000	14	0.00%	0.00%
Soy meal, at plant	4.081	14.1	25.38%	24.67%
Soybean residues, at field	10.873	12.5	67.64%	58.27%

Table 7
Comparison of algae-derived biodiesel and soy-derived biodiesel

	Climate change [kgCO_2eq]			
	Mass based allocation		Energy based allocation	
Fuel	GWP100	GWP20	GWP100	GWP20
Algae biodiesel, production, at plant	0.226	0.270	0.505	0.604
Soy biodiesel, production, at plant	0.004	0.004	0.010	0.011

6 Notes

1. In addition to the software discussed, in terms of biofuels, we recommend the reader investigates using Argonne National Laboratory's Greenhouse Gases, Regulated Emissions, and Energy Use in Transportation (GREET) software.

2. Energy use can be measured through multiplying the rated power by the time of operation, but this will lead to an overestimate. Using three-phase electricity meters to measure minute by minute power use is a far more accurate method. For our own projects we have built our own electricity meters, which cost approximately £500 in parts and labor.
3. It can be very complicated to find the appropriate process. You will have to consider the geographic factors, and method of construction.
4. Within GaBi there are <agg> and <u-so> processes. <agg> are aggregated systems, and for simplicity it is easiest to use these. However, if you want to work backward within a process, such as specifying a different energy source (like solar) for the factory, use <u-so>, and then search for the inputs for this process.

References

1. Hunt RG, Franklin WE (1996) LCA—how it came about. Int J Life Cycle Assess 1:4–7
2. Goedkoop M, Heijungs R, Huijbregts M, Schryver AD, Struijs J, van Zelm R. ReCiPe (2008) A life cycle impact assessment method which comprises harmonised category indicators at the midpoint and the endpoint level—report I: characterisation. Amersfoort, Leiden, Nijmegen, Bilthoven: PRé Consultants, University of Leiden (CML) Radboud University Nijmegen (RUN) National Institute for Public Health and the Environment (RIVM), 2013
3. Huijbregts MAJ SZ, Elshout PMF, Stam G, Verones F, Vieira MDM, Hollander A, Van Zelm R (2016) ReCiPe2016: a harmonized life cycle impact assessment method at midpoint and endpoint level. Report 2016-0104 ed. Bilthoven, The Netherlands
4. International Organization for Standardization (2006) ISO 14040: Environmental management—life cycle assessment—principles and framework. 20
5. International Organization for Standardization (2006) ISO 14044: Environmental management—life cycle assessment—principles and framework.
6. Wold M-A, Pant R, Chomkhamsri K, Sala S, Pennington D (2012) The international life cycle data system (ILCD) handbook, Brussels, Joint Research Centre—Institute for Environment and Sustainability
7. United Nations Environment Programme (2011) Global guidance principles for Life Cycle Assessment: a basis for greener processes and products
8. (2009) Directive 2009/28/EC of the European parliament and of the council of 23 April 2009 on the promotion of the use of energy from renewable sources and amending and subsequently repealing Directives 2001/77/EC and 2003/30/EC. Official Journal of the European Union, Brussels
9. Collet P, Spinelli D, Lardon L, Hélias A, Steyer J-P, Bernard O (2014) Chapter 13—life-cycle assessment of microalgal-based biofuels. In: Pandey A, Lee D-J, Chisti Y, Soccol CR (eds) Biofuels from Algae. Burlington MA, Elsevier, pp 287–312
10. Bradley T, Maga D, Antón S (2015) Unified approach to Life Cycle Assessment between three unique algae biofuel facilities. Appl Energy 154:1052–1061
11. Wang M, Huo H, Arora S (2011) Methods of dealing with co-products of biofuels in life-cycle analysis and consequent results with in the U.S.context. Energy Policy 39:5726–5736
12. Steubing GWB, Reinhard J, Bauer C, Moreno-Ruiz E (2016) The ecoinvent database version 3 (part II): analyzing LCA results and comparison to version 2. Int J Life Cycle Assess 21:1269–1281
13. Myhre G, Shindell D, Breon F-M, Collins W, Fuglestvedt J, Huang J et al (2013) Anthropogenic and natural radiative forcing. In: Stocker TF, Qin D, Plattner G-K, Tignor M, Allen SK, Boschung J et al (eds) Climate change: the physical science basis contribution of working group I to the fifth assessment report of the intergovernmental panel on climate change. Cambridge University Press, Cambridge, p 731

14. Forster P, Ramaswamy V, Artaxo P, Berntsen T, Betts R, Fahey DW et al (2007) Changes in atmospheric constituents and in radiative forcing. In: Solomon S, Qin D, Manning M, Chen Z, Marquis M, Averyt KB et al (eds) Climate change: the physical science basis contribution of working group I to the fourth assessment report of the intergovernmental panel on climate change. Cambridge University Press, Cambridge, p 212
15. Ciroth SMA, Weidema B, Lesage P (2016) Empirically based uncertainty factors for the pedigree matrix in ecoinvent. Int J Life Cycle Assess 21:1338–1348
16. Lee JCL, Lee AC (2013) Statistics for business and financial economics, 3rd edn. Springer, New York
17. Passell H, Dhaliwal H, Reno M, Wu B, Amotz AB (2013) Algae biodiesel life cycle assessment using current commercial data. J Environ Manag 129:103–111
18. National Renewable Energy Laboratory (2012) U.S. life cycle inventory database
19. Wernet G, Bauer C, Steubing B, Reinhard J, Moreno-Ruiz E, Weidema B (2016) The ecoinvent database version 3 (part I): overview and methodology. Int J Life Cycle Assess 21:1218–1230

Chapter 14

Assessment of Fuel Quality Parameters and Selection of Bacteria Using PROMETHEE–GAIA Algorithm

Sumathy Shunmugam, Manickam Gayathri, and Gangatharan Muralitharan

Abstract

Recently, biodiesel is gaining significant importance due to eco-friendly nature and development of large-scale production methodologies. Biodiesel is a mixture of mono-alkyl esters of fatty acids (FA). During transesterification, the long-chain FAs are combined with methanol to produce fatty acid methyl ester (FAME), the principle component of biodiesel. The biodiesel fuel properties are determined by structural components of FAs such as chain length, degree of unsaturation, and branching of the carbon chain. The fuel quality of biodiesel are evaluated by assessing the properties such as cetane number (CN), iodine value (IV), cold filter plugging point (CFPP), higher heating value (HHV), cloud point (CP), pour point (PP) etc., of FAME. The amount of lipid or fat produced may vary from organism to organism. A particular species may have high biomass with low lipid content and vice versa. So the selection of suitable species/ genus by decision analysis software is much needed. Besides various multi-criteria decision analyses, Preference Ranking Organization Method for Enrichment of Evaluation (PROMETHEE) and Graphical Analysis for Interactive Aid (GAIA) analysis is considered as the most promising tool in selecting the prominent biodiesel producing strain. Here we describe the method of evaluating the fuel quality parameters for the produced FAME and selecting the prominent strain through PROMETHEE–GAIA algorithm.

Key words Biodiesel properties, FAME profile, Fuel quality parameters, Multi-criteria decision analysis, PROMETHEE–GAIA, Strain selection

1 Introduction

Biodiesel is a renewable fuel that can be produced from edible and nonedible sources such as animal, plant as well as microbial oils [1, 2].These oils and fats are composed of triglycerides that consists of three long-chain fatty acids (FA), produces fatty acid methyl esters (FAME) after transesterification with methanol. These FAME molecules have reduced viscosity that is suitable for combustion in diesel engine [3]. Biodiesel is composed of mono alkyl esters of FA that can be used alternatively for petroleum diesel fuel. However, biodiesel fuel from any source or feed stock has to meet

Venkatesh Balan (ed.), *Microbial Lipid Production: Methods and Protocols*, Methods in Molecular Biology, vol. 1995, https://doi.org/10.1007/978-1-4939-9484-7_14,

the recognized quality standards for its commercial use as diesel engine fuel. The biodiesel fuel properties are assessed by the structural features of FA chain and the moiety derived from the alcohol [4]. It is imperative to know that these fuel properties can easily be predicted from various simple mathematical equations derived for this purpose. American Society for Testing and Materials (ASTM) D6751 of USA, EN14214 of Europe, and the Bureau of Indian standards (BIS) IS-15607 have specified permissible range of various fuel quality parameters for their country/region. The following are the critical fuel quality parameters to be analyzed for biodiesel before being considered for commercial application.

Iodine value (IV) determines the biodiesel oxidative stability. Higher IV prompts a higher rate of polymerization of glyceride that increases fuel thickness, resulting in the formation of deposits on engine fuel system and adversely affecting fuel injector spray patterns [5].

Saponification value (SV) indicates that milligrams of potassium hydroxide (KOH) required to neutralize the FAs resulting from complete hydrolysis of 1 g of oil. SV is inversely proportional to the molecular weight of FAs; higher the value of SV, the smaller will be the molecular mass.

Cloud point (CP), pour point (PP), and cold filter plugging point (CFPP) are the cold flow properties of biodiesel. CP is the temperature at which a cloud of wax crystals first appears when the fuel is cooled, whereas PP is the temperature at which wax formed fuel can flow, and CFPP is the least temperature at which a fuel portion will go through a standardized filtering device in a determined time [6]. Higher proportions of saturated fatty acids (SFAs) affect these properties and tend to crystallize the fuel, clogging the fuel filters of engine under colder climatic conditions [7].

Cetane number (CN) indicates the ignition quality of a fuel and identified with the ignition delay (ID) time, *i.e.* the time between injection of the fuel into the cylinder of a diesel engine and the onset of burning [4]. Higher CN leads to better combustion, improving engine motor efficiency and reducing noxious nitrogen oxides (NOx) [8].

Flash point (FP) is the lowest temperature at which the fuel will begin to vaporize to form an ignitable mixture when it comes in contact with the air [9]. Oxidation stability (OS) is the indication of degradation resistance of fuel due to oxidation during long-term storage [5]. The allylic position equivalent (APE), bisallylic position equivalent (BAPE), and degree of unsaturation (DU) predict and influence the OS [10, 11]. The higher heating value (HHV) indicates energy content in the fuel per unit mass. Long chain saturated factor (LCSF) is important in determining the cold response of the produced biodiesel and therefore influence CN, IV, OS, and CFPP.

Screening and selection of suitable microbial strains based on the various biodiesel quality parameters requires a multi-criterion decision analysis (MCDA). A review of the MCDA literature revealed that Preference Ranking Organisation Method for Enrichment of Evaluation (PROMETHEE) and Graphical Analysis for Interactive Aid (GAIA) have significant advantages (compared to other MCDA methods) because it facilitates rational decision making, *i.e.* the decision vectors stretch toward the preferred solution [12]. PROMETHEE is a computer-based multi-criteria decision tool to rank alternative solutions to a complex problem based on the weightings of selected preferences to determine positive and negative preference flows [13]. GAIA is a visual aid tool used with PROMETHEE that enables visualization and graphical representation of the analysis. Here we demonstrate how to calculate various fuel quality parameters from FAME profile of tested sample and the strategies of selecting the best suitable strain using PROMETHEE–GAIA algorithm [14–16].

2 Materials

1. Fatty acid composition (FAME) of different samples obtained from GC-MS (*see* Chapter 4).
2. Visual PROMETHEE software.

3 Methods

1. Make a table of FAME profiles obtained from samples by comparing the retention time of both standard and sample peaks (Table 1) (*see* Chapter 4).
2. Calculation of fuel properties.
   (a) Iodine value (IV)

$$\sum (254 \times DN)/M \quad (1)$$

   where D is the number of double/single bonds, M is the molecular weight of FA, and N is the area percentage of the FA obtained for the sample (*see* **Note 1**).

   (b) Saponification value (SV)

$$\mathrm{SV} = \sum (560 \times N)/M \quad (2)$$

   where M is the molecular weight and N is the area percentage of each FA obtained for the sample (*see* **Note 2**).

   (c) Cloud point (CP)

Table 1
Fatty acid (FA) profile of the samples with retention time (RT), area percentage (AP), and obtained FA for each sample

Sample 1			Sample 2			Sample 3		
RT	AP	FA (MW)$	RT	AP	FA	RT	AP	FA (MW)$
11.425	3.3166	C8:0 (144.216)	11.435	8.97	C8:0	11.424	2.96	C8:0
13.171	0.3341	C10:0 (172.27)	13.19	0.59	C10:0	13.17	0.193	C10:0
14.637	0.117	C11:0 (186.297)	13.531	10.51	C11:0	14.47	0.69	C11:0
15.718	0.6577	C12:0 (200.324)	15.499	0.68	C12:0	15.05	0.51	C12:0
17.215	0.1905	C13:0 (214.351)	17.239	0.43	C13:0	17.58	2.43	C13:0
18.561	0.2606	C14:0 (228.378)	18.285	1.17	C14:0	18.23	0.40	C14:0
20.008	2.2615	C14:1 (226.378)	20.029	5.9	C14:1	20.01	1.92	C14:1
21.991	23.4727	C16:0 (256.432)	21.968	14.39	C16:0	21.95	21.84	C16:0
23.171	4.2869	C16:1 (254.432)	23.201	2.26	C16:1	23.16	6.13	C16:1
23.617	3.887	C17:0 (270.459)	23.648	4.04	C17:0	23.62	3.89	C17:0
25.275	1.2331	C17:1 (268.459)	25.309	2.71	C17:1	25.28	1.42	C17:1
25.813	1.2393	C18:0 (284.486)	25.851	10.94	C18:0	25.82	0.54	C18:0
26.430	0.7182	C18:1n9c (282.486)	27.032	2.34	C180:1n9c	26.42	3.49	C18:1n9c
27.219	0.3159	C18:2n6t (280.486)	27.838	3.17	C18:2n6t	27.22	0.28	C18:2n6t
28.143	3.6240	C18:2n6c (280.486)	28.178	0.64	C18:2n6c	28.14	2.98	C18:2n6c
29.520	3.3360	C20:1n9 (310.54)	29.553	2.12	C20:1n9	29.52	4.67	C20:1n9
29.773	2.2755	C18:3n3 (278.486)	29.773	2.28	C18:3n3	29.77	2.10	C18:3n3 (278.486)
30.823	1.337	C20:2 (308.54)	30.787	0.66	C20:2	30.71	2.08	C20:2
31.909	1.2321	C22:0 (340.594)	33.193	0.52	C23:0	31.89	2.38	C22:0
33.193	0.5212	C23:0 (354.621)				33.21	0.42	C23:0
						33.89	0.42	C22:2 (336.594)
						35.94	0.18	C24:1 (366.648)

$Molecular Weight of the fatty acid is represented in parentheses

$$CP = (0.526 \times C16) - 4.992 \quad (\textit{see}\ \textbf{Note 3}) \qquad (3)$$

where C16 is the area percentage of C16 FA obtained for the sample.

(d) Pour point (PP)

$$PP = (0.571 \times C16) - 12.240 \text{ (}\textit{see}\text{ \textbf{Note 4})} \quad (4)$$

where C16 is the area percentage of C16 FA obtained for the sample.

(e) Cetane number (CN)

$$CN = 46.30 + (5458/SV) - (0.225 \times IV)\text{(}\textit{see}\text{ \textbf{Note 5})} \quad (5)$$

(f) Degree of unsaturation (DU)

$$DU = MUFA + (2 \times PUFA) \quad (6)$$

where MUFA—monounsaturated fatty acid (*see* **Note 6**), PUFA—polyunsaturated fatty acid (*see* **Note 7**) (in wt%)

(g) Long chain saturation factor (LCSF)

$$\begin{aligned} LCSF = {} & (0.1 \times C16) + (0.5 \times C18) + 1 \times C20) \\ & + (1.5 \times C22) + (2 \times C24) \text{ (}\textit{see}\text{ \textbf{Note 8})} \end{aligned} \quad (7)$$

where C16, C18, C20, C22, and C24 represent the area percentage of respective FA obtained for the sample.

(h) Cold filter plugging point (CFPP)

$$CFPP = (3.1417 \times LCSF) - 16.477 \quad (8)$$

(i) Allylic position equivalent (APE)

$$APE = 2(AC18:1 + AC18:2 + AC18:3) \quad \text{(}\textit{see}\text{ \textbf{Note 9})} \quad (9)$$

where A is the area percentage of corresponding FA obtained in the sample.

(j) Bisallylic position equivalent (BAPE)

$$BAPE = AC18:2 + 2\,(AC18:3) \quad (10)$$

where A is the area percentage of corresponding FA obtained in the sample.

(k) Higher heating value (HHV)

$$HHV = 49.43 - (0.041(SV) + 0.015\,(IV)) \quad (11)$$

where SV is the saponification value and IV is the iodine value calculated for the sample.

(l) Oxidative stability (OS)

$$OS = -0.03844 \times DU + 7.770 \quad (12)$$

where DU is the degree of unsaturation value calculated for the sample.

(m) Flash point (FP)

$$FP = 205.226 + 0.083 \times C16:0 - 1.723 \times C18:0 - 0.5717 \times C18:1 - 0.3557 \times C18:2 - 0.467 \times C18:3 - 0.2287 \times C22 \quad (13)$$

where C16:0, C18:0, C18:1, C18:2, C18:3, and C22 represent the area percentage of respective FAs obtained for the sample.

(n) Saturated fatty acid (SFA)

$$SFA = \text{Sum of area percentage of FA with no bond (see Note 10)} \quad (14)$$

3.1 Selection of Suitable Strains for Biodiesel Production Using MCDA–PROMETHEE

1. Download the Visual PROMETHEE academic edition in visual.promethee-gaia.net (Fig. 1).
2. Visual PROMETHEE is available for download either as a Windows installation file (VPsetup.msi) or an archive file (PROMETHEE.zip).
3. Double-click the VPsetup.msi file to launch the installation and then follow the instructions. Read and agree with Visual PROMETHEE license during the installation process.
4. Enter the organism name in the action tab and values of biodiesel properties in the criteria tab (*see* **Note 11**).
5. Set the preference function as linear; threshold as absolute; Q indifference as 0; P preference and min/max as mentioned in Table 2 and shown in Fig. 2 (*see* **Notes 12–14**).
6. Click the GAIA menu after entering all the corresponding values. It gives the graphical representation as shown in Fig. 3.
7. The fuchsia lines represent the biodiesel quality parameters, *i.e.* criteria; the aqua circles represent the samples analyzed, *i.e.* actions; and the red line represents the decision axis (*see* **Note 15**).
8. The strain along with the decision axis, *i.e.* Sample 2 is the most suitable candidate among the other strains.
9. Most of the biodiesel quality parameters lie adjacent to the decision axis (CP, OS, PP, IV, CFPP, CN, MUFA), while some lie opposite to the decision axis (HHV, FP) and having less influence on the decision axis, The highly deviated parameter from the center point influence the decision axis (*see* **Note 16**).
10. Click the PROMETHEE table in the menu bar, and it displays the ranking of the samples analyzed based on the phi score (Table 3) (*see* **Note 17**).

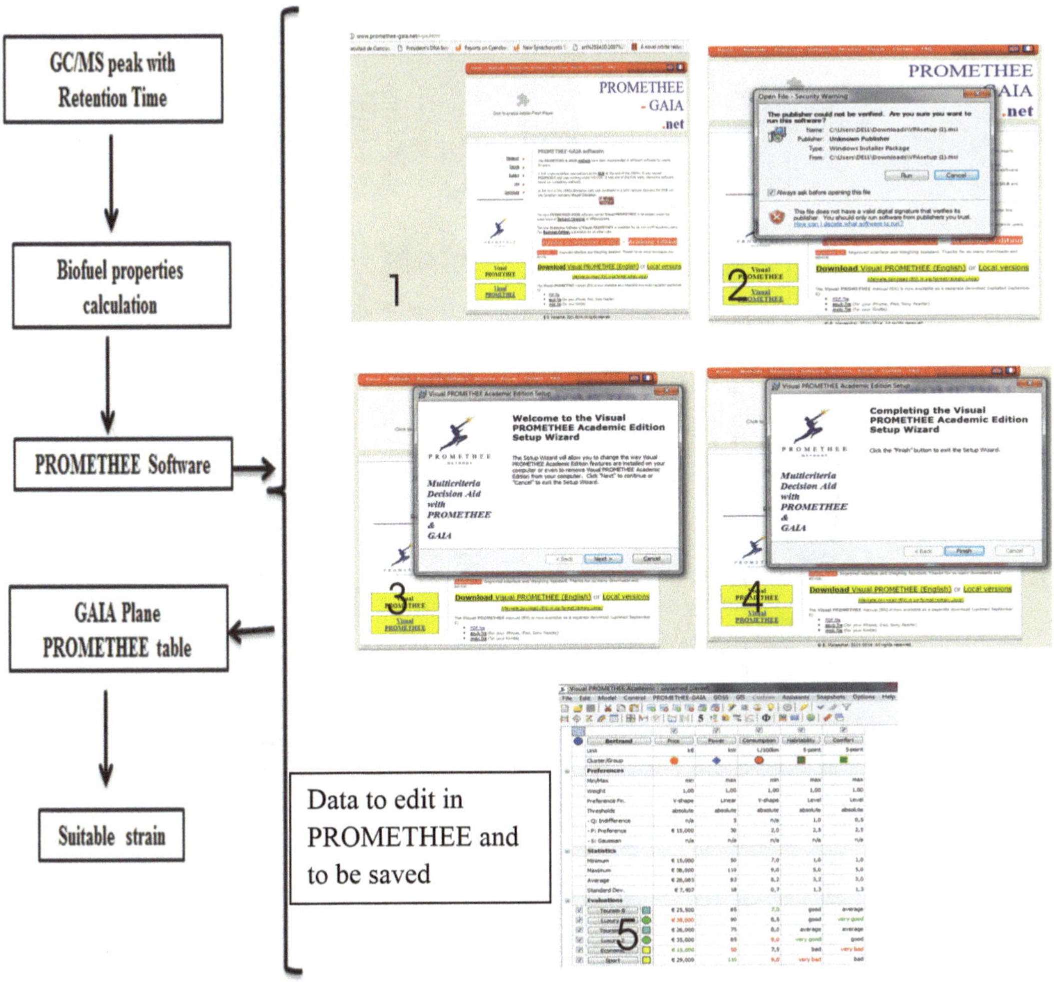

Fig. 1 Graphical Representation showing the method of PROMETHEE–GAIA software installation. Here, 1—Go to the PROMETHEE web page using the web address; 2—Run the set up file; 3—Click "next" option and continue; 4—Complete the wizard by select "finish" option; 5—Edit the data in the main window of PROMETHEE as mentioned in Table 2

Table 2
Calculated physical and chemical fuel properties of each sample FAME and the preference used in PROMETHEE–GAIA algorithm

Standards	IV (gI_2 100 g^{-1}fat)	SV (mg KOH g^{-1})	CP (°C)	PP (°C)	CN	DU (wt%)	LCSF (wt%)	CFPP (°C)	APE	BAPE	HHV (MJ kg^{-1})	OS (h)	FP (min)	SFA	P UFA	MUFA
Biodiesel standard EN 14214	≤120	–	–	–	≥51	–	–	≤5/−20	–	–	NA	≥6	120	–	–	–
Biodiesel standard ASTM D6751–02	NA	–	−3.0 to 12	−15.0 to 16.0	≥47	–	–	NA	–	–	NA	3	93	–	–	–
Biodiesel standard IS 15607	NA	–	12	3/15	≥51	–	–	6/18	–	–	NA	–	–	–	–	–
Biodisel standard EN 14213	–	–	–	–	–	–	–	–	–	–	35	–	–	–	–	–
Min/Max	Min	Min	Min	Min	Max	Min	Min	Min	Min	Min	Max	Max	Max	Max	Min	Max
Threshold value	120	–	12	16	51	–	–	18	–	–	35	6.0	120	–	–	–
S1	34.81	115.46	7.353	1.16	51	31.74	6.03	18	10.82	10.26	44.18	6.55	195.89	37.83	8.43	14.88
S2	28.93	184.72	2.58	−4.01	69.31	22.83	6.91	5.23	16.86	8.37	41.23	6.9	183.85	54.76	6.71	15.33
S3	36.19	145.32	17.4	0.2	75.7	37.49	6.72	4.66	24.9	14.74	42.93	6.33	199.15	36.92	7.99	21.51

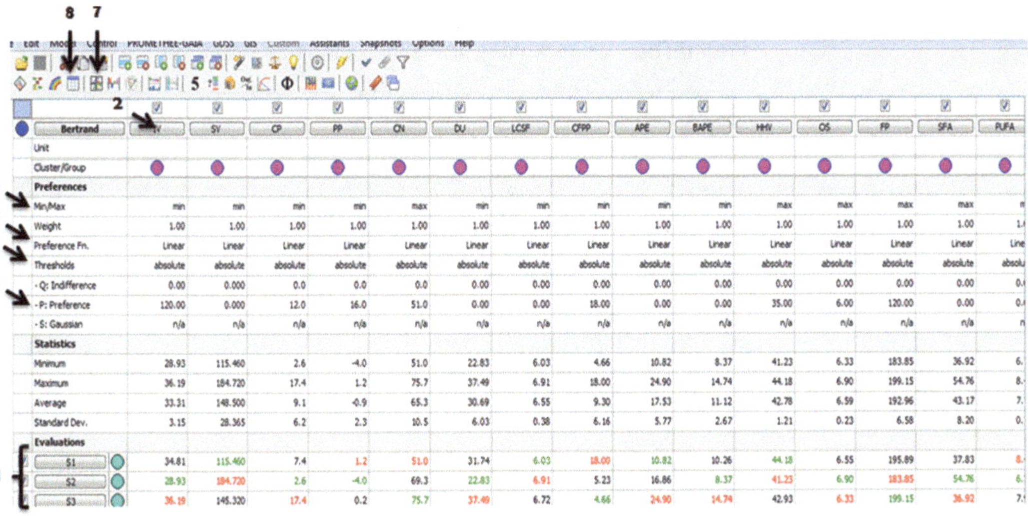

Bertrand	V	SV	CP	PP	CN	DU	LCSF	CFPP	APE	BAPE	HHV	OS	FP	SFA	PUFA
Unit															
Cluster/Group															
Preferences															
Min/Max	min	min	min	min	max	min	min	min	min	min	max	max	max	max	
Weight	1.00	1.00	1.00	1.00	1.00	1.00	1.00	1.00	1.00	1.00	1.00	1.00	1.00	1.00	
Preference Fn.	Linear	Linear	Linear	Linear	Linear	Linear	Linear	Linear	Linear	Linear	Linear	Linear	Linear	Linear	
Thresholds	absolute	absolute	absolute	absolute	absolute	absolute	absolute	absolute	absolute	absolute	absolute	absolute	absolute	absolute	
- Q: Indifference	0.00	0.000	0.0	0.0	0.0	0.00	0.00	0.00	0.00	0.00	0.00	0.00	0.00	0.00	
- P: Preference	120.00	0.000	12.0	16.0	51.0	0.00	0.00	18.00	0.00	0.00	35.00	6.00	120.00	0.00	
- S: Gaussian	n/a	n/a	n/a	n/a	n/a	n/a	n/a	n/a	n/a	n/a	n/a	n/a	n/a	n/a	
Statistics															
Minimum	28.93	115.460	2.6	-4.0	51.0	22.83	6.03	4.66	10.82	8.37	41.23	6.33	183.85	36.92	
Maximum	36.19	184.720	17.4	1.2	75.7	37.49	6.91	18.00	24.90	14.74	44.18	6.90	199.15	54.76	
Average	33.31	148.500	9.1	-0.9	65.3	30.69	6.55	9.30	17.53	11.12	42.78	6.59	192.96	43.17	
Standard Dev.	3.15	28.365	6.2	2.3	10.5	6.03	0.38	6.16	5.77	2.67	1.21	0.23	6.58	8.20	
Evaluations															
S1	34.81	115.460	7.4	1.2	51.0	31.74	6.03	18.00	10.82	10.26	44.18	6.55	195.89	37.83	
S2	28.93	184.720	2.6	-4.0	69.3	22.83	6.91	5.23	16.86	8.37	41.23	6.90	183.85	54.76	
S3	36.19	145.320	17.4	0.2	75.7	37.49	6.72	4.66	24.90	14.74	42.93	6.33	199.15	36.92	

1- Action; 2- Criteria; 3- Threshold value; 4- Threshold- absolute; 5- Preference function- Linear; 6- min/max; 7- GAIA plane; 8- Promethee table

Fig. 2 PROMETHEE display showing the action, criteria and preferences for evaluation of biodiesel quality

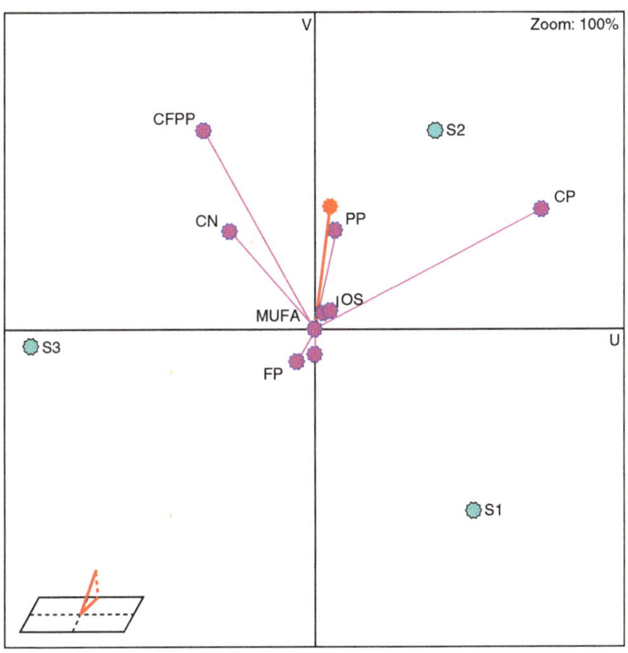

Fig. 3 PROMETHEE–GAIA plane showing the biodiesel properties of the tested samples S1, S2, and S3

Table 3
PROMETHEE table ranking the tested samples based on the Phi scores

Rank	Sample	Phi	Phi+	Phi−
1	S2	0.0875	0.1037	0.0162
2	S3	−0.0218	0.0514	0.0732
3	S1	−0.0657	0.0346	0.1002

4 Notes

1. From the *Sample 2* values in Table 1, IV is calculated as follows. Only the FAs of double/single bonds obtained for this sample is accounted for. For the molecular weight of each FA *see* Table 1 (values are given in parentheses against each FA obtained).

$$\text{C14:1} = 254 \times 1 \times 5.9/226.378 = 6.62$$

$$\text{C16:1} = 254 \times 1 \times 2.26/254.432 = 2.26$$

$$\text{C17:1} = 254 \times 1 \times 2.71/268.459 = 2.56$$

$$\text{C18:1n9c} = 254 \times 1 \times 2.34/282.486 = 2.1$$

$$\text{C18:2n6t} = 254 \times 2 \times 3.17/280.486 = 5.74$$

$$\text{C18:2n6c} = 254 \times 2 \times 0.63/280.486 = 1.14$$

$$\text{C20:1n9} = 254 \times 1 \times 2.12/310.540 = 1.734$$

$$\text{C18:3n3} = 254 \times 3 \times 2.28/278.486 = 6.24$$

$$\text{C20:2} = 254 \times 2 \times 0.66/308.540 = 0.54$$

$$\text{IV for Sample 2} = \sum(6.62 + 2.26 + 2.56 + 2.1 + 5.74 + 1.14 + 1.734 + 6.24 + 0.54)$$

$$\text{IV} = 28.93$$

2. From the *Sample 2* values in Table 1, SV is calculated as follows:

$$C8{:}0 = 560 \times 8.97/144.216 = 31.1$$
$$C10{:}0 = 560 \times 0.58/172.27 = 1.89$$
$$C11{:}0 = 560 \times 10.51/186.297 = 31.59$$
$$C12{:}0 = 560 \times 0.68/200.324 = 1.9$$
$$C13{:}0 = 560 \times 0.43/214.351 = 1.12$$
$$C14{:}0 = 560 \times 1.17/228.378 = 2.87$$
$$C14{:}1 = 560 \times 5.9/226.378 = 14.59$$
$$C15{:}0 = 560 \times 1.21/242.405 = 2.8$$
$$C16{:}0 = 560 \times 14.39/256.432 = 31.43$$
$$C16{:}1 = 560 \times 2.26/254.432 = 4.97$$
$$C17{:}0 = 560 \times 4.0384/270.459 = 8.36$$
$$C17{:}1 = 560 \times 2.71/268.459 = 5.65$$
$$C18{:}0 = 560 \times 10.936/284.486 = 21.53$$
$$C18{:}1n9c = 560 \times 2.34/282.486 = 4.64$$
$$C18{:}2n6t = 560 \times 3.17/280.486 = 6.33$$
$$C18{:}2n6c = 560 \times 0.64/280.486 = 1.28$$
$$C20{:}1n9 = 560 \times 2.12/310.54 = 3.82$$
$$C18{:}3n3 = 560 \times 2.28/278.486 = 4.58$$
$$C21{:}0 = 560 \times 1.31/326.567 = 2.25$$
$$C20{:}2 = 560 \times 0.66/308.54 = 1.2$$
$$C23{:}0 = 560 \times 0.52/354.621 = 0.82$$
$$\begin{aligned} SV = (&31.1 + 1.89 + 31.59 + 1.9 + 1.12 + 2.87 \\ &+ 14.59 + 2.8 + 31.43 + 4.97 + 8.36 \\ &+ 5.65 + 21.53 + 4.64 + 6.33 + 1.28 + 3.82 + 4.58 \\ &+ 2.25 + 1.2 + 0.82) \end{aligned}$$
$$SV = 184.72$$

3. CP for Sample 2 in Table 1 is $= (0.526 \times 14.39) - 4.992 = 2.58$.
4. PP for the Sample 2 is $= (0.571 \times 14.39) - 12.24 = -4.04$.
5. CN for Sample 2 $= 46.3 + (5458/189.23) - (0.225 \times 28.93) = 46.3 + (28.84) - (5.83) = 69.31$.
6. MUFA = Sum of area percentage of FAs with single bond. In Sample 2, C14:1, C16:1, C17:1, C18:1n9C, C20:1n9 are the

only single bond FAs obtained. Hence MUFA = 5.9 + 2.26 + 2.71 + 2.34 + 2.12 = 15.33.

7. PUFA = Sum of area percentage of FAs more than single bond. In Sample 2, C18:2n6t, C18:2n6c, C18:3n3, and C20:2 are the FAs with more than single bond. Hence PUFA = 3.17 + 0.64 + 2.28 + 0.66 = 6.71.DU = 15.33 + 2 × 6.71 = 28.75.
8. In Sample 2, only C16 and C18 FAs alone present. C20, C22, C24 are not detected in FAME. So LCSF for Sample 2 = (0.1 × 14.3947) + (0.5 × 10.936) = 6.91.
9. Add area percentage of all the C18:1 (C18:1n9t and C18:1n9c), C18:2 (C18:2n6t and C18:2n6c), and C18:3 (C18:3n3 and C18:3n6) obtained for the sample.
10. In Sample 2, C8:0, C10:0, C11:0, C12:0, C13:0, C14:0, C15:0, C16:0, C17:0, C18:0, C21:0, and C23:0 are the FAs with no bond. Hence, SFA = (8.97 + 0.59 + 10.51 + 0.68 + 0.43 + 1.17 + 1.21 + 14.39 + 4.04 + 10.94 + 1.31 + 0.52) = 54.76.
11. Add the new criterion/action by clicking the "new" option in each criteria/action tab. In the criterion, select numerical in scale option and edit the input name for each criteria/action as short form, *i.e.* for iodine value add IV for better visualization. The values are entered as mentioned in Table 2.
12. Enter the equal weight for all criteria, *i.e.* 1.0, which is the default value.
13. In the options tab, choose USA in general option for displaying values in decimal.
14. The icon for criteria and action can be changed to different colors and shapes by selecting the following option. Model> criteria groups/action categories> color or shape for easy visualization of each in GAIA plane.
15. In GAIA plane, the criteria that lie near to (±45°) are corresponded, while those lying in reverse bearing (135°–225°) are against related, and those in a generally orthogonal course have no or less impact. The quality of the GAIA plane should not be less than 75% which is displayed on the right side of the GAIA plane.
16. The fuel quality parameters CP, OS, PP, IV, CFPP, CN, and MUFA influence the decision axis and Sample 2 FAME profile has superior values compared to other samples (Table 2). IV should be at the maximum of 120 gI_2 100 g^{-1}fat. Flash point temperature should be at least 120 °C. OS should be higher or not lesser than 3–6 h. Higher CN is desirable for ensuring good cold start properties and reducing the development of white smoke. Whereas, lower CN may cause diesel knocking and enhancing the exhaust emissions. However, it is commonly

understood that biodiesels with low CFPP, CP, PP, and FP are better options for diesel engine fuels operating in cold weather condition.

17. Phi+ is the positive flow measuring the strength and Phi− is the negative flow measuring the weakness. Both Phi+ and Phi− can be used to rank the actions/parameters taken. The sum of the positive slices minus the sum of the negative ones is equal to the Phi net flow score of the action.

Acknowledgments

M.G. acknowledges Bharathidasan University authorities for the University Research Fellowship (05441/URF/K7/2013 dated 04.07.2013). The authors are thankful to DST-FIST programme (SR/FIST/LSI/-013/2012 dated 13.08.2012) for instrument facilities.

References

1. Ho DP, Ngo HH, Guo W (2014) A mini review on renewable sources for biofuel. Bioresour Technol 169:742–749
2. Dasgupta CN, Suseela MR, Mandotra SK, Kumar P, Pandey MK, Toppo K, Lone JA (2015) Dual uses of microalgal biomass: an integrative approach for biohydrogen and biodiesel production. Appl Energ 146:202–208
3. Ramos MJ, Fernández CM, Casas A, Rodríguez L, Pérez Á (2009) Influence of fatty acid composition of raw materials on biodiesel properties. Bioresour Technol 100:261–268
4. Knothe G (2005) Dependence of biodiesel fuel properties on the structure of fatty acid alkyl esters. Fuel Process Technol 86:1059–1070
5. Jahirul MI, Brown RJ, Senadeera W, Ashwath N, Rasul MG, Rahman MM, Hossain FM, Moghaddam L, Islam MA, O'Hara IM (2015) Physio-chemical assessment of beauty leaf (*Calophyllum inophyllum*) as second-generation biodiesel feedstock. Energ Rep 1:204–215
6. Jahirul MI, Brown RJ, Senadeera W, O'Hara IM, Ristovski ZD (2013) The use of artificial neural networks for identifying sustainable biodiesel feedstocks. Energies 6:3764–3806
7. Knothe G (2011) A technical evaluation of biodiesel from vegetable oils vs. algae. Will algae-derived biodiesel perform? Green Chem 13:3048–3065
8. Knothe G (2009) Improving biodiesel fuel properties by modifying fatty ester composition. Energy Environ Sci 2:759–766
9. Szybist JP, Song J, Alam M, Boehman AL (2007) Biodiesel combustion, emissions and emission control. Fuel Process Technol 88:679–691
10. Kumar M, Sharma MP (2015) Assessment of potential of oils for biodiesel production. Renew Sustain Energ Rev 44:814–823
11. Francisco ÉC, Neves DB, Jacob-Lopes E, Franco TT (2010) Microalgae as feedstock for biodiesel production: carbon dioxide sequestration, lipid production and biofuel quality. J Chem Technol Biotechnol 85:395–403
12. Brans JP, Mareschal B (1994) The PROMCALC & GAIA decision support system for multicriteria decision aid. Decis Support Syst 12:297–310
13. Behzadian M, Kazemzadeh RB, Albadvi A, Aghdasi M (2010) PROMETHEE: a comprehensive literature review on methodologies and applications. Eur J Oper Res 200:198–215
14. Anahas AMP, Muralitharan G (2015) Isolation and screening of heterocystous cyanobacterial strains for biodiesel production by evaluating the fuel properties from fatty acid methyl ester (FAME) profiles. Bioresour Technol 184:9–17
15. Wang M, Nie K, Yun F, Cao H, Deng L, Wang F, Tan T (2015) Biodiesel with low temperature properties: enzymatic synthesis of fuel alcohol fatty acid ester in a solvent free system. Renew Energ 83:1020–1025
16. Agarwal AK, Gupta T, Shukla PC, Dhar A (2015) Particulate emissions from biodiesel fuelled CI engines. Energ Convers Manage 94:311–330

Chapter 15

Production of High-Value Polyunsaturated Fatty Acids Using Microbial Cultures

Mingjie Jin, Rui Zhai, Zhaoxian Xu, and Zhiqiang Wen

Abstract

Microbes can produce not only commodity fatty acids, such as palmitic acid (16:0) and stearic acid (18:0), but also high-value fatty acids (essential fatty acids). Most high value fatty acids belong to long chain polyunsaturated fatty acids (PUFA), such as omega-3 fatty acids (e.g., eicosapentaenoic acid (EPA) and docosahexaenoic acid (DHA)) and omega-6 fatty acids (e.g., arachidonic acid (ARA) and γ-linolenic acid (GLA)). EPA (20:5n-3) is a 20-carbon fatty acid with five double bonds, and the first double bond is in the n-3 position. DHA (22:6n-3) is a 22-carbon fatty acid with 6 double bonds and the first double bond is also in the n-3 position. Both EPA and DHA play an essential role in cardiovascular health including prevention of atherosclerotic disease development (Zehr and Walker, Prostaglandins Other Lipid Mediat 134:131–140, 2018). ARA (20:4n-6) is a 20-carbon fatty acid with four double bonds, and the first double bond is in the n-6 position. GLA (18:3n-6) is an 18-carbon fatty acid with three double bonds, and the first double bond is in the n-6 position. ARA and GLA have multiple biological effects, such as lowering blood cholesterol, and lowering cardiovascular mortality (Poli and Visioli, Eur J Lipid Sci Technol 117 (11):1847–1852, 2015). This chapter provides details on microbial production of EAP, DHA, ARA, and GLA.

Key words Eicosapentaenoic acid (EPA), Docosahexaenoic acid (DHA), Arachidonic acid (RA), γ-Linolenic acid (GLA), Omega-3 fatty acids, Omega-6 fatty acids

1 EPA Production via Microbial Cultivation

1.1 Materials

1.1.1 Strains

Traditionally EPA was primarily recovered and purified from marine fish oils, which made it expensive due to the high production cost. Later, microorganisms, including microalgae, filamentous fungi, and marine bacteria, were found to be able to produce EPA. Based on these microbes, fermentation technology was developed for EPA production. With development of synthetic biology and metabolic engineering technology, genetic modified yeasts were constructed for EPA production and are now used for commercial production [1, 2]. The main EPA-producing strains are listed in Table 1.

Venkatesh Balan (ed.), *Microbial Lipid Production: Methods and Protocols*, Methods in Molecular Biology, vol. 1995, https://doi.org/10.1007/978-1-4939-9484-7_15, © Springer Science+Business Media, LLC, part of Springer Nature 2019

Table 1
EPA-producing strains and fermentation technologies

Organisms		Main carbon sources	Key production strategy	EPA content, yield, or productivity	Fermentation time	Main references
Microalgae	*Nannochloropsis* sp.	CO_2	Elevated the concentrations of CO_2	340 μg/L	4 days	[3]
	Nitzschia laevis	glucose	Subsequent fed-batch cultivation	695 mg/L	14 days	[4]
	Nannochloropsis oceanica CY2	CO_2	Long term semibatch operations	10.5 mg/L/d	NM	[5]
	Nannochloropsis oceanica CY2	CO_2	A novel photobioreactor with additional immersed light sources	14.4 mg/L/d	NM	[6]
	Nitzschia laevis UTEX 2047	Glucose	Perfusion culture	1112 mg/L	25 days	[7]
	Nannochloropsis gaditana B-3	CO_2	Flat-panel geometry application using medium recycling	30 mg/L/d	1 year	[8]
	Nannochloropsis sp	Glucose	Mixotrophic conditions with glucose as carbon source	22 mg/L	8 days	[9]
Filamentous fungi	*Pythium irregulare*	Crude glycerol	Use crude glycerol as the main carbon resource	89.8 mg/L	8 days	[10]
	Mortierella alpina	Glucose	Expressing heterologous the Δ17-desaturase gene under ordinary temperature	1.8 g/L	6 days	[11]
	Pythium ultimum strain #144	Glucose	25 °C for cell growth and 13 °C for EPA production	383 mg/L	10 days	[12]
	Mortierella alpina 20-17	Glucose	20 °C for cell growth and 12 °C for EPA production	490 mg/L	7 days	[13]
	Pythium irregulare (ATCC 10951)	Dry-milling derived thin stillage	Appling dry-milling derived thin stillage	243 mg/L	9 days	[14]

Marine bacteria	SCRC-8132	Glucose	High-producing strains isolation	20 mg/L	18 h	[15]
	Shewanella sp.717	Casein	Medium optimization strategy	350 mg/L	96 h	[16]
	SCRC-2738	Corn steep liquor	Medium optimization strategy	200 mg/L	18 h	[17]
	Synechococcus sp. NKBG15041c	CO_2	EPA synthesis gene cluster was cloned into *Synechococcus* sp. NKBG15041c and the recombinant was cultured at 23 °C	2.24 mg/L	48 h	[18]
Yeast	*Yarrowia lipolytica* Y5037	NA	Expression of a Δ9 elongase/Δ8 desaturase pathway	18.6% EPA to total lipid	NM	[1]
	Recombinant *Yarrowia lipolytica*	NA	Comprehensive metabolic engineering	At least 18.6% EPA to total lipid	NM	[19]
	Recombinant *Yarrowia lipolytica*	Glucose	Inactivation of the peroxisome biogenesis gene *PEX10* and other metabolic engineering	At least 56.6% EPA to total lipid	NM	[2]

NM not mentioned in the references

1.1.2 Fermentation Medium

The culture media for EPA production varies based on the EPA-producing strain. The carbon sources can be divided into two general types: (a) the organic carbon sources (e.g., glucose and glycerol) which are mainly applied for the culture of heterotrophic filamentous fungi, bacteria, yeasts, and some microalgae; (b) the inorganic carbon sources (e.g., CO_2 and Na_2CO_3) which are mainly applied for the culture of photoautotrophic microalgae (Table 1). For the nitrogen source, yeast extract, peptone, $NaNO_3$, KNO_3, and urea are often applied in EPA production. When the EPA-producing strain originated from the ocean, seawater and artificial seawater are commonly as well [5, 8, 20].

1.2 Method

1.2.1 Fermentation Protocol

Phaeodactylum tricornutum is used here as an example to illustrate fermentation protocol for EPA production since it is commonly used for EPA production. First, a slant of *P. tricornutum* was inoculated into 100 mL medium contained in a 500 mL flask, and incubated at 23 °C. The seed culture, when ready, was transferred into plastic carboys containing seawater enriched with F/2 culture medium under sterile conditions. It is then cultured at 23 °C for approximately 14 days. After cultivation, biomass is harvested and EPA is extracted. Cool white fluorescent lamps are set in a 24 h light regime of 60 μmol photons m^2/s [21]. To achieve high cell density, high yield and high productivity of EPA, advanced cultures strategies are also employed in EPA production, such as fed-batch and continuous cultivation, perfusion cultures, applying different illumination systems, and innovative bioreactor designs (Table 1).

1.2.2 Key Factors in Fermentation Process for EPA Production

Medium Components

Appropriate nutrients, including carbon, nitrogen, phosphorus, sulfur sources, and trace elements are vital for EPA production. Carbon sources are assimilated by the microorganisms to provide the energy and carbon skeletons for the cell growth and EPA biosynthesis. For photoautotrophic microalgae, CO_2 is sufficient as the carbon source, and in some cases EPA accumulation has been enhanced by elevated concentrations of CO_2 [3]. However, for heterotrophic microalgae and other microorganisms, they cannot utilize CO_2 directly and additional organic carbon sources are needed, including glucose, glycerol, fructose, sucrose, lactose, acetate, starch, and agricultural/forestry wastes. In particular, the addition of some oils can enhance EPA production significantly for some species because these oils contain abundant precursors for EPA biosynthesis [22]. Besides, the nitrogen source and concentration are also vital to cell growth and fatty acids composition produced by EPA-producing strains. Appropriate nitrogen source is critical for cell growth, lipid content, and EPA synthesis, and generally, a high C/N ratio favors EPA accumulation. Moreover, the consumption of nitrogen resource (e.g., the ammonium salt) often leads to the fluctuation of pH, which influences the growth and EPA production. Medium salinity can significantly affect the physiological properties of EPA-producing strains, especially for marine microalgae. Several microorganisms have been investigated to determine how salinity affects cell growth and EPA production. For example, in *Nannochloropsis* sp. cultivation, the highest EPA percentage was achieved using 13 g/L NaCl and the percentages of C16 fatty acids achieved their highest levels when NaCl was kept below 40 g/L [20].

Temperature

Temperature is another key factor affecting cell growth and lipid composition. In most cases, high temperature leads to high biomass concentration, but it has adverse influence on content of polyunsaturated fatty acids [23, 24]. Low temperature condition usually leads to a relatively high EPA content, and there is almost no report for wild microorganisms cultured above 25 °C. EPA production at low temperature is believed to be enhanced by the higher levels of intracellular O_2, because the enzymes responsible for the elongation and desaturation of EPA depend on the availability of O_2. In evolutionary terms, the generation of unsaturated fatty acids at low temperature could enhance fluidity and stability of cellular membranes. Due to the different temperature demands for cell growth and EPA biosynthesis, temperature shift strategies are often employed to enhance the overall production of EPA [12, 13]. Recently, Okuda et al. constructed a recombinant *Mortierella alpina* strain by heterologous expressing *Saprolegnia diclina* Δ17 desaturase gene in *M. alpina* ST1358, which allowed for EPA accumulation at 28 °C [11]. The higher temperature was advantageous because it reduced the amount of cooling water and associated energy.

pH

The culture pH also affects EPA production. An initial culture pH value at 6.0–8.0 is found to be optimal for EPA production by microalgae, fungi and marine bacteria [10, 15, 25]. For example, Sang et al. [26] investigated the influence of initial pH on the EPA production by *Pinguiococcus pyrenoidosus*, and results showed that cell growth was significantly inhibited at pH 5.0, and cell biomass increased as the initial pH was raised from 5.0 to 8.0. The maximum percentage of EPA, which was 23.13%, was achieved with an initial pH of 6.0, and the minimum percentage of EPA was observed at a pH of 5.0 [26]. For a marine bacteria SCRC-8132, the highest production of EPA was achieved when the culture pH was maintained at 7.0 [15].

Aeration and Agitation

Microorganisms require molecular oxygen for desaturation process in the EPA biosynthesis, and an increased oxygen concentration in the culture was observed to elevate unsaturated fatty acid content in *Chlorella sorokiniana* and *Gyronidium cohnii*

[27, 28]. Increased aeration and agitation improves dissolved oxygen content in medium. However, excessive agitation also leads to poor cell growth and low EPA production due to high shear stress. Typically, agitation is maintained at 120–550 rpm for EPA production [5, 12, 29].

Light Exposure

Most EPA-producing microalgal species are photoautotrophs, which require abundant light formation. As reported, the relationship between light intensity and polyunsaturated fatty acids (e.g., EPA) production varies with species: low light intensity enhanced formation and accumulation of polyunsaturated fatty acids in many diatoms and euglenids. However, for the green alga, and red alga, the effect of light intensity is reversed [25, 28]. Based on the relationship of EPA production and irradiance level, effective cultivation processes have been developed. For example, illumination with a 14 h light-on and 10 h light-off cycle resulted in a high EPA production rate and a high electricity conversion efficiency for *Chlorella vulgaris* [30]. Further investigation on the effect of different LED light sources on cell growth and EPA production by *Chlorella vulgaris* indicated LED-white leaded to the highest biomass production and LED-blue resulted in the highest EPA content [5].

1.2.3 Analytical Methods for EPA Content

The contents of EPA and other PUFAs are mainly analyzed with gas chromatography. First, 20 mg dried cells are suspended in 2 mL methanolic HCl (5%), and heated at 70 °C for 2 h in sealed tubes. Then, the fatty acid methyl esters are extracted from the cells with 0.6 mL hexane and then dried under nitrogen gas. The fatty acid methyl esters profiling is performed by gas chromatography (e.g., GC-2014 (Shimadzu) equipped with a flame ionization detector and SPTM-2560 capillary GC column (Sigma-Aldrich)) with helium as the carrier gas [5]. In addition, an internal standard is needed (e.g., nonadecanoic acid (C19:0)) to determine the amount of each fatty acid based on the areas of all peaks and the known concentration of the standard added.

In conclusion, the EPA production depends on the characteristics of the producing-strain and cultivation strategy. The medium components, temperature, pH, aeration, agitation, and light are all important parameters for EPA production, and the responses are interactive. For example, on the nitrogen-replete medium, increasing light intensity and salinity increased EPA content. In contrast, EPA productivity is relatively low under nitrogen starvation conditions with high salinity and light intensity [20]. Furthermore, with the metabolic pathways involved in EPA biosynthesis elucidated and the relevant genes identified, the application of genetically modified microorganisms is becoming a promising route for EPA production. The engineered strains have already been used to overcome restrictions associated with pH, temperature, and other culture conditions to a certain extent.

2 DHA Production via Microbial Cultivation

2.1 Materials

2.1.1 Strains

Schizochytrium is considered to be the most promising microorganisms for producing DHA. It is related to microalgae and can accumulate over 70% of its weight as lipids of which the DHA content can equal up to 35% of the total fatty acids [31]. Table 2 lists examples of DHA production using *Schizochytrium*.

2.1.2 Seed Culture Medium

The seed culture medium contained 40 g/L glucose and 0.4 g/L yeast extract, which are dissolved in artificial sea water [39], trace element solution (2 mL/L), and vitamin solution (2 mL/L). The artificial sea water is composed of (g/L) Na_2SO_4 10; $(NH_4)_2SO_4$ 0.8; KH_2PO_4 4; KCl 0.2; $MgSO_4$ 2; monosodium glutamate 20; $CaCl_2$ 0.1. The trace element solution contained the following components (g/L): Na_2EDTA 6, $FeSO_4$ 0.29, $MnCl_2{\cdot}4H_2O$

Table 2
Examples of DHA production by *Schizochytrium* in recent years

Strain	Scale (L)	Carbon source	Biomass (DCW g/L)	DHA (g/L)	DHA prod. (mg/L/h)	References
Schizochytrium sp. HX-308 (CCTCC M 209059)	1500	Glucose	71	17.5	119	[32]
Schizochytrium sp. HX-308 (CCTCC M 209059)	50	Glucose	92.72	17.7	111	[33]
Schizochytrium sp. S31	7.5	Glycerol	151.40	28.93	301	[34]
Schizochytrium sp. HX-308 (CCTCC M 209059)	1500	Glucose	71.28	14.16	101.1	[35]
Schizochytrium sp. HX-308 (CCTCC M 209059)	7000	Glucose	90.1	19.72	136.9	[35]
Schizochytrium sp. CCTCC M209059	10	Glucose	72.37	18.38	138.8	[36]
Schizochytrium sp. HX-308	0.5	Glucose	ND	8.9	148.3	[37]
Schizochytrium sp. S31	7.5	Glucose	123.04	21.26	177.2	[38]
Schizochytrium sp. HX-308 (CCTCC M 209059)	7000	Glucose	97.5	18.48	124.8	[39]
Schizochytrium sp. LU310	1	Glucose	88.6	24.74	241.5	[40]
Schizochytrium sp. HX-308 (CCTCC M 209059)	5	Glucose	84.34	26.40	220	[41]
Schizochytrium sp. B4D1	5	Glucose	107.91	30.7	284	[42]
Schizochytrium sp. HX-308 (CCTCC M 209059)	15	Glucose	108.09	21.42	187.9	[43]

0.86, $ZnSO_4$ 0.8, $CoCl_2{\cdot}6H_2O$ 0.01, $Na_2MoO_4{\cdot}2H_2O$ 0.01, $NiSO_4{\cdot}6H_2O$ 0.06, and $CuSO_4{\cdot}5H_2O$ 0.6. Vitamin solution is filter-sterilized (0.22 μm) and contained (mg/L) thiamine 50, biotin 1, and cyanocobalamin 10. All medium components are separately heat-sterilized (121 °C) in an autoclave.

2.1.3 Fermentation Medium

Same as seed culture with increased glucose concentrations (e.g., 120 g/L).

2.2 Methods

2.2.1 Fermentation Protocol

Laboratory scale: The culture preserved in the glycerine tube was inoculated into a 250-mL flask with 50 mL medium and cultivated for 24 h a shaker incubator set to 25 °C, 160 rpm. After three generation cultivation, the preculture was inoculated into a 10-L seed fermentor with an inoculum size of 1% (v/v) and cultivated for 24 h, and the seed culture (20%, v/v) was then transferred to a 50-L fermentor with a working volume of 35 L. In the fermentation culture, with initial glucose concentration 120 g/L and 5% (v/v) inoculum size, cells were grown at 25 °C, 160 rpm [32, 36].

Commercial scale:A frozen glycerol algae stock (e.g., −80 °C) is used for inoculation into a 250-mL flask with 50 mL medium and cultivated for 24 h in a shaker incubator set to 25 °C, 160 rpm. After three generations of cultivation, the preculture is transferred to a 150-L seed fermentor with an inoculum size of 1% (v/v) and cultivated for 24 h. Agitation speed is set at 220 rpm with aeration rate of 2.8 m^3/h. The seed culture (10%, v/v) is then transferred to a 1500-L seed fermentor with a working volume of 1000 L and cultivated for another 24 h. Agitation speed is set at 110 rpm with aeration rate of 28 m^3/h. Finally, the seed culture is transferred to a 7000-L fermentor with a working volume of 4500 L. Agitation speed is set at 90 rpm with aeration rate of 90 m^3/h. The temperature for all cultivations is set at 28 °C [39].

2.2.2 Effects of pH

The culture pH is a critical parameter for the production of DHA through fermentation using *Schizochytrium limacinum*. It directly affects both cell growth and intracellular metabolism for DHA production. It has been reported that the proper culture pH for *Schizochytrium* ranges from 7 to 3. Generally, neutral pH facilitated cell growth, while acidic pH is beneficial to DHA accumulation. Jia et al. [44] investigated the impact of pH as well as two-stage pH shifting strategies on cell growth and DHA production by *Schizochytrium limacinum*. It was found that maintaining pH at 3 in the first 4 days and then maintaining pH at 5 in the last 2 days resulted in the highest DHA yield [44].

2.2.3 Agitation and Aeration

As discussed above, oxygen concentration plays an important role in determining DHA production. Fatty acids fermentation is aerobic. High aeration is beneficial to cell growth by increasing cell growth rate. However, *Schizochytrium* species synthesize long chain PUFAs via a polyunsaturated fatty acid synthase that is O_2 independent [33]. Generally, dry cell weight and total fatty acids increased along with agitation and aeration, whereas DHA content decreases with increased agitation and aeration. Accordingly, aeration needs to be controlled appropriately to achieve high cell density, high lipid accumulation, and high DHA content [35]. Ren et al. proposed a stepwise aeration control strategy, in which the aeration rate was controlled at 0.4 volume of air per volume of liquid per minute (vvm) for the first 24 h, then shifted to 0.6 vvm until 96 h, and then switched back to 0.4 vvm until the end of the fermentation. This stepwise control scheme proved to be much more effective for DHA production than using a constant aeration rate [45].

2.2.4 Temperature

Temperature is one of the important environmental factors affecting the biosynthesis of DHA. It has been reported that the proper culture temperature for *Schizochytrium* ranged from 16 to 37 °C [46]. The optimal temperature for *Schizochytrium* is 27 °C. Generally, a low temperature slows cell growth, but promotes the accumulation of unsaturated fatty acids. Therefore, a two-step temperature profile is most suitable for DHA production. Zeng et al. investigated the effects of temperature and a stepwise temperature change on fatty acid production and DHA content in *Schizochytrium* sp. HX-308. It was found that the optimum strategy for DHA production was cultivation at 30 °C for 32 h before shifting temperature to 20 °C for the final 12 h of cultivation [45].

2.3 Medium Components

Schizochytrium sp., a thraustochytrid, is a heterotrophic marine fungus. Sufficient nutrients are required to support *Schizochytrium* growth and DHA synthesis.

2.3.1 Carbon Source

Schizochytrium use a broad spectrum of carbon sources for growth and DHA production. Monosaccharides glucose and fructose are good carbon sources for cell growth. Glucose is the best carbon source for biomass, lipid and DHA production. Disaccharides lactose, maltose, sucrose and polysaccharide, soluble starch, can all be used by *Schizochytrium* for cell growth and lipid production, but are not as good as monosaccharides for DHA production [47].

2.3.2 Nitrogen Source

Schizochytrium can utilize both complex and defined nitrogen sources. Among the complex nitrogen sources, such as yeast extract, peptone, and tryptone, yeast extract is the most effective for biomass, lipid, and DHA production. Among the defined nitrogen sources, monosodium glutamate and ammonium chloride performed much better than urea and sodium nitrate in DHA production [47].

2.3.3 Phosphorus

Phosphorus is an essential element required for cell growth, intracellular metabolism, and lipid biosynthesis. Phosphate-limitation strategy is commonly used in microbial fermentation process. Ren et al. investigated the effect of phosphate concentration on lipid and especially DHA synthesis in *Schizochytrium* sp. HX-308, and found that DHA yield and productivity in the medium with KH_2PO_4 concentration of 0.1 g/L were much higher than those in the media with KH_2PO_4 concentrations of 0.5, 1, 2, and 4 g/L [37].

2.3.4 Malate (NADPH Supply)

Malate is a key intermediate in the citrate/malate and transhydrogenase cycle that provides NADPH for DHA synthesis. It can be introduced into the fermentation system during the fast lipid accumulation stage to enhance NADPH supply. Zhang et al. investigated the effect of malate concentration on lipid and especially DHA synthesis in *Schizochytrium*, and found that adding 4 g/L malate was added during the rapid lipid accumulation stage significantly increased DHA production by 47%. Besides, a constant-concentration malate feeding strategy was also applied to enhance DHA titer, DHA yield and DHA productivity [42, 48].

2.3.5 Ethanol and Sodium Acetate (Acetyl-CoA Supply)

According to the metabolic pathway of DHA biosynthesis by *Schizochytrium* sp., the formation of acetyl-CoA in oleaginous microorganism is facilitated by the presence of ATP:citrate lyase. Acetyl-CoA in the cytosol of the cell is a necessary precursor for fatty acid synthetase (FAS) [49]. Ethanol and sodium acetate, two interesting alternative carbon sources for DHA production, can be directly converted into acetyl-CoA by acetyl-CoA synthetase in eukaryotes. Ren et al. investigated the effects of ethanol and sodium acetate concentration on lipid and especially DHA synthesis in the oleaginous fungi *Schizochytrium*, and found that 40 mL/L ethanol and 2 g/L sodium acetate resulted in the highest DHA yield among the tested concentrations (40, 80, 120, 160 mL/L ethanol; 2, 4, 6, 8 g/L sodium acetate) [48].

In summary, industrial production of DHA using Thraustochytrids has been applied since 20 years ago. During the last two decades, great progress has been made in DHA productivity and yield by optimization of medium and process engineering. However, development of genetically modified strain was hampered by limited genome sequence annotation and lack of genetic tools for gene transfer. In post-genomic era, as more sequences are becoming available, the metabolic pathways involved in DHA biosynthesis and targets for strain improvement can be identified. The synthetic biotechnology may provide promising approach for a system-level metabolic engineering for improved productivities.

3 ARA Production via Microbial Cultivation

3.1 Materials

3.1.1 Strains

ARA can be produced by various groups of microorganisms such as *Mortierella*, *Zygomycetes*, and *Oomycetes* [50]. Among these species, members belonging to the genus *Mortierella* have been widely studied because representatives have been observed to produce high amounts of ARA [50]. For example, *M. alpina* is able to generate a very high ARA content (more than 50% of total fatty acids) [51–53].

3.1.2 Fermentation Medium

Seed culture medium usually includes (g/L) glucose 30, KH_2PO_4 3, yeast extract 6, $MgSO_4{\cdot}7H_2O$ 0.5, and $NaNO_3$ 3. Fermentation medium usually contains (g/L) glucose 100, KH_2PO_4 3.8, yeast extract 11, $MgSO_4{\cdot}7H_2O$ 0.5, and $NaNO_3$ 3.4 [53]. Besides, urea is also used as a nitrogen source [54].

3.2 Methods

3.2.1 Fermentation Protocol

Seed culture is carried out in seed culture medium (50 mL) using 250 mL baffled flasks. The culture is grown at 25 °C under shaken condition (120 rpm) for 3 days. Subsequently, the fermentation medium (50 mL) is inoculated at 10% v/v in 250-mL baffled flask, which is then kept in a shaken incubator (130 rpm) for 6.5 days at 23 °C [53].

3.2.2 Key Factors Affecting ARA Fermentation

Fermentation Process

Many studies have mistakenly produced ARA using a constant shaking/agitation, which could damage mycelia and inhibit synthesis of ARA during the later stage of fermentation. To overcome this disadvantage, two-step fermentation processes were developed, which consist of a conventional high intensity aeration step to promote growth followed by a slow shaking agitation step. It was found that this process greatly enhanced ARA production. By using a two-step aeration scheme, an ARA yield (19.8 g/L), constituting 75% of total lipid, was reached after 11-day cultivation, which was 1.7 times higher than the yield obtained through one-step fed-batch cultivation [53]. This indicates that fermentation process control affects ARA yield and the two-step fermentation process is promising for production of ARA.

Temperature

Temperature affects cell growth, thus influencing the production of ARA. It has been shown that lower temperature resulted in a higher content of ARA in the lipid fraction. However, due to slow metabolic rate at low temperature, ARA productivity was not actually increased. It was found that the stain grew well at 30 °C, yielding a higher content of biomass and lipids, while the % ARA lipid content reached an optimum at 25 °C [55, 56].

pH and Cultivation Time

Another important factor is pH, which greatly affects ARA yield [56]. Although a relatively low pH (5–6) enhances the growth of cell biomass, the content of ARA is relatively low. It has been shown that ARA in lipids achieved the peak when the initial pH was 8.5 [56]. The culture time also affect the total amount of ARA produced by the microorganism. It was found that the lipids content increased continuously during the initial 6 days, and then decreased slightly. The ARA content increased with the cell growth, which also achieved maximum at around sixth day [55].

Dissolved Oxygen

Dissolved oxygen affects the rates of cell growth and ARA production [52, 54, 57]. Maintaining high levels of dissolved oxygen in the medium (e.g., 40%) can alleviate growth inhibition by high concentration of nutrients and lead to an increase of ARA content in the lipids.

Carbon Source

Carbon source affects both total lipid and ARS yields. It was found that among glucose, maltose, starch, sucrose, and glycerin, the growth of the strain was much faster when glucose was used. The strain growth was moderate with maltose and starch and slow with sucrose and glycerol [56]. In addition, carbon source concentration determines the amount of lipids that can be produced. For example, it was reported that ARA production increased with glucose concentration [55]. However, high carbon source concentration also has an inhibitory effect on stain growth, possibly due to osmotic stress [55]. Therefore, carbon source concentration should be maintained within an optimal range.

Nitrogen Source

The types of nitrogen source affect the amount of ARA produced by *M. alpina*. Yoshifumi Shinmen et al. examined the effect of nitrogen source on the production of ARA by *M. alpina* [58]. It was found that the yeast extract, corn steep liquor, and soybean were good nitrogen sources when used at optimum concentration. Among these nitrogen sources, the soybean resulted in the highest ARA concentration and ARA yield [59]. When evaluating the effect of nitrogen source on morphology, Park et al. found that the soybean led to formation of a rapidly growing morphology, which is favorable for ARA productivity [59].

C/N Ratio

The C/N ratio greatly affects cell growth and mycelial morphology, and thereby influences the production of ARA. It has been found that to achieve an optimum production of ARA, it is important to keep the ratio of C/N at around 15–20 [57]. As ARA is produced intracellularly, it is essential to achieve a high mycelial concentration in the fungal culture. A previous study has shown that an optimum C/N could facilitate the formation of feather-like morphology of mycelia with small size. Under this condition, nutrient supply helps the accumulation of ARA and improves ARA productivity

[57, 59]. As shown by Park et al., when the ratio of C/N was kept at 20, the amount of ARA production was proportional to the concentrations of carbon and nitrogen [59].

In summary, various factors such as pH, dissolved oxygen and temperature all play essential role in the fermentation for ARA production. Two-step fermentation process has been considered an effective way to produce ARA. Temperature and dissolved oxygen affect cell growth, thus influencing the production of ARA. In addition, carbon source and nitrogen source influence both total lipid and ARA yields. On top of that, the C/N ratio greatly affects cell growth and mycelial morphology, and thereby influences the production of ARA. Thus, to facilitate ARA production, all of these factors need to be considered and optimized.

4 GLA Production via Microbial Cultivation

4.1 Materials

4.1.1 Strains

Zygomycetes have been widely studied in the literature for their capability in producing GLA, which include the genera *Mucor*, *Mortierella*, *Rhizopus*, *Cunninghamella*, and *Zygorhynchus* [60]. The first microbial industrial process for producing GLA was developed in the UK using *Mucor circinelloides* [61]. Recently, *Mortierella* and *Mucor* have been successfully applied in Japan on the industrial scale to produce GLA [61].

4.1.2 Fermentation Medium

The fermentation medium usually includes a carbon source and a nutrient solution containing yeast extract, peptone, $NaNO_3$, $MgSO_4$, and K_2HPO_4. The carbon source varies considerably, but can be divided into two general types: those metabolized as C6/C5 sugars, such as glucose, xylose and polysaccharides, and those metabolized as C2, such as acetic acid and ethanol [62].

4.2 Methods

4.2.1 Fermentation Protocol

The fungal strains are maintained on Czapek-Dox agar slants at 4 °C. The spore inoculum is prepared by incubating the mycelium grown on Sabouraud agar with an aqueous solution of Tween 80 (0.1% w/v) for 4 days at 28 °C. Then, the mycelia are washed with sterile distilled water to obtain a concentrated spore suspension. The suspension is further diluted to obtain a final spore concentration of 10^5–10^8 spores/mL [61]. Subsequently, the spore suspension is used to inoculate the fermentation culture. A 500-mL Erlenmeyer flask filled with glucose and 60 mL nutrient solution containing $NaNO_3$, K_2HPO_4, $MgSO_4{\cdot}7H_2O$, and yeast extract is used for fermentation. The culture is incubated at 28 °C under shaken condition (120 rpm) for 6 days [61].

4.2.2 Key Factors Affecting Fermentation

Examples of GLA production using different species and cultivation conditions are shown in Table 3.

Table 3
Examples of GLA production using different species and cultivation conditions

Species	Carbon source	Nitrogen source	pH	Temperature [°C]	Duration	GLA content in lipids (%)	Dry weight (g/L)	References
Cunninghamella elegans CCF 1318	Barley	Yeast extract + peptone	NA	28	420 h	NA	15.8	[61]
Sporobolomyces carnicolor O33	Glucose	Urea	5.6	NA	10 h	NA	3.3	[63]
Mortierella isabellina ATHUM 2935	Xylose	Yeast extract + $(NH_4)_2SO_4$	5.2–6.3	28	216 h	3.90%	8.7	[62]
Cunninghamella echinulata ATHUM 4411	Xylose	Yeast extract + $(NH_4)_2SO_4$	5.2–6.0	28	192 h	6.60%	7.8	[64]
Cunninghamella echinulata ATHUM 4411	Orange peel	$(NH_4)_2SO_4$	4.1–6.4	28	240 h	5.20%	1.7	[64]

Temperature

Temperature influences growth, GLA production, lipid content, and fatty acid composition in *Zygomycetous* fungal cultures [65]. For example, when *C. elegans* was cultivated on barley, the optimum temperature for GLA production was 21 °C, yielding 12.6 mg GLA/g dry substrate after 9 days of cultivation [61]. Although the whole growth process including the metabolic processes of lipid accumulation and transformation was faster at higher temperatures, the GLA yield only achieved 9.3 mg GLA/g after 7 days of cultivation at 28 °C. The production yield could not be increased by extending incubation time [61]. Similarly, it was also found that the highest GLA content in *Mucor rouxii* was achieved at a low temperature and a reduced GLA proportion was observed when the culture temperature was raised (35 °C) [66]. The effect of temperature might be associated with the adaptive role of GLA in stabilizing membrane fluidity at low temperatures. GLA has a huge impact on the regulation of membrane fluidity within the microorganisms, thus possibly compensating for the decreased functionality of the biomembranes at low temperatures. Another explanation is that, at low temperature, more oxygen gets dissolved and become available in the culture medium for oxygen-dependent enzymes (needed for GLA production), thereby increasing the yield of more unsaturated fatty acids such as GLA [67]. Or it could be combination of both of these factors.

pH

In addition to temperature, pH has a great influence on GLA production. It has been reported that the GLA production was achieved when *Cunninghamella blakesleeana*-JSK2 was cultivated at pH 6. When the pH was adjusted to 4 or 8, the GLA accumulation was greatly reduced [65]. Besides, it has been also found that although more acidic pH is beneficial for the growth of some strains in submerged culture, an increase in initial pH was, in fact, able to stimulate the accumulation of metabolites [67, 68]. It is likely that pH affects the interaction of cells with various medium components such as salts and thus decreases the ability of the strain to utilize the available nutrients in the medium [67]. Another explanation for such phenomenon is that at different initial pH, the morphology of fungal mycelia varied in the medium, which might influence the metabolite formation [67, 68].

Carbon Source

According to previous studies, the carbon and nitrogen sources have great influence on the GLA production. It has been found that the GLA production is dependent on the types of sugars, following glucose > D-Mannitol = Fructose = maltose > starch > sucrose. It has been suggested that simple sugars such as glucose could serve as a great carbon sources for GLA production by using *Mucor* spp. [69], *Mortierella* spp. [70], and *Cunninghamella* spp. [71]. One of possible reasons for such phenomenon is related with the nature of *Zygomycetes*, which are generally saprophytic and more likely to grow and synthesize the metabolites by using simple sugars compared to complex sugars. However, a high concentration of glucose could also limit GLA production, as the cells are not tolerant to high concentrations of glucose due to the increase of osmotic potential of the medium [67].

Nitrogen Source

In addition to the carbon source, the choice of nitrogen source also affects the GLA production. It has been found that the exhaustion of nitrogen in the medium results in an increase in GLA accumulation because the cell proliferation is inhibited and the conversion of carbon to lipids is promoted [72]. In addition, it was found that the nitrogen source could regulate the lipogenesis and thus affect the lipid production [73]. One study showed that greater lipid production from *Cunninghamella blakesleeana*-JSK2 was achieved using potassium nitrate compared to ammonium sulfate or urea [65], while another study indicated that higher amount of GLA was produced by using urea as nitrogen sources for *M. rouxii* and *M.* spp. [69]. It is likely that different species prefer different nitrogen sources for the production of GLA.

In summary, for GLA production via microbial cultivation, factors such as temperature, pH, carbon source, and nitrogen source have been considered key factors. Based on the above discussion, optimized control of temperature and pH and suitable supplementation of carbon source and nitrogen source could improve cell growth and ultimately increase the production of GLA significantly.

Acknowledgments

This work was supported by "Natural Science Foundation of Jiangsu Province," Grant No. BK20170037. "National Key R&D Program of China," Grant No. 2016YFE0105400, and "The Fundamental Research Funds for the Central Universities," Grant No. 30916011202.

References

1. Damude HG, Gillies PJ, Macool DJ, Picataggio SK, Pollak DMW, Ragghianti JJ, Xue Z, Yadav NS, Zhang H, Zhu QQ (2011) High eicosapentaenoic acid producing strains of Yarrowia lipolytica. Google Patents
2. Xue Z, Sharpe PL, Hong S-P, Yadav NS, Xie D, Short DR, Damude HG, Rupert RA, Seip JE, Wang J (2013) Production of omega-3 eicosapentaenoic acid by metabolic engineering of Yarrowia lipolytica. Nat Biotechnol 31 (8):734–740
3. Hoshida H, Ohira T, Minematsu A, Akada R, Nishizawa Y (2005) Accumulation of eicosapentaenoic acid in Nannochloropsis sp. in response to elevated CO2 concentrations. J Appl Phycol 17(1):29–34
4. Wen Z-Y, Jiang Y, Chen F (2002) High cell density culture of the diatom Nitzschialaevis

for eicosapentaenoic acid production: fed-batch development. Process Biochem 37 (12):1447–1453

5. Chen C-Y, Chen Y-C, Huang H-C, Huang C-C, Lee W-L, Chang J-S (2013) Engineering strategies for enhancing the production of eicosapentaenoic acid (EPA) from an isolated microalga Nannochloropsis oceanica CY2. Bioresour Technol 147:160–167
6. Chen C-Y, Chen Y-C, Huang H-C, Ho S-H, Chang J-S (2015) Enhancing the production of eicosapentaenoic acid (EPA) from Nannochloropsis oceanica CY2 using innovative photobioreactors with optimal light source arrangements. Bioresour Technol 191:407–413
7. Wen Z-Y, Chen F (2002) Perfusion culture of the diatom *Nitzschia laevis* for ultra-high yield of eicosapentaenoic acid. Process Biochem 38 (4):523–529
8. Camacho-Rodríguez J, González-Céspedes A, Cerón-García M, Fernández-Sevilla J, Acién-Fernández F, Molina-Grima E (2014) A quantitative study of eicosapentaenoic acid (EPA) production by Nannochloropsis gaditana for aquaculture as a function of dilution rate, temperature and average irradiance. Appl Microbiol Biotechnol 98(6):2429–2440
9. Xu F, Hu H-h, Cong W, Z-l C, Ouyang F (2004) Growth characteristics and eicosapentaenoic acid production by Nannochloropsis sp. in mixotrophic conditions. Biotechnol Lett 26(1):51–53
10. Athalye SK, Garcia RA, Wen Z (2009) Use of biodiesel-derived crude glycerol for producing eicosapentaenoic acid (EPA) by the fungus Pythium irregulare. J Agric Food Chem 57 (7):2739–2744
11. Okuda T, Ando A, Negoro H, Muratsubaki T, Kikukawa H, Sakamoto T, Sakuradani E, Shimizu S, Ogawa J (2015) Eicosapentaenoic acid (EPA) production by an oleaginous fungus Mortierella alpina expressing heterologous the Δ17-desaturase gene under ordinary temperature. Eur J Lipid Sci Technol 117 (12):1919–1927
12. Gandhi S, Weete J (1991) Production of the polyunsaturated fatty acids arachidonic acid and eicosapentaenoic acid by the fungus *Pythium ultimum*. Microbiology 137 (8):1825–1830
13. Shimiziu S, Kawashima H, Shinmen Y, Akimoto K, Yamada H (1988) Production of eicosapentaenoic acid by Mortierella fungi. J Am Oil Chem Soc 65(9):1455–1459
14. Liang Y, Zhao X, Strait M, Wen Z (2012) Use of dry-milling derived thin stillage for producing eicosapentaenoic acid (EPA) by the fungus Pythium irregulare. Bioresour Technol 111:404–409
15. Yazawa K, Araki K, Okazaki N, Watanabe K, Ishikawa C, Inoue A, Numao N, Kondo K (1988) Production of eicosapentaenoic acid by marine bacteria. J Biochem 103(1):5–7
16. Ward AC, Glassey J (2014) Process development of eicosapentaenoic acid production. Biochem Eng J 82:53–62
17. Yazawa K (1996) Production of eicosapentaenoic acid from marine bacteria. Lipids 31(1): S297–S300
18. Yu R, Yamada A, Watanabe K, Yazawa K, Takeyama H, Matsunaga T, Kurane R (2000) Production of eicosapentaenoic acid by a recombinant marine cyanobacterium, Synechococcus sp. Lipids 35(10):1061–1064
19. Damude HG, Gillies PJ, Macool DJ, Picataggio SK, Pollak DMW, Ragghianti JJ, Xue Z, Yadav NS, Zhang H, Zhu QQ (2014) High eicosapentaenoic acid producing strains of Yarrowia lipolytica. Google Patents
20. Pal D, Khozin-Goldberg I, Cohen Z, Boussiba S (2011) The effect of light, salinity, and nitrogen availability on lipid production by Nannochloropsis sp. Appl Microbiol Biotechnol 90 (4):1429–1441
21. Pérez-López P, González-García S, Allewaert C, Verween A, Murray P, Feijoo G, Moreira MT (2014) Environmental evaluation of eicosapentaenoic acid production by *Phaeodactylum tricornutum*. Sci Total Environ 466:991–100
22. Shimizu S, Kawashima H, Akimoto K, Shinmen Y, Yamada H (1989) Conversion of linseed oil to an eicosapentaenoic acid-containing oil by Mortierella alpina 1S-4 at low temperature. Appl Microbiol Biotechnol 32(1):1–4
23. Sukenik A (1991) Ecophysiological considerations in the optimization of eicosapentaenoic acid production by Nannochloropsis sp. (Eustigmatophyceae). Bioresour Technol 35(3):263–269
24. Mitra M, Patidar SK, Mishra S (2015) Integrated process of two stage cultivation of Nannochloropsis sp. for nutraceutically valuable eicosapentaenoic acid along with biodiesel. Bioresour Technol 193:363–369
25. Wen Z, Chen S (2005) Prospects for eicosapentaenoic acid production using microorganisms. In: Cohen Z, Ratledge C (eds) Single cell oils, pp. 138–160. Champaign: AOCS Press, 2005
26. Sang M, Wang M, Liu J, Zhang C, Li A (2012) Effects of temperature, salinity, light intensity,

and pH on the eicosapentaenoic acid production of Pinguiococcus pyrenoidosus. J Ocean Univ China 11(2):181–186. (English Edition)

27. Chen F, Johns MR (1991) Effect of C/N ratio and aeration on the fatty acid composition of heterotrophic *Chlorella sorokiniana*. J Appl Phycol 3(3):203–209
28. Bajpai P, Bajpai PK (1993) Eicosapentaenoic acid (EPA) production from microorganisms: a review. J Biotechnol 30(2):161–183
29. Sevilla JF, Grima EM, Camacho FG, Fernandez FA, Perez JS (1998) Photolimitation and photoinhibition as factors determining optimal dilution rate to produce eicosapentaenoic acid from cultures of the microalga *Isochrysis galbana*. Appl Microbiol Biotechnol 50 (2):199–205
30. Chen CY, Yeh KL, Su HM, Lo YC, Chen WM, Chang JS (2010) Strategies to enhance cell growth and achieve high-level oil production of a *Chlorella vulgaris* isolate. Biotechnol Prog 26(3):679–686
31. Armenta RE, Valentine MC (2013) Single-cell oils as a source of Omega-3 fatty acids: an overview of recent advances. J Am Oil Chem Society 90(2):167–182
32. Ren L-J, Ji X-J, Huang H, Qu L, Feng Y, Tong Q-Q, Ouyang P-K (2010) Development of a stepwise aeration control strategy for efficient docosahexaenoic acid production by Schizochytrium sp. Appl Microbiol Biotechnol 87 (5):1649–1656
33. Qu L, Ji XJ, Ren LJ, Nie ZK, Feng Y, Wu WJ, Ouyang PK, Huang H (2011) Enhancement of docosahexaenoic acid production by Schizochytrium sp. using a two-stage oxygen supply control strategy based on oxygen transfer coefficient. Lett Appl Microbiol 52(1):22–27
34. Chang G, Gao N, Tian G, Wu Q, Chang M, Wang X (2013) Improvement of docosahexaenoic acid production on glycerol by Schizochytrium sp S31 with constantly high oxygen transfer coefficient. Bioresour Technol 142:400–406
35. Qu L, Ren L-J, Huang H (2013) Scale-up of docosahexaenoic acid production in fed-batch fermentation by Schizochytrium sp based on volumetric oxygen-transfer coefficient. Biochem Eng J 77:82–87
36. Qu L, Ren L-J, Sun G-N, Ji X-J, Nie Z-K, Huang H (2013) Batch, fed-batch and repeated fed-batch fermentation processes of the marine thraustochytrid Schizochytrium sp for producing docosahexaenoic acid. Bioprocess Biosyst Eng 36(12):1905–1912
37. Ren L-J, Feng Y, Li J, Qu L, Huang H (2013) Impact of phosphate concentration on docosahexaenoic acid production and related enzyme activities in fermentation of Schizochytrium sp. Bioprocess Biosyst Eng 36 (9):1177–1183
38. Chang G, Wu J, Jiang C, Tian G, Wu Q, Chang M, Wang X (2014) The relationship of oxygen uptake rate and k(L)a with rheological properties in high cell density cultivation of docosahexaenoic acid by Schizochytrium sp S31. Bioresour Technol 152:234–240
39. Ren L-J, Sun L-N, Zhuang X-Y, Qu L, Ji X-J, Huang H (2014) Regulation of docosahexaenoic acid production by Schizochytrium sp.: effect of nitrogen addition. Bioprocess Biosyst Eng 37(5):865–872
40. Ling X, Guo J, Liu X, Zhang X, Wang N, Lu Y, Ng IS (2015) Impact of carbon and nitrogen feeding strategy on high production of biomass and docosahexaenoic acid (DHA) by Schizochytrium sp LU310. Bioresour Technol 184:139–147
41. Sun X-M, Ren L-J, Ji X-J, Chen S-L, Guo D-S, Huang H (2016) Adaptive evolution of Schizochytrium sp by continuous high oxygen stimulations to enhance docosahexaenoic acid synthesis. Bioresour Technol 211:374–381
42. Zhang Y, Min Q, Xu J, Zhang K, Chen S, Wang H, Li D (2016) Effect of malate on docosahexaenoic acid production from Schizochytrium sp B4D1. Electron J Biotechnol 19:56–60
43. Zhao X, Ren L, Guo D, Wu W, Ji X, Huang H (2016) CFD investigation of Schizochytrium sp impeller configurations on cell growth and docosahexaenoic acid synthesis. Bioprocess Biosyst Eng 39(8):1297–1304
44. Jia J, Zhang Q, Lu X, Huang C, Ji J (2015) Impact of pH on docosahexaenoic acid (DHA) production in fermentation of Schizochytrium limacinum. J Biol 32(4):16–19
45. Zeng Y, Ji X-J, Lian M, Ren L-J, Jin L-J, Ouyang P-K, Huang H (2011) Development of a temperature shift strategy for efficient docosahexaenoic acid production by a marine fungoid protist, Schizochytrium sp HX-308. Appl Biochem Biotechnol 164(3):249–255
46. Zhu L, Zhang X, Ji L, Song X, Kuang C (2007) Changes of lipid content and fatty acid composition of Schizochytrium limacinum in response to different temperatures and salinities. Process Biochem 42(2):210–214
47. Wu S-T, Yu S-T, Lin L-P (2005) Effect of culture conditions on docosahexaenoic acid production by Schizochytrium sp. S31. Process Biochem 40(9):3103–3108
48. Ren L-J, Huang H, Xiao A-H, Lian M, Jin L-J, Ji X-J (2009) Enhanced docosahexaenoic acid

production by reinforcing acetyl-CoA and NADPH supply in Schizochytrium sp HX-308. Bioprocess Biosyst Eng 32 (6):837–843

49. Ping WEI, Lujing REN, Xiaojun JI, Qianqian T, Yun F, He H (2011) Effect of reinforcing acetyl-CoA supply in docosahexaenoic acid production by Schizochytrium sp. J Chinese Biotechnol 31(4):87–91
50. Eroshin V, Dedyukhina E, Chistyakova T, Zhelifonova V, Kurtzman C, Bothast R (1996) Arachidonic-acid production by species of Mortierella. World J Microbiol Biotechnol 12(1):91–96
51. Eroshin V, Satroutdinov A, Dedyukhina E, Chistyakova T (2000) Arachidonic acid production by Mortierella alpina with growth-coupled lipid synthesis. Process Biochem 35 (10):1171–1175
52. Hwang B-H, Kim J-W, Park C-Y, Park C-S, Kim Y-S, Ryu Y-W (2005) High-level production of arachidonic acid by fed-batch culture of Mortierella alpina using NH 4 OH as a nitrogen source and pH control. Biotechnol Lett 27 (10):731–735
53. Jin M-J, Huang H, Xiao A-H, Zhang K, Liu X, Li S, Peng C (2008) A novel two-step fermentation process for improved arachidonic acid production by Mortierella alpina. Biotechnol Lett 30(6):1087–1091
54. Higashiyama K, Yaguchi T, Akimoto K, Fujikawaa S, Shimizu S (1998) Effects of mineral addition on the growth morphology of and arachidonic acid production by Mortierella alpina 1S-4. J Am Oil Chem Soc 75 (12):1815–1819
55. Yamada H, Shimizu S, Shinmen Y (1987) Production of arachidonic acid by Mortierella elongata 1S-5. Agric Biol Chem 51 (3):785–790
56. Yuan C, Wang J, Shang Y, Gong G, Yao J, Yu Z (2002) Production of arachidonic acid by Mortierella alpina I~ 4~ 9-N~ 1~ 8. Food Technol Biotechnol 40(4):311–316
57. Koike Y, Cai HJ, Higashiyama K, Fujikawa S, Park EY (2001) Effect of consumed carbon to nitrogen ratio of mycelial morphology and arachidonic acid production in cultures of Mortierella alpina. J Biosci Bioeng 91(4):382–389
58. Shinmen Y, Shimizu S, Akimoto K, Kawashima H, Yamada H (1989) Production of arachidonic acid by Mortierella fungi. Appl Microbiol Biotechnol 31(1):11–16
59. Park EY, Koike Y, Higashiyama K, Fujikawa S, Okabe M (1999) Effect of nitrogen source on mycelial morphology and arachidonic acid production in cultures of Mortierella alpina. J Biosci Bioeng 88(1):61–67
60. Gema H, Kavadia A, Dimou D, Tsagou V, Komaitis M, Aggelis G (2002) Production of [gamma]-linolenic acid by Cunninghamella echinulata cultivated on glucose and orange peel. Appl Microbiol Biotechnol 58(3):303
61. Conti E, Stredansky M, Stredanska S, Zanetti F (2001) γ-Linolenic acid production by solid-state fermentation of Mucorales strains on cereals. Bioresour Technol 76(3):283–286
62. Kavadia A, Komaitis M, Chevalot I, Blanchard F, Marc I, Aggelis G (2001) Lipid and γ-linolenic acid accumulation in strains of Zygomycetes growing on glucose. J Am Oil Chem Soc 78(4):341–346
63. Ochsenreither K, Glück C, Stressler T, Fischer L, Syldatk C (2016) Production strategies and applications of microbial single cell oils. Front Microbiol 7
64. Fakas S, Papanikolaou S, Batsos A, Galiotou-Panayotou M, Mallouchos A, Aggelis G (2009) Evaluating renewable carbon sources as substrates for single cell oil production by Cunninghamella echinulata and Mortierella isabellina. Biomass Bioenergy 33(4):573–580
65. Sukrutha S, Adamechova Z, Rachana K, Savitha J, Certik M (2014) Optimization of physiological growth conditions for maximal gamma-linolenic acid production by cunninghamella blakesleeana-JSK2. J Am Oil Chem Soc 91(9):1507–1513
66. Jangbua P, Laoteng K, Kitsubun P, Nopharatana M, Tongta A (2009) Gamma-linolenic acid production of Mucor rouxii by solid-state fermentation using agricultural by-products. Lett Appl Microbiol 49(1):91–97
67. Nisha A, Venkateswaran G (2011) Effect of culture variables on mycelial arachidonic acid production by Mortierella alpina. Food Bioprocess Technol 4(2):232–240
68. Shu C-H, Lung M-Y (2004) Effect of pH on the production and molecular weight distribution of exopolysaccharide by Antrodia camphorata in batch cultures. Process Biochem 39 (8):931–937
69. Somashekar D, Venkateshwaran G, Sambaiah K, Lokesh B (2003) Effect of culture conditions on lipid and gamma-linolenic acid production by mucoraceous fungi. Process Biochem 38(12):1719–1724
70. Dyal SD, Bouzidi L, Narine SS (2005) Maximizing the production of γ-linolenic acid in Mortierella ramanniana var. ramanniana as a function of pH, temperature and carbon source, nitrogen source, metal ions and oil

supplementation. Food Res Int 38 (7):815–829

71. Chen HC, Chang CC (1996) Production of γ-linolenic acid by the fungus Cunninghamella echinulata CCRC 31840. Biotechnol Prog 12 (3):338–341
72. Wynn JP, Hamid AA, Li Y, Ratledge C (2001) Biochemical events leading to the diversion of carbon into storage lipids in the oleaginous fungi Mucor circinelloides and Mortierella alpina. Microbiology 147(10):2857–2864
73. Certik M, Megova J, Horenitzky R (1999) Effect of nitrogen sources on the activities of lipogenic enzymes in oleaginous fungus Cunninghamella echinulata. J Gen Appl Microbiol 45(6):289–293

Chapter 16

Screening for Oily Yeasts Able to Convert Hydrolysates from Biomass to Biofuels While Maintaining Industrial Process Relevance

Patricia J. Slininger, Bruce S. Dien, Joshua C. Quarterman, Stephanie R. Thompson, and Cletus P. Kurtzman

Abstract

Research has recently intensified to discover new oleaginous yeast strains able to function quickly and efficiently in low-cost lignocellulosic hydrolysates to produce high-quality lipids for use in biodiesel and chemicals. Detailed techniques are given here for ranking candidate yeast strains based on conversion of hydrolysate sugars to lipids and then optimizing cultivation conditions for best performers in a 96-well aerobic microcultivation format. A full battery of assays applicable to high throughput of small-volume samples are described for efficiently evaluating cell biomass production, lipid accumulation, fatty acid composition, and sugar utilization. Original data is additionally presented on the validation of the micro-technique for GC analysis of lipid composition in yeast since this application involved modification of a previously published assay for microalgae.

Key words Oleaginous yeast, Screen, High-throughput, Microculture, Triacylglycerols, Lipids, Biodiesel, Lipid assays, Bioconversion, Lignocellulose

1 Introduction

Lignocellulosic biomass is an abundant, renewable feedstock with potential for use as a substrate in the production of a host of bioproducts ranging from biofuels to plastics to antibiotics. Pretreatment and enzyme saccharification processes release sugars from the biomass, making them available for fermentation by yeast. However, traditional industrial *Saccharomyces* yeast do not ferment xylose (comprising up to 40% of plant sugars) and are not able to function in concentrated hydrolysates. Concentrated hydrolysates are needed to support economical product recovery, but they are laden with toxic by-products including organic acids, aldehydes, and phenols generated during pretreatment. While detoxification methods can render hydrolysates fermentable, they

Venkatesh Balan (ed.), *Microbial Lipid Production: Methods and Protocols*, Methods in Molecular Biology, vol. 1995, https://doi.org/10.1007/978-1-4939-9484-7_16,

are costly and generate waste disposal liabilities. Still lignocellulosic materials are potentially rich sources of low-cost renewable sugars that are convertible via microbial pathways to lipids, primarily in the form of triacylglycerols (TAGs). A number of reviews in the last decade have pointed to oleaginous yeasts as viable catalysts capable of producing bio-oils with fatty acid composition similar to vegetable oils, especially palm oil, which can undergo transesterification to high-quality diesel fuel [1–4]. Oleaginous yeasts by definition accumulate over 20% of their biomass as lipid [5], and they have a number of advantages over other microorganisms, especially algae and fungi, previously studied for lipid production. For example, they grow rapidly in traditional batch cultivation tanks with aeration, are capable of accumulating very high lipid titers, are often metabolically diverse utilizing a variety of carbon sources, and are frequently robust for industrial application at acidic pHs and moderate temperatures 25–30 °C [4]. By using conservative estimates for crop and bioprocess yields, it was determined that oleaginous yeasts can produce ~48 and ~190 gal oil per acre from corn stover and switchgrass, respectively, which compares favorably with the ~68 gallons of oil per acre produced from processing soybeans [6]. Converting all of the estimated 1.3 billion tons of annual biomass available to yeast-based biodiesel [7], would reduce current diesel consumption of 50 billion gallons by 62%. This estimate is based upon current technology and assumes 50% recoverable monosaccharides, 0.2 g oil/g sugar yield, 90% lipid recovery, and 7.68 lb/gal oil density [8–12]. Although biomass is likely to be converted to a variety of products in addition to lipids, this figure is noteworthy because it illustrates the potential of yeast to produce significant quantities of high quality bio-oil. Studies on yeast lipid production from biomass hydrolysates are relatively limited, and none have reported the expected yield of 0.2 g lipid/g sugar at volumetric productivity greater than 0.5 g/L/h, which are the projected minimum production parameters needed for economic feasibility [4, 6, 13, 14].

Research has recently intensified to discover new oleaginous yeast strains able to function quickly and efficiently in low-cost lignocellulosic hydrolysates. The choice of yeast strains to screen can be guided by literature and phylogenic analysis of clades containing known oleaginous yeast strains. Dien et al. [15] investigated 18 members of the *Lipomyces* and *Myxozyma* clade and demonstrated, however, that performances within this promising clade and even for a specific species was highly variable and random. While the highest ranking yeast strains on synthetic media were of the *Lipomyces* genus and had similar lipid accumulations—*L. tetrasporus* (21 g/L), *L. spencermartinsiae* (19.6 g/L), and *L. lipofer* (16.7 g/L)—the lipid accumulations of other strains of the *Lipomyces* genus were not so impressive—such as *L. tetrasporus* (2.7 g/L), *L. kononenkoae* (2.9 g/L), *L. starkeyi* (3.0 g/L), and

L. arxii (1.4 g/L). Furthermore, the strains performing well on hydrolysate were even fewer, and for example, the *L. lipofer* strain noted above accumulated only 0.7 g/L lipid on saccharified AFEX pretreated corn stover compared to 11.9 g/L lipid for the highest ranking *L. tetrasporus* strain [6]. Additionally, recognizing the intense interest in *Yarrowia lipolytica* as a model oleaginous yeast with a long history of safe commercial use and genetic tractability, Quarterman et al. [16] examined 57 members of the *Yarrowia* clade for growth and lipid production on dilute-acid switchgrass hydrolysate. A high level of variability was observed in robustness and lipid production and in particular one lesser characterized strain, *Candida phangngensis* Y-63743, accumulated nearly 9.8 g/L lipid, which is a threefold improvement over the frequently referenced control strain *Y. lipolytica* W29. Hence, there is need for an efficient screen methodology to screen large numbers of yeast strains for robustness and lipid production in concentrated lignocellulosic hydrolysates. Research to improve strains via strain engineering and targeted evolution will also require screening of numerous isolates to select mutants that have the greatest potential for success. Whether they are obtained from a culture collection, isolated from nature or improved by evolution or engineering procedures, the discovery of successful strains will require efficient screening strategies capable of identifying individuals with maximum utility to the biorefining and fuel industries.

Motivated by the relative scarcity of oleaginous strains, the even smaller subset of those able to function successfully in complex inhibitory hydrolysates of lignocellulosic biomass, and the variability within species, a high throughput microcultivation method has been developed [6]. The standard approach to screening for oleaginous strains has been to start with synthetic media, spiked in some cases with a few known metabolic inhibitors (typically furfural, HMF, and acetic acid) and then to transition best-performing strains to hydrolysate utilization. However, the more efficient approach to finding superior strains able to convert complex substrate mixtures in the presence of numerous synergistically acting inhibitors is to directly apply a variety of challenging industrially relevant hydrolysates in an initial microculture screen. This approach, as it is presented in the method below, promises to allow not only the efficient identification of lipid productive strains across genera on hydrolysates but also the efficient evaluation of robustness and performance optimization in response to hydrolysates varying in toxicity and nutritional content. A two-tier screening method for selection of strains is shown schematically in Fig. 1. It utilizes two well-studied hydrolysate types that differ in harshness in regard to effect on yeast survival and metabolism. The screen is designed with a less severely toxic hydrolysate for primary screening and a more toxic hydrolysate for a secondary screening to identify the most robust individuals. Yeast strains demonstrating the best

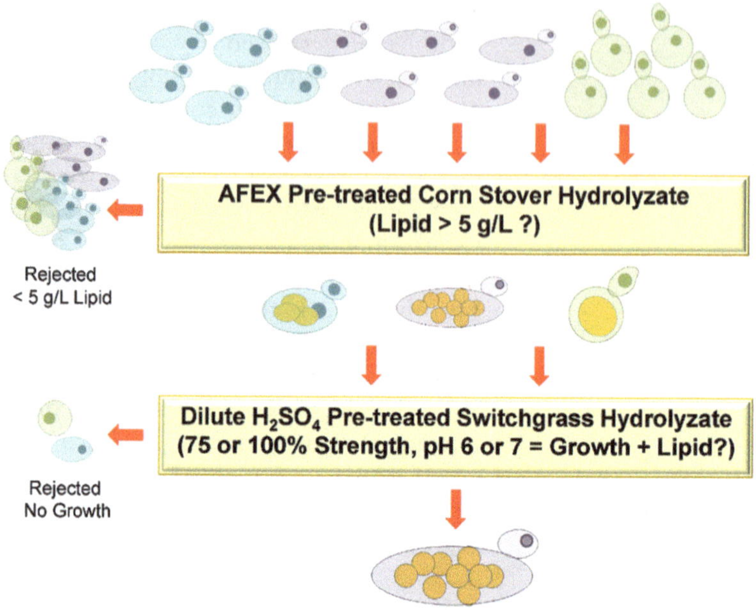

Fig. 1 Two tier screening concept to identify strains most robust for hydrolysate bioconversion to lipids. Putative oleaginous yeast strains are evaluated first on their ability to accumulate lipids in a less harsh alkaline hydrolyzate (enzyme-saccharified corn stover pretreated using ammonia fiber expansion (AFEX)). Strains accumulating over 5 g/L lipid in the primary screen are next evaluated in a secondary screen where they are challenged with a more harsh acidic hydrolysate (enzyme-saccharified dilute sulfuric acid-pretreated switchgrass) prepared at four severity levels ranging in hydrolysate strength and pH. Strains best able to grow and produce lipid under the most severe conditions (highest strength, lowest pH) are considered to be the most robust for use in hydrolysates

lipid production on both screens are predicted to have the greatest robustness for hydrolysate use. To provide a less harsh hydrolysate in the primary screen, an alkaline pretreatment process known as ammonia fiber expansion (AFEX) is used to pretreat corn stover prior to enzymatic saccharification, whereas for the secondary screen a dilute sulfuric acid pretreatment method is applied to pretreat switchgrass prior to saccharification to provide a less nitrogen-rich, more inhibitory hydrolysate (SGH) constituting a greater challenge to candidate strains most successfully passing the primary screen. Both of these hydrolysate production methods are industrially relevant, with dilute sulfuric acid pretreatment being more popular due to its relatively lower cost. The AFEX process has advantageous features, including high sugar yields at modest cellulase loadings and generating low-toxicity hydrolysates owing to low concentrations of acetic acid and furan aldehydes.

Best strains identified from the two-tiered screen are next subjected to a two-stage process to optimize lipid amplification and assess the potential of the yeast to accumulate ultrahigh lipid concentrations during conversion of nitrogen poor dilute acid pretreated switchgrass hydrolysates (SGH). This process-based screen applies four concepts recognized in prior literature. First, yeasts functioning well in hydrolysate are likely to induce expression of enzymes that reduce the furanaldehyde inhibitors generated during biomass pretreatment, especially when catalyzed with dilute acid [17]. Second, Crabtree-negative yeasts, such as *Lipomyces* and *Rhodosporidium* species and others, which were shown in our two-tier screen to be good lipid producers on hydrolysate [6], are likely superior because they use the pentose phosphate pathway and aerobic metabolism to oxidize diverse sugars, especially pentose sugars. In Crabtree negative yeasts, it is found that increasing sugar concentration increases triacylglyceride accumulation [18, 19]. Finally, shifting C:N within yeast cultures above 100:1 more strongly triggers lipid production [1, 18]. All four concepts are instrumental to designing the screen to identify yeast that could support an economical industrial process characterized with rapidly accumulating high lipid titers on hydrolysates of biomass. The 75% SGH at C:N of 60:1 in stage 1 primes yeast to detoxify inhibitors, provides N for abundant cell production, and supplies excess sugar to initiate lipid accumulation. The ultrahigh C:N in the second stage does not require N supplementation or dilution of SGH and provides high sugar concentration under severe N limitation for rapid lipid accumulation up to ~30 g/L in available yeast cells [6].

Nitrogen source manipulations may additionally be screened using additions of ammonia and relatively low cost commercially preferred sources of amino acids such as from soy, casein, or corn steep. The screening method can accommodate experimental designs to test the combined impact of strain, hydrolysate strength, nitrogen supply, and C:N. Depending on the type of biomass being targeted as the substrate of microbial conversion, other pertinent hydrolysates may be additionally incorporated into the screen. Dien et al. [13] extended the application of best oleaginous yeast strains previously identified in our two-tiered method to the successful conversion of acid sulfite (SPORL)-pretreated Douglas fir. This result suggests the robustness of the oleaginous strains discovered in the two-tiered screen using diverse enzymatic hydrolysates of alkaline pretreated corn stover and dilute acid pretreated switchgrass. However, since they may present different challenges than herbaceous biomass hydrolysates, inclusion of woody biomass hydrolysates among the array of screening substrates may ensure selection of optimal strains for bioconversion of forest products.

Features of the screening protocol include high throughput cultivations employing 96-well plates with deep square wells supporting high oxygen transfer rates and volume capacities commensurate with analytical methods to accommodate high-throughput data analysis including optical density assessment of cell growth, a colorimetric assay for lipid content, HPLC analysis of hydrolyzate (sugars, acetic acid and furans), and a streamlined GC-based method to analyze fatty acid composition. With this method superior strains are identified based on the following features: production of significant levels of lipid using two types of hydrolysate produced from enzyme saccharification of either dilute acid-pretreated switchgrass or alkaline-pretreated corn stover; ability to produce lipid on low-cost commercial sources of amino acids and ammonium to support growth in hydrolysates; and ability to quickly amplify lipid accumulation to >25 g/L in repitched cells when provided a diverse sugar-rich nitrogen-poor hydrolysate environment.

Numerous techniques are available for measuring lipid content of microorganisms [20]. Presently the most widely published methods include gravimetric analysis, fluorescent staining, colorimetric determination, time-domain based NMR, and GC detection of FAMEs. The standard method is to extract yeast lipids into a chloroform–methanol solution, evaporate the solvents, and weigh the extracted lipids [21]. This method is time-consuming and requires large sample volumes that are inconvenient for screening experiments. Fluorescence methods are based on staining yeast lipid granules with the lipid stains Nile Red [22] or more recently BODIPY [23] and measuring the resulting fluorescence signal. This method is convenient and amendable to use in a 96-well plate format. However, it requires a fluorescence plate reader, and staining varies widely with yeast strain and species. Time-domain NMR is convenient and a standard method for analysis of oil seeds [24]. The major disadvantages are that it requires a specialized and expensive instrument and requires dry yeast samples. In our laboratory, we prefer measuring lipids using the sulfophosphovanillin (SPV) colorimetric assay [2, 25–27]. This method appears to be consistent across various yeast species, does not require dry yeast cells, and is accurate with microquantities of yeast biomass. However, the response factor is expected to vary with fatty acid composition [15]. To circumvent this issue, we have also developed a GC method for measurement of fatty acid methyl esters (FAME). The major advantage is that lipid content is directly measured as fatty acid content, and consequently the fatty acid profile is simultaneously determined. The disadvantages are the requirement for a GC with a FID detector, albeit somewhat common in modern laboratories, and need for dry yeast cells. Beginning with the excellent in situ method developed by the US Department of Energy for measurement of lipids in algal cultures [28], we have modified the protocol for application to yeast and have validated its accuracy for lipid determinations by referencing to other methods.

Of these lipid quantitation methods, the SPV assay is preferred in our laboratory because it allows rapid screening of the numerous strains and conditions needed to find optimum lipid production. In principle sulfuric acid is believed to react with unsaturated lipid to form a carbonium ion that in turn reacts with the activated carbonyl group of phosphovanillin to produce a colored complex which absorbs maximally at 525 nm [27]. Lipid production is conveniently and sensitively measured using the SPV colorimetric assay applied to washed yeast cell pellets from 0.5–1 mL culture samples. The SPV assay method applied here has been adapted from the literature [29, 30] and is conveniently standardized using refined corn oil. Validation of this method to monitor yeast cultivated in hydrolysate is described in Dien et al. [15], where data showed that the lipid assay was robust to various yeast lipid and that interferences from nonlipid substrates should not be a concern if yeast cells are washed.

Though it is more time-consuming, the assessment of the fatty acid profile is also key because this will determine acceptability of the yeast lipid as a feedstock for biodiesel production. Consequently, we validated a streamlined approach to fatty acid profiling applicable to oleaginous yeast cultures. The traditional methodology is to release the fatty acids from the glycerol and transform them to fatty acid methyl esters (FAME) in one transesterification step, and then to measure the FAMEs by GC either equipped with a flame ionization detector (FID) or mass spectrometer. In one method, the yeast are dried and disrupted, lipids recovered from ground cells by solvent extraction, and the lipids transformed to FAME in the presence of excess methanol through base catalyzed transesterification [15]. Alternatively, in situ transesterification allows for direct production of FAME using dried yeast in the presence of excess methanol and an acid catalyst. A recently developed protocol by the Department of Energy allows for simultaneous determination of total lipid content and fatty acid composition of micro algal cells [28]. Herein this protocol has been modified and validated for use in oleaginous yeasts. The method demonstrates a robust linear range for lipid contents based upon a freeze-dried *Yarrowia lipolytica* culture (Fig. 2a). The lipid content of the *Y. lipolytica* cells was found by GC to be 28.6% (w/w) on a dry basis. Values determined by TD-NMR and the SPV assay were 26.5 ± 1.7% and 31.1 ± 5.6%, respectively. Furthermore, the profile of fatty acid composition as determined by GC-FAME was invariant to sample size (Fig. 2b). As a final validation, lipid titers were followed for triplicate *Y. lipolytica* cultures (Fig. 3a) by the three methods (Fig. 3b). Lipid titers were the same for GC-FAME and TD-NMR methods and followed the same pattern for the SPV assay. It is notable that the GC-FAME derived results matched those for TD-NMR, which is a standardized method for measuring lipids. Finally, the fatty acid profile was conveniently followed throughout cultivations by GC-FAME (Fig. 3c).

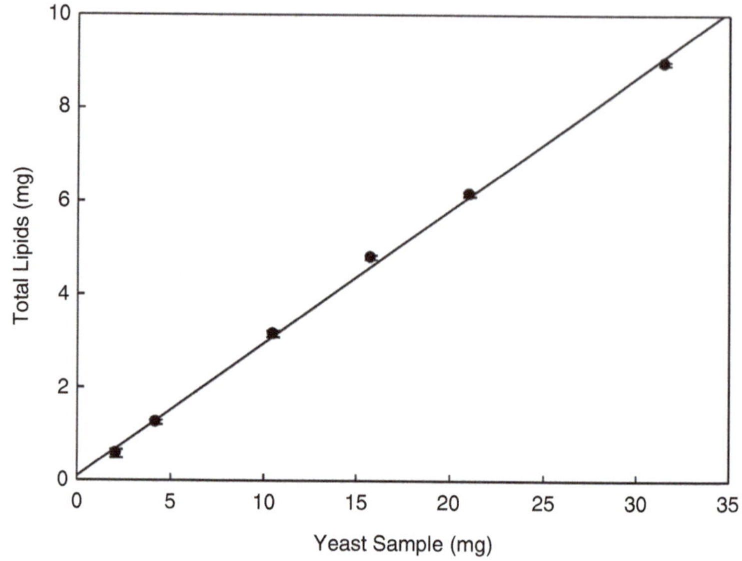

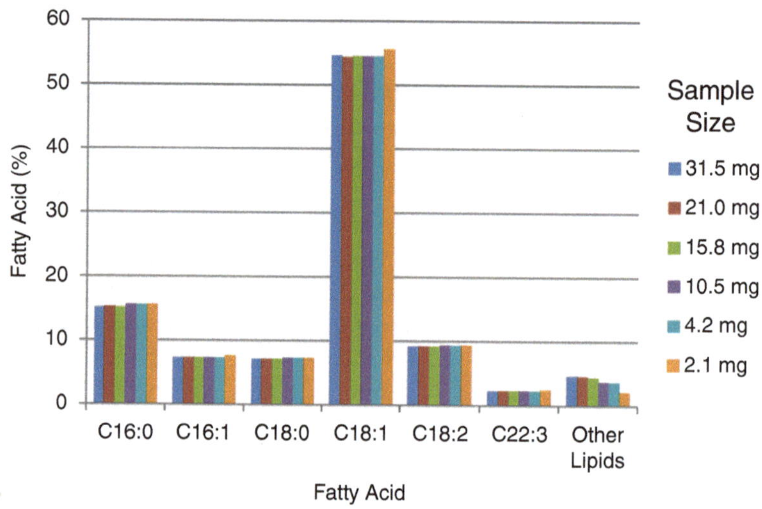

Fig. 2 Dependence of GC FAME fatty acid quantitation on yeast sample size (mg). (**a**) Lipid (mg) determined by the GC FAME method is a linear function of freeze-dried yeast mass (mg). The correlation coefficient of the linear regression line through the data is 0.998, and error bars represent the standard deviation. (**b**) As should be expected, the measured fatty acid profile is independent of yeast sample size (mg)

2 Materials

2.1 AFEX Corn Stover Hydrolysate at 6% Glucan (AFEX CSH)

1. Obtain corn stover, which is a readily available biomass, from a documentable source (*see* **Note 1**).
2. Prepare AFEX CSH at 6% glucan as previously described [31], filter-sterilize through a 0.22 μm filter unit, store frozen in

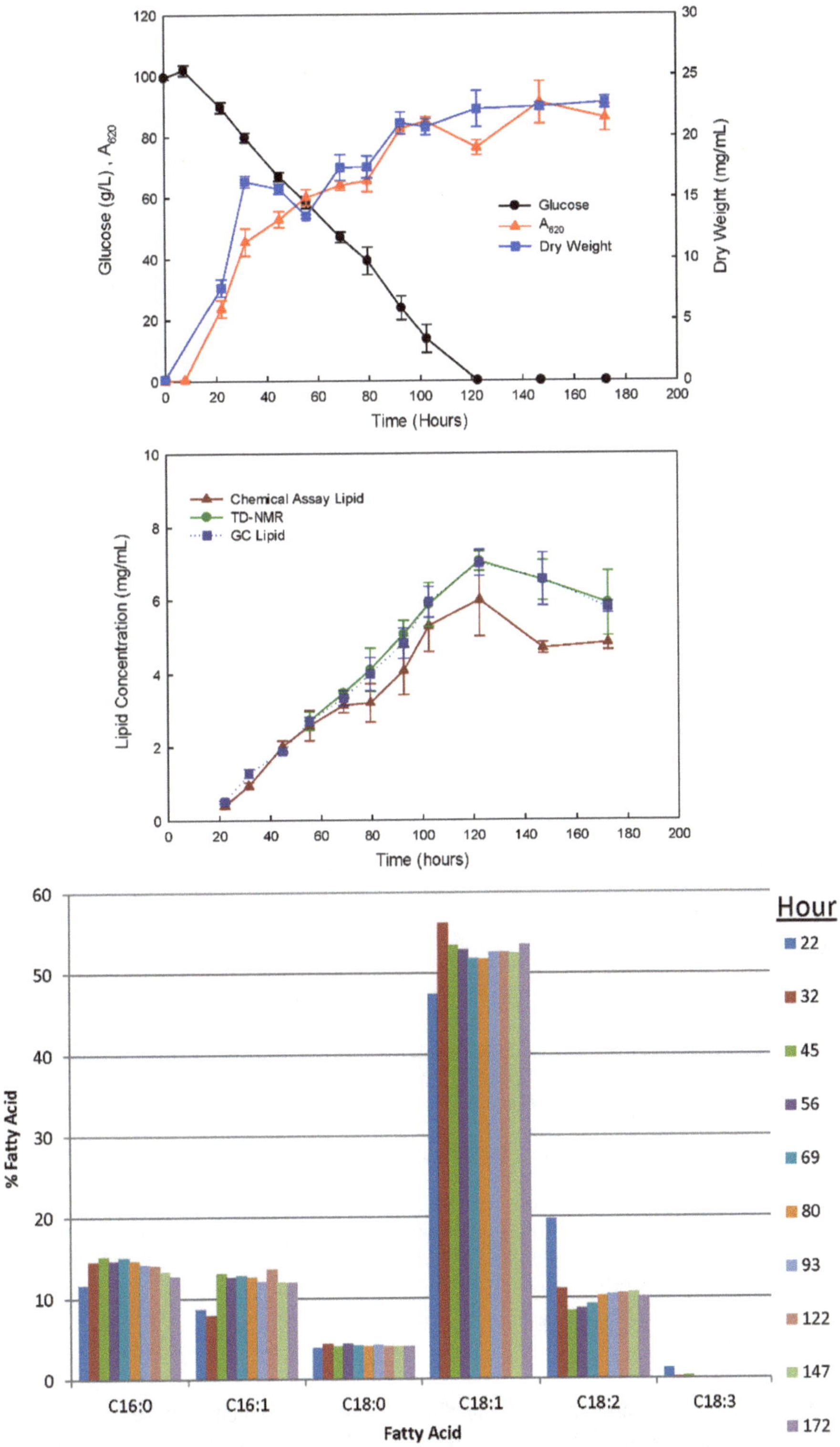

Fig. 3 Time course profiles of triplicate cultures of *Yarrowia lipolytica*, where error bars represent standard deviations. (**a**) Time course of yeast growth and sugar uptake. (**b**) Comparative lipid concentration time

aliquots at −20 °C. Also *see* Chapter 16 of this text for a detailed description of the AFEX process, variations on its configuration, and sources of the prepared AFEX CSH product for research. Alternatively, obtain prepared AFEX CSH from a supplying organization, such as Michigan State University, Department of Chemical Engineering or Glydia Biotech (Sugar Land, TX).

3. Thaw aliquots of hydrolysate in ice water, adjust pH to 5.8–6 as specified, filter-sterilize and store up to 3 days in the refrigerator (4 °C) until inoculation. Table 1 gives typical sugar, nitrogen, and inhibitor concentrations found in our prepared AFEX CSH.

Table 1
Compositions of hydrolysates used in lipid production screens (reproduced from Slininger et al. [6])

Component	AFEX CSH[a]	Dilute acid SGH (N-amended)[b]	Dilute acid SGH (Unamended)
Glucose (g/L)	59.3 ± 2.8	65.8 ± 5.1	66.4 ± 1.9
Xylose (g/L)	36.3 ± 2.2	49.8 ± 2.9	49.7 ± 2.3
Arabinose (g/L)	5.6 ± 0.5	6.9 ± 0.8	7.1 ± 0.8
HMF (mM)	0.6	4.8 ± 1.8	1.7 ± 0.9
Furfural (mM)	0.09	14.1 ± 4.1	15.0 ± 2.8
Acetic acid (g/L)	2.4 ± 0.3	5.1 ± 2.4	4.1 ± 1.3
PAN (mg N/L)	293	288 ± 65	76.8 ± 12.3
NH_3 (mg/L)	563	705 ± 115	41 ± 37
Urea (mg/L)	44	121 ± 114	14 ± 3
C:N[c]	62:1	62.4:1 ± 7.6	513:1 ± 116

Abbreviations: *HMF* 5-hydroxymethyl furfural, *PAN* primary amino nitrogen, *AFEX CSH* ammonia fiber expansion-pretreated corn stover (6% glucan loading), *SGH* enzyme-saccharified dilute acid-pretreated switchgrass hydrolysate (20% solids loading)

[a]A common batch of AFEX CSH was used for screening all strains and only variations of initial sugar and acetic acid concentrations are represented

[b]Nitrogen (N) amendments included amino acids and $(NH_4)_2SO_4$

[c]C:N is the molar carbon-to-nitrogen ratio based on total useable sugars (glucose, xylose, arabinose) and total available nitrogen (PAN, NH_3, and urea). The contribution of acetic acid to the carbon supply was only 2–4% of that arising from sugars and was ignored in this calculation

Fig. 3 (continued) courses determined by three different analysis methods. The GC Fame and TD-NMR methods give very similar results. TD-NMR, which is a standard method used to measure soybean lipids, showed a similar trend as measured by the phosphovanillin chemical assay. (**c**) The GC FAME method has the advantage of incorporating fatty acid analysis with lipid quantification

2.2 Dilute Acid-Pretreated Switchgrass Hydrolysate at 20% Solids Loading (SGH)

1. Prepare switchgrass hydrolysate at 20% solids loading (SGH) as previously described without detoxification [31]. Briefly, this involves pretreating the switchgrass with dilute sulfuric acid to partially delignify [11] and hydrolyze hemicellulose, adjusting pH of cooled material, then adding enzymes to saccharify the remaining cellulose, adjusting pH to accommodate fermentation, and filter sterilizing.
2. Obtain postfrost switchgrass, a herbaceous bioenergy crop from a documentable source (*see* **Note 1**).
3. Determine the dry biomass content of switchgrass that has been milled to pass through a 2-mm screen by delivering a known ~1 g weight of undried grass (W) to an aluminum pan, drying in an oven 105 °C for 24 h, determining the dry weight (D) of the sample and calculating the dry weight fraction ($f = D/W$). For each batch of switchgrass, this process is done at least in triplicate to arrive at an average value of f.
4. In our experiments a Mathis Labomat IR Dyer Oven with up to 12 reactors is used to carry out pretreatment. Since many groups may not have access to a Mathis Oven, *see* **Note 2** regarding guidance on alternative methods to accomplish the dilute acid pretreatment on laboratory scale samples. To load each stainless steel reactor vessel for pretreatment, weigh out the undried switchgrass needed to load 20 g of dry biomass ($20/f$), and add 80 mL of 0.936% (v/v) H_2SO_4 (i.e. 1.69% (w/v) H_2SO_4) plus 0.3 g of Pluronic F-268 surfactant, and cap tightly.
5. While mixing by tumbling at 50 rpm (60 s to left then 60 s right), heat the switchgrass in dilute acid to 160 °C and hold for 15 min, then cool to 40 °C.
6. After pretreatment and cooling, add 4.5 mL 1 M citric acid and ~7 mL of 15% $Ca(OH)_2$ to each reactor and continue mixing 15–20 min in the Mathis Labomat in order to adjust pH to 4.5 for subsequent saccharification of the cellulose. Add a little more or less calcium hydroxide as needed to achieve desired pH.
7. Transfer the pretreated switchgrass mixture from each vessel to a 250 mL Pyrex bottle, add 2.7 mL of Novozyme Cellic CTec and 0.5 mL of Cellic HTec, and cap tightly. Incubate the saccharification 72 h at 50 °C and 175 rpm (1″ diameter orbit).
8. After saccharification, pool the reactor vessel contents, filter-sterilize (0.22 μm Nalgene filter units), and refrigerate the SGH batch at 4 °C.
9. Just prior to use, SGH may be amended with nitrogen as required (*see* **Note 3**), diluted if necessary, and adjusted to pH 6–7 with 6 N NaOH as specified to support yeast

cultivation, then filter-sterilized and refrigerated briefly prior to inoculation.

10. Freeze a few milliliters of the hydrolysate prepared to assay available nitrogen as PAN, ammonia, or urea using enzyme-based kits according to manufacturer instructions (Megazyme International Ireland Ltd.) and to determine concentrations of sugar carbon sources and inhibitors including acetic acid, furfural and 5-hydroxymethylfurfural (HMF) using HPLC (*see* Subheading 3.6).
11. Calculate molar C:N ratio as the ratio of moles of carbon (C) available in usable sugars (glucose, xylose, arabinose) to the moles of usable nitrogen (N) available (ammonia, urea, amino acids).
12. The acetic acid consumption in our hydrolysates is less than 5% of available C, and so it can be ignored in the C:N calculation. Table 1 compares the concentrations of key C and N sources and inhibitors typically found in each type of switchgrass hydrolysate prepared.

2.3 Yeast–Malt–Peptone–Dextrose (YM) Agar

1. Prepare yeast–malt–peptone–dextrose (YM) agar by autoclaving a mix of 3 g/L yeast extract, 3 g/L malt extract, 5 g/L peptone, 10 g/L dextrose, and 20 g/L agar in a bottle.
2. Partially cool the bottle, and in a biocontainment hood pipet 15–20 mL of the still hot agar to each of 50–65 sterile disposable petri plates leaving lids ajar.
3. Warm plates are allowed to cool, and lids are closed on the room temperature plates.
4. Plates are inverted in plastic bags and stored until use in a closed cabinet.

3 Methods

Refer to Fig. 4 for a flow diagram of the general inoculation and screening process steps described in Subheadings 3.1–3.3.

3.1 Inocula Preparation for Screening

1. Obtain yeast cultures either by isolation of single colonies from nature or from an available culture collection, and store them in 20% (v/v) glycerol at −80 °C.
2. Streak yeast strains from glycerol stock cultures onto Yeast–Malt–Peptone–Dextrose (YM) agar plates (Subheading 2.3) and incubate at 25 °C for 48–96 h, then store refrigerated at 4 °C for up to a week.
3. Transfer a loop of cells from streak plates to fresh YM agar, then incubate at 25 °C 24 h prior to use as inoculum for hydrolysate precultures.

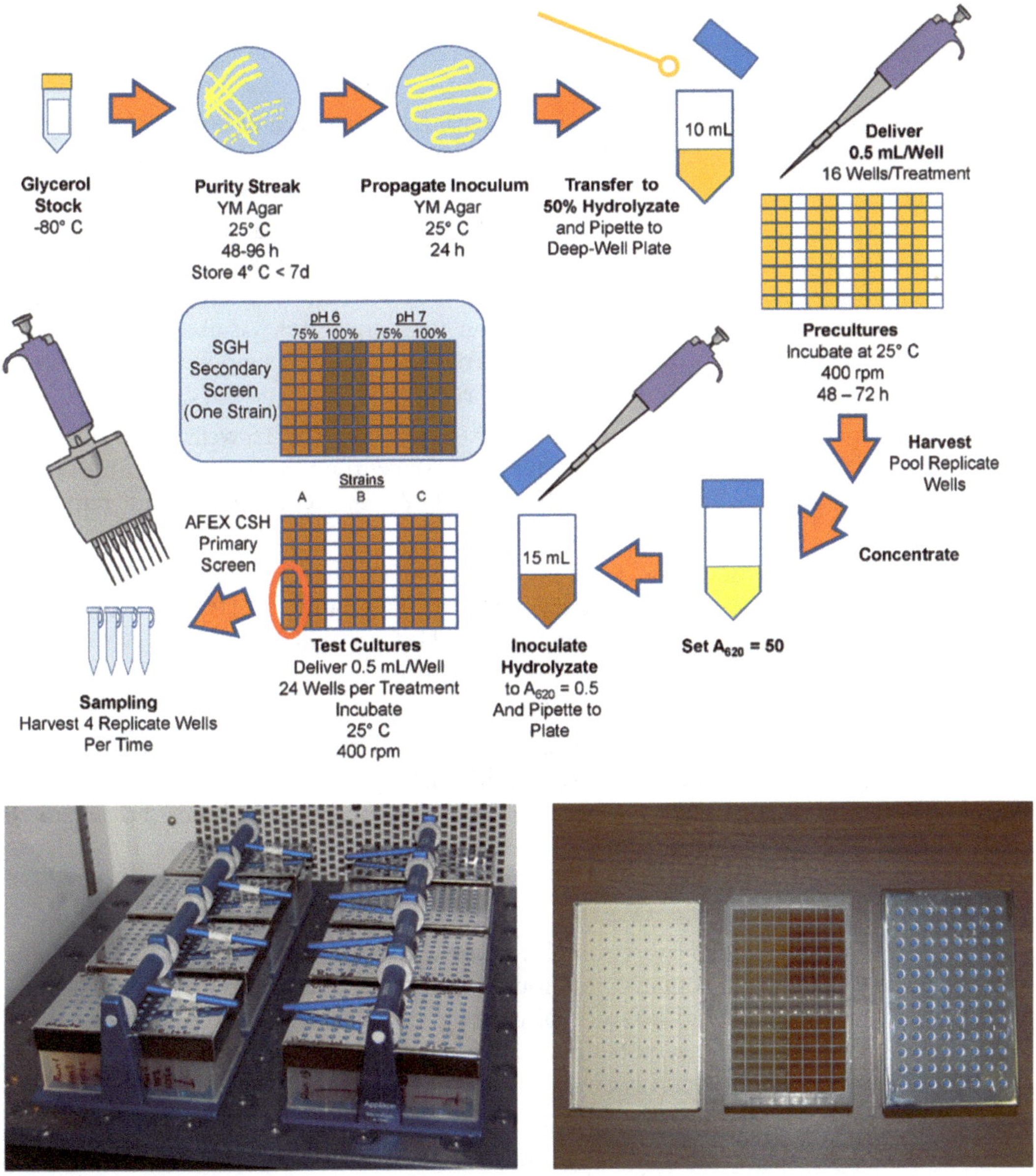

Fig. 4 Deep well plate screening cultivation process. (**a**) Flow diagram of the two-tier deep well plate screening cultivation procedure (*see* Subheadings 3.1–3.3) is shown with primary AFEX CSH and secondary SGH (inset) test plate designs. (**b**) The deep well plate incubation apparatus for high throughput screening of strains and culture conditions is shown. (**c**) Pictured is an open 96-well deep well plate (center) with steel lid (right) and standard inner sealing lining for maximal air transfer (left)

3.2 Primary Hydrolysate Screening of Yeast Strains on AFEX CSH

1. To prepare preculture medium for each strain, dilute 5 mL of AFEX CSH with 5 mL distilled water, and filter-sterilize the resulting 10 mL of 50% strength AFEX CSH. Multiply 10 mL by the number of strains to be tested in order to prepare enough for all.

2. With a loop of cells from a 24 h YM agar plate, inoculate a 10-mL aliquot of the preculture medium for each strain to be screened.
3. For each strain deliver 0.5 mL per each of 16 wells in a 96-well deep well plate. At least one empty row should separate each strain. Therefore four strains can be grown per plate (Fig. 4a).
4. Cover each preculture plate with a standard aeration cover and incubate at 25 °C with shaking (~400 rpm, 1″ diameter orbit) for 48 h. The Applikon System Duetz deep square well plate with standard aeration cover apparatus was used in our work to provide a high oxygen transfer rate to support cell growth and lipid accumulation (*see* **Note 4** and Figs. 4b, c).
5. To harvest preculture plates pool the 16 wells of 48 h precultures of each strain and concentrate the 8 mL to an absorbance of 50 at 620 nm (A_{620}) by centrifuging and removing supernatant.
6. To inoculate test cultures of each strain, deliver 0.15 mL of cell concentrate to 14.85 mL (1% volume) of 100% strength AFEX CSH (6% glucan) at pH 6 to achieve an A_{620} of 0.5 then distribute 0.5 mL of culture to each of 24,500-μL test cultures in 96-well deep well plates.
7. Cover each test culture plate with a standard aeration cover and incubate at 25 °C with shaking (~400 rpm) for up to 240 h. For test cultures, the 24-well strain blocks should be separated from others by leaving 1–2 empty rows between blocks to prevent cross contamination of different yeast strains (*see* Fig. 4a).
8. Harvest samples from each strain block by transferring the contents of four replicate microreactor wells containing 0.5 mL each of yeast culture to four 1.5-mL microfuge tubes (*see* **Notes 5** and **6**). (Thus, the 24-well blocks for each yeast strain allow for six sample points.)
9. Dilute a small aliquot from each of two replicate wells contained in microfuge tubes and measure absorbance (620 nm) (*see* **Note** 7), then centrifuge the microfuge tubes (15 min, 2287 × *g*) containing the remaining volumes from the two replicate wells, and freeze the cleared supernates at −20 °C for subsequent HPLC analysis of sugars, acetic acid, and furans (*see* Subheading 3.6).
10. Pool the other two 0.5 mL hydrolysate culture wells in a 1.5-mL microfuge tube for lipid analysis; centrifuge each sample 10 min at 4065 × *g*, and remove the supernate; wash the pellet two times with 1 mL volumes of deionized water, since the original volume of hydrolyzate in two wells was 1 mL, add 1 mL deionized water to the washed cell pellet and vortex to resuspend, then freeze at −20 °C and store the washed cell

suspension for subsequent lipid analysis (*see* Subheading 3.7 and **Note 6** for cell washing in multiple tubes).

11. Repeat **steps 8–10** for the six sample points: initially and daily, or every other day, pending the expected and observed rates of change.
12. Identify best strains as those showing significant lipid accumulation (>5 g/L, for example) on AFEX CSH, and subject them to a harsher secondary screen to further winnow down the number of strains for extended study. *See* example data collection results and best-performing strains (Table 2). For an alternative statistical strain ranking procedure (*see* **Note 8**).

3.3 Secondary Screening on SGH at Four Severity Levels: 75% or 100% Strength at pH 6 or 7

To further distinguish more robust strains, stress on yeasts may be increased by increasing severity. In this secondary screen example, a harsher type of hydrolysate, SGH, is applied at 75% and 100% of full-strength preparation in Subheading 2.2 at pH 7 and 6. Thus, each strain is tested on four SGH conditions increasing in harshness (75% SGH, pH 7 < 75% SGH, pH 6 < 100% SGH, pH 7 < 100% SGH, pH 6).

1. For precultures prepare 50% strength SGH (*see* Subheading 2.2) by adding N amendment, adjusting pH to 6 with 6 N NaOH, then diluting SGH 1:1 with distilled water, and filter sterilizing. For each strain to be screened, prepare 10 mL of 50% SGH, so the number of strains × 10 mL will be the volume of 50% SGH needed for all.
2. Inoculate a 10-mL aliquot of the preculture medium for each strain to be screened with a loop of cells from a 24 h YM agar plate.
3. For each strain deliver 0.5 mL per each of 16 wells in a 96-well deep well plate. At least one empty row should separate each strain. Therefore four strains can be grown per plate, as in Fig. 4a.
4. Incubate the inoculated precultures in deep, 96-well plates, as described above in Subheading 3.2, **step 4** for 72 h.
5. Prepare 15 mL of each of the following four SGH test culture media: 75% strength SGH (three parts SGH: one part water) or 100% SGH at pH 6 or 7. After 72 h incubation, pool the preculture wells of each strain, measure A_{620}, and concentrate to an A_{620} of 50 by centrifuging (2287 × *g*, 15 min) and removing supernate.
6. Inoculate test cultures to A_{620} of 0.5 by transferring 0.15 mL preculture to each 14.85-mL volume of test culture (a 1% volume transfer).
7. For each strain, prepare a 96 deep-well plate which contains 24 wells each of the four inoculated hydrolysate types at

Table 2
Comparative performance of isolates in the primary screen on AFEX CSH and use of Relative Performance Index (RPI)[a,b] to rank strains (*see* Note 8)

NRRL strain description			Maximum values			Relative Performance[a]		
Number	Genus	Species	A_{620} (–)	Oil (g/L)	Oil rate (g/L/h)	RPI_P	RPI_R	$RPI_{overall}$
YB-3399	*Chlorella*	Sp.	6.4	0.8	0.007	32.0	27.2	29.6
Y-1399	***Cryptococcus***	***aerius***	**85.5**	**9.8**	**0.089**	**100.2**	**91.3**	**95.8**
Y-552	*Geotrichum*	*candidum*	36.2	1.1	0.021	34.3	38.2	36.3
Y-17921	*Lipomyces*	*arxii*	NG	0	0	26.0	21.8	23.9
Y-27504	*Lipomyces*	*doorenjongii*	4	0.4	0.076	29.0	81.2	55.1
Y-7042	***Lipomyces***	***kononenkoae***	**69.4**	**11.3**	**0.081**	**111.5**	**85.1**	**98.3**
Y-11553	*Lipomyces*	*kononenkoae*	44.5	2.9	0.021	47.9	38.2	43.1
Y-11555	*Lipomyces*	*lipofer*	5.4	0.7	0.005	31.3	25.7	28.5
Y-17247	*Lipomyces*	*oligophaga*	2.8	0.1	0.001	26.7	22.6	24.6
Y-27493	*Lipomyces*	*starkeyi*	2.7	0.2	0.001	27.5	22.6	25.0
Y-27494	*Lipomyces*	*starkeyi*	3	0.2	0.002	27.5	23.3	25.4
Y-27495	*Lipomyces*	*starkeyi*	16.8	0.5	0.004	29.8	24.9	27.3
Y-11557	*Lipomyces*	*starkeyi*	60.8	3	0.022	48.7	39.0	43.8
Y-11562	***Lipomyces***	***tetrasporus***	**86.9**	**11.9**	**0.100**	**116.1**	**99.9**	**108.0**
Y-27496	*Lipomyces*	*tetrasporus*	50.2	0.5	0.003	29.8	24.1	26.9
Y-27497	*Lipomyces*	*tetrasporus*	11.8	3.5	0.017	52.5	35.1	43.8
Y-17252	*Myxozyma*	*geophila*	46	1.9	0.021	40.4	38.2	39.3
Y-17253	*Myxozyma*	*lipomycoides*	17.6	0.8	0.009	32.0	28.8	30.4
Y-11823	*Myxozyma*	*mucilagina*	42.1	1.8	0.021	39.6	38.2	38.9
Y-17387	*Myxozyma*	*udenii*	62.4	2.5	0.019	44.9	36.6	40.8
Y-17727	*Myxozyma*	*vanderwaltii*	62.7	2.4	0.026	44.2	42.1	43.1
Y-7903	*Pichia*	*nakazawae*	68.6	3.2	0.05	50.2	60.9	55.5
Y-1091	***Rhodosporidium***	***toruloides***	**93.8**	**8.8**	**0.095**	**92.6**	**96.0**	**94.3**
Y-63011	*Torulaspora*	*delbrueckii*	46.2	2	0.034	41.1	48.3	44.7
Y-7986	*Torulaspora*	*delbrueckii*	56.2	1.4	0.031	36.6	46.0	41.3
Y-1579	*Trigonopsis*	*variabilis*	77.5	2.4	0.035	44.2	49.1	46.6
YB-387	*Yarrowia*	*lipolytica*	97.2	2.5	0.053	44.9	63.2	54.1
YB-392	***Yarrowia***	***lipolytica***	**86.4**	**5.8**	**0.096**	**69.9**	**96.8**	**83.4**

(continued)

Table 2 (continued)

NRRL strain description			Maximum values			Relative Performance[a]		
Number	Genus	Species	A_{620} (–)	Oil (g/L)	Oil rate (g/L/h)	RPI_P	RPI_R	$RPI_{overall}$
YB-419	*Yarrowia*	*lipolytica*	89.2	3.2	0.058	50.2	67.1	58.7
YB-423	*Yarrowia*	*lipolytica*	77.4	3.4	0.069	51.7	75.7	63.7
YB-437	***Yarrowia***	***lipolytica***	**76**	**5.8**	**0.061**	**69.9**	**69.5**	**69.7**
YB-2330	***Saitoella***	***coloradoensis***	**67.4**	**8.5**	**0.071**	**90.3**	**77.3**	**83.8**
Y-17804	***Saitoella***	***complicata***	**74.3**	**7.4**	**0.067**	**82.0**	**74.1**	**78.1**
Average acroos strains[a]				2.5	0.032			
Standard deviation acroos strains[a]				2.6	0.028			

[a]Relative Performance Indices RPI_P and RPI_R designate, respectively, the RPI calculated for oil product accumulation and for oil production rate for each strain. The average of these two indices for each strain is designated $RPI_{overall}$. As described in **Note 8**, RPI are statistical ranking indices for each strain that are calculated based on the average and standard deviation of the data parameter across all strains

[b]Embolded text was used to highlight top ranked strains and associated date

0.5 mL/well, enough for six sampling time points. Since the same strain is cultivated on a plate there is no need to leave empty rows to separate the 24-well blocks, so use four 24-well blocks/per plate (*see* Fig. 4a, inset).

8. Incubate plates as described in Subheading 3.2, **step 7** and collect and prepare samples as described above in Subheading 3.2, **steps 8–10** for subsequent analysis.
9. Repeat the sampling **step 10** for a total of six time points: initially, daily, or every other day pending the expected and observed rates of change in the cultures.
10. Analyze results. *See* Fig. 5 example results.
11. The above procedure (Subheading 3.3) may be alternatively used to optimize conditions for growth with lipid accumulation for top-ranked strains (*see* **Note 9** and Fig. 6). Example results of an optimization experiment are shown in Fig. 6 from Slininger et al. [6].

3.4 Amplified Lipid Production Screen Using Two-Stage Cultivation on SGH in Deep-Well Plates

Strains showing best performance based on lipid accumulation in the secondary screen on SGH are further compared in a two-stage cultivation process at pH 7. In the first stage, yeast are grown on SGH at C:N 62 (Subheading 2.2, **items 1–9**) cut to 75% (v/v) strength with water. In the second stage, the yeast populations after first stage growth are transferred to unamended SGH at 100% strength with C:N 400–600.

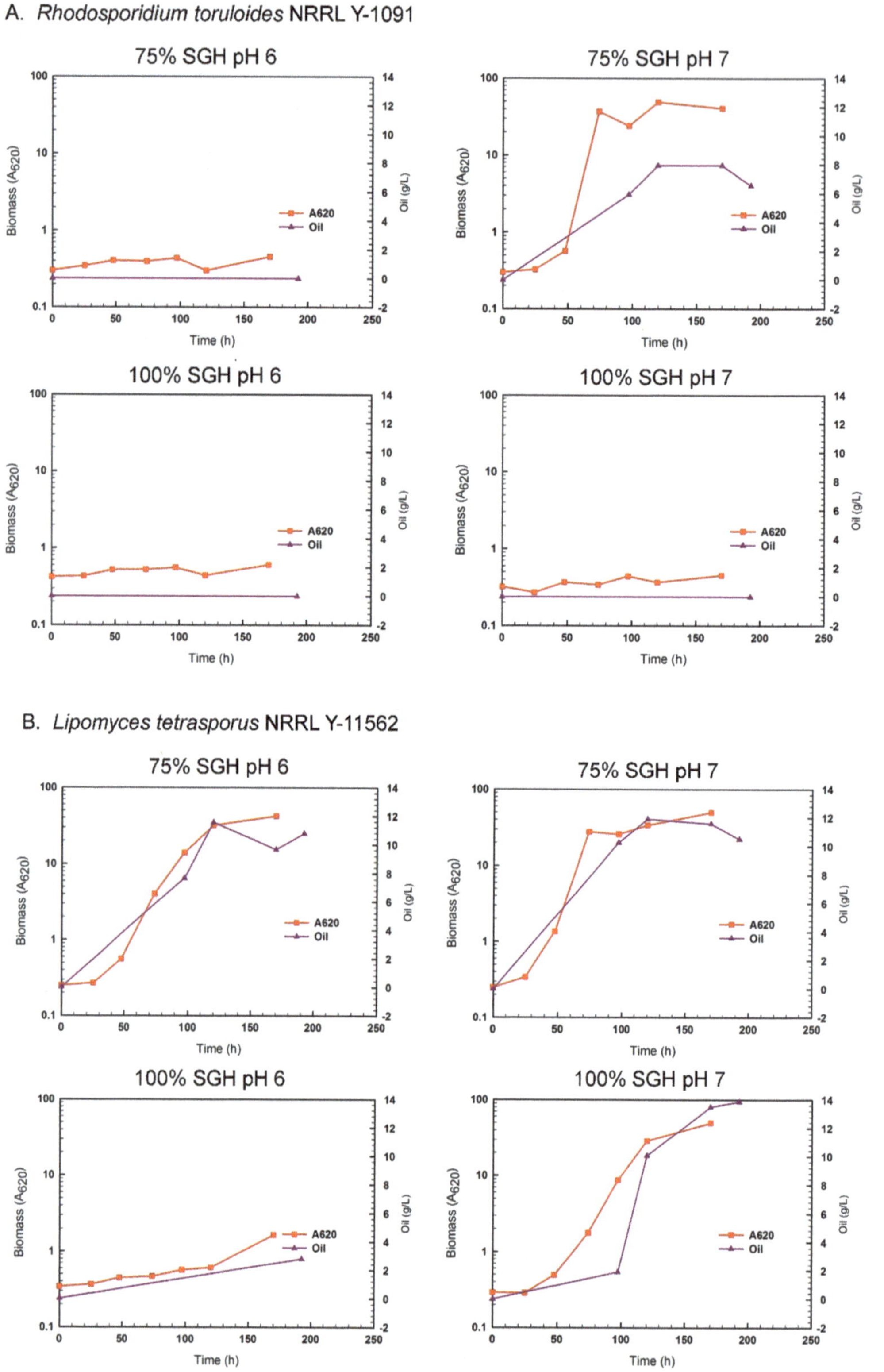

Fig. 5 Varying abilities of strains to grow and produce lipids when challenged with four severity levels of the enzyme-saccharified dilute acid-pretreated switchgrass at 20% solids loading (SGH). (**a**) Secondary screening

1. Prepare precultures on 50% SGH with C:N 62 (as in Subheading 2.2) and incubate as in Subheading 3.3, **steps 1–4**.
2. After 72 h growth, pool preculture wells for each strain and concentrate to $A_{620} = 50$ as in Subheading 3.3, **step 7** and use to inoculate test cultures.

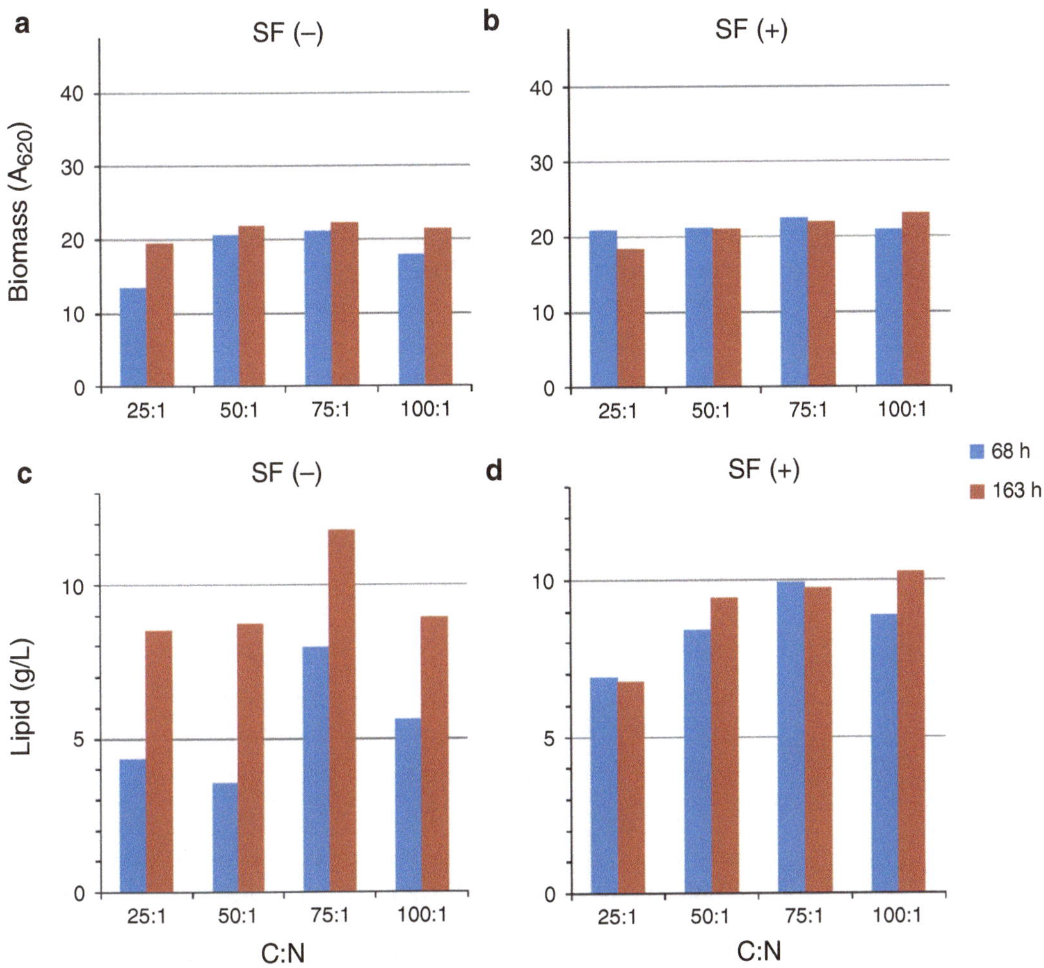

Fig. 6 In this example, a 2 × 4 level optimization of soyflour supplementation (SF) (presence (+) absence (−)) × carbon to nitrogen ratio (C:N) (25, 50, 75 and 100:1) showed the significant benefit of soy flour and optimal C:N for biomass (**a**, **b**) and lipid (**c**, **d**) accumulations of *Lipomyces tetrasporus* Y-11562 (right) in 75% strength switchgrass hydrolysate (SGH) during deep-well plate cultivation (*see* Subheading 3.3 and **Note 9**). Blue bars represent data at 68-h and red bars at 163-h sample points (figure excerpt from Slininger et al. [6])

Fig. 5 (continued) of *Rhodosporidium toruloides* NRRL Y-1091 on SGH showed growth and lipid production only on the mildest of the four hydrolysate conditions (*see* Subheading 3.3). (**b**) Secondary screening of *Lipomyces tetrasporus* NRRL Y-11562 on SGH showed robust growth and good lipid production on all but the harshest of the four conditions

3. For each strain to be tested, transfer a 1% preculture volume (0.45 mL) to inoculate 45 mL of the first stage growth culture medium (75% SGH C:N 62 at pH 7) to an initial A_{620} of 0.5.
4. For each strain, use the inoculated first stage growth culture to fill 40 wells of each of two separate 96-deep well plates: 40 wells on the first plate are to sample and monitor the growth stage, and the other group of 40 on the second plate are used for the second stage resuspension. Two different yeast strains can be grown on each plate when they are separated by dual empty 8-well rows.
5. After preparing plates in this way for all strains to be tested, incubate them as described in Subheading 3.2, **step 7**.
6. Sample plates initially, daily or every other day as appropriate for the conversion rates being observed. Prepare samples as described above in Subheading 3.2, **steps 8–10**. Store prepared samples frozen at −20 °C for subsequent analysis.
7. Once the first stage sugars are consumed by a strain, combine the yeast microcultures from the unsampled 40-well growth plate, pellet the cells by centrifuging (15 min, 2287 × *g*), and resuspend the cells in 20 mL of 100% strength SGH prepared with no nitrogen source amendments such that C:N is 400–600. Repeat for all strains being tested as sugars become depleted.
8. Distribute the 20 mL of resuspended culture 0.5 mL per each of 40 wells of the stage 2 plate and incubate as described in Subheading 3.2, **step 7**.
9. Sample immediately after resuspension and daily using the sampling procedure as described in **step 6**.
10. Analyze results. *See* example data plot in Fig. 7a.

In addition, the above procedure (Subheading 3.4) may be used alternatively to optimize conditions for growth and to explore how such conditions may also impact second stage lipid amplification for top-ranked strains (*see* **Note 10**). Example data are shown in Fig. 7a, b comparing the impact of 75% and 100% hydrolyzate strengths used in stage 1.

3.5 Amplified Lipid Production Using Two-Stage Cultivation on SGH at Flask Scale

The two-stage lipid production strategy may be carried out in flasks in triplicate to confirm potential for scalability of top oleaginous yeast strains and also to prepare larger samples for other purposes. A comparison of Figs. 7a and 8 indicates the scalability of deep-well plate results to flask applications, that is, from 0.5 to 100 mL working volume, respectively.

1. For cultivation of each strain to be tested in triplicate, prepare 300 mL of SGH at pH 7 with nitrogen supplementation by soy flour and ammonium sulfate (C:N 62) and 350 mL of SGH at

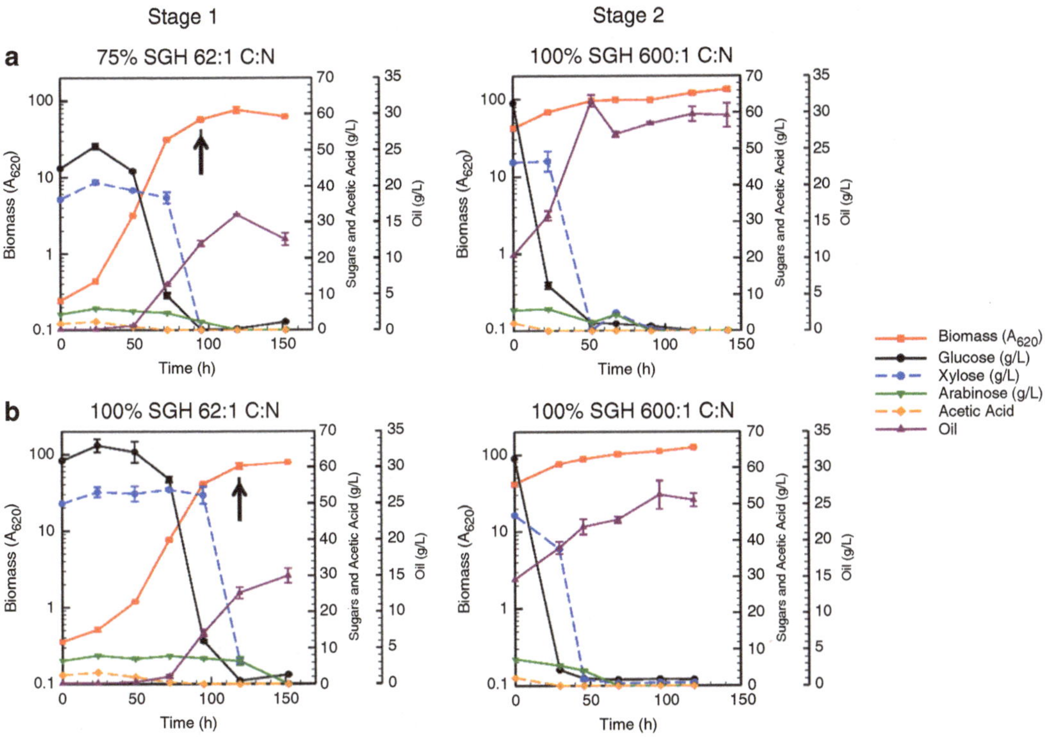

Fig. 7 Example showing an optimization result of the two-stage lipid amplification process carried out in deep well plates (*see* Subheading 3.4 and **Note 10** on optimization). Compare growth and lipid accumulation when the yeast *Lipomyces tetrasporus* NRRL Y-11562 is cultivated in the first growth stage (62:1 C:N) on 75% (**a**) versus 100% (**b**) strength switchgrass hydrolysate (SGH), and note the subsequent impact of this growth condition on continued lipid accumulation when the yeast is transferred to the second lipid amplification phase. The arrow indicates the point of yeast transfer from stage 1 growth to stage 2 lipid amplification (figure excerpt from Slininger et al. [6])

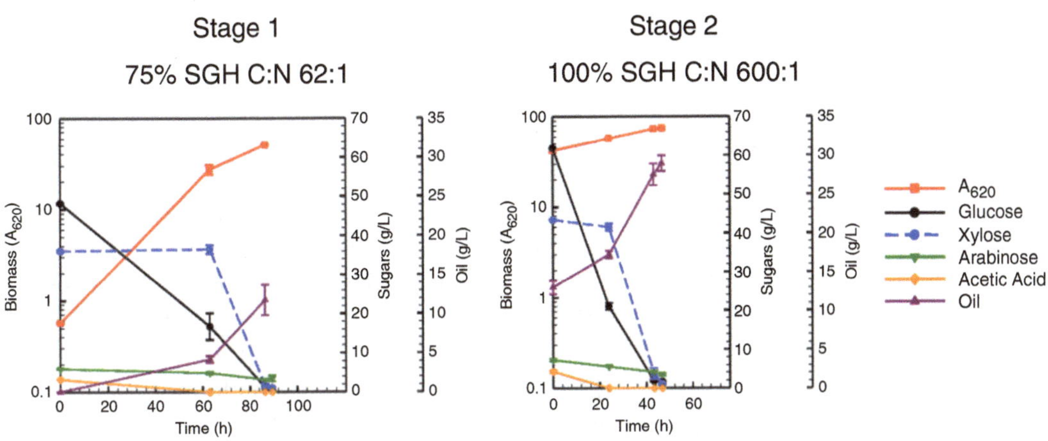

Fig. 8 Example showing the two stage lipid amplification process catalyzed by the yeast *Lipomyces tetrasporus* NRRL Y-11562 carried out in flasks with 100 mL working volume of switchgrass hydrolysate (SGH) (*see* Subheading 3.5). The similarity of the cell growth and lipid accumulation kinetics in stages 1 and 2 of this flask experiment with that shown for the same process carried out in deep well plate microcultures (Fig. 7a) indicates their reliable scalability to flask applications (figure excerpt from Slininger et al. [6])

pH 7 without nitrogen supplementation (C:N 400–600) as described in Subheading 2.2 and Table 1.

2. To a sterile 250-mL baffled DeLong flask (such as Bellco product SKU 2540-00250), add 50 mL of 0.2 μm filter-sterilized 50% SGH C:N 62. Inoculate one preculture for each strain by loop (5 loops per 50 mL), close flasks with silicone closures (such as Bellco product SKU 2004-00005), and incubate at 25 °C at 300 rpm (1″ orbit) for 72 h.
3. Remove 50 μL from the mature preculture, dilute and measure cell density (A_{620}), then centrifuge the preculture 15 min at 2287 × *g* and remove supernate to prepare a cell concentrate with A_{620} at 50.
4. Use a 1% (v/v) inoculum to initiate yeast growth at A_{620} 0.5 in 100 mL of 75% strength SGH C:N 62 at pH 7. Incubate stage 1 "growth" cultures in triplicate in 500-mL baffled DeLong flasks (such as Bellco product SKU 2540-00500) as in **step 2**.
5. Sample by drawing 0.75 mL from each culture for analysis of cell density (A_{620}), sugars, inhibitors, and lipids.
6. Pipet each 0.75 mL sample into a 1.5 mL microfuge tube and chill on ice. Move 0.5 mL of the sample into a separate tube to be prepared for lipid analysis.
7. Dilute an aliquot of the remaining 0.25 mL in the original tube into a cuvette and measure the cell density absorbance (A_{620}) using a spectrophotometer.
8. Centrifuge each lipid sample 10 min at 4065 × *g*, and remove the supernate to a 1.5-mL microfuge tube to be frozen for subsequent HPLC analysis of sugars, acetic acid, and furans. Wash the remaining pellet two times with 0.5 mL volumes of deionized water (*see* **Note 6**). Bring the final volume to 0.5 mL in deionized water. Freeze and store the washed cell suspension for subsequent lipid analysis.
9. Repeat **steps 5–8** initially, at 48 h and daily until sugar consumption has finished.
10. At the end of sugar consumption, collect the growth culture cells from each of the three flasks by centrifugation 15 min at 2287 × *g*, resuspend each in 100 mL of pH 7 100% strength SGH C:N 400–600 with no N amendment, and take an initial sample as described in **steps 5–8**.
11. Incubate the triplicate stage 2 "lipid amplification cultures" at high cell density in 500-mL baffled DeLong flasks at 25 °C and 300 rpm (2.54 cm orbit). Sample daily as in **steps 5–8** until all sugars are consumed.

12. At the end of sugar consumption draw a 1-mL sample from each flask and dispense it to a 1.5-mL microfuge tube for analysis of the dry cell concentration (*see* Subheading 3.8).
13. Next deliver a 1-mL sample to a 10-mL conical glass screw cap tube for determination of the neutral lipid composition and freeze (−20 °C) until analysis (*see* Subheading 3.9).
14. Optionally, cells from an additional sample may be pelleted, ball milled and extracted to isolate carotenoids from ruptured cells for analysis (*see* **Note 11**).

3.6 HPLC Analyses of Culture Concentrations of Sugar, Furfural, HMF, and Acetic Acid

1. Thaw frozen samples for HPLC analysis by spreading the vials out in a rack in cool tap water, leaving at least one space open between vials to equalize thawing rates of all samples.
2. Dilute samples 1:1 by adding 0.05 mL of sample and 0.05 mL of deionized water in 1-mL vials (8 mm × 40 mm from Sun Sri, Cat. #500-602) with 50–100 μL inserts (Thomas Scientific, Cat. #4005BS-425).
3. Install vials of samples and control solutions of each analyte on the autosampler tray and analyze all for glucose, xylose, arabinose, xylitol, acetic acid, furfural, and HMF.
4. Set the autosampler (such as Waters 717 Plus) to chill samples at 10 °C and to inject 10 μL from each sample onto the chromatography column.
5. Operate an Aminex HPX-87H ion exclusion column (125-0140) fitted with a Micro-guard Cation H Micro-Guard Cartridge (125-0129) precolumn heated at 60 °C and eluted with 15 mM nitric acid mobile phase flowing at 0.6 mL/min (such as via a Waters 590 Pump). Alternatively an HPX-87P column may be used to enhance hydrolysate sugar separations (*see* **Note 12**).
6. Detect separated analytes as they leave the column by refractive index change such as can be measured using a Waters 2414 RI Detector with the following settings: sensitivity at 16, time constant at 1 s, 410 mode, and RC (resistor capacitor) filter.
7. Apply chromatography data software to identify and quantitate analytes relative to control solutions which occur at retention times less than 25 min.
8. If desired, assess the furan inhibitors HMF and furfural at longer retention times, ~33.5 and ~51.6 min, respectively, using UV detection at 215 nm.

3.7 SPV Assay of Culture Lipid Concentration

The SPV assay serves as the primary method applied to process large numbers of microsamples taken during screening experiments where high throughput is required. For selected samples, the GC analysis of lipid composition (*see* Subheading 3.9) and time-domain

NMR (*see* Subheading 3.10) may be applied along with the SPV assay if additional lipid measurements are needed to understand the fatty acid composition of lipid accumulated by promising strains or to confirm SPV assay consistence [32, 33], respectively. *See* **Note 13** for variations of the SPV assay found useful to save time and resources, using a plate reading spectrophotometer if available in the laboratory.

1. On the same day as the assays are to be run, prepare vanillin-- phosphoric acid solution by dissolving 0.12 g vanillin in 20 mL deionized H_2O and adjusting the volume to 100 mL by adding 80 mL 85% o-phosphoric acid.
2. Prepare three standard stock solutions in volumetric flasks, by dissolving 62.5 mg refined corn oil in 25 mL of 2:1 chloroform–methanol (2.5 mg/mL) and similarly 125 mg of oil in 25 mL of the solvent (5.0 mg/mL) and a third flask with solvent only (0 mg/mL). Weigh flasks before and after oil addition to obtain the exact weight of oil, then dilute to volume with solvent. Cap with glass stopper, seal with Parafilm, and set aside.
3. Thaw frozen yeast samples as described in Subheading 3.6, **step 1** above. In a dry heating bath with 40 wells, 34 samples (or 2 replicates each of 17 culture samples) and 6 controls (2 reps each of 0, 125, and 250 μg of oil per 50 μL) may be processed at a time.
4. Add 1 mL of 18 M sulfuric acid to each of 40 test tubes (13 × 100 mm 9 mL Pyrex™).
5. With a Hamilton syringe, transfer 50 μL of control solution or sample washed yeast suspension to each tube.
6. Transfer the tubes to a dry heating bath, and heat at 100 °C for 10 min.
7. Cool the hot tubes for 5 min in a room temperature water bath.
8. To each tube add 2.5 mL of the vanillin–phosphoric acid solution and allow the reaction to proceed for 15 min in an incubator or dry bath at 37 °C.
9. Cool the test tubes for 10 min in a water bath at room temperature.
10. Transfer developed sample reactions to disposable cuvettes (1 cm) and read the absorbance of each reaction at 530 nm against a reference sample prepared with 50 μL of the 0 mg/ mL standard (solvent only) in place of the yeast sample suspension.
11. Dilute samples as necessary to maintain lipid within range of the lipid controls used to standardize the assay (i.e., 0–5 μg/μ L). Note that the mass (μg) of oil added to the assay is (50 μL)

(oil concentration of the control, μg/μL), and as described here, the control oil additions (x) of 0, 125, and 250 μg should generate a linear colorimetric response. Therefore, plot the control absorbances (A_{530}) versus mass (μg) of oil added to the assay (x). Perform linear regression to obtain the slope (m) and intercept ($A_{530,i}$), such that $A_{530} = mx + A_{530,i}$ and so $x = (A_{530} - A_{530,i})/m$.

12. Convert the sample absorbance measurements to lipid concentration in the culture (C_{oil}) by accounting for the volume (v) of sample added to the assay ($v = 50$ μL in this case) and any dilution factor (D) that was applied to the original culture sample to maintain the sample lipid level within the control range: $C_{oil} = D(x/v)$.
13. Repeat **steps 1–12** for each sample set with controls.

3.8 Determination of Dry Cell Weight Concentration

1. Deliver the 1-mL culture samples to 1.5 mL microfuge tubes.
2. Centrifuge the tubes to pellet the cells (15 min at 2287 × *g*).
3. Remove the supernates from each pellet, and then pipet 1-mL distilled water to each microfuge tube pellet.
4. Vortex to resuspend pellets. If available, for processing efficiency, insert the microfuge tubes in a multitube shaker (such as Heidolph Multi Reax shaker), and vortex at maximum speed for 5–10 min depending on the tightness of the pellet characteristic of the strain.
5. Repeat **steps 2–4** for a second washing (*see* **Note 6** for enhanced processing efficiency).
6. Finally resuspend twice-washed cells in 1-mL distilled water, and pipet the cell suspension to a preweighed aluminum weigh pan predried to constant weight at 105 °C.
7. Rinse the tube with 0.5 mL water to move the last traces of cells to the weigh pan.
8. Dry all pans at 105 °C for 24 h.
9. Calculate the dry cell mass concentration (g per 1-mL culture sample basis) = (final weight of the pan with dried cells) – (dry weight of the pan).

3.9 GC Analysis of Lipid Composition

Using gas chromatography, the total lipid content and fatty acid composition are simultaneously measured in dried samples of oleaginous yeast. The following procedure is modified from the Department of Energy method developed for the determination of lipid composition of micro algae [28], and original data included in Subheading 1 were presented to show its validation for application to oleaginous yeasts. All chemicals used are of analytical grade.

Sample Preparation and Freeze Drying:

1. Transfer 0.5–1.0 mL of yeast culture to a 10 mL conical borosilicate disposable glass centrifuge tube closed with a Teflon-coated black phenolic threaded screw cap which has been pre-weighed. The estimated lipid in the tube should be between 1 and 20 mg. Samples can be kept at −20 °C until they are submitted to freeze drying.
2. Centrifuge to remove culture media and wash cells twice with 1.0 mL of deionized (DI) water (purity of 18.2 MΩ, 25 °C).
3. Add 1.0 mL of DI water. Vortex to resuspend cells and store at −80 °C for up to at least a week.
4. Turn on the freeze dryer (e.g., Freezone 1, Labconco) and when temperature reaches −50 °C, manually turn on the vacuum pump.
5. When the freeze drier is ready, take samples out of the freezer and place them on dry ice. Remove caps and place two layers of Kimwipes™ on top of each tube and secure with rubber bands. Set the cap back on top of each tube without twisting it on. Caps need to be loose to allow evacuation.
6. Leave tubes on dry ice until ready to place them in a 900 or 1200 mL freeze dryer glass vessel. Connect vessel(s) immediately to the freeze dryer and dry for 18–24 h.
7. Reweigh tubes to obtain the yeast dry weight.
8. Tubes can be stored in a −20 °C freezer until ready for extraction.

Transesterification In Situ:

9. Add 0.4 mL of 2:1 chloroform–methanol (v/v) and 0.6 mL of 5% (v/v) concentrated HCl in methanol. Then add 20 μL of 24.5 mg/mL C15:0 triglyceride standard (Nu-Chek Prep, T-145) prepared in 2:1 chloroform–methanol (v/v). Note that once 2 mL of hexane is added in **step 12**, the C15:0 triglyceride standard will have a concentration of 245 μg/mL. Solvents are added using a positive displacement pipettor (such as Pos-D Rainin MR-1000, Mettler Toledo) or glass syringes, and the standards are added using a repeating dispenser (such as Hamilton PB6000-1) equipped with a 1 mL glass syringe. Typical plastic pipette tips suitable for air displacement pipettors lead to the appearance of spurious peaks on the GC chromatograms.
10. Vortex and heat at 85 °C for 1 h in a dry bath. Briefly vortex every 15 min.
11. Allow to cool to ambient temperature (e.g., for 20 min).
12. Add 2 mL of hexane and 20 μL of 10.8 mg/mL C13:0 methyl ester standard prepared in hexane.

13. Vortex and separate the hexane layer by centrifugation for 5 min at 1240 × max *g*.
14. Using a 5¾″ Pasteur glass pipette, remove 1.2 mL of the top hexane layer and place in a new 10 mL centrifuge tube. Samples may be stored overnight at −20 °C. For ease of handling, the hexane layer is first removed to a dram vial, then a 1 mL glass syringe is used to transfer precisely 1.2 mL (2 × 0.6 mL) to the new container.
15. Wash hexane by adding 0.4 mL of 0.3% NaCl (w/v), mix with vortex, and separate hexane layer by centrifugation for 5 min at 1240 × max *g*.
16. Using a Pasteur pipette, transfer hexane (upper layer) to a 1.5 mL vial for GC analysis. Exact volume is not critical because internal standards are used to correct for differences in volume.
17. Prepare external standards such that each external standard mixture contains 100 mg total FAMEs. Dilute standards to 50 mL final volume in analytical grade hexane using a calibrated volumetric flask. Dispense 1 mL aliquots into GC vials and store at −20 °C. Inject 1 μL onto GC column for separation (*see* below for GC method). External FAME standards include FAME reference mixture suitable for peanut oil analysis comprising C14:0, C16:0, C18:0, C18:1, C18:2, C18:3, C20:0, C22:0, C22:1, C24:0 (catalog #17A, Nu-Chek Prep, Inc., Elysian, MN, USA) and FAME reference mixture suitable for palm oil analysis comprising C14:0, C16:0, C16:1, C18:0, C18:1, C18:2, C18:3 (catalog #20A, Nu-Chek Prep, Inc.).
18. Inject 1 μL onto the gas chromatography (GC) column for separation under the following conditions:

    Run time: 10 min.

    Column: Agilent HP-88 (112-8837) 30 m × 250 μm × 0.2 μm nominal.

    Injector: Splitless, 1 μL injection.

    Inlet: 240 °C, 7.5 psi.

    Oven: Initially 140 °C, then ramp 15 °C/min to 240 °C, and hold 2.5 min (total temperature ramp time is 9.2 min).

    Detector: FID, 280 °C, H_2 flow at 35 mL/min, air flow at 400 mL/min, N_2 make up flow at 5 mL/min.
19. Use the external standards in **step 17** for peak identification based upon retention times.
20. Calculate total FAME using the C13:0 methyl ester internal standard (C13:0 IS) and peak areas of individual FAME analytes (FA) as follows:

$$\text{FA } [\mu g/\text{mL}] = (1/R) \times (\text{peak area FA/peak area C13:0}) \times \text{concentration C13:0 IS},$$
$$\text{Total lipid } [\mu g/\text{mL}] = \text{sum of FA of all individual fatty acids present},$$

where R represents the relative response factor, which relates the peak area per unit FAME analyte concentration (FA) to peak area per unit internal standard concentration as follows:

$$R[-] = (\text{peak area FA/concentration FA})/(\text{peak area C13:0 IS/concentration C13:0 IS}).$$

21. Correct the total lipid concentration measured by GC based on the fraction recovery of the C15:0 triglyceride internal reaction standard as follows:

$$\text{Corrected total lipids} = \text{total lipid, GC} \times \text{C15:0 IS,i/C15:0 IS, GC},$$

where C15:0 IS, GC is the GC measured concentration, and C15:0 IS,i is the initial concentration based on the known addition of C15:0 triglyceride to the assay in **step 9** prior to transesterification.

3.10 Time Domain NMR Method to Determine Lipid Concentration

After each yeast cell sample has been added to a 1.65 or 2 mL microfuge tube, washed, and freeze-dried as specified in Subheading 3.9, **steps 1–8**, then proceed as follows:

1. Cut off the cap and place the entire tube inside an 8 × 18 mm glass test tube, and close with a size 0 rubber stopper. Samples should contain 20–100 mg of freeze-dried yeast cells. Note the exact weight of dried cells added to the sample tube in order to determine lipid content later.
2. Prepare external standards by placing 2–20 mg at intervals of 2 mg refined corn oil (or other appropriate reference oil) onto Kimwipes™ inserted into the glass test tubes so the oil is approximately 10–30 mm from the bottom. Tare the tubes and tissue ahead of time to determine the exact amount of oil added to each. Cap tubes with size 0 rubber stoppers.
3. Heat samples to 40 °C by incubating in a dry heating bath for 1 h.
4. Run samples using recommended operating conditions for the time domain NMR instrument as specified in the manual. For example, the following operating parameters are recommended for the Minispec mq20 TD-NMR spectrometer (Bruker Optics Inc., Billerica, MA): 8 mm tube size (OD), 60 MHz resonance frequency, 40 °C probe temperature, 3.48 μs 90° pulse length, 7.52 μs 180° pulse length, 16 total scans, 2 s recycle delay, 64 dB gain.

4 Notes

1. It is useful to have plant biomass field source data since post-harvest processing can be impacted by the field management history and season. For example, for our first tier screen corn hybrid Pioneer 36H56 triple stack (a corn borer/rootworm/ Roundup Ready variety) was harvested in September 2008 from the Arlington Research Station, Wisconsin, USA. Additionally for our second tier studies, the switchgrass was obtained post-frost from the Kanlow N1 field, Mead, Nebraska, USA grown by the USDA ARS Grain, Forage, and Bioenergy Research Unit, Lincoln, Nebraska. Pertinent to this study, post-frost harvested switchgrass has a low protein content [10], and this impacts the availability of amino acids to support yeast growth and the C:N ratio involved in triggering lipid production.
2. Alternatively, the dilute acid pretreatment of switchgrass or other herbaceous biomass may be carried out using an autoclave at 121 °C or a pipe reactor at 150 °C in a fluidized sand bath according to the method presented in Dien et al. [11]. These methods were applied to small 20 g samples for screening purposes but could be scaled up to larger volumes commensurate with the Mathis oven. Although autoclaves are common to most research groups doing biological research, the relatively low reaction temperature limit of 121 °C leads to a less severe pretreatment such that the sugar yields obtainable are significantly reduced, making it less representative of an expected industrial process. Another very practical alternative would be to obtain dilute acid pretreated biomass from an industrial collaborator involved in producing hydrolysates or else through a suitable research-associated source, such as from the Integrated Biorefinery Research Facility at the National Renewable Energy Laboratory. Golden, Co. The latter works with research customers to produce hydrolysates at the laboratory and pilot scales.
3. For nitrogen amendment as specified by experiment designs, the SGH may be supplemented with 2.18 g/L ammonium sulfate and 30 g/L soy flour (Toasted Nutrisoy Flour Product, Archer Daniels Midland, code 063160) to bring the primary amino nitrogen (PAN) content of the SGH to 211 mg N/L. Soy flour has long been recognized as the lowest cost and most abundant source of essential amino acids [34] and as such is commonly used to fortify industrial cultivations (at <$0.25/ lb). Alternatively a vitamin-free casamino acids mix supplying 5.25 g/L Bacto Difco Vitamin Assay Casamino Acids (Becton Dickinson, no. 228830), 0.053 g/L D,L-tryptophan (Sigma,

T330-100G), and 0.209 g/L L-cysteine (Sigma, C7352-100G) may be added to SGH to supply an equivalent PAN content (211 mg N/L) to that achieved with the soy flour. Despite its higher cost, the casamino acids mix above is recommended for its consistent quality as a good starting point to begin optimization and design studies aimed at developing future delivery of most cost-effective amino acids to industrial processes [35].

4. Based on manufacturer data collected at 30 °C and 300 rpm, the oxygen transfer rate of this system would be estimated to be at least 12 mM oxygen/L/h under our incubation conditions.
5. Emptied wells can be filled with sterile water to maintain humidity and dissolve culture media from the wells for easier cleaning later.
6. Multiple samples can be efficiently processed at one time by using an electronic programmable multichannel variable tip spacing pipette (such as an Integra Viaflo Voyager) and a multi-tube shaker (such as Heidolph Multi Reax) for cell resuspension and mixing. Sample mixing in deep wells can be achieved with the programmable pipette set to draw and expel sample volume approximately three times before dispensing to microfuge tubes.
7. Measure biomass concentration by either analyzing culture absorbance at 620 nm (A_{620}) in 1 cm path length cuvettes using a spectrophotometer (for example, Evolution 60) or in 96-well microplate wells filled to 200 μL using a plate reader (such as BioTek Powerwave XS) if higher throughput is needed. Dilute samples as needed to obtain absorbance in the linear range for each instrument. Correlate the absorbance readings from microplate wells to equivalent cuvette values using the appropriate conversion factor to adjust for the difference in light path lengths for the cuvettes and wells.
8. When strains are being screened to determine the best performers in hydrolysates, a statistical ranking process can be applied which utilizes dimensionless Relative Performance Index (RPI) to order strains from best to worst. Pertinent formulas to calculate RPI are: $RPI = (2 + F)(25)$, where $F = (X - X_{avg})/s_x$ and X is the value of a measured parameter for a specific strain that characterizes its performance, such as product titer, yield, or productivity. X_{avg} and s_x are the average and standard deviation, respectively, of the values of X observed across all of the strains tested. If the observed values of X are normally distributed across the strains tested, then F will by definition range from −2 to +2, and, here, to scale the value of RPI to 0–100 (worst to best), the scaling factor 25 is applied. For example, to order strains relative to product lipid

accumulation, P (g/L) in AFEX CSH, the RPI_P for each strain applied is calculated as follows: $RPI_P = (25)[2 + ((P - P_{avg})/s_P)]$. RPI_P values calculated for our screen of oleaginous yeasts on AFEX CSH are provided in Table 2. An overall RPI can be based on performance relative to various options, including various kinetic traits in a specific hydrolysate or various hydrolysates, or other trait deemed to be of importance to successful performance in the target industrial environment. Since dimensionless, an overall RPI can be calculated for each strain by averaging the RPIs for each performance trait. For example, if we consider lipid accumulation (P) and lipid productivity (R) for a given hydrolysate, then for each yeast isolate tested on it, the $RPI_{overall} = (RPI_P + RPI_R)/2$. In this example, equal weight is given to the performance traits, lipid accumulation and productivity, in the calculation of average $RPI_{overall}$. Note that in general, yield and rate parameters can be given different weights if they are deemed unequal in importance. An example ranking result using equally weighted RPI values is shown in the Table 2. *See* additional applications of RPI to ranking isolates based on performance [6, 31].

9. Full factorial (or other) experiment designs may be used as a modification of Subheading 3.3 to examine the impact of conditions, such as SGH strength (50–100%), amino acids source (casamino acids versus soy flour or others) and pH (5–7), on growth and lipid production of top ranked strains. Various amino acids sources can be incorporated to supply equivalent PAN levels for comparison while the balance of nitrogen is supplied as ammonium sulfate. Urea can replace ammonium sulfate to reduce pH drift if urea is utilized by the strain of interest. Such optimizations of lipid production were performed for our best lipid producing strains (*Lipomyces tetrasporus* and *Rhodosporidium toruloides*) and are given in full detail in Slininger et al. [6]. The ranges of variation are initially chosen based on the industrially preferred settings of SGH strength high for more concentrated products, amino acids sources of lowest cost and concentration, and acidic pHs to reduce contamination issues. These would be modified as data are collected to progress toward optimal production kinetics. Kennedy and Krouse [36] provide a useful review of experiment design techniques for optimization which are pertinent to microbial processes. Additionally, Webb [37] gives a useful listing of partial factorial designs which can be applied in combination with analysis of variance to study the impact of multiple variables (3–8) in a single experiment using a minimum number of treatment combinations.
10. Alternatively, for top-ranked strains the culture conditions may be optimized relative to cell growth and lipid accumulation in

the first stage and also relative to further lipid accumulation in the second stage under ultra high molar C:N ratio. Ultra high C:N ratios may be achieved at >400:1 without supplementing SGH and are likely to strongly induce lipid metabolism [1, 18]. First stage conditions to optimize might include pH (6–7), molar C:N (25:1–100:1), SGH strength (50–75%), and soy flour amino acids (presence, absence or continuously variable) while second stage variables might be likely to include SGH strength and pH. C:N ratios can be adjusted based on assays of the usable nitrogen sources PAN, ammonia, and urea. If PAN is variably supplemented for first stage growth, the molar N availability is maintained constant with added ammonium sulfate. Certainly variable shifts in stage 2 are expected to impact its performance, but it is also expected that culture conditions applied to stage 1 may impact stage 2 performance. Both stage 1 and 2 cultures are optimized in 96-deep well plates for high throughput as described above (*see* Subheading 3.4, **steps 1–9**). Using these experiment design concepts together with appropriate statistical analysis methods (*see* **Note 9**), variable impacts on yeast and lipid accumulation during first and second stage processes may be determined. An application of this type of optimization is shown in Slininger et al. [6] where the performance of preferred strains in a two-stage process was optimized as a function of first stage growth culture variables (SGH strength, PAN amendment, pH) in combination with the second stage variables (pH, SGH strength at C:N > 400).

11. Using solvent extraction of cell contents after ball milling, it may also be desired to assess carotenoid composition in addition to lipids. Carotenoids occur with the lipid layer upon solvent extraction [21] and can be analyzed using spectrophotometry [38, 39] and gradient HPLC chromatography [40, 41] to indicate if there is potential for carotenoid recovery [6]. The recovery of carotenoids as a high value product along with lipids is expected to greatly enhance the economics of microbial biodiesel [14].
12. For hydrolysate compositional analysis with enhanced separation of hydrolyzate sugars and sugar alcohols, a Bio-Rad Aminex HPX-87H carbohydrate analysis column may be optionally replaced with the HPX-87P (125-0098) column, which is used with Deashing cartridge (125-0118) and Carbo-P Micro-Guard Cartridge (125-0119) installed as precolumns. This column is heated at 80 °C and eluted with water as mobile phase at 0.6 mL/min.
13. Time and resource-saving modifications of the SPV assay have been successfully applied in our laboratory. The first modification involves cutting all of the reagent volumes in Subheading

3.7, **steps 3–8** by a factor of 2.5 to have the following: 0.4 mL 18 M H_2SO_4, 1 mL vanillin–phosphoric acid solution, and 20 μL yeast sample. The smaller reaction volume is additionally consistent with the use of a 96-well microplate for absorbance readings provided that a plate-reading spectrophotometer is available in the laboratory. In this modification, 200 μL aliquots of fully reacted samples, following cooling in Subheading 3.7, **step 9**, are loaded into a 96-well microtiter plate for rapid simultaneous absorbance readings. While in the tubes, the reaction continues to proceed slowly even in cool water, and drift error between first and last sample readings may be eliminated by using a microplate reader. Due to the reduced light pathlength of the 200 μL sample depth in plates, the standard curve is linear up to ~10 μg/μL oil. Hence, a broader range of oil standards from 0 up to 6–8 μg/μL may be used with samples diluted to fall in this range.

Acknowledgments

The authors would like to thank Maureen Shea-Andersh for her valuable technical assistance in support of the oleaginous yeast project.

Disclaimer: The mention of trade names or commercial products in this chapter is solely for the purpose of providing specific information and does not imply any recommendation or endorsement by the US Department of Agriculture. USDA is an equal opportunity provider and employer.

References

1. Ageitos JM, Vallejo JA, Veiga-Crespo P, Villa TG (2011) Oily yeasts as oleaginous cell factories. Appl Microbiol Biotechnol 90 (4):1219–1227
2. Leiva-Candia D, Pinzi S, Redel-Macías M, Koutinas A, Webb C, Dorado M (2014) The potential for agro-industrial waste utilization using oleaginous yeast for the production of biodiesel. Fuel 123:33–42
3. Sitepu IR, Garay LA, Sestric R, Levin D, Block DE, German JB, Boundy-Mills KL (2014) Oleaginous yeasts for biodiesel: current and future trends in biology and production. Biotechnol Adv 32(7):1336–1360
4. Jin M, Slininger PJ, Dien BS, Waghmode SB, Moser BR, Sousa LC, Orjuela A, Balan V (2015) Microbial lipid based lignocellulosic biorefinery: feasibility and challenges. Trends Biotechnol 33(1):43–54
5. Thorpe R, Ratledge C (1972) Fatty acid distribution in triglycerides of yeasts grown on glucose or n-alkanes. J Gen Microbiol 72:151–163
6. Slininger PJ, Dien BS, Kurtzman CP, Moser BR, Bakota EL, Thompson SR, O'Bryan PJ, Cotta MA, Balan V, Jin M (2016) Comparative lipid production by oleaginous yeasts in hydrolyzates of lignocellulosic biomass and process strategy for high titers. Biotechnol Bioeng 113:1676–1690
7. US Department of Energy (2011) U.S. billion-ton update: biomass supply for a bioenergy and bioproducts industry (R.D. Perlack and B.J. Stokes (Leads)), Oak Ridge, TN: ORNL/TM-2011/224. Oak Ridge National laboratory, p 227
8. Dugar D, Stephanopoulos G (2011) Relative potential of biosynthetic pathways for biofuels

and bio-based products. Nat Biotechnol 29 (12):1074–1078
9. Ratledge C, Cohen Z (2008) Microbial and algal oils: do they have a future for biodiesel or as commodity oils? Lipid Technol 20 (7):155–160
10. Bates G, Keyser P, Harper C, Waller J (2008) Using switchgrass for forage (https://utextension.tennessee.edu/publications/documents/SP701-B.pdf). UT Biofuels Initiative, University of Tennessee Institute of Agriculture
11. Dien BS, Jung H-J, Vogel K, Casler M, Lamb J, Iten L, Mitchell R, Sarath G (2006) Chemical composition and response to dilute-acid pretreatment and enzymatic saccharification of alfalfa, reed canarygrass, and switchgrass. Biomass Bioenergy 30(10):880–891
12. Dien BS, O'Bryan P, Hector R, Iten L, Mitchell R, Qureshi N, Sarath G, Vogel K, Cotta M (2013) Conversion of switchgrass to ethanol using dilute ammonium hydroxide pretreatment: influence of ecotype and harvest maturity. Environ Technol 34 (13–14):1837–1848
13. Dien BS, Zhu J, Slininger PJ, Kurtzman CP, Moser BR, O'Bryan PJ, Gleisner R, Cotta MA (2016) Conversion of SPORL pretreated Douglas fir forest residues into microbial lipids with oleaginous yeasts. RSC Adv 6(25):20695–20705
14. Koutinas A, Chatzifragkou A, Kopsahelis N, Papanikolaou S, Kookos I (2014) Design and techno-economic evaluation of microbial oil production as a renewable resource for biodiesel and oleochemical production. Fuel 116:566–577
15. Dien BS, Slininger PJ, Kurtzman CP, Moser BR, O'Bryan PJ (2016) Identification of superior lipid producing Lipomyces and Myxozyma yeasts. AIMS Environ Sci 3(1):1–20
16. Quarterman J, Slininger PJ, Kurtzman CP, Thompson SR, Dien BS (2017) A survey of yeast from the Yarrowia clade for lipid production in dilute acid pretreated lignocellulosic biomass hydrolysate. Appl Microbiol Biotechnol 101(8):3319–3334
17. Liu ZL, Slininger PJ, Dien BS, Berhow MA, Kurtzman CP, Gorsich SW (2004) Adaptive response of yeast to furfural and 5-hydroxymethylfurfural and new chemical evidence for HMF conversion to 2,5-bis-hydroxymethylfuran. J Ind Microbiol Biotechnol 31:345–352
18. Weete JD (1980) Chapter 2: Fungal lipids. In: Weete JD (ed) Lipid biochemistry of fungi and other organisms. Plenum Press, New York, pp 9–48
19. Sitepu IR, Sestric R, Ignatia L, Levin D, Bruce GJ, Gillies LA, Almada LA, Boundy-Mills KL (2013) Manipulation of culture conditions alters lipid content and fatty acid profiles of a wide variety of known and new oleaginous yeasts species. Bioresour Technol 144:360–369
20. Hounslow E, Noirel J, Gilmour DJ, Wright PC (2017) Lipid quantification techniques for screening oleaginous species of microalgae for biofuel production. Eur J Lipid Sci Technol 119(2):1500469
21. Bligh EG (1959) A rapid method of total lipid extraction and purification. Can J Biochem Physiol 37(8):911–917
22. Sitepu I, Ignatia L, Franz A, Wong D, Faulina S, Tsui M, Kanti A, Boundy-Mills K (2012) An improved high-throughput Nile red fluorescence assay for estimating intracellular lipids in a variety of yeast species. J Microbiol Methods 91:321–328
23. Govender T, Ramanna L, Rawat I, Bux F (2012) BODIPY staining, an alternative to the Nile Red fluorescence method for the evaluation of intracellular lipids in microalgae. Bioresour Technol 114:507–511
24. Krygsman P, Barrett A (2004) Simple methods for measuring total oil content by bench-top NMR. In: D.L. Luthira (ed), Oil extraction and analysis, AOCS Press, Champagin, Illinois pp 152–165
25. Cheng Y-S, Zheng Y, VanderGheynst JS (2011) Rapid quantitative analysis of lipids using a colorimetric method in a microplate format. Lipids 46(1):95–103
26. Johnson K, Ellis G, Toothill C (1977) The sulfophosphovanillin reaction for serum lipids: a reappraisal. Clin Chem 23(9):1669–1678
27. Knight JA, Anderson S, Rawle JM (1972) Chemical basis of the sulfo-phospho-vanillin reaction for estimating total serum lipids. Clin Chem 18(3):199–202
28. Laurens LM, Quinn M, Van Wychen S, Templeton DW, Wolfrum EJ (2012) Accurate and reliable quantification of total microalgal fuel potential as fatty acid methyl esters by in situ transesterification. Anal Bioanal Chem 403 (1):167–178
29. Izard J, Limberger RJ (2003) Rapid screening method for quantitation of bacterial cell lipids from whole cells. J Microbiol Methods 55 (2):411–418
30. Wang J, Li R, Lu D, Ma S, Yan Y, Li W (2009) A quick isolation method for mutants with high lipid yield in oleaginous yeast. World J Microbiol Biotechnol 25(5):921–925

31. Slininger PJ, Shea-Andersh MA, Thompson SR, Dien BS, Kurtzman CP, Balan V, da Costa Sousa L, Uppugundla N, Dale BE, Cotta MA (2015) Evolved strains of Scheffersomyces stipitis achieving high ethanol productivity on acid-and base-pretreated biomass hydrolyzate at high solids loading. Biotechnol Biofuels 8(1):60
32. Gao C, Xiong W, Zhang Y, Yuan W, Wu Q (2008) Rapid quantitation of lipid in microalgae by time-domain nuclear magnetic resonance. J Microbiol Methods 75(3):437–440
33. Moreton R (1989) Yeast lipid estimation by enzymatic and nuclear magnetic resonance methods. Appl Environ Microbiol 55 (11):3009–3011
34. Ravimannan SP (2016) Soy flour as alternative culture media for yeasts. Global J Sci Front Res 16(3):75–78
35. Zabriski DW, Armiger WB, Phillips DH, Albano PA (1980) Fermentation media formulation. In: Trader's guide to fermentation media formulation. Trader's Protein Memphis, TN, pp 1–39
36. Kennedy M, Krouse D (1999) Strategies for improving fermentation medium performance: a review. J Ind Microbiol Biotechnol 23 (6):456–475
37. Webb SR (1971) Small incomplete factorial experiment designs for two- and three-level factors. Technometrics 13(2):243–256
38. Donaldson M (2012) Development of a rapid, simple assay of plasma total carotenoids. BMC Res Notes 5(1):521
39. Tinoi J, Rakariyatham N, Deming R (2005) Simplex optimization of carotenoid production by *Rhodotorula glutinis* using hydrolyzed mung bean waste flour as substrate. Process Biochem 40(7):2551–2557
40. Hart DJ, Scott KJ (1995) Development and evaluation of an HPLC method for the analysis of carotenoids in foods, and the measurement of the carotenoid content of vegetables and fruits commonly consumed in the UK. Food Chem 54(1):101–111
41. Rodriguez-Amaya DB (2001) A guide to carotenoid analysis in foods. ILSI Press, International Life Sciences Institute, Washington, DC

Chapter 17

Conversion of Microbial Lipids to Biodiesel and Basic Lab Tests for Analysis of Fuel-Quality Parameters

Annaliese K. Franz and Cody Yothers

Abstract

This chapter describes lab-scale procedures for the direct conversion of microbial lipids to fatty acid methyl esters (FAMEs) for use as biodiesel fuel. Methods for the gas chromatography analysis of FAME profiles and equations to predict several fuel-quality parameters are detailed herein. This chapter also provides a complete list summarizing each of the fuel quality tests (e.g., sample size and equipment) that are required by ASTM International D6751 regulations for pure biodiesel fuel (B100) or blend stock. Recommendations for the decolorization of microbial lipid sources containing pigments are also included. This resource should provide a guide to basic conversion and characterization of microbial-derived biodiesel fuels and a roadmap for more-detailed testing required to assess commercial feasibility.

Key words Biodiesel, Lipids, Fatty acid, Transesterification, Fatty acid methyl ester, Mono-alkyl ester, Ethyl ester, Fuel quality, Biofuel, Microbial lipids, Microalgae, Yeast, Cyanobacteria, ASTM

1 Introduction

Biodiesel, as defined by ASTM D6751, is the mono-alkyl esters of fatty acids from biological sources [1]. Microbial lipids represent a viable biodiesel feedstock because they can be directly converted to mono-alkyl esters for biodiesel fuels and blend stocks [2]. Microbial lipids are hydrocarbon constituents in the cell that primarily exist in esterified forms such as triglycerides. The direct conversion of isolated lipids to fatty acid methyl or ethyl esters is accomplished by transesterification. Free fatty acid constituents also exist in microbial cells and can be simultaneously converted to fatty acid alkyl esters using esterification [3]. Conversion from triglycerides to mono-alkyl esters for biodiesel fuel significantly decreases viscosity and improves volatility and cold flow properties when compared with unaltered lipids, thereby improving operation in diesel engines [4].

Microbial lipid sources include yeast, bacteria, fungi, and algae, all of which can be from either genetically modified sources or

Venkatesh Balan (ed.), *Microbial Lipid Production: Methods and Protocols*, Methods in Molecular Biology, vol. 1995, https://doi.org/10.1007/978-1-4939-9484-7_17,

native species and may have widely varying lipid profiles [2]. The various structural features, such as chain length and degree of unsaturation, of the fatty acid components derived from microbial lipids will influence key properties of biodiesel (i.e., viscosity and oxidative stability) [5]. Microbial-derived biodiesel fuels have special considerations compared with petrodiesels due to the degree of unsaturation and the chain length of microbial lipid hydrocarbons. The higher levels of polyunsaturated fatty acids (PUFAs) that are often found in microbial lipids can have a negative impact on the oxidative stability of the resultant biodiesel. Longer chain lengths can result in higher energy content and lubricity, but may reduce volatility or cause increased engine deposits [6]. In addition, lipid sources from algae and cyanobacteria can contain chlorophyll and other pigments, which can be considered a detrimental contaminant and may obscure fuel quality tests. One recourse is to decolorize FAME biodiesel samples, so a decolorization protocol is provided in Subheading 3.5 of this chapter.

The process for conversion of microbial lipids to biodiesel typically utilizes isolated lipids that have been extracted from biomass. Methods for the extraction of lipids from biomass are addressed in Chapter 19 of this volume. Subheading 3.1 of this chapter provides a microscale method for acid-catalyzed transesterification of lipid extract to fatty acid methyl esters (FAMEs) (*see* **Note 1**). A microscale conversion and analysis of the FAME profile of microbial lipid sources using gas chromatography (GC) (*see* Subheading 3.2) is a simple and cost-effective method to estimate several fuel-quality parameters based set of empirical correlations. In Subheading 3.3, equations are provided to approximate cetane number, viscosity, density, and energy content of biodiesel samples based on their FAME profile.

The commercial use of microbial lipid-derived biodiesel necessitates adherence to standardized fuel-quality regulations put forth by ASTM International or the European Standardization Organizations. Subheading 4 of this chapter presents a comprehensive list summarizing each fuel-quality parameter governed by the relevant ASTM International regulation D6751, "Standard Specification for Biodiesel Fuel Blend Stock (B100) for Middle Distillate Fuels" (*see* **Note 2**) [1]. Tables 1 and 2 provide a list of all D6751 fuel parameters and regulatory limits, as well as the approximate sample size required for a certified laboratory submission (*see* **Note 3**). Many of these fuel quality tests require specialized equipment, and the broad nature of these techniques means some laboratories choose to outsource this analysis. Often, the cost and amount of material required for a full test panel is prohibitive for many research-scale operations (*see* **Note 4**). A gram-scale transesterification process is provided in Subheading 3.4 for the preparation of bulk FAME samples for fuel-quality analysis. Readers should consider the summaries in Subheading 4 and assess which tests are necessary for their particular research questions.

Table 1
Category 1 specifications (related to the molecular composition of fuels) for biodiesel fuel blend stock (B100) for middle distillate fuels from ASTM D6751

Fuel parameter, unit of measure	Test method	Limit (min or max)	Sample size	SE	Biodiesel-related qualities
Cetane number, unitless	D613	47 (min)	1 L	Y	Carbon chain length, degree of unsaturation
Cloud point, °C	D2500	Application specific	75 mL	N	Degree of unsaturation
Cold soak filterability, s	D7501	200 s (max, 1-B grades) 360 s (max, 2-B grades)	300 mL	N	Degree of unsaturation
Distillation temperature, 90% recovery, °C	D1160	360 °C (max)	200 mL	Y	Carbon chain length, residual alcohol
Flash point, °C	D93	93 °C (min)	75 mL	Y	Carbon chain length, residual alcohol
Kinematic viscosity at 40 °C, mm^2/s	D445	1.9–6.0 mm^2/s	15 mL	N	Carbon chain length, degree of unsaturation
Oxidative stability, h	EN 15751	3 h (min)	50 mL	Y	Degree of unsaturation
Sulfur, ppm	D5453	15 ppm (max, S15 grades) 500 ppm (max, S500 grades)	5 mL	Y	Limited sulfur expected
Phosphorus content, ppm	D4951	10 ppm (max)	100 mL	Y	Unconverted lipids; combined with measurements of Ca, Mg, Na, and K

SE indicates the need for specialized equipment to complete the relevant test method

2 Materials

Unless otherwise stated, use deionized water and ACS-grade reagents. Prepare and store all reagents at room temperature. Diligently follow all waste disposal regulations when disposing of waste materials. Wear appropriate personal protective equipment (PPE) when handling any chemicals and solvents.

2.1 Microscale Conversion of Lipid Extracts to FAMEs

1. Analytical balance.
2. Culture tubes with threaded caps (100 mm × 13 mm culture tube Thomas Scientific CS1000 Catalog #: 9215D32).
3. Nitrogen manifold (Reacti-Vap III #TS-18826 Evaporation unit).

Table 2

Category 2 specifications (related to the production and downstream purification of fuels) for biodiesel fuel blend stock (B100) for middle distillate fuels from ASTM D6751

Fuel parameter, unit of measure	Test method	Limit (min or max)	Sample size	SE	Biodiesel related qualities/ Notes
Metals, ppm					
Ca and Mg (combined), ppm	EN 14538	5 ppm (max)	100 mL		Catalyst residues; combined with measurements of Na, Mg and P
Na and K (combined), ppm	EN 14538	5 ppm (max)	100 mL		Catalyst residues; combined with measurements of Ca, Mg and P
Acid number, mg KOH/g	D664	0.05 (max)	50 mL	Y	Residual fatty acids
Alcohol control (one of the following must be met)					
Flash point, °C	D93	130 °C (min)	75 mL	Y	Carbon chain length, residual alcohol
Methanol content, % mass	EN 14110	0.2% mass (max)	20 mL	Y	Residual methanol
Carbon residue, % mass	D4530	0.5% mass (max)	20 mL	Y	Residual fatty acids and glycerides
Copper strip corrosion, 3 h at 50 °C, standard comparison	D130	Class 3 (dark tarnish, no corrosion)	30 mL	Y	Downstream processing, residual fatty acids
Glycerin (one method measures all three simultaneously)					
Total glycerin, % mass	D6584	0.240% mass (max)	20 mL	Y	Residual glycerin
Free glycerin, % mass	D6584	0.020% mass (max)	20 mL	Y	Residual glycerin
Monoglyceride, % mass	D6584	0.40% mass (max, 1-B grades)	20 mL	Y	Residual glycerin
Sulfated ash, % mass	D874	0.02% mass (max)	100 mL	N	Trace metal contaminants, catalyst residues
Water and sediment, % volume	D2709	0.050% vol (max)	100 mL	N	Downstream processing, residual water

SE indicates the need for specialized equipment to complete the relevant test method

4. Dry block heater (DuPont Qualicon BAX® System).
5. Aluminum heat blocks (13-mm vial insets).
6. Thermometer (0–150 °C).
7. Cooler and crushed ice.
8. 1.5-mL microcentrifuge tubes.
9. Microcentrifuge.
10. Centrifuge with capacity for 13-mm culture tubes (optional).
11. Vortex mixer.
12. Chemical reagents and solvents: toluene, hexanes, concentrated hydrochloric acid, and methanol.

13. Internal standard solution: Dissolve 84.9 mg of internal standard (17:0 triglyceride, trihexadecanoin) in 10 mL of chloroform. Internal standards can be stored at −20 °C for up to 1 year (*see* **Notes 5** and **6**).
14. 8% v/v HCl–MeOH solution: Add 80 μL of concentrated HCl to 920 μL of methanol. This reagent should be prepared fresh for each procedure. A volume of 600 μL is required for each sample, so this solution should be scaled accordingly.

2.2 GC Analysis of FAMEs

1. Gas chromatograph equipped with flame ionization detector (FID) (Perkin Elmer Clarus 500 GC).
2. Nitroterephthalic acid-modified polyethylene glycol (PEG) column of high polarity for the analysis of volatile FAs and FAMEs (DB-FFAP Agilent part 122-3232, length 30 m, diameter 0.25 mm, film 0.25 μm) (*see* **Note 7**).
3. Amber GC vials.
4. Amber GC vial caps.
5. 150-μL GC vial inserts.
6. Hexanes (mixture of isomers), for wash and blank cycles.
7. Solution of FAME standard mix with internal standard: Create a stock solution FAME mix by dissolving 1 mg of each standard in 1 mL of chloroform. Standard mixes can be scaled up and stored at −20 °C for up to 1 year. An example of an effective FAME standard mix is: 8:0, 10:0, 12:0, 13:0, 14:0, 14:1, 15:0, 16:0, 16:1, 17:0, 18:0, 18:1 cis, 18:2 ω-6, 18:3 ω-6, 18:3 ω-3, 20:0, 20:1 ω-9, 20:2 ω-6, 20:3, 20:4 ω-6, 20:3 ω-3, 20:5 ω-3, 22:0, 22:1, 22:2, 22:5 ω-6, 22:5 ω-3, and 22:6 (*see* **Note 8**).

2.3 Gram-Scale Conversion of Microbial Lipids to FAMEs

1. 500-mL round-bottom flask.
2. Hot plate with magnetic stir function.
3. Magnetic stir bar.
4. 500-mL round-bottom flask heat block, or Pyrex hot bath and metal heating beads.
5. 1-L separatory funnel.
6. Amber glass or corrosion resistant plastic vials for sample storage.
7. Rotary evaporator.
8. Chemical reagents: methanol, chloroform, and concentrated sulfuric acid.

2.4 Decolorization of FAME/Biodiesel Sample

1. 100-mL Erlenmeyer flask.
2. Hot plate with magnetic stir function.
3. Magnetic stir bar.

4. Montmorillonite K10 powder.
5. Celite.
6. Buchner funnel and 250-mL vacuum filter flask.
7. Rotary evaporator.
8. Amber glass or corrosion resistant plastic bottles for sample storage.

3 Methods

Carry out all procedures at room temperature, unless otherwise specified. The methods below are described for one sample, but replicates (at least $n = 3$ per microbial lipid sample) are generally recommended.

3.1 Microscale Conversion of Lipid Extracts to FAMEs for GC Profiling (See Note 9) [7, 8]

1. Weigh 30 mg of lipid extract into a 100 × 13 mm culture tube (*see* **Note 10**).
2. Add 100 μL (1 mmol) of trihexadecanoin (C17:0 triglyceride) solution as an internal standard to lipid extract (*see* **Note 11**).
3. Evaporate the sample to dryness under nitrogen using a Reacti-Vap manifold (*see* **Note 12**).
4. Add 400 μL of toluene to the culture tube, cap tightly, and vortex vigorously for 3 s (*see* **Note 13**).
5. Add 3 mL of methanol to the culture tube, cap tightly, and vortex vigorously for 3 s (*see* **Note 14**).
6. Add 600 μL of the 8% HCl–methanol reagent to the culture tube, cap tightly, and vortex vigorously for 5 s.
7. Incubate tube for 60 min in a dry heating block set to 90 °C (*see* **Note 15**).
8. Transfer the hot tube to a cooler packed with ice and cool for 5 min.
9. Add 1 mL of hexanes to the tube, cap tightly, and vortex vigorously for 3 s.
10. Add 1 mL of deionized water to the tube, cap tightly, and vortex vigorously for 3 s (*see* **Note 16**).
11. Let tube stand undisturbed for about 10–15 min until the aqueous and organic layers have separated (*see* **Note 17**).
12. Transfer the hexanes (upper) layer to a 1.5-mL microcentrifuge tube containing 450 μL of deionized water and vortex for 3 s (*see* **Note 18**).
13. Centrifuge microcentrifuge tube for 2 min at 15,000 × *g*.
14. If continuing to GC analysis, transfer the hexanes layer to a second 1.5-mL microcentrifuge tube and evaporate to dryness

under nitrogen; for storage, transfer the hexanes layer to a 1-dram amber vial, purge with nitrogen, and store at −80 °C until analysis (*see* **Note 19**).

3.2 GC Analysis and Calculation of the Mole Fraction for FAMEs

1. Reconstitute microbial FAME sample from **step 14** of Subheading 3.1 (approximately 15–30 mg of FAMEs) in 150 μL hexanes and transfer to a GC vial with a 150-μL insert, being careful to avoid any debris.
2. Run microbial FAME sample on the GC equipped with FID (*see* **Note 20**). Reconstitute and rerun samples if necessary (*see* **Note 21**).
3. Run FAME standard mixture on the GC during the sample run.
4. To identify GC peaks, compare the retention times of the microbial FAME sample to the standard mix (*see* **Note 22**). Figure 1 shows a sample standard mix chromatograph and a sample FAME chromatograph isolated from microalgae lipid extract.
5. To quantify GC peaks for the microbial FAME sample, export the peak integrations from sample runs into a spreadsheet format such as excel or CSV.
6. Calculate moles of each microbial FAME peak using Eq. 1 based on your internal standard. One individual GC chromatograph FAME peak is defined as the *i*th peak (*see* **Note 23**).

$$\text{Moles of } i\text{th FAME} = \left(\frac{\text{Moles of internal standard}}{\text{Area of internal standard peak}}\right) * \text{area of } i\text{th peak} \quad (1)$$

7. Calculate the mole fraction (z_i) of each microbial FAME peak using Eq. 2 (*see* **Note 24**).

$$z_i = \frac{\text{Moles of } i\text{th FAME}}{\text{Total moles}} \quad (2)$$

3.3 Equations for Estimating Fuel Parameters Using GC Data [9]

The mole fractions and chemical structure of each microbial FAME calculated from the GC data can be used to approximate several fuel quality parameters based on the following series of equations. First, the molecular weight and degree of unsaturation (number of double bonds) of each FAME in the profile should be compiled. Next, use Eqs. 3–6 to calculate the fuel property of each constituent FAME (the *i*th FAME). For these equations, M_i is the molecular weight of the *i*th FAME, and N refers to the number of double bonds in that FAME. Finally, use Eq. 7 to approximate each fuel quality parameter of your biodiesel sample by summation of the values for each FAME constituent multiplied by their respective molar fractions (z_i, determined in Subheading 3.2, Eq. 2).

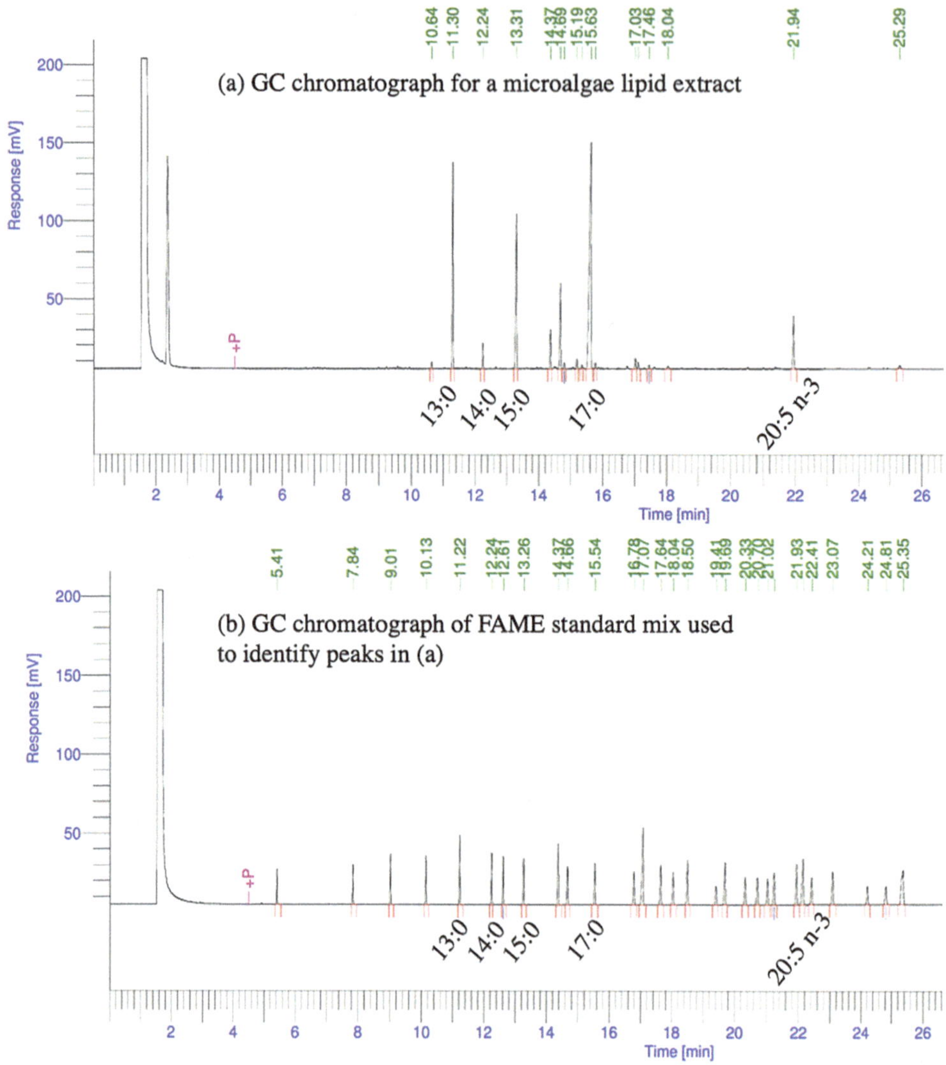

Fig. 1 Gas chromatographs of (**a**) FAME profile of a microalgae lipid extract, and (**b**) the FAME standard mix used to identify peaks

Equations 3–6 should be applied to find values for each FAME constituent individually, whereas Eq. 7 combines those values to provide an estimate of the overall biodiesel fuel quality parameter based on the FAME profile.

1. Cetane number: The ignition quality of a fuel is measured by cetane number (*also see* Subheading 4.1.1). High cetane numbers indicate smooth combustion and rapid engine starting; however, cetane number requirements vary and an increase over specific design requirements does not necessarily improve engine performance. Cetane number generally increases with

molecular weight (i.e., chain length) of the FAME, and decreases with more degrees of unsaturation. The cetane number ($\emptyset_i$) of the *i*th FAME can be estimated with empirical equation 3 (*see* **Notes 25** and **26**).

$$\emptyset_i = -7.8 + 0.302 \cdot M_i - 20 \cdot N \tag{3}$$

2. Viscosity: Viscosity of a fuel is relevant to the pumping and injection performance of engines (*also see* Subheading 4.1.6). Viscosity limits will depend on engine design and size; however, the maximum limit for biodiesel viscosity is higher than that of most petrodiesel fuels. Viscosity at 40 °C generally increases directly with fatty acid chain length (due to increased van der Waals forces) and decreases with more degrees of unsaturation. The kinematic viscosity (η_i, mm^2/s) at 40 °C of the *i*th FAME can be estimated with empirical equation 4 (*see* **Note 27**).

$$\ln \eta_i = -12.503 + 2.496 \cdot \ln M_i - 0.178 \cdot N \tag{4}$$

3. Density: Density of a fuel is important for proper fuel injection system performance, and can be used in conjunction with other properties to determine energy content and other fuel properties (*also see* Subheading 4.3.1). The density of biodiesel fuels is generally higher than that of petrodiesel fuels; however, the increased levels of PUFAs found in many microbial lipids can also lead to a decrease in the density of microbial-derived biodiesel fuel relative to other biodiesel fuels. Density at 20 °C generally decreases as the number of carbon atoms increases; however, this effect is diminished as the chain gets longer. Increasing the degrees of unsaturation also decreases density. Density of FAMEs can be expressed using Eq. 5, where ρ_i is the density at 20 °C of the *i*th FAME in g/cm^3 (*see* **Note 28**).

$$\rho_i = 0.8463 + \frac{4.9}{M_i} + 0.0118 \cdot N \tag{5}$$

4. Energy content: The measure of the energy content of a given unit of fuel is represented by its higher heating value (HHV) and considered a good indicator of fuel value and utility (*also see* Subheading 4.3.2). HHV can be used to calculate the mass heat of combustion, and is important when considering the thermal efficiency of engines. In biodiesel FAME sources, HHV increases with chain length, but decrease with each degree of unsaturation (PUFA content). Higher heating value can be estimated from Eq. 6, where δ_i is the HHV of the *i*th FAME in MJ/kg (*see* **Note 29**).

$$\partial_i = 46.19 - \frac{1794}{M_i} - 0.21 \cdot N \quad (6)$$

5. The fuel properties determined for each FAME in Eqs. 3–6 can be combined with the molar fraction to estimate each fuel property of the biodiesel sample. In Eq. 6, f_b is a function of one physical property of the biodiesel sample (i.e., δ_b, ρ_b, $\emptyset_b$, or $\ln(\eta_b)$). f_i refers to the function (or calculated value) of the *i*th FAME, and z_i is the mole fraction of *i*th FAME, which was calculated in Subheading 3.2. Using Eq. 7, determine the value (f_b) for each fuel quality parameter of the biodiesel sample.

$$f_b = \sum_{i=1}^{n} z_i \cdot f_i \quad (7)$$

3.4 Gram-Scale Conversion of Microbial Lipids to FAMEs [10]

To perform in-house fuel quality tests or to submit samples for external analysis, a larger quantity of biodiesel sample is required. This section provides a method for a simple acid-catalyzed transesterification to convert microbial lipids to FAME biodiesel that is suitable for up to 50 g of starting lipid extract (*see* **Note 30**).

1. Dissolve 25 g of microbial lipid extract (about 30 mL) in 200 mL of 1:1 methanol–chloroform (~6× volume of the lipid extract) in a 500-mL round-bottom flask with a magnetic stir bar (*see* **Note 31**).
2. Add 2.720 mL of concentrated sulfuric acid to the flask, which equates to 20% w/w of the original 25-g lipid sample (*see* **Note 32**).
3. Attach a 60-cm Liebig condenser to the flask and heat the mixture at 90 °C in a heating block for 1 h (*see* **Note 33**).
4. Let the reaction cool to room temperature.
5. Pour mixture into a 1 L separatory funnel and add 500 mL (~2× volume) of water.
6. Shake separatory funnel to mix and release gas through the stopcock. Then allow separatory funnel to rest until phases separate.
7. Drain the organic (lower) layer out of the separatory funnel, discard the aqueous layer, and repeat wash with the organic layer (*see* **steps 5** and **6**).
8. Drain the organic (lower) layer into a preweighed 500-mL round-bottom flask.
9. Remove organic solvents using a rotary evaporator and determine product mass. Store FAME biodiesel in amber glass or other corrosion resistant vessel in the dark at 4 °C.

3.5 Decolorization of FAME/Biodiesel Sample for Analysis [11, 12]

Decolorization is required when pigmentation prevents determination of the cloud point during biodiesel fuel quality testing (*see* Subheading 4.1.2). Decolorization is particularly relevant when working with algae- or cyanobacteria-derived feedstocks because chlorophyll will not be excluded by most lipid extraction methods.

1. Heat 15 g of biodiesel in a 100-mL Erlenmeyer flask containing a magnetic stir bar on a hot plate to 60 °C.
2. Add 3.0 g (20% w/w of biodiesel) of montmorillonite K10 powder and stir for 1 h.
3. Add 6 g of Celite to a 60-mm–diameter Buchner funnel attached to a 250-mL vacuum filtering flask.
4. Filter decolorized sample through Celite, then use several portions of hexanes to rinse the flask and Celite to recover all material. Collect the filtrate and all hexanes rinses in a 500-mL round-bottom flask; light vacuum can be used if necessary.
5. Remove hexanes using a rotary evaporator.
6. Store decolorized sample at 4 °C for at least 24 h to allow settling of any residual montmorillonite K10 powder.
7. Before analysis, let sample return to room temperature and decant to remove any residual sorbent (*see* **Note 34**).

4 Summary of Laboratory Tests for Each Fuel Parameter Governed by ASTM (D6751)

Fuel parameters for biodiesel B100 are governed by ASTM D6751. Biodiesel B100 must consist primarily of mono-alkyl esters of long chain fatty acids, as described above, and must conform to the standard specifications presented in Tables 1 and 2 (*see* **Note 35**). Biodiesel samples should also be inspected visually to check that there is limited turbidity and a lack of phase separation or precipitation. Samples for analysis can be taken using either manual sampling (ASTM D4057) or automatic sampling (ASTM D4177) procedures [13, 14].

There are two general categories for the specifications of fuel-quality parameters. Category 1 includes fuel parameters that are based on the specific molecular composition of the biodiesel feedstocks, such as the chain length and degrees of unsaturation (Table 1). Category 2 includes fuel parameters that are considered to be more related to the production and downstream purification of fuels (Table 2). Tests in category 1 are more relevant to lab-scale research for the fuel-quality parameters of microbial lipid-derived biodiesel.

4.1 Summary of Category 1 Fuel Quality Tests

4.1.1 Cetane Number (D613) [15]

The ignition quality of a fuel is measured by cetane number. High cetane numbers generally indicate smooth combustion and rapid engine starting; however, cetane number requirements vary based on engine design, size, vehicle speed, and load. Therefore, an increase in cetane number over specific design requirements does not necessarily improve engine performance, and commercial agreements often supplant the D6751 requirements.

The cetane number of a biodiesel sample is measured by comparing the combustion performance of the sample biodiesel to standard diesel blends with known cetane numbers in a specialized test engine under standard operating conditions. The biodiesel is injected into a specialized engine chamber, and each injection produces a single compression ignition combustion cycle. Ignition delay is measured and averaged over several tests, then compared to performance of two standard fuel blends. Approximately 1 L of biodiesel is required to perform this test.

4.1.2 Cloud Point (D2500) [16]

The cloud point of a fuel sample is the temperature at which a precipitate (cloud or haze of wax crystals) appears in the oil under test conditions. This point represents the lowest temperature for its utility as a fuel in a specific application (i.e., the lowest temperature it can be pumped through an engine). ASTM D2500 is a manual method for cloud point determination; however, automated methods have also been developed using visible spectroscopy [17]. Biodiesel samples from pigment-containing microbial sources may be opaque or darkened liquids, which can prohibit the determination of the cloud point. Decolorization and purification methods for biodiesel have been developed; however, if the biodiesel is processed prior to measuring the cloud point, then all tests should be performed on the decolorized biodiesel sample.

To measure the cloud point, the biodiesel sample is first heated above the expected cloud point, then cooled at a specific rate. The sample is periodically examined for cloudiness, and the temperature at which cloudiness appears is recorded as the cloud point. There is no regulatory limit for cloud point; however, the reported value can influence the application of the fuel based on expected conditions of operation or blending ratio. Approximately 75 mL of biodiesel is required to perform this test.

4.1.3 Cold Soak Filterability (D7501) [18]

This method predicts precipitation in biodiesel during storage and cold weather that can cause engine filter plugging. This method measures substances that may precipitate out of samples upon cooling to temperatures above the cloud point or after extended periods of time in storage or fuel tanks. Method D7501 is an accelerated means of determining the presence of these contaminants in biodiesel samples using glass fiber filters. Filtration time is measured, with a quicker elapsed time indicating a more satisfactory fuel operation.

A biodiesel sample is stored at 5 °C for 16 h, then allowed to warm to room temperature. The sample is then filtered through a 0.7-μm glass filter under controlled vacuum. The cold soak filterability is recorded as the filtration time in seconds. Approximately 300 mL of biodiesel is required to perform this test.

4.1.4 Distillation (D1160) [19]

The distillation temperature of a biodiesel sample is a measure of volatility, which is the major indicator of the ability of a hydrocarbon mixture to produce explosive vapors. This parameter has implications related to both safety and performance of biodiesel fuels. High-boiling point components can significantly increase the formation of engine deposits. Biodiesel will generally have a determinable boiling point, rather than a distillation curve that is measured for petrodiesel. The regulatory limit for B100 is 360 °C, and most biodiesel has a distillation temperature measured in the range of 330–357 °C.

The distillation temperature is measured by distilling a biodiesel sample at reduced pressure (0.13–6.7 kPa) under single theoretical plate fractionation conditions. A maximum boiling point is obtained after 90% recovery of the initial sample is obtained (*see* **Note 36**). Approximately 200 mL of biodiesel is required to perform this test.

4.1.5 Flash Point (D93) [20]

Flash point is important for the safe storage and handling of fuel, but is not directly related to engine performance. Flash point measures the temperature at which a spark causes vapors of a sample under specific test conditions to display a flame large enough to also catch the liquid fuel sample on fire. Most biodiesel fuels have flash points well above petrodiesel, which is a beneficial consideration in biodiesel's application to fuel blending. The regulatory limit for flash point of biodiesel is significantly higher than that of petrodiesel fuels and should be considered an important measurement. Flash point increases generally with FAME chain length. The flash point can decrease rapidly with the presence of residual alcohols in the biodiesel sample; this correlation provides an indicator of residual methanol in the FAME mixture, and a higher flash point limit validates acid-control regulations.

The flash point is measured using a manual or automatic Pensky–Martens closed cup (PMCC) device with a range of 40–360 °C. The biodiesel sample is heated at a slow, continuous ramp rate and regularly charged with a spark. The flash point is recorded as the lowest temperature at which the spark ignites the vapor above the sample. Approximately 75 mL of biodiesel is required to perform this test.

4.1.6 Kinematic Viscosity (D445) [21]

Viscosity of a fuel is relevant to the pumping and injection performance of engines. Lower limits on viscosity protect against power loss caused by injection pump/injector leakage. Maximum viscosity is limited due to effects on injection spray and volume flow through pumps; however, the true limits will depend on engine design and size. The maximum limit for biodiesel viscosity is higher than that of most petrodiesel fuels, and thus blending of high-viscosity biodiesel fuels must be monitored so that the resulting fuel does not exceed the upper limit of the resultant fuel product.

To determine the kinematic viscosity, a calibrated viscometer is used to measure the time it takes for a volume of biodiesel sample to flow under gravity through the viscometer at a controlled temperature. The kinematic viscosity is the product of the measured flow time and calibration constant of the viscometer. Selection of the viscometer range depends on the viscosity of the sample, and calibration of a viscometer requires the use of certified viscosity reference standards. Approximately 15 mL of biodiesel is required to perform this test.

4.1.7 Oxidative Stability (EN 15751) [22]

The oxidative stability test of biodiesel samples is primarily of concern for fuel storage and gum formation; however, correlation of this test with gum formation changes under different fuel storage conditions. The oxidation products of biodiesel, such as polymers or acids, can cause fuel system deposits and filter clogging, which is generally mitigated by application of stabilizing additives in a final fuel product. Due to the presence of PUFAs in many microbial lipid feedstocks, oxidative stability is often one of the more troublesome fuel parameters for biodiesel fuels to meet ASTM specifications. This method can measure oxidative stability of B100 or blended biodiesel.

The analysis method for oxidative stability is split into two phases: the induction period and the tainted odor and flavor phase. In the first phase, an oxygen-rich airstream passes through the sample at specific temperature. The air and any released vapors are captured in a flask of water with a conductivity electrode. The induction period is defined as the elapsed time from the start to the point where conductivity increases rapidly. The measurement is terminated during the second phase when the conductivity curve plateaus. This method requires specialized equipment, for which there are only two vendors, the Rancimat Model 73 (Metrohn AG) and the OSI Instrument (Omnion, Inc.). Approximately 50 mL of biodiesel is required to perform this test.

4.1.8 Sulfur Content (D5453) [23]

Although most microbial B100 fuel samples have been shown to be essentially sulfur free, the measurement of sulfur content in biodiesel fuels is important to illustrate the value of biodiesel for blending applications and environmental impact. More stringent sulfur limits have been imposed at state levels for environmental reasons.

To measure sulfur content, a biodiesel sample is inserted into a quartz combustion tube at high temperature where sulfur is oxidized to sulfur dioxide in an aerobic test atmosphere. The combustion tube provides oxygen, carrier gas, and water removal. Combustion gas is then subjected to UV light, exciting sulfur dioxide and generating a fluorescence signal. Approximately 5 mL of biodiesel is required to perform this test.

4.1.9 Phosphorus Content (D4951) [24]

The measurement of phosphorous content in fuel is important because phosphorous contamination can deactivate the exhaust catalysts in catalytic converters. Phosphorous is present in all microbe and plant oils, primarily in phospholipids. Although most contemporary biodiesel sources have been shown to have low phosphorous content (<1 ppm, which is well below the specified limit of 10 ppm), this regulation targets alternative biodiesel sources such as microbial lipids, which may contain higher levels of phosphorous.

A biodiesel sample is weighed and diluted with xylenes (or other solvent) before introduction to an inductively coupled plasma-atomic emission spectrometry (ICP-AES) system. The use of an internal standard and calibration standards is required; the wavelengths (nm) designated for analysis include 213.7, 214.9, and 253.4. Approximately 100 mL of biodiesel is required to perform this test. Although D4951 is currently the official method for phosphorous content determination, ASTM D7111 is a more sensitive alternative (*see* **Note 37**) [25]. The same test will also measure calcium, magnesium, sodium, and potassium content (*see* Subheading 4.2.1).

4.2 Summary of Category 2 Fuel Quality Tests

4.2.1 Calcium, Magnesium, Sodium, and Potassium Content (EN 14538) [26]

Alkali and alkali earth metal contamination of fuels can cause fuel system corrosion and degrade the exhaust catalysts in catalytic converters. The primary source of these elements in biodiesel samples results from the fuel production process, such as catalyst residues. Regulatory limits consider the combined content of sodium and potassium and the combined content of calcium and magnesium. These metal contaminants can be efficiently removed using modern fuel refinery practices.

A biodiesel sample is weighed and diluted with kerosene in a 1:1 weight ratio. This mixture is directly injected in an ICP-AES instrument. The wavelengths (nm) designated for analysis of each element include but are not limited to: calcium (422.6), magnesium (279.6), sodium (589.0), and potassium (796.9). Approximately 100 mL of biodiesel is required to perform this test. The same test will also determine phosphorous content (*see* Subheading 4.1.9).

4.2.2 Acid Number (D664) [27]

Acids can lead to increased engine deposits and fuel system corrosion. The acid number is defined as the quantity of base (mg of KOH) required to titrate a dilute fuel sample to its endpoint. In biodiesel fuels, this method is relevant to determine the level of residual free fatty acids or volatile fatty acids, such as acetic acid and formic acid.

A biodiesel sample is dissolved in a mixture of solvents (toluene, isopropanol, and water), and then a potentiometric titration is performed with ethanolic KOH to the endpoint of a glass indicator electrode. The protocol specifically developed for biofuels D664 B requires the use of an automatic titrator with endpoint-seeking functionality. Often, multiple inflection points are observed during analysis, and this protocol requires reporting the last well-defined endpoint. Approximately 50 mL of biodiesel is required to perform this test.

4.2.3 Alcohol Control (D93 and EN 14100)

The presence of alcohol contamination in biodiesel can cause engine corrosion, low lubricity and affect fuel injection. In biodiesel, unreacted alcohol may remain from the transesterification process or other solvent exposure. The regulatory parameter of alcohol control can be met by either a demonstrably higher flash point than the minimum flash point requirement, 130 °C (D93, *see* Subheading 4.1.5), or by the specific determination of residual methanol content (EN 14110, 0.2% mass maximum) [28]. Approximately 20 mL of biodiesel is required to perform method EN 14110.

4.2.4 Carbon Residue (D4530) [29]

The carbon residue is used to approximate the tendency of a biofuel to create engine deposits. The carbon residue is defined as the carbonaceous residue left over after the vaporization of volatile compounds in the absence of air. In biodiesel, the carbonaceous residue mostly results from downstream processing in terms of residual glycerides, free fatty acids, remaining catalyst, or other contaminants. Microbial-derived biodiesel with high concentrations of polyunsaturated FAMEs may have an increased carbon residue.

A biodiesel sample is placed in a glass vial and heated to 500 °C under anaerobic conditions. The volatiles formed are removed by a consistently flowing nitrogen atmosphere, and the carbon residue remaining is reported as a percent mass compared to the original sample weight. A specialized cooking oven with a circular heating chamber and exhaust for inert atmosphere purging is typically used for this method. Approximately 20 mL of biodiesel is required to perform this test.

4.2.5 Corrosion of Copper Strip (D130) [30]

This test measures the presence of compounds (e.g., sulfur compounds and acids) in biodiesel that can cause the corrosion of fuel systems containing bronze, brass, and copper. Method D130 is operationally simple and does not require specialized equipment; however, this method requires the purchase of ASTM standard copper strips and sample corrosion standards.

An ASTM standard copper strip is polished, placed in a sample of biodiesel, and subjected to sustained heat (for at least 3 h at 50 °C). At the end of the test, the copper strip is removed and washed in a hydrocarbon solvent (i.e., hexanes). The color and tarnish of the copper strip is compared by visual inspection to the ASTM copper corrosion standard, and a class is assigned based on that standard (i.e., Class 3, dark tarnish, no corrosion). Approximately 30 mL of biodiesel is required to perform this test.

4.2.6 Total Glycerin, Free Glycerin, and Monoglyceride Content (D6584) [31]

The presence of glycerin in biodiesel can cause issues with long-term fuel storage and lead to fouling of engine injector components. Glycerin is a byproduct of the transesterification process. Glycerin content is a quality measure of biodiesel samples and is primarily dependent on the transesterification and fuel-polishing processes. D6584 is a single GC method that simultaneously measures for three separate fuel-quality parameters: monoglyceride, free glycerin, and total glycerin content.

A biodiesel sample is treated with *N*-methyl-*N*-trimethylsilyl trifluoroacetamide to silylate any glycerin or monoglyceride before the sample is injected into a GC system equipped with FID. Samples are quantified using internal standards and identified by retention time in comparison with reference materials. A 5% phenylpolydimethylsiloxane column is recommended. Approximately 5–20 mL of biodiesel is required to perform this test.

4.2.7 Sulfated Ash (D874) [32]

Sulfated ash is used to approximate trace metal impurities, such as barium, magnesium, sodium, potassium, calcium, tin, and zinc, that can contribute to engine wear and deposits. Trace metal contaminants in biodiesel can result from unremoved catalysts, soluble metallic soaps, and abrasive solids. Biodiesel samples are generally very low in trace metals, often to the point of being below the detection limit of the D874 protocols.

A biodiesel sample is burned aerobically in an electric muffle furnace until the sample is completely combusted. After cooling, the remaining ash residue is treated with sulfuric acid. The sample is then heated in the furnace over several cycles until there is <1.0 mg difference in weight between heating cycles (to constant weight). The remaining sulfated ash is weighed and reported as a percent of the original sample weight. Approximately 100 mL of biodiesel is required to perform this test.

4.2.8 Water and Sediment (D2709) [33]

Residual water and sediment can cause significant fuel system and engine damage and is generally a concern for the downstream processing of biodiesel. D2709 is a simple and cost-effective screening to determine high levels of free water and sediment, but cannot measure low levels of water dissolved in biodiesel. In storage, natural autoxidation processes of biodiesel containing polyunsaturated FAMEs can cause a precipitate and form a gel-like mass that can be measured by this test. This method is only valid for biodiesel with viscosities between 1 and 4.1 mm^2/s (at 40 °C) and densities in the range of 770–900 kg/m^3 (at 15 °C).

A biodiesel sample is placed in a graduated centrifuge tube and subjected to 800 relative centrifugal force (rcf) for 10 min at 20–32 °C. The combined volume of free water and sediment that has accumulated in the bottom of the centrifuge tube is read and reported as the volumetric percent of the total sample. Approximately 100 mL of biodiesel is required to perform this test.

4.3 Additional Fuel Quality Tests Not Governed by D6751

Several fuel parameters that are not required by D6751 are useful to determine biodiesel quality and to compare with other biofuels and petrodiesels.

4.3.1 Density, Relative Density, or API Gravity (D1298) [34]

The density of biodiesel is generally higher than that of petrodiesel, but decreases with increasing PUFA concentration. Test method D1298 is a relatively simple laboratory test, so it has relevance for lab-scale biofuel research.

A biodiesel sample is brought to a specific temperature and placed into a hydrometer vessel. A hydrometer is lowered into the sample and allowed to settle; after a temperature equilibrium has been reached, the hydrometer is read and temperature is noted. Values are corrected for thermal glass expansion and meniscus effects, then converted to density values at 15 or 60 °C using a series of standard tables and calculations provided in D1298. The sample size required for this test depends on hydrometer selection, but generally less than 100 mL of biodiesel is required.

4.3.2 Energy Content (Higher Heating Value) (D240) [35]

Higher heating value (HHV), or the gross calorific value, is a measure of the heat released when a fuel sample is combusted and the products return to 25 °C. HHV is a measure of the energy content of a given unit of fuel and is considered a good indicator of the fuel value and utility. HHV can be used to calculate the mass heat of combustion, and is important when considering the thermal efficiency of engines. In biodiesel FAME sources, HHV is increased as chain length increases and decreased as PUFA content increases [36].

A small sample of biodiesel is burned aerobically in a bomb calorimeter under controlled conditions. The heat of combustion is computed before, during, and after the burn. The mass percentage of hydrogen in the sample must be known to calculate the heat of combustion and the HHV. Approximately 5 mL of biodiesel is required to perform this test.

4.3.3 Iodine Number (D1541) [37]

The iodine number reflects the stability of biodiesel to oxidation, which is directly related to the units of unsaturation (number of double bonds) present in the biodiesel mixture. Total iodine value is the measure of the total unsaturation in biodiesel samples. This method has generally been replaced with oxidative stability in ASTM D6751 (*see* Subheading 4.2.7); however, iodine number is still useful to predict biodiesel properties and is often reported in biodiesel characterization reports (*see* **Note 38**).

A biodiesel sample is dissolved in isooctane or hexanes in a 250-mL iodine determination flask in a dark room. Then, mercuric acetate and Rosenmund–Kuhnhenn reagent (a mixture of bromine, sulfuric acid, glacial acetic acid, and pyridine) are added to the flask, followed by addition of a small amount of potassium iodide. The flask is sealed and mixed, then stored in the dark at room temperature. After 1 h, water and more potassium iodide are added to the mixture, which is then titrated with $Na_2S_2O_3$ using a starch indicator. An iodine value is reported based on the titrant volume used and amount of initial sample. Approximately 50 mL of biodiesel is required to perform this test.

5 Notes

1. Several alternative methods (e.g., base-catalyzed) and various catalysts for transesterification are known and may impact the resultant quality of the FAME biodiesel (*see* Subheadings 3.1 and 3.4).
2. Specific ASTM test methods, complete with materials lists and notes, are available via ASTM International. The test methods for many of these fuel characteristics were not developed for biofuels, and several have not yet (at the time of writing this chapter) been updated to include biodiesel in the scope of the method; thus, it is prudent to reference and utilize the most contemporary ASTM methods possible.
3. D6751 governs four grades of B100 biodiesel blend stock for use in middle distillate fuels with slight differences. "2-B" grades are for general purpose use, whereas "1-B" grades indicate special purpose B100 that are intended for low temperature use and are sensitive to glyceride contamination. Each of these grades are further differentiated due to sulfur content as either "S15," with a maximum of 15 ppm sulfur content, or "S500," with a maximum of 500 ppm sulfur content.

4. The cost of a full ASTM 6751 biodiesel fuel quality test ranges from \$1000 to \$1500 and requires about 2 L of unadulterated biodiesel. Most individual tests cost between \$50 and \$150, with the notable exception for measuring Cetane number, which typically costs >\$400 and requires a 1-L sample of biodiesel.
5. We purchase our standards from Nu-Check Prep (Elysian, MN). Standards should be stored at −80 °C.
6. Chloroform is suitable for esterified internal standards; when using a free fatty acid standard, a 1:1 chloroform–methanol mixture should be used.
7. There are several other appropriate columns that can be used; some of these options are described in biofuel technology reviews [38–40].
8. There are premade FAME standard mixes available; mixes should be chosen to reflect the expected FAME profile of biodiesel samples.
9. Additionally, this transesterification method can be used directly on dried biomass samples. When using biomass, starting with approximately 50 mg of dried biomass is usually sufficient.
10. When performing transesterification with lipid extract, a sample containing about 10–30 mg of material is appropriate to yield enough FAMEs for GC analysis. Less material (<10 mg) can be used if manual injection into the GC is being used. Lipid extracts should be transferred to the culture tubes in chloroform. We have found a Folch extraction to work well with microbial samples that are rich in lipids.
11. A triglyceride standard should be used that contains a fatty acid constituent that is not found in your sample lipid profile and used in the amount of about 10–15% of your total fatty acid content. Often lipids consisting of odd-chain fatty acids are acceptable standards, for example, we use the triglyceride trihexadecanoin (C17:0 TG). If both free and esterified fatty acids (e.g., triglycerides) are present in your sample, which is generally the case for microbial lipid sources, then an esterified standard should be used.
12. While nitrogen gas is generally preferred because of its low cost and inert chemistry, argon gas can also be used for drying samples. The use of compressed air should be avoided for drying samples because this method can cause oxidation of PUFAs present in the lipid extract, which would suppress their detection on the GC-FID.
13. Toluene is used to promote solubilizing of esterified lipids into the methanol reaction mixture due to its partial miscibility. In

addition, toluene protects the lipid sample (e.g., PUFAs) from oxidation. Alternative nonpolar solvents that are miscible in methanol could be used for this step, such as methyl *tert*-butyl ether (MTBE).

14. Molar excess of methanol increases the FAME yield; the molar ratio of methanol to oil should be at least 3:1 for both acid- and base-catalyzed homogenous methods. Acid-catalyzed routes are often performed at an even higher methanol-to-oil ratio (>12:1).
15. Ensure caps are screwed on "finger tight" before placing on heating block. If they are loose, excess boiling and loss of solvent can occur. However, the culture tube threads tend to shear if overtightened; if this happens, transfer contents to a new culture tube before incubation and replace any lost methanol.
16. The addition of water serves to deactivate the acid catalyst. Notably, even slight water contamination (<0.5 wt%) in earlier steps can significantly decrease the FAME yield of this method.
17. If the organic and aqueous phases do not separate after about 15 min, vials can be centrifuged at low speed to assist with separation.
18. The upper layer will be the organic hexanes layer. Be careful not to contaminate the hexanes layer with the lower water layer; contamination with water will become apparent when attempting to dry samples. Recovering the entire hexanes layer is not necessary, as the internal standard will account for any process loss through the extraction steps.
19. When using this protocol with dried biomass samples, or if residual debris is observed in the hexane layer after **step 14**, use the following additional procedure instead of drying and reconstituting the sample.
    (a) Centrifuge for 5 min at 15,000 × *g* to pellet debris, and transfer the supernatant to a third 1.5-mL microcentrifuge tube.
    (b) Evaporate to dryness under nitrogen, and then reconstitute with 150 μL hexanes. Vortex for 3 s.
    (c) Centrifuge for 5 min at 15,000 × *g*. Use the supernatant from this step for GC analysis (Subheading 3.2).
20. Suggested GC protocol and equipment:

    Instrument: PerkinElmer Clarus 500 gas chromatograph with FID and auto sampler.

    Column: Agilent Technologies DB-FFAP (nitroterephthalic acid–modified PEG) 30 m × 0.25 mm inner diameter × 0.25 μm film thickness).

Injector temp.: 300 °C.

Detector temp.: 250 °C.

Initial oven temp.: 50 °C hold for 2 min, ramp 10 °C/min to 180 °C, hold for 5 min, ramp 5 °C/min to 240 °C, hold for 5 min. Total run time of 37 min.

Carrier gas: helium (1.3 mL/min).

Blanks: hexanes should be used as blanks before and after each analysis run and between every ten samples.

21. After GC analysis, you will be able to see if samples are too concentrated or too diluted. Samples can be evaporated and reconstituted in the appropriate amount of hexanes to achieve a clean and clear GC chromatograph devoid of overlapping peaks. If running multiple samples and using an auto sampler, adding 200 μL of hexanes to the bottom of the GC vial (below the insert) can reduce sample concentration resulting from evaporation.

22. Retention times recorded for standards and samples will change with time, particularly during instrument maintenance. Standard mixes should be run after each maintenance event and/or before large studies.

23. Moles of internal standard must be corrected for the resultant quantity of mono-alkyl esters; for example, moles of the triglyceride trihexadecanoin (C17:0) in a sample must be multiplied by three because the triglyceride will convert to three mono-alkyl esters.

24. When calculating the molar fraction, do not include unidentified peaks from GC analysis. Unidentified peaks may make up a significant portion of the profile (>10%) and will affect the accuracy of any biodiesel property estimates. Depending on the percent of unknown compounds in the sample, it may be necessary to add more samples to your standard mix, use database resources, or use other analytical methods to determine the unknown components.

25. The average absolute deviation (AAD) was reported by Ramírez-Verduzco et al. for each of these empirical equations compared with experimentally determined values [9]. The AAD was calculated via Eq. 8, considering experimentally determined values (exp) and calculated values (cal) for individual FAMES and biodiesel samples.

$$\mathrm{AAD} = \left(\frac{1}{n}\right)\left(\sum_{j=1}^{n}\left|\frac{\left(f_{j,\exp} - f_{j,\mathrm{cal}}\right)\cdot(100)}{f_{j,\exp}}\right|\right) \tag{8}$$

26. AAD for cetane number: 5.95%.

27. AAD for kinematic viscosity at 40 °C: 2.57%.
28. AAD for density at 40 °C: 0.11%.
29. AAD for HHV: 0.21%.
30. There are several alternative methods to the homogeneous HCl-catalyzed transesterification method presented in this section. Generally, scaling these reactions should use transesterification and downstream processing methods that are as close as possible to those that would be used on a commercial scale for the biodiesel sample in question [39, 41, 42] Sulfuric and phosphoric acids are homogeneous acid alternatives [43]. Two phase methods, without the use a cosolvent are also suitable alternatives [44]. Base-catalyzed methods are the most commonly used commercially because of the lower cost compared with acid-catalyzed methods, but the disadvantage of base catalysis (particularly for microbial lipids) is its sensitivity to free fatty acids, which deactivate the catalyst through the formation of water and soap [45]. In commercial scale applications, acid catalysis is often used to pretreat oil samples, converting free fatty acids to their respective esters before base-catalyzed treatment of triglycerides completes the process [3, 41]. Additionally, there are several heterogeneous solid-supported catalyst methods including metal oxides, zeolites, and lipase enzymes [46].
31. The volume of microbial lipids will depend on their density; most densities will be in the range of 0.8–0.95 mg/cm^3.
32. This method can be safely scaled to a 50-g sample of microbial lipids. Reaction flasks should be selected so that the total volume of the mixture made in **steps 1** and **2** constitutes less than 60% of the total flask volume. Molar excess of methanol will help drive the reaction to completion; 6–12x the lipid extract volume of the 1:1 methanol–chloroform mixture should be used.
33. Condenser selection should result in a complete reflux of reaction solvents, loss of solvent volume will result in an incomplete conversion to FAMEs.
34. This test is scalable. About 90% w/w mass recovery of biodiesel sample can be expected. Fuel properties may change upon decolorization, so if decolorization is required, then this method should be used consistently before all fuel quality tests.
35. The governing ASTM test methods are presented in this section; however, there are acceptable alternative test methods for most parameters with some qualifications. A handbook comparing and discussing all accepted biofuel test methods is available from ASTM International [47].

36. The reduced-pressure distillation apparatus is difficult to set up and operate manually, so many automated and atmospheric pressure methods have also been developed. Although D1160 (manual distillation) is the official method, D7344 "Distillation at atmospheric pressure mini method" is also accepted [48]. D7344 is an automatic method that uses an instrument solely available from Grabner Industries (Vienna, Austria).
37. This ICP-AES method is designed for the analysis of lubricants and other materials that have significantly higher phosphorous levels than biodiesel; therefore, the sensitivity of this method is not rated for measuring the 10 ppm maximum regulatory threshold [47].
38. The American Oil Chemists' Society produces a contemporary method for total iodine number determination, Cd 1-25 [49]. ASTM D1541 has not been updated since its expiration in 2006.

References

1. ASTM D6751-15ce1 Standard Specification for Biodiesel Fuel Blend Stock (B100) for Middle Distillate Fuels (2015) ASTM International, West Conshohocken, PA
2. Li Q, Du W, Liu D (2008) Perspectives of microbial oils for biodiesel production. Appl Microbiol Biotechnol 80(5):749–756
3. Meher LC, Vidya Sagar D, Naik SN (2006) Technical aspects of biodiesel production by transesterification—a review. Renew Sust Energ Rev 10(3):248–268
4. Bello EI, Mogaji TS, Agge M (2011) The effects of transesterification on selected fuel properties of three vegetable oils. J Mech Eng Res 3(7):218–225
5. Knothe G (2005) Dependence of biodiesel fuel properties on the structure of fatty acid alkyl esters. Fuel Process Technol 86 (10):1059–1070
6. Ramos MJ, Fernández CM, Casas A, Rodríguez L, Pérez Á (2009) Influence of fatty acid composition of raw materials on biodiesel properties. Bioresour Technol 100 (1):261–268
7. Ichihara K, Fukubayashi Y (2010) Preparation of fatty acid methyl esters for gas-liquid chromatography. J Lipid Res 51(3):635–640
8. Richardson CE, Hennebelle M, Otoki Y, Zamora D, Yang J, Hammock BD, Taha AY (2017) Lipidomic analysis of oxidized fatty acids in plant and algae oils. J Agric Food Chem 65(9):1941–1951
9. Ramírez-Verduzco LF, Rodríguez-Rodríguez JE, Jaramillo-Jacob AR (2012) Predicting cetane number, kinematic viscosity, density and higher heating value of biodiesel from its fatty acid methyl ester composition. Fuel 91 (1):102–111
10. O'Neil GW, Knothe G, Williams JR, Burlow NP, Culler AR, Corliss JM, Carmichael CA, Reddy CM (2014) Synthesis and analysis of an alkenone-free biodiesel from Isochrysis sp. Energy Fuel 28(4):2677–2683
11. O'Neil GW, Knothe G, Williams JR, Burlow NP, Reddy CM (2016) Decolorization improves the fuel properties of algal biodiesel from Isochrysis sp. Fuel 179:229–234
12. Issariyakul T, Dalai AK (2010) Biodiesel production from greenseed canola oil. Energy Fuel 24(9):4652–4658
13. ASTM D4057-12 Standard Practice for Manual Sampling of Petroleum and Petroleum Products (2012) ASTM International, West Conshohocken, PA
14. ASTM D4177-16e1 Standard Practice for Automatic Sampling of Petroleum and Petroleum Products (2016) ASTM International, West Conshohocken, PA
15. ASTM D613-17b Standard Test Method for Cetane Number of Diesel Fuel Oil (2017) ASTM International, West Conshohocken, PA
16. ASTM D2500-16b Standard Test Method for Cloud Point of Petroleum Products and Liquid

Fuels (2016) ASTM International, West Conshohocken, PA
17. ASTM D5771-17 Standard Test Method for Cloud Point of Petroleum Products and Liquid Fuels (Optical Detection Stepped Cooling Method) (2017) ASTM International, West Conshohocken, PA
18. ASTM D7501-12a Standard Test Method for Determination of Fuel Filter Blocking Potential of Biodiesel (B100) Blend Stock by Cold Soak Filtration Test (CSFT) (2012) ASTM International, West Conshohocken, PA
19. ASTM D1160-15 Standard Test Method for Distillation of Petroleum Products at Reduced Pressure (2015) ASTM International, West Conshohocken, PA
20. ASTM D93-16a Standard Test Methods for Flash Point by Pensky-Martens Closed Cup Tester (2016) ASTM International, West Conshohocken, PA
21. ASTM D445-17a Standard Test Method for Kinematic Viscosity of Transparent and Opaque Liquids (and Calculation of Dynamic Viscosity) (2017) ASTM International, West Conshohocken, PA
22. EN 15751 Automotive Fuels – Fatty Acid Methyl Ester (FAME) Fuel and Blends with Diesel Fuel – Determination of Oxidation Stability by Accelerated Oxidation Method (2014) European Committee for Standardization
23. ASTM D5453-16e1 Standard Test Method for Determination of Total Sulfur in Light Hydrocarbons, Spark Ignition Engine Fuel, Diesel Engine Fuel, and Engine Oil by Ultraviolet Fluorescence (2016) ASTM International, West Conshohocken, PA
24. ASTM D4951-14 Standard Test Method for Determination of Additive Elements in Lubricating Oils by Inductively Coupled Plasma Atomic Emission Spectrometry (2014) ASTM International, West Conshohocken, PA
25. ASTM D7111-16 Standard Test Method for Determination of Trace Elements in Middle Distillate Fuels by Inductively Coupled Plasma Atomic Emission Spectrometry (ICP-AES) (2016) ASTM International, West Conshohocken, PA
26. EN 14538 Fat and Oil Derivatives – Fatty Acid Methyl Ester (FAME) – Determination of Ca, K, Mg and Na Content by Optical Emission Spectral Analysis with Inductively Coupled Plasma (ICP OES) (2006) European Committee for Standardization
27. ASTM D664-11ae1 Standard Test Method for Acid Number of Petroleum Products by Potentiometric Titration (2011) ASTM International, West Conshohocken, PA
28. EN 14110 Fat and Oil Derivatives – Fatty Acid Methyl Esters (FAME) – Determination of Methanol Content (2003) European Committee for Standardization
29. ASTM D4530-15 Standard Test Method for Determination of Carbon Residue (Micro Method) (2015) ASTM International, West Conshohocken, PA
30. ASTM D130-12 Standard Test Method for Corrosiveness to Copper from Petroleum Products by Copper Strip Test (2012) ASTM International, West Conshohocken, PA
31. ASTM D6584-13e1 Standard Test Method for Determination of Total Monoglycerides, Total Diglycerides, Total Triglycerides, and Free and Total Glycerin in B-100 Biodiesel Methyl Esters by Gas Chromatography (2013) ASTM International, West Conshohocken, PA
32. ASTM D874-13a Standard Test Method for Sulfated Ash from Lubricating Oils and Additives (2013) ASTM International, West Conshohocken, PA
33. ASTM D2709-16 Standard Test Method for Water and Sediment in Middle Distillate Fuels by Centrifuge (2016) ASTM International, West Conshohocken, PA
34. ASTM D1298-12b Standard Test Method for Density, Relative Density, or API Gravity of Crude Petroleum and Liquid Petroleum Products by Hydrometer Method (2012) ASTM International, West Conshohocken, PA
35. ASTM D240-17 Standard Test Method for Heat of Combustion of Liquid Hydrocarbon Fuels by Bomb Calorimeter (2017) ASTM International, West Conshohocken, PA
36. Sanz-Tejedor MA, Arroyo Y, San José J (2016) Influence of degree of unsaturation on combustion efficiency and flue gas emissions of burning five refined vegetable oils in an emulsion burner. Energy Fuel 30(9):7357–7366
37. ASTM D1541-97 Standard Test Method for Total Iodine Value of Drying Oils and Their Derivatives (Withdrawn) (2006) ASTM International, West Conshohocken, PA
38. Knothe G, Krahl J, Van Gerpen J (2015) The biodiesel handbook. Elsevier, Amsterdam
39. Mittelbach M, Remschmidt C (2004) Biodiesel—the comprehensive handbook. Martin Mittelbach, Graz, Austria. There is no corresponding record for this reference
40. Knothe G (2001) Analytical methods used in the production and fuel quality assessment of biodiesel. Trans ASAE 44(2):193
41. Gerpen JV (2005) Biodiesel processing and production. Fuel Process Technol 86 (10):1097–1107
42. Van Gerpen J, Knothe G (2005) Basics of the transesterification reaction. In: The biodiesel

handbook. AOCS Press/Academic Press, London, pp 26–41

43. Van Gerpen J, Canakci M (1999) Biodiesel production via acid catalisis. Trans ASAE 42:1203–1210
44. Ataya F, Dubé MA, Ternan M (2007) Acid-catalyzed transesterification of canola oil to biodiesel under single-and two-phase reaction conditions. Energy Fuel 21(4):2450–2459
45. Anastopoulos G, Zannikou Y, Stournas S, Kalligeros S (2009) Transesterification of vegetable oils with ethanol and characterization of the key fuel properties of ethyl esters. Energies 2 (2):362–376
46. Thanh LT, Okitsu K, Boi LV, Maeda Y (2012) Catalytic technologies for biodiesel fuel production and utilization of glycerol: a review. Catalysts 2(1):191–222
47. Nadkarni R (2016) Analysis of biofuels – a laboratory resource, MNL77-EB. ASTM International, West Conshohocken, PA
48. ASTM D7344-17 Standard Test Method for Distillation of Petroleum Products and Liquid Fuels at Atmospheric Pressure (Mini Method) (2017) ASTM International, West Conshohocken, PA
49. American Oil Chemists S, Firestone D (2004) Official methods and recommended practices of the American Oil Chemists' Society. AOCS, Champaign

Chapter 18

Impact of Culture Conditions on Neutral Lipid Production by Oleaginous Yeast

Irene Fakankun, Maryam Mirzaei, and David B. Levin

Abstract

Oleaginous yeasts have the ability to accumulate and store triacylglycerides (TAGs) to more than 20% of their cell mass. Oleaginous yeasts have advantages over oil seed plants and microalgae because they grow much faster (doubling time is usually less than an hour), accumulate cell mass to much higher densities, and are less affected by seasonal or weather conditions. The TAGs synthesized by oleaginous yeasts are often rich in polyunsaturated fatty acids and can be used either for biodiesel production or as edible oils. "Red" yeasts are oleaginous yeasts that can synthesize and accumulate high concentrations of TAGs. Many factors affect the growth of red yeasts and subsequent yields of TAGs. These factors include carbon and nitrogen sources, their concentrations, the C/N ratio, temperature, pH, aeration rate, mineral elements, inorganic salts, and inhibitors. The effect of each factor varies with the yeast strain and its growth phase. *Rhodosporidium diobovatum* is a "red" yeast that can utilize low-cost substrates, such as waste glycerol derived from biodiesel production as a carbon source, and can synthesize and accumulate high concentrations of both TAGs and carotenoids.

Key words Oleaginous yeasts, Red yeasts, Single cell oils, Biodiesel production, *Rhodosporidium diobovatum*, Triacylglycerides

1 Introduction

Nonrenewable fossil fuels are presently ubiquitous because they are abundant, transportable, and low cost compared to other alternatives. These factors drive the continuous use of fossil fuels as the main source of energy, but at a cost to environmental and climate stability. Environmental concerns along with the long-term availability of fossil fuels have encouraged research into alternative fuel sources, which includes biofuels [1, 2]. It has been projected that at the current rate of consumption, all current oil reserves may be completely depleted within a few centuries [3]. Although a few centuries seem to be a long period of time, the need to develop alternative sources of transportation fuels is urgent because of the alarming rate of carbon dioxide (CO_2) accumulation in the

Venkatesh Balan (ed.), *Microbial Lipid Production: Methods and Protocols*, Methods in Molecular Biology, vol. 1995, https://doi.org/10.1007/978-1-4939-9484-7_18,

atmosphere due to fossil fuel combustion, and its contribution to global climate change [4].

Biodiesel derived from plant oils has been demonstrated as a potential alternative to petroleum diesel. It is produced by transesterification of triacylglycerols (TAGs), which yields mono-alkyl esters, also known as fatty acid methyl esters (FAMEs) and fatty acid ethyl esters (FAEEs) [5]. Biodiesel has several advantages over petroleum-derived diesel, such as, lower carbon monoxide emissions, lower sulfur content, higher combustion efficiency, higher cetane number, and higher biodegradability [6].

Production of biodiesel is not without its own challenges: TAG feedstocks required for biodiesel synthesis are primarily derived from vegetable oils, so biodiesel production competes with their use for human food; their production requires large land areas for cultivation; and the biodiesel produced may not meet existing quality standards for transportation fuels, such as cold weather performance [7, 8]. To address the "food/feed versus fuel" issue, research on the use of second generation feedstock, that is, nonedible oils, for biodiesel production was intensified [9]. However, even the use of nonedible oil feedstocks for biodiesel production has several limitations: production of nonedible oil crops still requires large land areas for cultivation and is highly dependent on weather conditions, as well as the geopolitics climate [10].

Researchers have therefore identified another promising alternative for the production of TAGs for biodiesel production, called single cell oils, which are derived from "oleaginous" microorganisms, which can accumulate intracellular TAGs [11]. To effectively produce high quality single cell oils for biodiesel production from oleaginous microorganisms, a detailed understanding of the concept of oleaginicity and how culture conditions affect the yield and lipid profile is required.

2 Oleaginous Yeasts

2.1 What Are Oleaginous Yeasts?

All microorganisms synthesize structural lipids, but only a few can accumulate TAGs and be classified as oleaginous. A microorganism is said to be oleaginous if it is capable of synthesizing and accumulating lipids in excess of 20% of its cell dry weight [12]. These microorganisms include yeasts, bacteria, and microalgae. Oleaginous yeasts have several advantages over other oleaginous microorganisms, such as high growth rate, high rate of TAG synthesis and accumulation, the low cost of the media needed for their growth, and the ease of controlling the culture conditions [13]. Oleaginous yeasts are generally also able to grow on a great variety of carbon substrates [14], ranging from pure glucose to waste-derived substrates (Table 1).

Table 1
Oleaginous yeasts grow and accumulate lipids on waste-derived substrates

Oleaginous yeast	Substrate in culture medium	Lipid accumulation (% cell dry weight)	Reference
Crytococcus curvatus	Spent yeast from brewery industry	37.7	[11]
Lipomyces starkeyi	Spent yeast cell mass hydrolysates	30.8	[15]
Rhodotorula glutinis	Crude glycerol from biodiesel plant	60.7	[16]
Rhodosporidium babjevae	Biodiesel derived waste glycerol	24.2	[17]
Rhodosporidium diobovatum	Biodiesel-derived waste glycerol	50.3	[17]
Yarrowia lipolytica	Palm oil mill effluent	48–61	[18]
Yarrowia lipolytica	Serum latex	8–13	[18]
Yarrowia lipolytica	Biodiesel-derived waste glycerol	31.0	[19]

2.2 The Importance of Oleaginous Yeast

Triacylglycerols (TAGs) make up 90% of the total lipids accumulated by oleaginous yeasts. These TAGs are useful in the production of biodiesel [10] and cocoa butter equivalent material. Oleaginous yeasts have the capability to produce TAGs for biodiesel production from lignocellulosic biomass [20]. They are also capable of producing polyunsaturated fatty acids useful for nutritional and medical purposes [21]. Oleaginous yeasts are also important organisms for lipid biochemistry research.

In addition, "red" oleaginous yeasts are useful in the production of carotenoids such as β-carotene, torularhodin, and torulene [22], which are important in the nutraceuticals and food industry. Mutants of oleaginous *Rhodotorula* yeast species have been engineered to produce increased amounts of these useful carotenoids [23]. In addition to carotenoid production, oleaginous yeasts are a promising source of ubiquinone Q_{10}, which has been found useful in cosmetology and medicine [24].

2.3 Screening of Oleaginous Yeasts

Oleaginicity is not a clade specific property, but varies among and within species [25]. Oleaginicity can vary from very low lipid-accumulating yeast to very high lipid-accumulating strains within the same species [26]. This makes screening for oleaginicity in a given yeast population an important procedure. There are a number of oleaginous yeast species that have been identified by employing varying screening methods. For example, according to Kicha and Cheirsilp [27], 889 yeast strains were isolated from different locations in the south of Thailand and screened for their oleaginicity, but only 23 strains were identified as oleaginous. Tanimura et al. [28] researched the oleaginicity of 500 yeast strains isolated from locations in Japan and found that only

31 of these yeast strains were lipid producers. It is useful to note that the culture conditions affect the oleaginicity of these yeasts [25, 27]. For example, Sriwongchai et al. [29] screened eight known oleaginous yeast strains belonging to the genera *Cryptococcus*, *Rhodotorula*, *Rhodosporidium*, and *Yarrowia*. They were cultured using glycerol as the sole carbon source. The results showed that *Cryptococcus albidus* was unable to utilize glycerol as its sole carbon source, while *Yarrowia lipolytica* strains tested showed the highest growth yield.

2.4 Oleaginous Yeasts

More than 40 of the 1600 known species of yeasts have been determined to be oleaginous [25]. They belong to the genera *Yarrowia*, *Candida*, *Rhodotorula*, *Rhodosporidium*, *Crytococcus*, *Trichosporon*, and *Lipomyces* [30]. A few species have been well characterized, including *Cryptococcus albidus*, *Cryptococcus curvatus*, *Lipomyces starkeyi*, *Rhodotorula glutinis*, *Rhodotorula graminis*, and *Yarrowia lipolytica*. The lipid content and fatty acid profiles of these oleaginous yeast species have been elucidated in different researches under different growth conditions [31].

Lipid accumulation and FAME profiles vary between species. For example, *Cryptococcus curvatus* is capable of growing in media containing glycerol and producing biomass up to 118 g/L [32]. It can accumulate TAGs to greater than 58% (w/w) of its dry cell weight (cdw), of which 28% is palmitic acid (C16:0), 15% is stearic acid (C18:0), and 48% is oleic acid (18:1) [32]. *Lipomyces starkeyi* is capable of utilizing xylose, ethanol, L-arabinose, and glucose as its carbon source and can accumulate TAGs up to 63% cdw, of which 33% is palmitic acid and 55% is oleic acid [33]. *Rhodotorula glutinis* can produce biomass as high as 180 g/L, and accumulate TAGs to 72% cdw using glucose as its main carbon source [30]. *Rhodosporidium toruloides* is capable of biodegrading corn hydrolysate as a carbon source to lipid and accumulating TAGs up to 76% cdw, of which about 24% is palmitic acid and 55% is oleic acid [33]. Munch et al. [17] compared the lipid accumulation of *Rhodosporidium babjevae* and *Rhodosporidium diobovatum* with 40 g/L of glycerol as the carbon source. *R. babjevae* accumulated TAGs to about 35% cdw, of which 24% was palmitic acid, 38% was oleic acid, 16% was linoleic acid (C18:2). *R. diobovatum* accumulated TAGs to about 64% cdw, of which 28% was palmitic acid, 33% was oleic acid, and 20% was linoleic acid. Finally, in 40 g/L of glycerol as substrate, *Y. lipolytica* can accumulate TAGs up to 31% cdw, of which 55% was oleic acid, 11% was palmitic acid, 10% was linoleic acid and 9% was oleic acid [19].

3 Important Parameters for Cellular Activity and Lipid Production

Many factors that affect the yeast productivity have been studied. These factors include carbon and nitrogen sources, their concentrations, the C/N ratio, temperature, pH, aeration rate, mineral elements, inorganic salts, and inhibitors. The effect of each factor varies with the yeast strain and its growth phase [17, 31, 34–38].

3.1 Carbon Sources

To optimize the lipid accumulation process, it is important to select a suitable carbon source. Oleaginous yeasts can grow and synthesize TAGs on a wide variety of carbon sources. Sugars, organic acids, alcohols, phenyl alkanes, n-alkanes, oils, and coal have been reported as carbon sources in the literature [25, 39–46]. However, to keep the costs of production low, it is desirable to utilize "waste" carbon sources derived from agriculture or industry by-product streams (Table 1).

Glucose is the most common carbohydrate used as a carbon source for yeast growth and productivity [31, 47]. However, many studies have reported the use of carbon sources other than glucose, such as arabinose [48, 49], mannose [16, 31, 50], mannitol [50], ethanol [51], xylose [41, 47, 52–54], lactose [55], galactose [31], cellobiose [31, 52], whey [56, 57], and starch [28, 58]; waste glycerol derived from biodiesel production is an important carbon source for the yeast oil production [59, 60]. Waste glycerol produced as a by-product of industrial processes may be used as a carbon source by yeasts synthesize many chemicals that can be used as precursors for synthesis of other chemicals or as final products [60–67].

3.2 Carbon-to-Nitrogen Ratio

Depletion of one or more nutrients, such as nitrogen, in the presence of excess of carbon in the culture medium can trigger the transition of yeast cells from growth phase to a TAG production phase [5, 35, 38, 68–70]. As the nitrogen source is exhausted during exponential growth phase, carbon will be consumed and stored as fatty acids within intracellular lipid bodies [71, 72]. Most oleaginous yeast species generate higher total fatty acids by increasing in the C/N ratio to a certain point, for example, as was reported by Kraisintu et al. [73]. Lipid accumulation by *Rhodosporidium toruloides* DMKU3-TK16 was higher at a C/N ratio of 140 compared to 115, 90, and 65 with 4.23, 4.13, 3.24, and 2.71 g/L, respectively [73]. However, this is not a general trend and has a limit, as high carbon concentrations can inhibit the growth of cells [27, 71, 73–81]. Kraisintu et al. [73] reported that the lowest biomass (6.79 g/L) resulted at C/N ratio of 140 while the highest amount resulted at C/N ratio of 65 (10.52 g/L) [73].

3.3 Nitrogen Sources

Different nitrogen sources have various influences on total TAG synthesis accumulation, as well as on the fatty acid profiles of the TAGs [25, 82, 83]. Sitepu et al. [25] have shown that greater TAG yields can be induced by nitrogen limitation in oleaginous yeasts. It has been widely reported that inorganic nitrogen sources for instance ammonium sulfate and ammonium chloride play a good role in cell growth while they are not suitable for lipid production. In contrast, organic nitrogen sources such yeast extract and peptone appear to stimulate lipid production and have less effect on the growth of yeast cells [78, 84, 85].

3.4 Minerals and Inorganic Salts

It has been known for quite some time that mineral elements and inorganic salts, such as calcium, iron, magnesium, phosphorus, potassium, sodium, sulfur, and zinc are essential for yeast cellular activity and TAGs production [71, 86, 87]. In addition to nitrogen limitation, limiting concentrations of phosphorus, sodium, or sulfate concentrations can also induce oleaginous yeasts to synthesize TAGs as energy storage molecules [36, 63, 88, 89].

3.5 Aeration Rate

Aeration is well known as an essential factor for yeast growth and TAG accumulation [37]. However, the influence of aeration (oxygen concentration) varies with yeast species [36, 90, 91]. In some yeast strains, oxygen limitation can enhance the proportion of saturated fatty acids incorporated into TAGs, while low oxygen concentration in some yeasts result in an increase in the incorporation of unsaturated fatty acids into the TAGs [91, 92]. Thus, the relationship between oxygen concentration, oxygen demand, and fatty acid synthesis must be considered an important factor in the overall productivity of TAGs [70, 90].

3.6 Temperature

Different oleaginous yeast species have different optimum temperatures for cell growth, and temperature can also influence TAG accumulation, and the fatty acids profile of the TAGs produced. Mostly, the optimum temperature is in the range of 25–30 °C, and temperatures that are too high or too low affect cell productivity. In general, depending on yeast species, high cultivation temperatures enhance the incorporation of saturated fatty acids in the TAGs, whereas lower cultivation temperatures favor incorporation of unsaturated fatty acids [93–97]. However, some oleaginous yeasts do not grow well at low temperatures, but warming large cultures in bioreactors to increase growth rate can increase the costs of production [25, 71, 83].

3.7 pH

Selection of an oleaginous yeast strain that is naturally resistant to different pH levels during growth culture is highly desirable [31, 70, 77, 83, 98]. Many yeast strains with low pH tolerance are now used to convert agro-industrial wastewaters with low pH to lipid products that are very useful in both biofuel production and food industry. An advantage of this strategy is that lower pH can inhibit bacterial growth in mixed populations, but the low pH does not affect the growth of the yeast cells (which have low pH tolerance), nor their ability in biodegradation of agro-industrial waste products to TAGs [99].

3.8 Effects of Inhibitors on Oil Productivity

Some of the by-products of lignocellulosic biomass pretreatment, with acid and/or steam explosion, to release fermentable sugars contain toxic compounds that can affect the cellular activity and TAG accumulation of oleaginous yeasts [31]. Resistance to various types of inhibitors has been known as one of the main factors for selection of oleaginous yeasts for biotechnological production in industry [100–103]. Chemical agents such as 5-(hydroxymethyl) furfural (HMF), furfural, acetic acid, formic acid, levulinic acid, lignin derivatives (sodium lignosulfonate and vanillin) [104], and p-benzaldehyde [51, 105–107] are known to be toxic to yeasts and can inhibit cell growth and, consequently, TAG synthesis [51, 73, 100, 101, 108].

3.9 Example Experiment

Here, using *Rhodosporidium diobovatum* 08-225 as the oleaginous yeast strain of choice, we show the impact of varying culture conditions (glucose versus glycerol as carbon source) on the total lipid yield and fatty acid profile. Based on the fact that glucose (dextrose; 180.16 g/mol) contains twice the number of carbon atoms of glycerol (92.09 g/mol), we conducted experiments to determine if the rate of *R. diobovatum* growth and/or the kinetics and yields of lipid synthesis are influenced by carbon availability. Precultures of *R. diobovatum* in nitrogen-limiting glucose–yeast extract–malt (GYM) medium containing 217 mM of dextrose (glucose) or 434 mM of glycerol were inoculated into 1000 mL Erlenmeyer flasks containing 300 mL of the same media, and cultured at 30 °C on a rotary shaker (150 rpm) for 120 h [17]. The pH was adjusted to 5.5 before autoclaving. Samples were collected as described below.

The total lipid yield was not much different at the end of 120 hours of cell growth—46.29% dcw in glucose media versus 44.57% dcw in glycerol media. However, total lipid accumulation was approximately 40% dcw within 72 h for glucose fed cultures, while it was approximately 30% dcw for glycerol fed cultures. This showed that the rate of total lipid production is influenced by the carbon source.

4 Materials

4.1 Yeast Strain and Culture Medium

The yeast *R. diobovatum* 08-225 used for this experiment was obtained from the Phaff Yeast Culture Collection, at the University of California at Davis (UCDFST). Maintain yeast strain on yeast extract–peptone–dextrose (YPD) with dextrose agar plate containing 20 g/L dextrose, 10 g/L yeast extract, 20 g/L peptone, and 15 g/L agar grown at 30 °C and store at 4 °C (*see* **Note 1**).

4.2 GMY Media

Nitrogen-limiting medium (GMY) should be freshly prepared using 3 g/L yeast extract, 8 g/L KH_2PO_4, 0.5 g/L $MgSO_4{\cdot}H_2O$, and 40 g/L glucose (dextrose) (*see* **Note 2**).

5 Methods

5.1 Culture Conditions and Analysis of Cell Growth

1. Revive yeast cultures from previously stored stocks at −80 °C by growing yeast cells on YPD agar plates at least for 2 days and then restreak on fresh YDP agar plates two times.
2. Pick single colonies from a streaked YDP agar plate and inoculate into 100 mL nitrogen-limiting medium (GMY) media. Precultures are grown aerobically at 30 °C for 24 h in a rotary shaker (at 150 rpm) using 500 mL baffled Erlenmeyer flasks (*see* **Note 3**).
3. Measure the medium pH and adjust to pH 5.5 before inoculation.
4. Carry out the experiments in 1000 mL Erlenmeyer flasks (*see* **Note 2**) with 300 mL of autoclaved GMY medium with a volume of inoculum equivalent to an initial optical density of 0.02 at a wavelength of 600 nm (OD_{600}).
5. Take samples (4 mL) to measure cell growth, dry cell weight (dcw), glucose and glycerol consumption, and ammonium ion (NH_4^+) concentration every 6 h for the initial 30 h, and then at the time points of 48, 72, 96, and 120 h postinoculation (hpi). Take 10 mL samples for lipid analysis every 24 hpi (*see* **Note 4**).
6. Monitor cell growth at OD_{600}, and harvest cell biomass by the centrifugation. Collect and store supernatant at −20 °C until substrate and ammonium ion concentration analysis. Wash pelleted cells with deionized water, centrifuge again, and decant the water. Dry all the cell pellets at 60 °C for 24 h and weigh to determine the dry cell weight (dcw).

5.2 Analytical Methods

1. There are various methods of measuring the ammonium ion concentration (*see* **Note 5**).
2. Vortex sample and measure ammonium ion concentrations (*see* **Note 6**).
3. Determine dextrose and glycerol concentrations at each time point by High Performance Liquid Chromatography (HPLC) (*see* **Note 7**).

5.3 Lipid Extraction and Fatty Acid Composition

The lipid extraction and transesterification of fatty acids was carried out as described by Bligh and Dyer [109], with modifications by Sestric et al [19].

1. Extract lipids from lyophilized biomass by boiling 10–20 mg dried biomass in 2 mL of 15% concentrated H_2SO_4 in methanol and 2 mL of a solution of 0.5 mg/mL methylheptadecanoic acid (C17) internal standard (Nu-Chek Prep Inc., USA) in chloroform for 5–6 h.
2. Add 1 mL of deionized water and leave to stand under the fume hood overnight to ensure a separation of the organic layer.
3. Extract 1 mL aliquot of fatty acid methyl esters (FAMEs) in chloroform and transfer into glass vials for FAMEs analysis. Analyze FAME composition by gas chromatography (GC) (*see* **Note 8**).

6 Notes

1. Analytical grade reagents and ultrapure water should be used to prepare all solutions.
2. All media should be autoclaved before inoculation. Avoid autoclaving the glucose with the media to avoid caramelization of the sugar at high temperatures. The glucose should be sterilized by filtration through a 0.2 μm filter and then added to the autoclaved media. The easiest way is to dissolve glucose in a portion of autoclaved media and then filter into the second portion of autoclaved media containing all other reagents.
3. 500 mL baffled Erlenmeyer flasks or larger ones should be used to grow cell cultures depending on the volume of the culture media. This is because aeration in the bottles influences the growth rate of the yeast cells.
4. It is important to measure parameters such as the ammonium ion concentration, dry cell weight, and substrate consumption during the time course of the experiment because it helps to give a clearer picture of what is going on within the cells while they produce lipids.
5. We quantified ammonium ion concentrations using a Lachat Quick Chem 8500 instrument (Ontario, Canada), according to ammonia method 10-107-06-1-I. Dilute the supernatants previously stored for this purpose two times with deionized water. Further dilutions with deionized water may be necessary as the instrument is not able to measure ammonium ion concentrations greater than 25 mg/mL and lower than

0.2 mg/mL. Dilutions should be carefully done to fit into the range of the instrument.

6. We used the Lachat Quick Chem 8500 instrument. Process Data was processed using Omnion software. It may be necessary to centrifuge again and transfer substrate into a new collection tube if the sample appears cloudy. The Lachat Quick Chem 8500 instrument has very sensitive collection needle and therefore requires that samples for ammonium ion concentration be clear liquid.
7. We used a Waters Breeze HPLC, equipped with a 300 mm × 7.8 mm HPX-87H Ion Exclusion Column (Aminex) maintained at 45 °C with 5 mM H_2SO_4 as the mobile phase at a flow rate of 0.6 mL/min. Glucose and glycerol standards should first be prepared with concentrations ranging from 50 to 0.1 mM. Dilute all liquid samples 1:10 or 1:20 with deionized water to prevent overloading of the column and to produce clear peaks. Load about 250 μL of standard and sample into a 96-well plate. Cover plate and run the HPLC system. Data were analyzed using the Waters Breeze 2 software.
8. FAMEs in chloroform were analyzed using an Agilent 7890A gas chromatograph (GC) with a split-splitless injector, and a flame ionization detector with a split ratio of 1:10. Separation of target molecules was achieved using a DB-23 capillary column (30 m × 320 μm × 0.25 μm; Agilent, California, USA). The carrier gas was Helium, at a flow rate of 1.8 mL/min. The initial oven temperature was set at 75 °C for 4 min and was then increased from 75 to 250 °C with increments of 20 °C/min, with a final hold at 250 °C for 4 min. A Supelco 37 component FAME mix (Sigma-Aldrich) was used as standard for peak identification and calculation different fatty acids.

Acknowledgments

This work was supported by the Natural Sciences and Engineering Research Council (NSERC) of Canada through an NSERC Discovery Grant (award #: RGPIN 05961-16).

References

1. Ridley CE, Clark CM, Leduc SD, Bierwagen BG, Lin BB, Mehl A, Tobias DA (2012) Biofuels: network analysis of the literature reveals key environmental and economic unknowns. Environ Sci Technol 46:1309–1131
2. Koroneos C, Spachos T, Moussiopoulos N (2003) Exergy analysis of renewable energy sources. Renew Energy 28:295–231
3. Shafiee S, Topal E (2009) When will fossil fuel reserves be diminished? Energy Policy 37 (1):181–118
4. Abdullah M, Mehmet IY (2015) World energy outlook and place of renewable resources. Int J Sci Technol Res 1:10–17
5. Meng X, Yang J, Xu X, Zhang L, Nie Q, Xian Q (2009) Biodiesel production from

oleaginous microorganisms. Renew Energy 34:1–5. https://doi.org/10.1016/j.renene.2008.04.01

6. Saxena V, Sharma CD, Bhagat SD, Saini VS, Adhikari DK (1998) Lipid and fatty acid biosynthesis by *Rhodotorula minuta*. J Am Oil Chem Soc 75(4):501–505
7. Muniyappa PR, Brammer SC, Noureddini H (1996) Improved conversion of plant oils and animal fats into biodiesel and co-product. Bioresour Technol 56:19–24
8. Vicente G, Bautista LF, Rodriguez R, Gutierrez FJ, Sadaba I, Ruiz-Vazquez RM, Torres-Martinez S, Garre V (2009) Biodiesel production from biomass of an oleaginous fungus. Biochem Eng J 48(1):22–27
9. Chhetri AB, Tango MS, Budge SM, Watts CK, Islam RM (2008) Non-edible plant oils as new sources for biodiesel production. Int J Mol Sci 9(2):169–180
10. Donot F, Fontana A, Baccou JC, Strub C, Schorr-Galindo S (2014) Single cell oils (SCOs) from oleaginous yeasts and moulds: production and genetics. Biomass Bioenergy 68:135–150
11. Ryu BG, Kim J, Kim K, Choi YE, Han JI, Yang JW (2012) High-cell-density cultivation of oleaginous yeast Cryptococcus curvatus for biodiesel production using organic waste from the brewery industry. Bioresour Technol 135:357–364
12. Ratledge C (1992) Microbial lipids: commercial realities or academic curiosities. In: Kyle DJ, Ratledge C (eds) Industrial applications of single cell oil. AOCS, Illinois, pp 1–15
13. Li Y, Zhao Z, Bai F (2007) High-density cultivation of oleaginous yeast *Rhodosporidium toruloides* Y4 in fed-batch culture. Enzym Microb Technol 41(3):312–317
14. Papanikolaou S (2012) Oleaginous yeasts: biochemical events related with lipid synthesis and potential biotechnological applications. Ferment Technol 1:e103
15. Yang X, Jin G, Gong Z, Shen H, Song Y, Bai F, Zhao ZK (2014) Simultaneous utilization of glucose and mannose from spent yeast cell mass for lipid production by *Lipomyces starkeyi*. Bioresour Technol 158:383–387
16. Saenge C, Cheirsilp B, Suksaroge TT, Bourtoom T (2010) Potential use of oleaginous red yeast *Rhodotorula glutinis* for the bioconversion of crude glycerol from diesel plant to lipids and carotenoids. Process Biochem 46:210–218
17. Munch G, Sestric R, Sparling R, Levin DB, Cicek N (2015) Lipid production in the under-characterized oleaginous yeasts, Rhodosporidium babjevae and *Rhodosporidium diobovatum* from biodiese-derived waste glycerol. Bioresour Technol 185:49–55
18. Cheirsilp B, Louhasakul Y (2013) Industrial wastes as a promising renewable source for production of microbial lipid and direct transesterification of lipid to biodiesel. Bioresour Technol 142:329–337
19. Sestric R, Munch G, Cicek N, Sparling R, Levin DB (2014) Growth and neutral lipid synthesis by *Yarrowia lipolytica* on various carbon substrates under nutrient-sufficient and nutrient-limited conditions. Bioresour Technol 164:41–46
20. Gong Z, Wang QH, Hu C, Jin G, Zhao ZK (2013) Co-fermentation of cellobiose and xylose by *Lipomyces starkeyi* for lipid production. Bioresour Technol 117:20–24
21. Cohen Z, Ratledge C (2005) Single cell oils. AOCS Press, Champaign, IL
22. Sperstad S, Lutnaes BF, Stormo SK, Liaaen-Jensen S, Landfald B (2006) Torularhodin and torulene are the major contributors to the carotenoid pool of marine Rhodosporidium babjevae (Golubev). J Ind Microbiol Biotechnol 33:269–273
23. Moline M, Libkind D, van Broock M (2012) Production of torularhodin, torulene, and B-carotene by Rhodotorula yeasts. In: Barredo J (ed) Microbial carotenoids from fungi: methods and protocols, Methods in molecular biology, vol 898. Springer, New York, pp 275–284
24. Yurkov AM, Vustin MM, Tyaglov BV, Maksimova IA, Sineokiy SP (2008) Pigmented basidiomycetous yeasts are a promising source of carotenoids and ubiquinone Q10. Microbiology 77(1):5–10
25. Sitepu IR, Sestric R, Ignatia L, Levin D, German JB, Gillies LA, Boundy-Mills KL (2013) Manipulation of culture conditions alters lipid content and fatty acid profiles of a wide variety of known and new oleaginous yeast species. Bioresour Technol 144:360–369
26. Sitepu I, Ignatia L, Franz A, Wong D, Faulina S, Tsui M, Kanti A, Boundy-Mills K (2012) An improved high-throughput Nile red fluorescence assay for estimating intracellular lipids in a variety of yeast specie. J Microbiol Methods 91:321–328
27. Kitcha S, Cheirsilp B (2011) Screening of oleaginous yeasts and optimization for lipid production using crude glycerol as a carbon source. Energy Procedia 9:274–282
28. Tanimura A, Takashima M, Sugita T, Endoh R, Kikukawa M, Yamaguchi S, Sakuradani E, Ogawa J, Shima J (2014)

Selection of oleaginous yeasts with high lipid productivity for practical biodiesel production. Bioresour Technol 153:230–235

29. Sriwongchai S, Pokethitiyook P, Kruatrachue M, Bajwa PK, Lee H (2013) Screening of selected oleaginous yeasts for lipid production from glycerol and some factors which affect lipid production by Yarrowia Lipolytica strains. J Microbiol Biotechnol Food Sci 2(5):2344–2348
30. Ageitos JM, Vallejo JA, Veiga-Crespo P, Villa TG (2011) Oily yeasts as oleaginous cell factories. Appl Microbiol Biotechnol 90:1219–1227
31. Sitepu I, Selby T, Lin T, Zhu S, Boundy-Mills K (2014) Carbon source utilization and inhibitor tolerance of 45 oleaginous yeast species. J Ind Microbiol Biotechnol 41 (7):1061–1070
32. Iassonova DR, Hammond EG, Beattie SE (2008) Oxidative stability of polyunsaturated triacylglycerols encapsulated in oleaginous yeast. J Am Oil Chem Soc 85:711–716
33. Li Q, Du W, Liu D (2008) Perspectives of microbial oils for biodiesel production. Appl Microbiol Biotechnol 80:749–756
34. Ratledge C (2002) Regulation of lipid accumulation in oleaginous microorganisms. Biochem Soc Trans 30(6):1047–1050
35. Rattray J (1988) Yeast. In: Ratledge C, Wilkinson S (eds) Microbial lipids, vol 1. Academic Press, London, pp 555–697
36. Rattray J, Scheibeci A, Kidby D (1975) Lipids of yeasts. Bacteriol Rev 39:197–231
37. Thiru M, Sankh S, Rangaswamy V (2011) Process for biodiesel production from Cryptococcus curvatus. Bioresour Technol 102 (22):10436–10440
38. Woodbine M (1959) Microbial fat: microorganisms as potential fat producers. Prog Ind Microbiol 1:181–245
39. Alvarez HM, Mayer F, Fabritius D, Steinbüchel A (1996) Formation of intracytoplasmic lipid inclusions by *Rhodococcus opacus* strain PD630. Arch Microbiol 165:377–386
40. Alvarez H, Steinbüchel A (2002) Triacylglycerols in prokaryotic microorganisms. Appl Microbiol Biotechnol 60:367–376
41. Freitas C, Parreira TM, Roseiro J, Reis A, da Silva TL (2014) Selecting low-cost carbon sources for carotenoid and lipid production by the pink yeast *Rhodosporidium toruloides NCYC* 921 using flow cytometry. Bioresour Technol 158:355–359
42. Füchtenbusch B, Wullbrandt D, Steinbüchel A (2000) Production of polyhydroxyalkanoic acids by *Ralstonia eutropha* and *Pseudomonas oleovorans* from an oil remaining from biotechnological rhamnose production. Appl Microbiol Biotechnol 53(2):167–172
43. Kavadia A, Komaitis M, Chevalot I, Blanchard F, Marc I, Aggelis G (2001) Lipid and γ-linolenic acid accumulation in strains of *Zygomycetes* growing on glucose. J Am Oil Chem Soc 78(4):341–346
44. Makula RA, Lockwood PJ, Finnerty WR (1975) Comparative analysis of the lipids of *Acinetobacter* species grown on hexadecane. J Bacteriol 121(1):250–258
45. Olukoshi ER, Packter NM (1994) Importance of stored triacylglycerols in Streptomyces: possible carbon source for antibiotics. Microbiology 140(4):931–943
46. Xian M, Yan J, Kang Y, Liu J, Bi Y, Zhen K (2001) Production of γ-linolenic acid by *Mortierella isabellina* grown on hexadecanol. Lett Appl Microbiol 33(5):367–370
47. Zhao X, Kong X, Hua Y, Feng B, Zhao ZK (2008) Medium optimization for lipid production through co-fermentation of glucose and xylose by the oleaginous yeast *Lipomyces starkeyi*. Eur J Lipid Sci Technol 110 (5):405–412
48. Kurtzman C, Fell J, Boekhout T (2011) The yeasts: a taxonomic study, 5th edn. Elsevier, Amsterdam
49. Wiebe MG, Koivuranta K, Penttilä M, Ruohonen L (2012) Lipid production in batch and fed-batch cultures of *Rhodosporidium toruloides* from 5 and 6 carbon carbohydrates. BMC Biotechnol 12(1):1
50. Hansson L, Dostálek M (1986) Effect of culture conditions on fatty acid composition in lipids produced by the yeast *Cryptococcus albidus var. albidus*. J Am Oil Chem Soc 63 (9):1179–1184
51. Zhang J, Zhu Z, Wang X, Wang N, Wang W, Bao J (2010) Biodetoxification of toxins generated from lignocellulose pretreatment using a newly isolated fungus, *Amorphotheca resinae* ZN1, and the consequent ethanol fermentation. Biotechnol Biofuels 3(26):1–15
52. Gong Z, Wang Q, Shen H, Hu C, Jin G, Zhao ZK (2012) Co-fermentation of cellobiose and xylose by *Lipomyces starkeyi* for lipid production. Bioresour Technol 117:20–24
53. Heredia L, Ratledge C (1988) Simultaneous utilization of glucose and xylose by *Candida curvata* D in continuous culture. Biotechnol Lett 10(1):25–30
54. Hu C, Wu S, Wang Q, Jin G, Shen H, Zhao ZK (2011) Simultaneous utilization of glucose and xylose for lipid production by

Trichosporon cutaneum. Biotechnol Biofuels 4 (1):1

55. Daniel HJ, Otto RT, Binder M, Reuss M, Syldatk C (1999) Production of 4 sophorolipids from whey: development of a two-stage process with *Cryptococcus curvatus* 5 ATCC 20509 and *Candida bombicola* ATCC 22214 using deproteinized whey concentrates as 6 substrates. Appl Microbiol Biotechnol 51 (1):40–45
56. Liang MH, Jiang JG (2013) Advancing oleaginous microorganisms to produce lipid via 34 metabolic engineering technology. Prog Lipid Res 52(4):395–408
57. Moon N, Hammond E, Glatz B (1978) Conversion of cheese whey and whey permeate to oil and single-cell protein. J Dairy Sci 61:1537–1547
58. Guerzoni M, Lambertini P, Lercker G, Marchetti R (1985) Technological potential of some starch degrading yeasts. Starch 37 (2):52–57
59. Morita T, Konishi M, Fukuoka T, Imura T, Kitamoto D (2007) Microbial conversion of glycerol into glycolipid biosurfactants, mannosylerythritol lipids, by a basidiomycete yeast, *Pseudozyma antarctica* JCM 10317T. J Biosci Bioeng 104:78–81
60. Zhu C, Chen B, Fang B (2013) Pretreatment of raw glycerol with activated carbon for 1,3-propanediol production by *Clostridium butyricum*. Eng Life Sci 13:376–384
61. Leoneti AB, Aragão-Leoneti V, Oliveira SVWB (2012) Glycerol as a by-product of biodiesel production in Brazil: alternatives for the use of unrefined glycerol. Renew Energy 45:138–145
62. Papanikolaou S, Aggelis G (2009) Biotechnological valorization of biodiesel derived glycerol waste through production of single cell oil and citric acid by *Yarrowia lipolytica*. Lipid Technol 21:83–87
63. Papanikolaou S, Muniglia L, Chevalot I, Aggelis G, Marc I (2002) *Yarrowia lipolytica* as a potential producer of citric acid from raw glycerol. J Appl Microbiol 92:737–744
64. Panikolaou S, Fakas S, Fick M, Chevalot I, Galiotou-Panayotou M, Komaitis M, Marc I, Aggelis G (2008) Biotechnological valorisation of raw glycerol discharged after bio-diesel (fatty acid methyl esters) manufacturing process: production of 1,3-propanediol, citric acid and single cell oil. Biomass Bioenergy 32:60–71
65. Santos EO, Michelon M, Furlong EB, Burkert JFM, Kalil SJ, Burkert CAV (2012) Evaluation of the composition of culture medium for yeast biomass production using raw glycerol from biodiesel synthesis. Braz J Microbiol 43:432–440
66. Santos EO, Michelon M, Gallas JA, Kalil SJ, Burkert CAV (2013) Raw glycerol as substrate for the production of yeast biomass. Int J Food Eng 9:413–420
67. Silva CM, Borba TM, Burkert CAV, Burkert JFM (2012) Carotenoid production by *Phaffia rhodozyma* using raw glycerol as an additional carbon source. Int J Food Eng 8 (4):1556–3758
68. Lundin H (1950) Fat synthesis by microorganisms and its possible applications in industry. J Inst Brew 56:17–28
69. Ratledge C (1989) Biotechnology of oils and fats. In: Ratledge C, Wilkinson S (eds) Microbial lipids, vol 2. Academic Press, London, pp 567–668
70. Turcotte G, Kosaric N (1989) Lipid biosynthesis in oleaginous yeasts. In: Bioprocesses and engineering. Springer, Heidelberg, pp 73–92
71. Beopoulos A, Cescut J, Haddouche R, Uribelarrea JL, Molina-Jouve C, Nicaud JM (2009) *Yarrowia lipolytica* as a model for bio-oil production. Prog Lipid Res 48:375–387
72. Galafassi S, Cucchetti D, Pizza F, Franzosi G, Bianchi D, Compagno C (2012) Lipid production for second generation biodiesel by the oleaginous yeast *Rhodotorula graminis*. Bioresour Technol 111:398–403
73. Kraisintu P, Yongmanitchai W, Limtong S (2010) Selection and optimization for lipid production of a newly isolated oleaginous yeast, *Rhodosporidium toruloides* DMKU3-TK16. Kasetsart J Nat Sci 44:436–445
74. Angerbauer C, Siebenhofer M, Mittelbach M, Guebitz GM (2008) Conversion of sewage sludge into lipids by *Lipomyces starkeyi* for biodiesel production. Bioresour Technol 99:3051–3056
75. Chang YH, Chang KS, Hsu CL, Chuang LT, Chen CY, Huang FY, Jang HD (2013) A comparative study on batch and fed-batch cultures of oleaginous yeast *Cryptococcus sp.* in glucose-based media and corncob hydrolysate for microbial oil production. Fuel 105:711–717
76. Chatzifragkou A, Makri A, Belka A, Bellou S, Mavrou M, Mastoridou M (2011) Biotechnological conversions of biodiesel derived waste glycerol by yeast and fungal species. Energy 36:1097–1108
77. Karatay SE, Dönmez G (2010) Improving the lipid accumulation properties of the yeast cells

for biodiesel production using molasses. Bioresour Technol 101:7988–7990

78. Kumar SV, Kumutha K, Krishnan PS, Gopal H (2010) Influence of nitrogen on lipid and biomass production by oleaginous yeast culture. Asian J Biol Sci 5:87–91
79. Pedersen TA (1961) Lipid formation in *Cryptococcus terricolus*. Acta Chem Scand 15:651–662
80. Pedersen T (1962) Lipid formation in *Cryptococcus terricolus*. III. Extraction and purification of lipids. Acta Chem Scand 16:374–382
81. Raimondi S, Rossi M, Leonardi A, Bianchi MM, Rinaldi T, Amaretti A (2014) Getting lipids from glycerol: new perspectives in biotechnological exploration of *Candida freyschussii*. Microb Cell Factories 13:83
82. Evans CT, Ratledge C (1984) Effect of nitrogen source on lipid accumulation in oleaginous yeasts. Microbiology 130 (7):1693–1704
83. Fei Q, Chang HN, Shang LA, Choi JDR, Kim N, Kang J (2011) The effect of volatile fatty acids as a sole carbon source on lipid accumulation by *Cryptococcus albidus* for biodiesel production. Bioresour Technol 102:2695–2701
84. Huang JZ, Shi QQ, Zhou XL, Lin YX, Xie BF, Wu SG (1998) Studies on the breeding of *Mortierella isabellina* mutant high producing lipid and its fermentation conditions. Microbiology 25:187–191
85. Husain SS, Hardin MM (1952) Influence of carbohydrate and nitrogen sources upon lipid production by certain yeasts. J Food Sci 17 (1–6):60–66
86. Li YH, Liu B, Zhao ZB, Bai FW (2006) Optimized culture medium and fermentation conditions for lipid production by *Rhodosporidium toruloides*. Chin J Biotechnol 22:650–656
87. Spencer JFT, Spencer DM, Figueroa LIC (1997) Yeasts as living objects: yeast nutrition. In: Yeasts in natural and artificial habitats. Springer, Berlin, pp 68–79
88. Somashekar D, Venkateshwaran G, Sambaiah K, Lokesh BR (2003) Effect of culture 27 conditions on lipid and gamma-linolenic acid production by *mucoraceous* fungi. Process Biochem 38(12):1719–1724
89. Wu S, Hu C, Jin G, Zhao X, Zhao ZK (2010) Phosphate-limitation mediated lipid production by *Rhodosporidium toruloides*. Bioresour Technol 101:6124–6129
90. Choi SY, Ryu DD, Rhee JS (1982) Production of microbial lipid: effects of growth rate and oxygen on lipid synthesis and fatty acid composition of *Rhodotorula gracilis*. Biotechnol Bioeng 24(5):1165–1172
91. Duarte SH, Ansolin M, Maugeri F (2014) Cultivation of *Candida* sp. LEB-M3 in glycerol: lipid accumulation and prediction of biodiesel quality parameters. Bioresour Technol 161:416–422
92. Valero E, Millan C, Ortega JM (2001) Influence of oxygen addition during growth phase on the biosynthesis of lipids in *Saccharomyces cerevisiae* (M(3)30-9) in enological fermentations. J Biosci Bioeng 92:33–38
93. Amaretti A, Raimondi S, Sala M, Roncaglia L, De Lucia M, Leonardi A, Rossi M (2010) Single cell oils of the cold-adapted oleaginous yeast *Rhodotorula glacialis* DBVPG 4785. Microb Cell Factories 9:73
94. Bass A, Hospodka J (1952) Biosynthesis of fats by yeast. II. Composition of fat at various temperatures. Chem List 46:243–245
95. Kates M, Baxter R (1962) Lipid composition of mesophilic and psychrophilic yeasts (Candida species) as influenced by environmental temperature. Can J Biochem Physiol 40 (9):1213–1227
96. Kates M, Paradis M (1973) Phospholipid desaturation in *Candida lipolytica* as a function of temperature and growth. Can J Biochem 51(2):184–197
97. McMurrough I, Rose AH (1973) Effects of temperature variation on the fatty acid composition of a psychrophilic Candida species. J Baceriol 114(1):451
98. Kessell RHJ (1968) Fatty acids of *Rhodotorula gracilis*: fat production in submerged culture and the particular effect of pH value. J Appl Bacteriol 31(2):220–231
99. Miller JJ, Webb NS (1954) Isolation of yeasts from soil with the aid of acid, rose bengal, and oxgall. Soil Sci 77(3):197–204
100. Chen W, Zhang C, Song L, Sommerfeld M, Hu Q (2009) A high throughput Nile red method for quantitative measurement of neutral lipids in microalgae. J Microbiol Methods 77:41–47
101. Hu C, Zhao X, Zhao J, Wu S, Zongbao K (2009) Effects of biomass hydrolysis by-products on oleaginous yeast *Rhodosporidium toruloides*. Bioresour Technol 100:4843–4847
102. Yu X, Zheng Y, Dorgan KM, Chen S (2011) Oil production by oleaginous yeasts using the hydrolysate from pretreatment of wheat straw with dilute sulfuric acid. Bioresour Technol 102:6134–6140
103. Sitepu IR, Garay LA, Sestric R, Levin D, Block DE, German JB, Boundy-Mills KL

(2014) Oleaginous yeasts for biodiesel: current and future trends in biology and production. Biotechnol Adv 32(7):1336–1360
104. Zhao X, Peng F, Du W, Liu C, Liu D (2012) Effects of some inhibitors on the growth and lipid accumulation of oleaginous yeast Rhodosporidium toruloides and preparation of biodiesel by enzymatic transesterification of the lipid. Bioprocess Biosyst Eng 35 (6):993–1004
105. Palmqvist E, Hahn-Hägerdal B (2000) Fermentation of lignocellulosic hydrolysates. I: inhibition and detoxification. Bioresour Technol 74(1):17–24
106. Palmqvist E, Hahn-Hägerdal B (2000) Fermentation of lignocellulosic hydrolysates. II: inhibitors and mechanisms of inhibition. Bioresour Technol 74(1):25–33
107. Parawira W, Tekere M (2011) Biotechnological strategies to overcome inhibitors in lignocellulose hydrolysates for ethanol production: review. Crit Rev Biotechnol 31(1):20–31
108. Qi F, Kitahara Y, Wang Z, Zhao X, Du W, Liu D (2014) Novel mutant strains of *Rhodosporidium toruloides* by plasma mutagenesis approach and their tolerance for inhibitors in lignocellulosic hydrolyzate. J Chem Technol Biotechnol 89(5):735–742
109. Bligh E, Dyer WJ (1959) A rapid method of total lipid extraction and purification. Can J Biochem Physiol 37:911–917

Chapter 19

Producing Oleaginous Microorganisms Using Wastewater: Methods and Guidelines for Lab- and Industrial-Scale Production

Kayla M. Rude, Tyler J. Barzee, and Annaliese K. Franz

Abstract

Cultivation of oleaginous microorganisms on wastewater provides alternative biofuel options while also acting as a remediation technique for alternative wastewater treatment. This chapter describes guidelines and methods for the production of oleaginous microorganisms—with a focus on microalgae—using wastewater as a growth medium while considering a variety of general challenges for both lab- and industrial-scale production. Cultivation techniques described here range in scale from microplates with 10-mL working volumes, up to multigallon, industrial-scale microorganism cultivation, with a focus on microalgae. This chapter includes guidelines for the preparation of wastewater and selection of oleaginous microorganisms combined with methods for the production of oleaginous microorganisms cultivated using wastewater.

Key words Oleaginous microorganism, Microalgae, Cultivation, Wastewater, Remediation, Biodiesel

1 Introduction

The microorganisms of interest in this chapter includes various oleaginous species of microalgae, bacteria, and yeast that are capable of being used as a feedstock for biofuel production. To be commercially competitive on the fuel market, biofuels prepared from oleaginous microorganisms must have a minimum fuel selling price (MFSP) of around \$3/gal [1, 2]. The cost of cultivating and harvesting these oleaginous microorganisms is often the limiting factor in being able to reach the goal MFSP. A recent technoeconomic assessment (TEA) performed by the National Renewable Energy Laboratory (NREL) found that between 65 and 85% of the overall biofuel production costs is the cost of biomass production [2].

Wastewater is a readily available and inexpensive resource that can provide nutrients such as nitrogen, phosphorous, carbon, potassium, and various trace metals to promote microorganism

Venkatesh Balan (ed.), *Microbial Lipid Production: Methods and Protocols*, Methods in Molecular Biology, vol. 1995, https://doi.org/10.1007/978-1-4939-9484-7_19,

growth. Using wastewater to cultivate oleaginous microorganisms has the dual purpose of remediating the high nutrient levels in wastewater while serving as a growth medium for the affordable production of biofuel feedstock. However, there are inherent challenges in using wastewater as a growth medium such as ammonia toxicity, heavy metals, and high chemical oxygen demand. There are many kinds of wastewater that can be used for culturing different strains of oleaginous microorganisms. This chapter provides several guidelines and methods for growing oleaginous microorganisms using a wastewater medium. The focus is on lab-scale production but also general guidelines for engineering larger systems for industrial production. For a specific discussion of cultivation of oleaginous yeast on sewage and industrial wastewater, refer to Chapter 24.

2 Guidelines

There are numerous wastewater streams that can be considered for use as a medium for microorganism growth including dairy, swine, anaerobic digestate, municipal, and industrial. These wastewaters have significant compositional variety which must be considered to provide a supportive environment for microorganism production. The following guidelines aim to help address issues of wastewater composition and selection of microorganism based on the wastewater source.

2.1 Collection, Sampling, and Characterization of Wastewater

To characterize wastewater composition, it is necessary to obtain representative samples of the wastewater that will be used as a culturing medium. Depending on the wastewater source, this can be accomplished in several ways including buckets, sampling ports of tank pipes, and sampling devices. The sampling method should be selected based on the source and availability of the wastewater of interest (*see* Subheading 4.1). For all methods it is critical that wastewater is mixed to ensure a representative sampling process for compositional analysis.

Characterization of nutrients and other wastewater parameters (pH, chemical oxygen demand, temperature, etc.) is necessary to consider microalgae strains and culturing conditions. Due to the limits of many analysis methods, a combination of several techniques and appropriate dilutions should be utilized for accurate measurements (*see* Subheadings 2.2 and 4.2).

Standardized methods are governed by Title 40 of the Code of Federal Regulations. Table 1 describes the limits of some of these standardized methods, the sample volume needed, and the interferences that must be considered and dealt with.

Table 1
Federally approved methods of analyzing wastewater samples for various chemical parameters. Details obtained from the Standard Methods for the Examination of Water and Wastewater (2017) [3]

Wastewater parameter	Analysis	Limits	Volume needed	Relevant interferences (*see* Note 1)
Nitrate	Cadmium reduction method	<1.0 mg NO_3^-/L	25 mL	• Suspended matter • Several mg/L of iron, copper, or other metals • Oil and grease • Residual chlorine
Nitrite	Colorimetric method	<1.0 mg NO_2^-/L	50 mL	• Sb^{3+}, Au^{3+}, Bi^{3+}, Fe^{3+}, Pb^{2+}, Hg^{2+}, Ag^{2+}, chloroplatinate, metavanadate • Copper ions • Suspended solids • Colored ions
Organic nitrogen	Total Kjeldahl nitrogen method	<100 mg N/L	25 mL	• Nitrate in excess of 10 mg/L • Inorganic salts and solids • Chloride ions • Large amounts of organic matter
Ammonia	Distillation and titration method	>5 mg NH_3/L	500 mL	• Volatile alkaline compounds (e.g., hydrazine and amines) • Residual chlorine • Glycine, urea, glutamic acid, cyanates, and acetamide
Phosphorous	Digestion and ascorbic acid method	<1.3 mg P/L	50 mL	• ≥0.1 mg arsenate/L • Cr(VI) • Sulfide: <1.0 mg/L • Silicate: <10 mg/L
Total organic carbon (TOC)	Combustion-infrared method	≥1 mg C/L	• If inorganic carbon present: 10–15 mL sample to purge • Analysis volume varies by instrument	• Inorganic carbon • Artifacts from filtration
Chemical oxygen demand (COD)	Closed-reflux, Colorimetric method	>50 mg O_2/L	2.5 mL	• >2000 mg Cl^-/L • Volatile straight/chain aliphatic compounds • Nitrite (1.1 mg O_2/mg NO_2^-)

2.2 Filtration, Pretreatment, and Sterilization of Wastewater

Excessive levels of total suspended solids (TSS) can cause interferences with photosynthetic microorganisms' ability to photosynthesize and can interference with analysis methods, so lowering TSS is essential for an optimized culturing design.

For laboratory and larger scale experiments, particle removal can be accomplished using numerous methods including centrifugation, sedimentation, flocculation, flotation, and filtration (*see* Subheading 2.9). Filtration of wastewater with a coarse material such as cheesecloth can reduce some of the large particles. However, wastewaters with large quantities of fine solids should be further filtered with membranes or otherwise treated to reduce TSS. For laboratory-scale investigations (<5 L wastewater), it is generally desirable to sterilize the wastewater by autoclaving or sterile filtration (0.22 μm) into an autoclaved bottle.

For large scale operations, treating the wastewater with hypochlorite followed by thiosulfate neutralization or ultraviolet irradiation are common initial sterilization techniques. Maintaining sterile conditions requires a closed-culture operation; therefore, sterilization may not be implemented as part of industrially sized algae production unless the product value justifies the added expense.

2.3 Wastewater Storage

When working at lab scale with small volumes, the wastewater should be stored at 4 °C in the dark (e.g., in walk-in cooler) in capped glass Pyrex bottles or plastic five-gallon carboys. When storing larger volumes, wastewater is usually stored in warehouses or outdoors where higher temperature may occur. If stored above room temperature, the wastewater should be sampled to monitor pH, total nitrogen, ammonia, phosphorus, etc. over time to account for possible compositional changes.

2.4 Strain Selection and Inoculum Preparation

To grow an organism on wastewater, it is useful to know the chemical tolerances and relative growth rates. Important characteristics to note about common oleaginous microorganism species can be referenced in Table 2. Both the strain and the species should be considered when designing procedures based on literature and when comparing results to reference values. Axenic strains can be purchased from culture collections (*see* **Note 2**). Native strains can be isolated from ponds, lakes, and rivers in the local area or selected by placing a sample of the wastewater outdoors to observe and isolate any natural microbial growth. Small-scale culturing and experimentation may benefit from starting with an axenic species purchased and cultured in synthetic media or sterilized dilutions of wastewater to establish a viable, sterile inoculum. In contrast, the sterility of strains and the surrounding conditions is not always feasible for large scale culturing and experimentation, and therefore contamination is less controlled, if at all.

Table 2
Common oleaginous microorganism species grown in wastewater systems and their heavy metal tolerances (Adapted from Kumar et al. [4])

Oleaginous microorganism	Type	Synthetic media	Potential wastewater substrates	Heavy metal removal [4–7]	Notes
Chlorella sorokiniana	Microalgae (Freshwater)	BG-11, JM, BBM	Anaerobically digested cattle manure effluent [8], digested food waste [9], municipal wastewater [9]	Cd(II), Ni(II)	• Fast growing • Can grow in a wide temperature range
Chlorella vulgaris	Microalgae (Freshwater)	BG-11, JM	Sewage pump effluent [10], agro-industrial effluent of the dairy and piggery industry [11], dairy industry effluent [12]	Cd(II), Cr(IV), Cu(II), Ni(II), Fe(III), Hg(II), Pb(II), Zn(II), U(II), Au(II)	• Can grow in some saline conditions [13] • Potential for high lipid content [14]
Scenedesmus obliquus	Microalgae (Freshwater)	JM, BBM	Municipal wastewater [15], olive oil industry effluent [16], brewery industry effluent [17]	Cd(II), Cr(IV), Cu(II), Ni(II), Zn(II), U(II)	• Can grow at higher temps [18] • Tolerant to high CO_2 conc [19]
Chlamydomonas reinhardtii	Microalgae (Freshwater)	TAP	Municipal wastewater [20]	Cd(II), Cr(IV), Cu(II), Hg(II), Ni(II), Pb(II), Mo(I)	• Widely studied as a model organism
Spirulina platensis	Cyanobacteria (Marine)	Modified BG-11[a]	Anaerobically digested piggery waste effluent [21], dairy effluent [22]	Cd(II), Cu(II), Pb(II), Zn(II)	• Can grow in a wide range of temps (20–40 °C) [23, 24]
Rhodococcus opacus	Bacteria	Trypticase soy broth, LB	Synthetic dairy wastewater [25], biomass gasification wastewater [26]	Al(III), Ni(II), Cr(III)	• Tolerant of many organic compounds including phenol [26] • Efficient lipid producer [26]
Cryptococcus curvatus	Yeast	YPD	Distillery wastewater [27], food waste wastewater [9], municipal wastewater [9]	N/A	• Accumulates lipids on a broad range of substrates [28] • Efficient lipid producer (>60%) [29]

JM Jaworski's medium, *BBM* bold basal medium, *LB* lysogeny broth, *YPD* yeast peptone dextrose agar, *BG-11* blue-green medium, *TAP* tris acetate phosphate
[a]Modified BG-11 made with saltwater instead of freshwater

2.5 Recognizing Nutrient Toxicity and/or Limitation—Dilutions and Supplementations

Excessive nutrient levels (e.g., nitrogen in the form of ammonia) along with high chemical oxygen demand (COD) can induce cellular stress and inhibit microbial growth. Similarly, high-salt concentrations in certain wastewater streams can negatively affect nonhalophilic freshwater organisms. To overcome inhibitory effects, the wastewater can be diluted to reduce the salt and nutrient concentrations to tolerable levels (*see* Subheadings 2.2 and 4.2). Some toxicities, such as those associated with heavy metal-polluted wastewater, cannot be overcome using dilutions; therefore, chemical remediation and bioremediation methods may be considered (*see* Table 2). Certain wastewaters, such as anaerobic digestates, may be dark in color and in many cases, the color can interfere with growth and chemical analysis.

Wastewater may also be limited in some nutrients and require supplementation to achieve an optimal nutrient ratio (e.g., C:N:P ratio) (*see* **Note 3**), for an organism's optimal growth [30, 41]. Possible supplementations include carbon (e.g., addition of CO_2, acetate, glucose, or sodium bicarbonate) and phosphorous (e.g., potassium phosphate or calcium phosphate).

2.6 Designing a Cultivation System

Wastewater characteristics, growth characteristics of the selected organism, scale and environmental conditions (indoors, outdoors, covered, etc.) are all critical factors to design a productive cultivation system (*see* **Note 4**). Standard parameters such as light cycles, temperature, aeration, and water current must also be considered in the design.

2.6.1 Mixing

With lab-scale experiments, mixing can be performed using a stir bar and stir plate, bubbling, or shaker plates. For larger scale systems, mixing can be achieved using methods such as submerged impellers, paddle wheels, or bubble diffusers.

2.6.2 Air and Carbon Dioxide Suffusion

Suffusion of a culture with oxygen and CO_2 is commonly necessary to promote microorganism growth [42]. The type and rate of suffusion depends on the wastewater, the microorganism tolerance and the size/design of the system.

Air Suffusion

Air suffusion enhances mixing and can also reduce the high chemical oxygen demand (COD) and biochemical oxygen demand (BOD) characteristic of many wastewater streams (*see* Subheading 4.5). However, a high rate of aeration can increase oxidative stress levels, causing growth and productivity to be limited.

Carbon Dioxide (CO_2) Suffusion

Supplementation with carbon dioxide can significantly increase production of microalgae in wastewater systems while sequestering ammonia-nitrogen [10]. The cost of CO_2 supplementation varies depending of the system design and source of wastewater. When carbon dioxide is created as a by-product of the wastewater production and treatment process (e.g., exhaust) [32], supplementation contributes to a cost effective, environmentally friendly, and efficient wastewater/microorganism culturing design.

2.7 Microplate-Scale Optimization Studies

Multiple experiments can be performed with replicates using a 6-well microplate design (10-mL working volume per well), including microorganism screenings, wastewater dilution studies, light cycle and temperature testing, nutrient studies, toxicity studies, and supplementation depending on the goals of the overall design. Conditions identified in the microplate screening can then be scaled up to achieve data that better represents a large-scale process. This method is acts as a preliminary assessment of variables and microplate results must be supported by further investigation at larger scales.

2.8 Measuring the Growth of Oleaginous Microorganisms

Common methods to measure microbial growth include spectrophotometric readings of cell density, pigment analysis and quantification, cell counting using a hemocytometer, flow cytometry, and measuring volatile suspended solids (VSS) (*see* **Note 5** and Subheadings 4.9–4.12).

2.9 Harvesting Oleaginous Organisms

Once microalgae cultures reach the desired biomass density (*see* **Note 6**), microalgae must be harvested. Common methods for harvesting include flocculation, gravity and centrifugation sedimentation, flotation, and filtration. These techniques are commonly used in combination or in successive iterations to obtain a desired biomass concentration.

2.9.1 Flocculation

This technique relies on the ability to coagulate particles into flocs that settle quickly using coagulant chemicals and natural flocculation (*see* **Note 7**). This technique is relatively low-cost and quick, but environmental concerns can arise from downstream users of the algae biomass that may be affected by increased mineral contents associated with the coagulant chemicals. Alternatives to multivalent salts are synthetic and bio-derived polymers and cationic polymers (*see* **Note 8**). Other methods of flocculation can involve cocultivation of bacteria, algae, and/or fungi (*see* **Note 9**).

2.9.2 Sedimentation

In this technique, the force for separation is being supplied by either gravity or a centrifuge. It is possible to increase the rate of particle sedimentation by increasing the particle size, such as by flocculation, increasing the particle's density relative to the surrounding fluid, decreasing the fluid's viscosity, or increasing the acceleration experienced by the particle. Implementing gravity-based sedimentation may be preferred for large scale operations (*see* Subheading 4.14 and 4.15).

Table 3
Categories of membrane filtration processes including rejected particle types and required operational pressures [33]

	Microfiltration	Ultrafiltration	Nanofiltration	Reverse osmosis
Pore size (nm)	100–10,000	2–100	0.5–2	<0.5
Pressure (psi)	1.5–30	1.5–70	40–300	70–1700
Example of substances removed/collected	Bacteria, algae	Smaller bacteria, viruses	Macromolecules, organic molecules	Monovalent ions

2.9.3 Filtration

This technique relies on using pressure to separate materials through a membrane. Membrane filtration processes are generally grouped into four categories depending on the size of particles they are designed to remove/collect. In order of decreasing rejected particle size, these categories are: microfiltration, ultrafiltration, nanofiltration, and reverse osmosis (*see* **Note 10**). The four processes are briefly summarized in Table 3.

2.10 Remediation Analysis, and Recycling of Remediated Wastewater and Wastewater Nutrients

After microalgae cultivation, the treated wastewater should be analyzed prior to discharge or transfer to other operations/facilities. In the USA, the Environmental Protection Agency (EPA) authorizes the regulation of discharged waters to individual state agencies. The EPA states minimum 30-day average requirements for secondary treatment as 30 mg/L for BOD_5 and TSS with a pH between 6 and 9 [34]. State agencies frequently regulate ammonia and/or nitrate levels as well as bacterial content (*see* **Note 11**).

In order to minimize inputs of clean freshwater to the cultivation system, the treated wastewater can be recycled as dilution water or used to make up for evaporative losses in cultivation ponds. Extensively recycled water can eventually accumulate salts or other inhibitory organic molecules that may impact the algae cultivation efficiency [35]. Inhibitory effects of recycled wastewater can generally be resolved by disposing a certain portion of the water in a process called "blowdown disposal" [36]. Recycling of certain harvesting chemicals (i.e., coagulant chemicals) is also possible [37].

3 Materials

All reagents and media can be stored at room temperature (25 °C), unless otherwise indicated. Although wastewater is the culturing media, proper waste disposal guidelines should be followed to dispose of the wastewater accordingly. In order to maintain sterile conditions, prepare all reagent solutions using ultrapure water and analytical grade reagents inside a laminar flow hood and/or fume hood when necessary (*see* **Note 12**).

3.1 Wastewater

Collect wastewater from designated source and store at room temperature or 4 °C for lab use and testing (*see* Subheading 4.1)

1. Long-handled dipper (e.g., submersible bucket).
2. Column sampler with check valve.
3. Plastic or glass sample bottles.

3.2 Reagents

Ultrapure water is used to make all reagents and for wastewater dilutions, unless otherwise indicated.

1. Deionized water when indicated.
2. 98% H_2SO_4 in water and 1.0 M HCl in water for pH adjustments.
3. 50% NaOH in water and 1.0 M NaOH in water for pH adjustments.
4. Synthetic media for preparation of inoculum and synthetic controls for comparison: Jaworski's medium (JM), BG-11 medium, and F/2 medium (*see* **Note 13**).
5. Coagulant chemical (ferric chloride, alum, cationic polymers, etc.).
6. Activated carbon.

3.3 Lab Equipment

3.3.1 Sampling and Filtration

1. 2-L Glass Pyrex bottles for sampling and storage of the wastewater.
2. Submersible bucket for sampling open-channel wastewater sources.
3. Long, cylindrical device equipped with a check valve on the bottom.
4. Three-way fitting and valve attached to a line for dispensing wastewater.
5. Buchner funnel and 1-L vacuum flask.
6. Whatman No. 1 filter papers to the dimension of the Buchner funnel.
7. 0.45 μm filter paper.
8. Sterile filtration unit (0.22 μm).
9. Bench scale pump-driven filtration unit.

3.3.2 Culturing

1. T-5 grow lights connected to a timer (*see* **Note 14**).

Microplate-Scale Design (For Culturing <10 mL)

1. 6-well, polystyrene, flat-bottomed, clear microplates with lid.
2. Shaker equipped with plate holders.

Small-to-Medium Lab-Scale Designs (For Culturing ≤150 mL)

1. 250-mL volumetric flasks with rubber filter caps.
2. Corning Pyrex media bottles (500-mL to 2-L) with filter caps and aeration set-up.
3. Shaker plate or stir plate and TFE-coated stir bars.
4. Nalgene PVC tubing (I.D. × O.D. × Wall: 1/8 × 3/16 × 1/32 in.; 50 ft) (*see* **Note 15**).
5. Air pump or CO_2 source (*see* **Note 16**).
6. Gas flow regulator.
7. Flow meters.
8. Autoclave (optional).
9. Graduated pipettes (5-mL, 25-mL, or 50-mL sizes).
10. Automatic pipettor or pipette bulb.
11. Graduated cylinder (100-mL, 200-mL sizes).
12. Micropipettes and pipette tips.
13. 1-mL cuvettes.

3.4 Microorganism Strains

There are a variety of oleaginous bacteria, fungi, microalgae, and cyanobacteria species available to choose from depending on the wastewater source *see* Table 2.

3.5 Wastewater Characterization Supplies

For a complete list of lab equipment and reagents needed to run the federally regulated laboratory methods to characterize the wastewater, refer to Standard Methods for the Examination of Water and Wastewater (2017) [3].

3.5.1 Lab Equipment

1. Buchner funnel and 1-L vacuum flask.
2. Whatman No.1 filter papers to the dimension of the Buchner funnel.
3. Aluminum dishes or porcelain crucibles.

3.5.2 Instrumentation

1. UV-Vis Spectrophotometer with a large working range of wavelengths and capable of holding 16-, 20-, or 25-mm culture tubes or ampules (*see* **Note 17**).
2. Total organic carbon analyzer.
3. pH meter.
4. Oven.
5. Muffle furnace.

3.5.3 Hach Kits (See ***Note 18****)*

1. Ultrapure water to use as diluent.
2. Hach reactor block (*see* **Note 19**).
3. Hach benchtop spectrophotometer.
4. Analysis kits including: Total nitrogen (HR) (Product # 2714100), Total Kjeldahl nitrogen (Product # 2495300), Total phosphates (LR) (Product # 2742645), Total organic carbon (Product # 2760445), Ammonia (HR) (Product # 2606945), and Chemical oxygen demand (Product # TNT82206).

4 Methods

Carry out all procedures at room temperature, unless otherwise specified. Acid-wash all glassware to avoid contamination of residual compounds of interest. Use ultrapure water unless otherwise indicated (*see* **Note 20**). Some sample preservation techniques cannot be utilized if running a certain analysis or experimental procedure. Reagent blanks should be carried through all procedures and accounted for in calculations.

4.1 Sampling of Wastewater

Obtain wastewater samples with submersible buckets (*see* **Note 21**). Water column samples may also be obtained with a long, cylindrical device equipped with a check valve on the bottom as described.

4.1.1 Well-Mixed, Open-Channel Wastewater Sources with Easy Access to the Liquid Surface

1. Lower the device through the water column to the bottom.
2. Pull up to engage the check valve to hold in the liquid.
3. With stratified water column in cylinder:
   - Sample directly at a desired height.
   - Empty into a sample bottle (*see* **Note 22**).

4.1.2 For Wastewater Samples Contained in Large Tanks Accessible Only by Pipes

1. Locate or construct a sampling port with a manual valve along the pipe (*see* **Note 23**).
2. Flush the pipe to clear any settled solids or other materials that would produce nonrepresentative samples (*see* **Note 24**).
3. To determine the flush duration, open the valve and take samples of the flow over time.
4. Analyze the solids contents of these samples and determine the time at which the samples became stable and use that as a starting point for any samples for material used in algae cultivation.

4.1.3 For Wastewater Samples Contained in Smaller Tanks

1. Mix wastewater using hydraulic mixing with a pump at the bottom on the container until complete mixing has been achieved (*see* **Note 25**).
2. Divert flow from container to sample bottle.
3. Collect desired volume of sample.

4.2 Preparation of Wastewater for Characterization Methods

Dilute all wastewater samples using ultrapure water for chemical analysis. Prepare dilutions using graduated or volumetric pipettes with an automatic pipettor or pipette bulb.

1. Filter 2 L of wastewater through a 0.45 μm filter (*see* **Note 26**).
2. Allow any possible remaining particulate matter to settle for at least 15 min. This will vary according to the viscosity and turbidity of the wastewater.
3. Prepare a 1% wastewater dilution by pipetting 1 mL of wastewater supernatant into 99 mL of ultrapure water in a glass beaker or flask; mix.
4. Repeat **step 1** with the appropriate volumes to prepare a 10% and 25% wastewater dilution.
5. Analyze the dilution series utilizing the desired analytical method and adjust accordingly to obtain values within the range of each method (*see* **Note 27**).

4.2.1 Decolorization of Wastewater

1. Add approximately 20 g/L activated carbon to the wastewater solution in shaker flasks (*see* **Note 28**).
2. Shake on shaker plate at constant temperature until color is visibly removed.
3. Filter decolorized wastewater through Whatman No. 1 filter paper followed by filtration using a Whatman No. 5 filter paper (*see* **Note 26**). Collect and store at 4 °C in the dark until analyzed.

4.3 Utilizing Commercial Chemical Analysis Kits

Each commercial analysis kit, such as those from Hach Co., follows a specified method provided by the company (*see* Table 4).

4.3.1 Preparing Wastewater for Analysis Using a Commercial Analysis Kit

1. Filter samples and create a dilution series as described in Subheading 4.2.
2. Identify possible interferences as listed in Table 4 and further described in each listed Hach method in order to properly test wastewater using the kits (*see* **Note 1**).

4.4 Federally Regulated Laboratory Tests for Examining Each Wastewater Parameter (See Note 29)

4.4.1 Nitrogen [3]

Federally regulated methods separate nitrogen analysis into the different species nitrogen: nitrate, nitrite, ammonia, and TKN, but they are limited given the concentrations limits of the assays. Federally approved AHPA method summaries for nitrogen are described for each of the listed parameters.

1. *Ammonia (AHPA 4500 NH_3 B/C) method summary*: A 500-mL wastewater sample must first be dechlorinated using a 0.022 M $Na_2S_2O_3 \cdot 5\ H_2O$ solution. One milliliter of dechlorinating reagent removes 1 mg/L residual chlorine. The dechlorinated sample is distilled, and the volatilized

Table 4
Available Hach kits and associated methods including the individual kits' working ranges and interferences that must be considered

Test	Hach kit method	Limits for testing kit (mg/L) (*see* Note 27)	Relevant interferences (*see* Note 1)
Total nitrogen	*HR*: Method 10072 *LR*: Method 10071	*HR*: 2–150 *LR*: 0–25	Calcium: 1200 mg/L Chromium (III): 2 mg/L Iron: 8 mg/L Lead: 26.4 μg/L Magnesium: 2000 mg/L Organic Carbon: 600 mg/L Phosphorous: 400 mg/L Chloride: > 3000 mg/L Direct sunlight
Total Kjeldahl nitrogen (TKN)	Method 8075	1–150	N/A
Ammonia	*HR*: Method 10031 *LR*: Method 10023	*HR*: 0.4–50 *LR*: 0.02–2.50	Iron: All levels Nitrate: 5000 mg/L Nitrite: 600 mg/L pH: adjust samples to pH of 7 Phosphate: 5000 mg/L Sulfate: 6000 mg/L Turbidity and color may intensify indicator color
Phosphorous	*HR*: Method 10127 *LR*: Method 8190	*HR*: 1–100 *LR*: 0.02–1.10	Turbidity A pH > 1 after addition of the acid to the sample Iron: >100 mg/L
Total organic carbon (TOC)	*HR*: Method 10128 *LR*: Method 10129	*HR*: 100–700 *LR*: 0.3–20	Copper: 10 mg/L Iron (II/III): 10 mg/L Nitrite: 500 mg/L Phosphate: 3390 mg/L Sulfates: 5000 mg/L Sulfide: 20 mg/L Sulfite: 50 mg/L Zinc: 5 mg/L
Chemical oxygen demand (COD)	*HR/LR*: Method 8000	*HR*: 20–1500 *LR*: 3–150	Chloride: >1000 mg/L

HR high range, *LR* low range

ammonia is caught in 50 mL of 0.32 M boric acid receiving solution. At least 200 mL of distillate is diluted to 500 mL and titrated using a standardized 0.02 N H_2SO_4 titrant. A blank is run through the entire process and Eq. 1 can be used to determine the concentration of ammonia.

$$\text{mg}\ (\text{NH}_3 - \text{N})/\text{L} = \frac{(A - B) \times 280}{\text{mL sample}} \quad (1)$$

Where:
A = volume of H_2SO_4 titrated for sample, mL, and
B = volume of H_2SO_4 titrated for blank

2. *Nitrites (AHPA 4500 NO_2^- B) method summary*: A 50-mL filtered wastewater sample is adjusted to a pH of 5 using 1 M HCl or NH_4OH, and color reagent consisting of water, H_3PO_4, sulfanilamide, and *N*-(1-napthyl)-ethylenediamine dihydrochloride is added to the sample. The sample is mixed, reacted, and the absorbance of the mixture is measured at 543 nm. The concentration of nitrite is calculated by preparing a standard curve using a stock nitrite standard solution and using the linear regression equation to relate the characteristic absorbance to the nitrite concentration.
3. *Nitrates (AHPA 4500 NO_3^- E) method summary*: A cadmium reduction column is prepared. A 25-mL aliquot of sample is combined with 75 mL of an NH_4Cl–EDTA solution and the 100 mL wastewater solution is adjusted to a pH of between 7 and 9 using 1.0 M HCl or 1.0 M NaOH. The wastewater + NH_4Cl–EDTA mixture is run through the cadmium column. The eluate is collected, and the color reagent used in the nitrites assay is added and mixed. After a 10 min incubation at room temperature (25 °C), the sample is measured at 543 nm. The concentration of nitrate is calculated by preparing a standard curve using a stock nitrate standard solution and using the linear regression equation to relate the characteristic absorbance to the nitrate concentration.
4. *Total Kjeldahl Nitrogen (AHPA 4500 N_{org} B) method summary*: A 25 mL sample volume is filtered, dechlorinated, and added to a Kjeldahl flask along with NaOH and borate buffer solutions to remove ammonia. The sample is digested in H_2SO_4 containing K_2SO_4 and $CuSO_4$ catalysts and cooled. After the addition of 50 mL of sodium hydroxide-thiosulfate, the ammonia is distilled from the alkaline medium and absorbed into 0.32 M boric acid. The ammonia can be measured as described in the ammonia section (AHPA 4500 NH_3 C).

4.4.2 Phosphorous (AHPA 4500 P B5/E) [3]

A persulfate digestion followed by an ascorbic acid colorimetric method is a federally approved method for measuring total phosphorous and is detailed in method AHPA 4500 P B5/E.

AHPA method summary: A persulfate digestion is performed on a 50-mL wastewater sample to oxidize organic matter releasing the phosphorous and orthophosphate. The digested wastewater is transferred to an Erlenmeyer flask where one drop of a phenolphthalein indicator is added along with 8 mL of prepared combined reagent containing sulfuric acid, potassium antimonyl tartrate, ammonium molybdate, and ascorbic acid solution. After a 10 min incubation period, the sample absorbance is measured at 880 nm. The concentration of phosphorus is calculated using a linear regression equation provided by running a standard curve of phosphorus standard solutions.

4.4.3 Total Organic Carbon (AHPA 5310 B) [3]

Total organic carbon (TOC) can be used as an indication toward the biochemical oxygen demand (BOD) and chemical oxygen demand (COD). The federally approved method for TOC is a combustion method and is detailed in method AHPA 5310 B.

AHPA method summary: About 10–15 mL of wastewater sample is homogenized to remove any remaining gross solids or insoluble matter. If the sample is preserved using acid, the wastewater sample must be purged using any CO_2-free gas. The sample is injected into a total organic carbon analyzer instrument and the organic carbon is calculated by comparing the corrected resulting carbon peaks to a standard curve prepared with both organic and inorganic carbon standards. The organic carbon content should be reported at non-purgable organic carbon if the purging step was performed to account for the loss of volatile organic carbons during the purging of the sample.

4.4.4 Chemical Oxygen Demand (AHPA 5220 D) [3]

The federally approved method for COD is a spectrophotometric method and is detailed in method AHPA 5220 D.

AHPA method summary: A 2.5-mL wastewater sample is placed in a 16 × 100 mm culture tube with 1.5 mL of a digestion solution containing $K_2Cr_2O_7$, ultrapure water, H_2SO_4, and $HgSO_4$. Slowly, 3.5 mL of 0.018 molal Ag_2SO_4 sulfuric acid solution is added to form an acid layer and the tubes are capped and inverted to mix. The samples are placed in a block digester or oven set at 150 °C for 2 h and then cooled to room temperature (25 °C). The tubes are inverted again to dislodge the solids from the sides of the tube before measuring. The absorbance of each sample is measured at 600 nm, and the COD concentration is determined by comparing the sample absorbance to a calibration curve prepared from potassium hydrogen phthalate standard solutions.

4.5 Cultivating Oleaginous Microorganisms

All cultivation procedures can be performed on the benchtop at room temperature (25 °C) using ultrapure water unless otherwise indicated.

4.5.1 Creating a Microorganism Stock Culture

1. Pipet or use a graduated cylinder to dispense 250 mL of synthetic media into a culturing vessel (*see* **Notes 12** and **30**) with a magnetic stir bar and aeration tubing.

2. If the sample is maintained on an agar plate or slant, use a pipette tip to gently scrape the cells from the culturing agar and transfer to the media in the culture flask. Pipet up and down until the sample is pushed off the pipette tip into the culture.
3. If the sample is maintained in liquid media, inoculate the synthetic media with the organism at a 10% inoculation volume (*see* **Note 31**); mix thoroughly.
4. Place the culture flask on a stir plate with a 16:8 light–dark cycle with constant stirring and aeration.
5. Measure the absorbance at 680 nm using a UV/Vis every other day to identify when the culture is in exponential phase (*see* Fig. 1) and ready to be transferred for culturing in wastewater (*see* **Note 32**).

4.5.2 Inoculation of Stock Culture into Wastewater

Prepare and transfer wastewater dilutions into the culturing vessel using graduated pipettes and an automatic pipettor or bulb or graduated cylinder (*see* **Note 33**).

1. Inoculate the wastewater with at least a 10% inoculation volume of the stock microorganism culture in exponential phase (*see* Fig. 1).
2. Provide aeration and mixing based on the selected design of the culturing system.

4.6 Microplate Studies for Screening Oleaginous Microorganism Growth on Wastewater

All wastewater dilutions for microplate studies should use ultrapure water or filtered seawater. Using 6-well plates provides an experimental design with either three or five replicates. Optical density measurements are as described in Subheading 4.9.

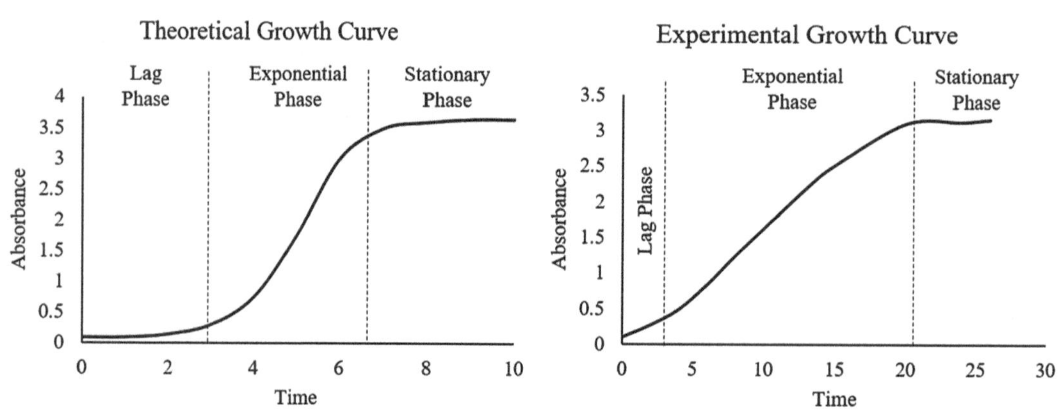

Fig. 1 Theoretical (right) and experimental (left) growth curves for microalgae. The arrows on each graph indicate the absorbance at which the stock culture should be used to inoculate wastewater samples. Time can be in hours or days depending on the type of microorganism being used

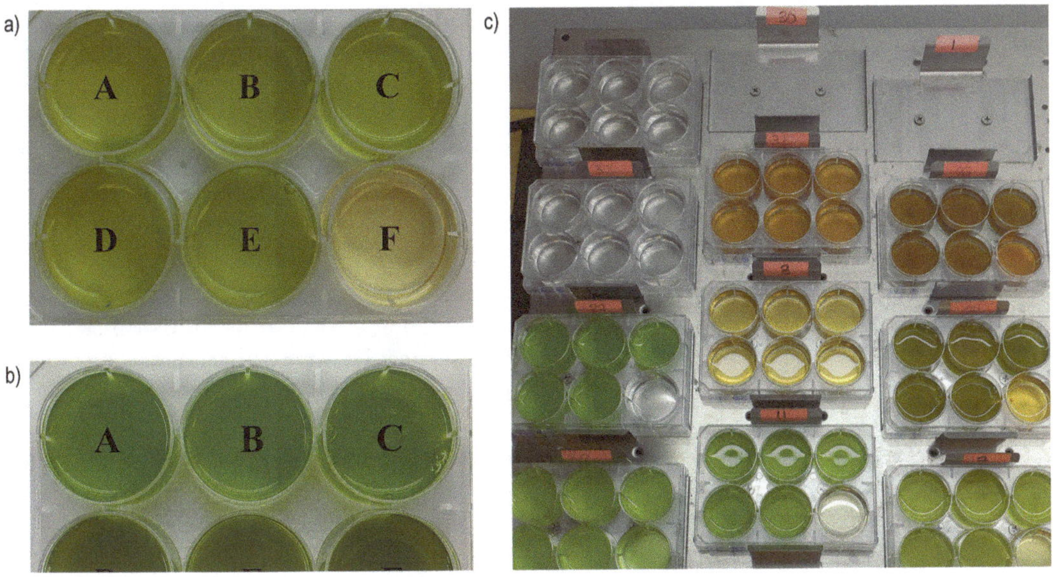

Fig. 2 (**a**) 6-well microplate design with five replicates; (**b**) 6-well microplate design 3 replicates; (**c**) Example microplate setup using 6-well microplates in an automated shaker

1. Pipet 9 mL of diluted wastewater media (or synthetic media for controls) into each preassigned well of the 6-well plate.
2. As shown in Fig. 2, label each well to identify the conditions that are being tested in those wells. Microplates can be prepared for a 5-replicate or 3-replicate culturing design:
   (a) *5-replicate microplate culturing design:* Wells A through E will be one condition (treatment, dilution, etc.) and F will be the medium blank. Prepare a separate plate (s) containing the controls for each experiment (*see* **Note 34** and Fig. 2a).
   (b) *3-replicate microplate culturing design:* Wells A through C will be one condition (treatment, dilution, etc.) and C through F will be a separate condition. Prepare separate plates containing blanks (at $n = 1$) and controls (at $n = 3$) for all conditions being tested (*see* Fig. 2b).

4.7 Batch-Scale Studies

All wastewater dilutions for batch-scale studies should be perform with deionized water or filtered seawater.

1. Pipet 225 mL of diluted wastewater media (or synthetic media for controls) into each labeled 500-mL culture flask (*see* **Note 12**).
2. Pipet 25 mL of the prepared microorganism culture into the 500-mL culture flask with the media and mix thoroughly.
3. Provide the culture with constant suffusion and mixing using caps equipped for aeration and a stir bar on a stir plate.

4.8 Controlling Culturing Conditions

1. *Lab-scale operations*: Light cycles and temperature can be controlled using grow lights set on a constant light–dark cycle (e.g., 16:8 or 12:12) using a timer and keeping the cultures in a temperature-controlled room (*see* **Note 35**). Photometers should be used to measure the intensity of light to promote optimal growth.
2. *Large-scale operations*: Light is often supplied solely from the sunlight (*see* **Note 36**). Depending on the climate and time of year, algae reactors can be heated or cooled to maintain optimal conditions (*see* **Note 37**). Large scale operations can be constructed in-ground so the heat transfer from outside conditions is limited, allowing for more stable temperature conditions.

4.9 Measuring Optical Density with a Spectrophotometer

The optical density of a culture is measured using a spectrophotometer to measure the absorbance. Cultures composed of more cells or larger cells will have a higher absorbance compared to cultures with less or smaller cells (*see* **Note 38**).

If using microplate culturing design (*see* **Note 39**):

1. Use a microplate reader to take efficient absorbance measurements.
2. Create an experimental method in the instrument's program to read at wavelengths characteristic to the organism of interest (e.g., 680 nm and 750 nm for microalgae) and select 6-well microplate in the method.
3. Place the microplate on the tray of a microplate reader and run the plate.
4. Use a microplate containing the appropriate blanks for each sample.

If using a batch culturing design:

1. Aliquot the necessary volume of sample into a cuvette.
2. Read the absorbance of the sample using a UV-Visible spectrophotometer at the indicated wavelength(s).
3. Use appropriate blanks for each sample.

4.10 Counting Cells Using a Hemocytometer

Counting cells using a hemocytometer is a way to measure an approximate number of cells per milliliter of media. A trained individual can physically visualize and identify between different kinds of cells resulting in more detail regarding cell growth patterns (*see* **Note 40**).

1. Aliquot 1 mL of well-mixed sample from the culture. The sample should be diluted to allow for a count of roughly 100 cells per red square on the hemocytometer (*see* **Note 41** and Fig. 3).

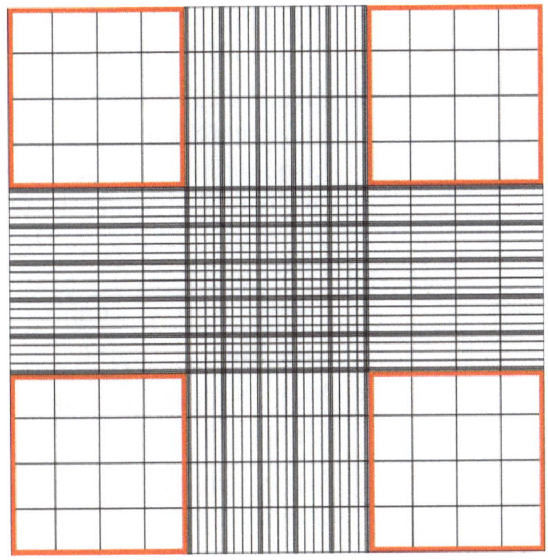

Fig. 3 Grid of a hemocytometer used to count cells (Adapted from: Zephyris at the English language Wikipedia, CC BY-SA 3.0, https:/commons.wikimedia.org/w/index.php?curid=10553213)

2. Add 10 μL of the diluted sample into the hemocytometer using a micropipette and capillary action. Figure 3 demonstrates the layout of a hemocytometer.
3. The number of cells in each of the 16 squares within each red square should be summed together to achieve an overall red square count (*see* **Note 42**).
4. **Step 1** should be repeated for at least three out of the four red squares. This provides you with at least three counts that can be averaged, and an overall cell count can be calculated.
5. Average the red square counts (*see* Fig. 3).
6. Calculations using the average of the red square counts:

$$\text{Cell density} = \text{Average\#of cells} \times 10,000 \times \text{Dilution factor} \quad (2)$$

4.11 Measuring Total Suspended Solids and Volatile Suspended Solids in Sample Culture

Total suspended solids (TSS) or the similar measure, sans ash, of volatile suspended solids (VSS) is a valuable measure of biomass concentration.

1. Pass a microbe-containing sample through a glass fiber filter and dry at 105 °C in an oven. The remaining mass on the filter is the TSS consisting of algae biomass and any ash contained within or from other particles or microorganisms.
2. Combust the filter paper at 550 °C using a muffle furnace and the remaining mass is only the fixed suspended solids (FSS) contained in the larger particles.

3. Calculate the VSS by taking the difference between the TSS mass and FSS mass.
4. Create a calibration curve as a plot of VSS versus the absorbance at 680 nm and add a line of best fit with an associated R-squared value. Use the equation of the linear regression line on the calibration curve to calculate VSS from their associated absorbance values (*see* **Note 43**).

4.12 Performing Pigment Extractions on Chlorophyll-Producing Microorganisms

1. Aliquot 1 mL of sample into a 2-mL Eppendorf tube.
2. Centrifuge at 13,450 × *g* for 10 min.
3. Remove supernatant and resuspend pellet in 1 mL of ultrapure water.
4. Repeat centrifugation step and remove supernatant for a total of two washes.
5. Centrifuge and remove supernatant.
6. Flash-freeze pellets in liquid nitrogen for 3 min.
7. Remove, allow to thaw, and then place on ice.
8. Add 1.5 mL of 100% methanol at 4 °C to each tube.
9. Vortex for at least 30 s and ensure the pellet is completely resuspended.
10. Place at 4 °C for at least 24 h or until all the pigments are extracted from the biomass (i.e., biomass pellet is no longer pigmented).
11. Centrifuge at 13,450 × *g* for 10 min and pipet 300 μL extract into a 96-well plate with three replicates; use methanol as the blank; do not resuspend.
12. Read on plate reader at the following spectral ranges:
    - 468–472 nm (step size: 2 nm).
    - 631–633 nm (step size: 1 nm).
    - 647–653 nm (step size: 1 nm).
    - 663–668 nm (step size: 1 nm).
    - 300–800 nm (step size: 5 nm).
13. *Calculations:* [38].

    Mackinney equation to calculate chlorophyll concentration:

$$\frac{\mu g_{\text{chlorophyll}}}{\text{mL}_{\text{medium}}} = \frac{(13.43(\text{Abs}_{665})(v))}{lV} \tag{3}$$

Where Abs_{665} = absorbance at 665 nm − absorbance at 750 nm; v = volume of solvent (mL), l = spectrophotometric path length, V = sample volume (mL).

Lichtenthaler equation to calculate carotenoid concentration:

$$\frac{\mu g_{\text{carotenoids}}}{\text{mL}_{\text{medium}}} = \frac{((1000)(\text{Abs}_{470}) - (44.76)(\text{Abs}_{666}))}{221} \quad (4)$$

Where Abs_{xxx} = absorbance at xxx nm − absorbance at 750 nm.

4.13 Evans Blue Viability Assay (See Note 44)

1. Aliquot 1 mL of sample into a 1.5-mL Eppendorf tube.
2. Centrifuge at 7250 × *g* for 5 min.
3. Remove supernatant and resuspend pellet in 100 μL 1% Evans blue dye.
4. Incubate at room temperature (25 °C) for 10–15 min.
5. Centrifuge at 7250 × *g* and wash pellet twice with ultrapure water.
6. Resuspend pellet in 100 μL of ultrapure water.
7. Use a hemocytometer to count the blue cells (i.e., dead cells) and record the number (*see* **Note 45**) (*see* Subheading 4.11 for details on how to count using a hemocytometer).
8. Calculation:

$$\frac{\text{Cells}}{\text{mL}} = \left(\frac{\text{Total \# cells counted}}{3}\right)(1 \times 10^4)(\text{DF}) \quad (5)$$

$$\%\text{Viability} = \left(\frac{\#\text{ dead}\frac{\text{cells}}{\text{mL}}\text{observed}}{\#\text{ total}\frac{\text{cells}}{\text{mL}}}\right) \times 100 \quad (6)$$

4.14 Harvesting Using Coagulation, Flocculation, and Sedimentation

1. Aliquot representative and well-mixed samples from the algae reactor of interest in equal volumes to multiple beakers or test tubes (*see* **Note 46**).
2. Prepare a 5–10 g/L stock solution of coagulant (e.g., $FeCl_3$ and $Al_2(SO_4)_3$) in sufficient volume for all experimental vessels.
3. Pipet the stock solution into each beaker or test tube to achieve a desired coagulant concentration in solution (*see* **Note 47**).
4. Vigorously mix each vessel during coagulant loading or immediately following it (preferably during loading). Vigorously mix for 1 min and follow with slower mixing for up to 30 min (*see* **Note 48**).
   - *For test tubes*: cap tightly with Parafilm or rubber stopper and shake vigorously for 1 min before letting settle for 1 h.
   - *For larger vessels* (e.g., *beakers*): Use stir bars or vertical impellers to accurately control mixing speeds.

5. Allow each vessel to settle for 1 h before taking a sample of the supernatant.
6. Measure the absorbance of each vessel at 680 nm and calculate the harvesting efficiency using the formula:

$$\text{Harvesting efficiency (\%)} = \frac{OD_i - OD_f}{OD_i} \times 100 \quad (7)$$

where OD_i and OD_f is the optical density at 680 nm initially and finally, respectively.
7. Repeat the process as needed to determine the lowest coagulant or flocculant dosage that produces acceptable harvesting efficiencies (at least 95%) (*see* **Note 49**).
8. Once an optimal coagulant dosage is determined, the chemical can be added in batches or on a continuous basis to the reactors of interest for larger scale harvesting.

4.15 Harvesting Using Centrifugation

1. Decant the supernatant and distribute equal volumes of the algae solution to appropriately sized centrifuge tubes.
2. Activate the centrifuge to rotate at appropriate speeds for its size and the volume of the centrifuge tubes. If possible, control the temperature inside the centrifuge to 4 °C. Spin for 20–30 min or until a tight algae pellet is formed.
3. Remove the tubes taking care to not disturb the algae pellet.
4. Decant the supernatant and collect the algae pellet for use in subsequent experiments or freeze-dry the algae pellet and store dried biomass at −20 °C.

5 Notes

1. Wastewater samples are typically diluted so interferences such as color and turbidity are reduced; therefore, the dilution factor should be used when determining if the concentration of a certain analyte may affect the assay.
2. Desired strains can be purchased from companies such as the Culture Collection of Algae at the University of Texas at Austin (UTEX), the Culture Collection of Algae and Protozoa (CCAP), and the Culture Collection of Algae at Göettingen University (SAG).
3. Using marine and freshwater microalgae does not necessarily follow the Redfield ratio (106:16:1 C/N/P) when grown on wastewater, so optimization of the practical nutrient ratio may be necessary [30].

4. The design of a proper cultivation system is also important to avoid creating an anaerobic, stagnant body of wastewater that poses a hazard to the environment.
5. Make note of the following regarding biomass measurements:
   - Bacterial colonies can cause interference when measuring biomass spectrophotometrically.
   - VSS is a nonspecific measure, so other microbes, such as bacteria, are also included.
6. The desired density usually lies between 0.3 g biomass/L and 5.0 g biomass/L.
7. Algae cells normally have a net negative charge and tend to repel each other, so the addition of cationic salts neutralize the surface charge and encourage floc formation.
8. In many cases cationic polymers have been shown to be effective at lower loadings than multivalent salts [39]. Various biopolymers such as chitosan and cationic starch have also been demonstrated for this purpose but may be cost-prohibitive on larger scale systems at the current time.
9. Bioflocculation techniques may reduce the risk of biomass contamination from synthetic or inorganic materials but limited information is currently confirmed about other advantages.
10. As the size of the rejected particles decreases, the pressure and energy required increase.
11. For each project, state agency requirements must be consulted, and the project designed accordingly.
12. Sterile conditions are recommended when making a sterile inoculum and when you are making growth media. Sterile conditions are also recommended if the experiments being run require that no other contamination besides what is present in the wastewater such as the case with an experimental control if necessary. To provide sterile conditions and to maintain an axenic stock culture, the media and culturing vessel should be autoclaved for at least 30 min at 150 psi and 121 °C and the culture should be prepared in a laminar flow hood to avoid contamination.
13. These are a few common synthetic medias available. A variety of culture mediums can be purchased from companies such as the Culture Collection of Algae and Protozoa (CCAP) and Sigma-Aldrich.
14. Grow lights are necessary for photosynthetic microorganisms and should be chosen based on the microorganism of interest. Variations of the grow light listed here are available from a variety of vendors.

15. The size of tubing used may vary depending on the aeration setup of the system, but the type/material of the tubing should be the same as that stated here.
16. At small scale, CO_2 and compressed air gas canisters may be used as the source and can further be attached to a gas blender to maintain a constant and accurate CO_2/air mixture.
17. The capability to hold the culture tubes and/or ampules is required only if regulated analysis methods are being used for wastewater analysis. Attachments can be purchased for spectrophotometers if needed, but if using a Hach kit for analysis, it is recommended to use a Hach spectrophotometer.
18. There are various other types of Hach kits available including variations of the ones described in this chapter. Choose the analysis kits that are most applicable to the possible wastewater composition that is being analyzed. Specifically, consider the working range of each assay as described in their associated method. There are high range (HR) and low range (LR) variations of some of these kits (*see* Table 4). Dilutions less than 1% are not desired.
19. Adaptors can be purchased from Hach Co. to accommodate smaller test tubes such as those used in the COD analysis kit.
20. If ultrapure water is not available, proper distillation or ion exchange procedures must be performed to obtain ultrapure water. This can be done using method AHPA 4500-NH_3 B.
21. This method is sensitive to mixing conditions in the tank. Therefore, samples of the entire column of water should be obtained and analyzed for solids contents, at least; relative differences less than 1% indicate acceptable mixing.
22. For both open channel and water column wastewater sources, it is desirable to obtain and compare samples at different time points to ensure that steady state conditions are being obtained in the reactor.
23. Construction of a sampling port and manual valve should be performed by qualified technicians. The sampling port should be a similar diameter opening as the pipe itself to encourage representative samples.
24. The duration of pipe flushing is system dependent and depends on the frequency of material movement. Systems with intermittent material transfer may be at a greater risk for solids settling than those with continuous movement.
25. This should be established by measuring the solids content as described when discussing large tank systems accessible only by port valves.

26. A variety of filters may be used to remove the particulate matter. Whatman No. 1 filters are adequate at removing large particulate matter while smaller pore sizes are desired when filtering out the smaller particulate matter. A filter with a 0.45 μm pore size is a common filtration pore size when aiming to remove the remaining particulate matter for analysis purposes. Further, sterile filtration can be used if native bacterial species are not desired when growing the microorganisms on the wastewater. Filtration may take more than one pass through a filter to obtain a more fully filtered sample.
27. The dilution factors vary widely depended on the wastewater being used, so these can pose as a possible initial dilution and a dilution series can be created around it and adjusted as necessary. Common dilutions for each test include:
    - TN = 10–25%.
    - COD = 10–25%.
    - TP = 1–2.5%.
    - NH_3 = 1–2.5%.
28. Small-scale isotherm equilibrium experiments may be run with varying amounts of activated carbon to determine the optimal loading for color removal. Biochar may also be used in place of activated carbon.
29. There a variety of other tests accepted by federal regulations not mentioned here but can be found in the Standard Methods for the Examination of Water and Wastewater [3]. There are also various regulated methods to measure other trace elements in wastewater such as iron (AHPA 3500-Fe) and calcium (AHPA 3500-Ca B). Many of these ions are best measured using instrumental methods such at atomic absorption, IC, or ICP/MS.
30. The volume of media needed to start a culture depends on the strain. Cyanobacterial strains may have to be initially cultured from a plate into 10 mL of synthetic media. On the other hand, microalgae strains such as *Chlorella sorokiniana* and *Phaeodactylum tricornutum* can be cultured from a plate into over 2 L of synthetic media. It is recommended that a culture initially be prepared using 250 mL of media and adjusted accordingly.
31. A 10% inoculation volume is adequate for oleaginous yeast and algae species, but a 1% inoculation volume can be used for fast growing bacteria.
32. The wavelength and measurement intervals may vary depending on species of microorganism being used. For example, yeast growth may be measured every 2 h rather than every other day as is the case for slower growing microorganisms such as microalgae.

33. When culturing in wastewater, an initial dilution of at most 50% wastewater or lower should be used. Subsequent culturing experiments can be adjusted accordingly depending on if the species/strain being cultured grows.
34. When using a 5-replicate microplate design, the blank for that condition is in the same plate as the sample replicates. Because of this, it is easier for the blank to become contaminated with the sample, so it is necessary to monitor the blank for this contamination throughout the experiment.
35. The majority of algae species grow well between 16 and 22 °C [40].
36. While light control and supplementation can be considered, this will add considerable cost in large-scale setups.
37. An opportunity to minimize additional temperature control costs could be realized through efficiently capturing waste heat from other unit operations.
38. When measuring cell density, it is assumed that only the microorganisms of interest are being measured and that the cells are healthy and physically the same size and shape. Other measurements may be necessary to support the absorbance data such as an Evans Blue assay (*see* Subheading 4.14) to measure cell death or with microscopy imaging using a light microscope or a transmission electron microscope to understand the viability and characteristics of the cells.
39. Synergy HI Multi-Mode Plate Reader is an adequate example of the instrument to use for measuring optical density with a microplate design. A hemocytometer could also be used as a method to monitor biomass growth.
40. This method is subjective and highly prone to human error, so a standardized method of counting cells must be established within each respective laboratory to ensure repeatability and remove possible bias. If available, a flow cytometer could be a less subjective option to measure cell growth compared to individual cell counting.
41. A dilution factor is likely necessary depending on the concentration of the cells in the sample.
42. If the cells are on the border of the squares, only count those on the bottom and left borders of each of the 16 squares making up the red square to avoid double counting cells and to maintain consistency.
43. Be sure to only use this relationship for absorbance values measured within the range of the calibration curve. Values outside of the range require interpolation and they may be outside of the linear region resulting in inaccurate results.

44. Optimization must be performed to ensure the effectiveness and accuracy of its assay.
45. The sample should be diluted to allow for a count of roughly 100 cells/square.
46. Larger volumes of algae samples are preferred, when possible. Jar tests are usually performed with volumes as high as 1 L to better simulate the process on larger scales and decrease inaccuracies in loading that can be caused from working with small volumes.
47. Use the relationship $C_1V_1 = C_2V_2$ where C_1 is the concentration of the stock solution, V_1 is the volume of stock to load to the beaker, C_2 is the desired coagulant concentration in each test tube or beaker, and V_2 is the final total volume of solution in each test tube or beaker.
48. Mixing times and speeds should be controlled in the laboratory environment to be as similar as possible to those expected on larger scales. "Gang stirrer" style machines are very useful to ensure each reactor experiences exactly the same mixing conditions.
49. A useful strategy initially is to use a large coagulant range with few test vessels to create a general idea of the optimal dosage. Then, repeat the loading experiments with finer resolution on both sides of the first dosage determined.

References

1. Davis R, Kinchin C, Markham J, et al (2014) Process design and economics for the conversion of algal biomass to biofuels: Algal biomass fractionation to lipid and carbohydrates-derived fuel products. NREL/TP-5100-62368, Golden, CO.
2. IEA Bioenergy (2017) State of technology review–algae bioenergy. Report for IEA Bioenergy funded by IEA Bioenergy: Task 39, Golden, CO.
3. American Public Health Association (ed) (2017) Standard methods for the examination of water and wastewater,. 23rd ed. American Public Health Association, Washington, D.C.
4. Kumar KS, Dahms H, Won E et al (2015) Microalgae–a promising tool for heavy metal remediation. Ecotoxicol Environ Saf 113:329–352
5. Akpor OB (2014) Heavy metal pollutants in wastewater effluents: sources, effects and remediation. Adv Biosci Bioeng 2:37
6. Clausen CA (2000) Isolating metal-tolerant bacteria capable of removing copper, chromium, and arsenic from treated wood. Waste Manag Res 18:264–268
7. Ahluwalia SS, Goyal D (2007) Microbial and plant derived biomass for removal of heavy metals from wastewater. Bioresour Technol 98:2243–2257
8. Kobayashi N, Noel EA, Barnes A et al (2013) Characterization of three *Chlorella sorokiniana* strains in anaerobic digested effluent from cattle manure. Bioresour Technol 150:377–386
9. Chi Z, Zheng Y, Jiang A, Chen S (2011) Lipid production by culturing oleaginous yeast and algae with food waste and municipal wastewater in an integrated process. Appl Biochem Biotechnol 165:442–453
10. Nayak M, Karemore A, Sen R (2016) Performance evaluation of microalgae for concomitant wastewater bioremediation, CO_2 biofixation and lipid biosynthesis for biodiesel application. Algal Res 16:216–223
11. González LE, Cañizares RO, Baena S (1997) Efficiency of ammonia and phosphorus removal from a Colombian agroindustrial wastewater by the microalgae *Chlorella vulgaris*

and *Scenedesmus dimorphus*. Bioresour Technol 60:259–262

12. Choi H-J (2016) Dairy wastewater treatment using microalgae for potential biodiesel application. Environ Eng Res 21:393–400
13. Sharma KK, Schuhmann H, Schenk PM (2012) High lipid induction in microalgae for biodiesel production. Energies 5:1532–1553
14. Chen B, Wan C, Mehmood MA et al (2017) Manipulating environmental stresses and stress tolerance of microalgae for enhanced production of lipids and value-added products–a review. Bioresour Technol 244:1198–1206
15. Wenbiao SH, Renjie J, Abd T (2016) Optimization of aeration for biodiesel production by *Scenedesmus obliquus* grown in municipal wastewater. Bioprocess Biosyst Eng 39:1073–1079
16. Hodaifa G, Martínez ME, Sánchez S (2008) Use of industrial wastewater from olive-oil extraction for biomass production of *Scenedesmus obliquus*. Bioresour Technol 99:1111–1117
17. Mata TM, Melo AC, Meireles S et al (2013) Potential of microalgae *Scenedesmus obliquus* grown in brewery wastewater for biodiesel production. Chem Eng Trans 32:901–906
18. Hodaifa G, Martínez ME, Sánchez S (2010) Influence of temperature on growth of *Scenedesmus obliquus* in diluted olive mill wastewater as culture medium. Eng Life Sci 10:257–264
19. Kaewkannetra P, Enmak P, Chiu T (2012) The effect of CO_2 and salinity on the cultivation of *Scenedesmus obliquus* for biodiesel production. Biotechnol Bioprocess Eng 17:591–597
20. Kong QX, Li L, Martinez B et al (2010) Culture of microalgae *Clamydomonas reinhardtii* in wastewater for biomass feedstock production. Appl Biochem Biotechnol 160:9–18
21. Chaiklahan R, Chirasuwan N, Siangdung W et al (2010) Cultivation of *Spirulina platensis* using pig wastewater in a semi-continuous process. J Microbiol Biotechnol 20:609–614
22. Kulkarni SD, Auti T, Saraf S (2016) Bioremediation study of dairy effluent by using *Spirulina platensis*. Res J Life Sci Bioinform Pharm Chem Sci 1:317–326
23. Panyakampol J, Cheevadhanarak S, Sutheeworapong S et al (2015) Physiological and transcriptional responses to high temperature in *Arthrospira* (*Spirulina*) *platensis* C1. Plant Cell Physiol 56:481–496
24. De Oliveira MACL, Monteiro MPC, Robbs PG, Leite SGF (1999) Growth and chemical composition of *Spirulina maxima* and *Spirulina platensis* biomass at different temperatures. Aquac Int 7:261–275
25. Gupta N, Manikandan NA, Pakshirajan K (2017) Real-time lipid production and dairy wastewater treatment using *Rhodococcus opacus* in a bioreactor under fed-batch, continuous and continuous cell recycling modes for potential biodiesel application. Biofuels 7269:1–7
26. Goswami L, Tejas Namboodiri MM, Vinoth Kumar R et al (2017) Biodiesel production potential of oleaginous *Rhodococcus opacus* grown on biomass gasification wastewater. Renew Energy 105:400–406
27. Gonzalez-Garcia Y, Hernandez R, Zhang G et al (2013) Lipids accumulations in *Rhodotorula glutinis* and *Cryptococcus curvatus* growing on distillery wastewater as culture medium. Environ Prog Sustain Energy 32:69–74
28. Meesters PAEP, Huijberts GNM, Eggink G (1996) High-cell-density cultivation of the lipid accumulating yeast *Cryptococcus curvatus* using glycerol as a carbon source. Appl Microbiol Biotechnol 45:575–579
29. Meng X, Yang J, Xu X et al (2009) Biodiesel production from oleaginous microorganisms. Renew Energy 34:1–5
30. Pidou M, Ometto F, Whitton R et al (2016) Influence of microalgal N and P composition on wastewater nutrient remediation. Water Res 91:371–378
31. U.S. Environmental Protection Agency (2010) Part 437–the centralized waste treatment point source category. U.S. Environmental Protection Agency, Washington, DC, pp 359–380
32. Kumar K, Banerjee D, Das D (2014) Carbon dioxide sequestration from industrial flue gas by *Chlorella sorokiniana*. Bioresour Technol 152:225–233
33. Van Der Bruggen B, Vandecasteele C (2003) Removal of pollutants from surface water and groundwater by nanofiltration: overview of possible applications in the drinking water industry. Environ Pollut 122:435–445
34. U.S. Environmental Protection Agency (2010) Technology-Based Effluent Limitations. In: National Pollutant Discharge Elimination System (NPDES) Permit Writers' Manual. Washington, DC, pp 1–49
35. Bacellar Mendes LB, Vermelho AB (2013) Allelopathy as a potential strategy to improve microalgae cultivation. Biotechnol Biofuels 6:152
36. (ABO) Algae Biomass Organization (2015) Industrial algae measurements. (ABO) Algae Biomass Organization, Preston, MN

37. Das P, Thaher MI, Abdul Hakim MAQM et al (2016) Microalgae harvesting by pH adjusted coagulation-flocculation, recycling of the coagulant and the growth media. Bioresour Technol 216:824–829
38. Henriques M, Silva A, Rocha J (2007) Extraction and quantification of pigments from a marine microalga: a simple and reproducible method. Commun Curr Res Educ Top Trends Appl Microbiol 2:586–593
39. Milledge J, Heaven S (2013) A review of the harvesting of microalgae for biofuel production. Rev Environ Sci Biotechnol 12:165–178
40. Coutteau P, Lavens P, Sorgeloos P (1996) Manual on the production and use of live food for aquaculture. FAO Tech Pap 1–47
41. Beuckels A, Smolders E, Muylaert K (2015) Nitrogen availability influences phosphorus removal in microalgae-based wastewater treatment. Water Res 77:98–106
42. Ji MK, Abou-Shanab RAI, Kim SH et al (2013) Cultivation of microalgae species in tertiary municipal wastewater supplemented with CO_2 for nutrient removal and biomass production. Ecol Eng 58:142–148

Chapter 20

Volatile Fatty Acid Production from Anaerobic Digestion of Organic Residues

Sibel Uludag-Demirer, Wei Liao, and Goksel N. Demirer

Abstract

Short chain volatile fatty acids (VFAs) from acetic acid (C2) to valeric acid (C5) are important starting chemicals for chemical industry. The production of VFAs from rejected resources (organic residues) using self-sustaining technologies has an exciting potential in supporting the US chemical industry to achieve the goal that 20% of chemicals produced in the USA will be bio-based. Acidogenic anaerobic digestion as a robust, well-established, and versatile biological technology can be applied as an alternative approach for the valorization of organic residues (municipal, agricultural, and industrial wastes) by the production of VFAs. In a typical acidogenic anaerobic digestion operation, residue type, pretreatment, reactor operation, and VFA recovery are the key factors that influence VFA production. This chapter discusses these factors and provides an experimental approach of VFA production from organic residues.

Key words Acidogenic anaerobic digestion, Organic residue, Volatile fatty acids, Liquid–liquid extraction

1 Introduction

Anaerobic digestion is an established technology for the treatment of domestic, industrial and agricultural residues, which also generates value added products such as methane, hydrogen, and organic acids [1]. Many classes of bacteria play important role in this complex biological process, which is composed of sequential steps, namely; hydrolysis, acidogenesis, acetogenesis, and methanogenesis. These steps can be engineeringly separated into two phases to enhance digestion efficiencies and produce different chemical products. In the first phase, anaerobic bacteria generate VFA metabolites as value-added chemicals via steps of hydrolysis, acidogenesis, and acetogenesis. The VFA intermediates are further converted to methane and carbon dioxide by archaea in the second phase (methanogenesis, or "methane fermentation") to achieve waste stabilization.

Venkatesh Balan (ed.), *Microbial Lipid Production: Methods and Protocols*, Methods in Molecular Biology, vol. 1995, https://doi.org/10.1007/978-1-4939-9484-7_20,

It is apparent that acidification phase of the anaerobic digestion is the target stage for VFA production [2]. During the acidification phase, the hydrolysis first takes place to convert various carbon substrates to energy (ATP), reducing agents, and monomers. It has been widely reported that *Bacteroidetes* are a key microbial group that plays a substantial role in the hydrolysis stage to degrade primary carbon polymers [3, 4]. The intermediate compounds from the hydrolysis step are then fermented by acidogens and acetogens through the glycolytic pathway to produce short chain VFAs (acetic, propionic, and butyric acids). *Synergistetes* and *Clostridium* are the representative bacteria in anaerobic digestion process that carry out such transformations [3, 5, 6]. However, during conventional anaerobic digestion process, VFAs as the substrates for archaeal metabolism are used for methane production. In order to accumulate them, digestion conditions need to be adjusted to inhibit or eliminate methanogens during the digestion.

1.1 Effects of Organic Wastes and Their Pretreatment in VFA Production

A large variety of organic residues has been investigated for VFA production. They include both solid and liquid streams, particularly the organic fraction of municipal solid waste, municipal sludge (both primary and waste activated sludge), food and food processing wastes, crop residues, and livestock wastes. For instance, organic wastes from the sugar industry (beet pulp) and wastewater mixture was processed in a continuously mixed anaerobic acidification process, which resulted in 43.8–52.9% acidification of initial chemical oxygen demand (COD) [2]. The sulfite spent liquor from the pulp and paper industry was used by a completely mixed stirred tank reactor (CSTR) to produce VFAs [7]. The maximum VFA concentration reached 7.0 g COD equivalent/L with the degree of acidification of 36%. Various mixtures of different organic wastes had been studied aiming to promote a higher production of VFA. Dogan et al. [8] and Erguder and Demirer [9] evaluated the potential of high-rate anaerobic acidification of high solids containing substrates (organic fraction of municipal solid waste and cow manure) for the VFA production in leaching bed reactors (LBRs). The maximum VFA concentration (mainly acetic and butyric acids) was around 6 g/L (acetic acid equivalent). Ma et al. [10] mixed potato peel waste, food waste, and waste activated sludge to run a cofermentation to improve anaerobic acidification of VFA production. The results revealed that acetate was the most dominant VFA, making up 41.3–57.6% of the total VFAs produced during acidification under alkaline condition. The concentrations of propionate and valerate decreased dramatically once more potato peel waste was added. It was also observed that starch in the food wastes favored butyric acid accumulation, while increase of lipids and protein in the feed enhanced the synthesis of valerate and propionate.

Due to the complex composition of organic residues and wastes such as sludge, food waste, and organic fraction of municipal solid waste, hydrolysis is the rate-limiting step for the anaerobic digestion. Various pretreatment methods have been explored to enhance the solubilization of these wastes, which include chemical, biological, thermal, physical, and combined approaches [11]. Chemical methods use acid and alkali, or strong oxidizing agent (i.e., ozone and hydrogen peroxide) to achieve the solubilization of polymeric substances in organic residues and wastes. Biological pretreatment applies enzymes and microbes that degrade cell walls and large polymer and consequently realize the solubilization of the residues and wastes. Compared to chemical and biological pretreatment, thermal treatment is a simple and economical method to achieve the solubilization by breaking hydrogen and other chemical bonds of the substances in the residues and wastes. Ultrasonic treatment as a physical process has also been intensively studied to break down polymeric structure of substances. In addition, the pretreatment efficiency of the organic wastes was improved using the combination of different pretreatment methods. For example, alkali and ultrasonics were integrated to treat food wastes; hydrogen peroxide and microwave were combined to treat municipal solid wastes.

1.2 VFA Accumulation by Acidogenic Anaerobic Digestion

Anaerobic acidification relies on the establishment of environmental and operational conditions that favor growth of acidogenic microbes, while preventing methanogenic activity. There are many methods to achieve anaerobic acidification, such as chemical inhibition of methanogens, heat treatment of sludge, inoculation of bacteria, and adjustment of digestion parameters [11]. Considering both technical soundness and economical feasibility, the adjustment of digestion parameters is a practical approach to realize anaerobic acidification. It has been reported that pH, solids retention time (SRT), temperature, and organic loading rate (OLR) are the key operational parameters that influence anaerobic VFA production (Table 1) [11, 12].

Preferred pH values for the VFA production vary with the type of wastes and are in the range of 5.25–11.0 [13]. Alkaline pH (from 8 to 11) is preferred once municipal sludge is used as the feed [14]. The alkaline condition enhances hydrolysis of the sludge through ionization of the charged groups (e.g., carboxylic groups) of the extracellular polymeric substances in the sludge. Neutral pH values (around 7) lead to high solubilization of carbohydrates, proteins, and lipids, and consequently good VFA production from kitchen wastes [15]. pH control influences the composition of VFA mixtures produced in anaerobic acidification [16]. pH 6–6.5 favored accumulation of acetic and butyric acids from dairy wastewater, while propionic acid production was observed in the pH range of 4–4.5 [17, 18]. As for glucose-rich medium, acidic and alkaline conditions were favorable for accumulation of acetic and butyric acids, respectively [19].

Table 1
VFA concentrations and production yields (modified from Lee et al. [11] and Cavinato et al. [12])

Feedstock	HRT (d)	OLR (kg VS/m^3·d)	Temperature (°C)	VFA Conc. (g VFA/L)	VFA yield (g VFA/kg COD)	Reference
Cheese whey	2.1	–	37	0.84	–	[54]
OFMSW	3.5	16	55	13.8	263	[36]
Food waste	3.0	16.8	55	12.3	221	[34]
Fruit/vegetable waste	1–11	11–102	37	7.6–28.5	33–279	[30]
Food waste + sludge	8.92	8.31	35	24.1	–	[55]
OFMSW+WAS	3.3	18	55	8.1	137	[56]
Palm oil mill effluent	4.0	–	30	15.3	–	[57]
OFMSW	6	4.1	23	9.5	127	[26]
Gelatin-rich proteinaceous Wastewater	0.5	–	37	1.57	–	[31]
WAS	4.6–5.9	1.2–1.9	35	3.2–7.5	207–325	[29]
Sugar industry wastewater + pressed beet pulp	2	–	35	3.64	–	[2]
Olive oil mill effluent	1.4	–	25	10.7	–	[58]

It has been reported that reducing solids retention time (SRT) can actively promote the production of VFA [20, 21]. Shorter SRTs limit the growth of methanogens in anaerobic digestion, so that acidogenic bacteria dominate the culture resulting in accumulation of VFAs [22, 23]. Numerous studies demonstrated that 2–15 days are the preferred SRTs on a variety of feedstocks to increase VFA production [2, 16, 18, 24, 25].

Temperature has a significant impact on anaerobic acidification. It has been reported that VFA production from waste occurred at a wide temperature range from psychrophilic (4–20 °C) [26–28] to mesophilic (20–50 °C) [7, 9, 17, 28–31] to thermophilic (50–60 °C) [32–34] and extreme/hyperthermophilic (60–80 °C) [24, 35] conditions. It has been demonstrated that increasing the temperature within the psychrophilic and mesophilic conditions is generally beneficial for the accumulation of VFAs.

Organic loading rate (OLR) is another important operational parameter for anaerobic VFA production [17, 22, 36] and it is linked with type of organic residues [11]. The VFA production from starch-based wastewater was increased proportionally with the increase of OLR from 1 g COD/L/d to 32 g COD/L/d

[37]. While, the VFA production from olive oil mill residue in the OLR range of 3.2–15.1 g COD/L/d demonstrated that the highest VFA concentration was obtained at the OLR of 12.9 g COD/L/d [38]. It has also been reported that feeding frequency influences the VFA production [11].

The aforementioned operational parameters have been combined to synergistically enhance VFA accumulation. Sans et al. [39] studied the effects of different OLRs (22.4–85.2 kg VS/m^3/d), retention times (8 h to 8 days), and sludge recirculation (related with SRT) on mesophilic (37 °C) anaerobic acidification of solid food wastes. The results showed that the highest VFA concentration of 23.1 g/L was reached with retention time of 6 days and high OLRs without sludge recirculation. Acetic and butyric acids are the major VFAs in this study. Dogan and Demirer [25] investigated the combined effects of OLR and pH on VFA production from organic fraction of municipal solid waste using completely stirred tank reactors (CSTRs). The study concluded that the maximum VFA production was observed at OLR of 20 g VS/L/d and HRT of 2 days with a pH value of 5.5. Meanwhile, they also demonstrated that selective production of organic acids was possible during the digestion by controlling the OLR and pH. Higher OLRs and lower pH enhanced the acetic acid production and reduced butyric acid production. Jankowska et al. [16] investigated the impact of pH (4–12) and retention times (5–15 days) on anaerobic acidification of primary sludge and waste activated sludge from a municipal wastewater treatment plant. The conditions of the short retention time (5 days) with low pH (4) and the long retention time (15 days) with high pH (10) were in favor of VFA accumulation. The highest VFA concentration was achieved from the condition of 15 days retention time at pH 10 (0.62 g/g VS). Jiang et al. [17] studied the effects of pH, temperature, and organic loading rate (OLR) on anaerobic acidification of food waste. The results indicated that under the culture conditions of low OLR (5 g TS/L/d), slightly acidic pH (6), and mesophilic temperature (35–45 °C), the acidification of food waste produced 39.5 g/L of VFAs with a yield of 0.504 g VFAs/g VS fed. Acetic and butyric acids were the predominant acids accounting for 77% of total VFAs.

1.3 Recovery of VFA from the Anaerobic Digestion Broth

Due to the complicated composition of the broth of anaerobic acidification, the major technical and economic barrier of using VFAs is their recovery, separation, and purification [40–43]. Numerous studies have been conducted to address this challenge [41, 44]. Techniques such as electrodialysis [45], ion exchange [42], adsorption [46], and liquid–liquid extraction [47] have been intensively investigated to recover VFAs from the broth. Liquid–liquid extraction among these methods is widely considered an efficient, economical, and environmentally sound method for the recovery of carboxylic acids [48]. The efficiency of VFA recovery during the liquid–liquid extraction highly depends on the nature of the VFAs, the concentration of the solvent (extractant), the type of diluent [49], and pH [50]. Alcohols, ketones, ethers, aliphatic hydrocarbons, and organophosphates are the solvents widely used for the extraction. Compared to the other solvents, organophosphates such as trioctylphosphine oxide (TOPO) and tri-n-butyl phosphate (TBP) have much larger distribution coefficient for carboxylic acids [51], so they are the better solvents for VFA extraction and recovery. Alkaya et al. [52] investigated the anaerobic acidification of sugar beet processing wastes and subsequent liquid–liquid extraction of the resulted VFAs. The influence of pH and the extractant (TOPO in kerosene) concentrations on the recovery of VFAs from fermentation broth were evaluated. The results indicated that TOPO concentration and pH were the crucial factors that influence VFA extraction efficiency. pH 2.5 was determined as the optimal pH for the extraction. Increase of TOPO concentration in kerosene significantly improved the extraction efficiency. Higher VFA recoveries (60.7–97.6%) were achieved at 20% TOPO in kerosene at pH 2.5.

2 Materials

2.1 Feedstock

The wastes and wastewater could be animal manure/wastewater, food wastes, municipal sewage sludge, and industrial organic wastes/wastewater. The feedstocks need to have a moisture content 85% or more.

2.2 Basal Medium (BM)

The BM contained the following chemicals to supply the necessary nutrients for optimum anaerobic microbial growth (concentrations are given in parentheses):NH_4Cl (1200 mg/L), $MgSO_4{\cdot}7H_2O$ (400 mg/L), KCl (400 mg/L), $Na_2S{\cdot}9H_2O$ (300 mg/L), $CaCl_2{\cdot}2H_2O$ (50 mg/L), $(NH_4)_2{\cdot}HPO_4$ (80 mg/L), $FeCl_2{\cdot}4H_2O$ (40 mg/L), $CoCl_2{\cdot}6H_2O$ (10 mg/L), KI (10 mg/L), $MnCl_2{\cdot}4H_2O$ (0.5 mg/L), $CuCl_2{\cdot}2H_2O$ (0.5 mg/L), $ZnCl_2$ (0.5 mg/L), $AlCl_3{\cdot}6H_2O$ (0.5 mg/L), $NaMoO_4{\cdot}2H_2O$ (0.5 mg/L), H_3BO_3 (0.5 mg/L), $NiCl_2{\cdot}6H_2O$ (0.5 mg/L), $NaWO_4{\cdot}2H_2O$ (0.5 mg/L), and cysteine (10 mg/L) [53].

3 Methods

VFA production from organic residues requires bench or pilot scale tests to evaluate the acidification performance under different digestion conditions (pH, temperature, organic loading rate, and hydraulic and solids retention time). Typical steps for such a feasibility and performance study are described as follows:

3.1 Characterization of Feedstock

1. Homogenize the feedstock (liquid) using a commercial blender to vigorously mix prior to its compositional analyses (*see* **Note 1**).
2. Analyze the feedstock to determine the total and soluble COD, total and total volatile solids (TS and TVS), total Kjeldahl nitrogen (TKN), total phosphorus (TP), total ammonium nitrogen (TAN), pH, and alkalinity using the standard methods [24] (*see* **Note 2**).
3. Store the feedstock at −20 °C until it is used.

3.2 Inoculation

1. Obtain a mixed anaerobic culture from an anaerobic digester operating in full scale (*see* **Note 3**).
2. Allow the culture to settle down for 24 h to increase the number of microorganisms.
3. Discard the supernatant and use the settlings to inoculate the digester.
4. Use the basal medium as a part of daily feeding in addition to the feedstock (substrate) to provide micronutrients and macronutrients, minerals, and vitamins for microorganism activities.
5. Make the total solids content of the solution in the digester and the feed at 5% (w/w).

3.3 Operation of Anaerobic Reactors and Monitoring

1. Decide the operational mode (batch, semicontinuous, or continuous) of the digester and decide a feeding schedule based on the operational mode.
2. Specify the reactor operation such as, hydraulic retention time, pH, organic loading rate, temperature, mixing rate (*see* **Note 4**).
3. Monitor biogas production, pH, VFA concentration, soluble COD, TS, and TVS during the test (*see* **Note 5**).
4. Establish a steady-state condition of the digestion following the start-up period regarding stabilization of biogas production and sCOD reduction.
5. Monitor sCOD reduction to VFAs (degree of acidification) in the steady-state digestion to evaluate performance of the acidification (*see* **Note 6**).

3.4 Liquid–Liquid Extraction of VFA from Anaerobic Broth

1. Prepare 20% by weight trioctylphosphine oxide (TOPO) solution in kerosene at 30 °C.
2. Collect the supernatant from the anaerobic digester and adjust pH to 2.5.
3. Mix the collected supernatant from the anaerobic digester with TOPO in kerosene with a volumetric ratio of 1:3.
4. Mixed vigorously the supernatant and solvent for 5 min and then allowed to separate by gravity for 3 min (This cycle needs to repeat three times to improve the extraction efficiency).

5. Analyze VFAs left in the aqueous phase after the extraction using gas chromatography (GC) (*see* the **Note 5**).
6. Calculate the partition of VFAs in aqueous and organic solvent and report the extraction efficiency (*see* **Note 7**).

4 Notes

1. The blended waste and wastewater will be resuspended to pass a screen (1–2 mm opening). The solution collected after the screening is used as the substrate to carry out the culture.
2. Standard methods for wastewater analyses are adopted in this part. The total and soluble CODs are measured by a closed reflux colorimetric method (Method no 5220-D). TS and TVS concentrations are measured gravimetrically following Methods 2540B and 2540E, respectively. Total Kjeldahl nitrogen (TKN) is measured using Method 4500-N_{org}-B, total phosphorus (TP) is measured by Method 4500-P, and total ammonium nitrogen (TAN) concentration is measured by Method 4500-NH_3. pH is measured by electrochemical method (Method 4500-H+), and alkalinity is measured by titration method (Method 2320).
3. Anaerobic digestate from municipal/industrial wastewater treatment plants can be used as the seed.
4. All reactors should be operated in triplicates for statistical interpretations of the results.
5. The gas production in the digesters is measured by water displacement setup. The biogas compositions are measured via GC equipped with Thermal Conductivity Detector (TCD). Methane (CH_4), nitrogen (N_2), carbon dioxide (CO_2), oxygen (O_2), and hydrogen (H_2) are separated with two series of columns with injector and detector temperature of 50 °C and 80 °C, respectively. Helium gas is used as the carrier at 100 kPa constant pressure 45 °C of oven temperature. VFA measurements are carried out using GC with flame ionization detector (FID). The samples are filtered through 0.45 μm online filters, and their pH are lowered to 2.5 by formic acid prior to their injection to GC column. The GC oven temperature is increased from 100 °C to 200 °C by 8 °C/min ramping. Inlet and detector temperatures are 250 and 280 °C respectively. Helium is again used as a carrier gas with a flow rate of 6 mL/min.
6. Degree of acidification = $(S_f/S_i)\times 100$, where S_i is the initial concentration of sCOD (mg/L), S_f is the produced VFAs expressed as theoretical equivalents of sCOD (mg/L) concentrations.

7. Partition of VFAs in organic solvent and aqueous phases at equilibrium can be calculated using the equation of $K_d = ([VFA]_{org})/([VFA]_{aq})_{eq}$, where $[VFA]_{org}$ and $[VFA]_{aq}$ are the concentrations of VFA in organic solvent and water, respectively. K_d is the partition coefficient without a unit.

References

1. Baumann I, Westermann P (2016) Microbial production of short chain fatty acids from lignocellulosic biomass: current processes and market. Biomed Res Int 2016:8469357
2. Alkaya E, Demirer GN (2011) Anaerobic acidification of sugar-beet processing wastes: effect of operational parameters. Biomass Bioenergy 35:32–39
3. Chen R, Murillo-Roos M, Zhong Y, Marsh T, Bustamante M, Hernandez W, Uribe L, Uribe-Lorio L, Kirk D, Reinhold D, Miranda A, Baudrit-Buiz D, Aguilar J, Rodriguez W, Srivastava A, Liao W (2016) Responses of anaerobic microorganisms to different culture conditions and corresponding effects on biogas production and solid digestate quality. Biomass Bioenergy 85:84–93
4. Yang F, Chen R, Yue Z, Liao W, Marsh T (2016) Phylogenetic analysis of anaerobic co-digestion of animal manure and corn stover reveals linkages between bacterial communities and digestion performance. Adv Microbiol 6:879–897
5. Kondo K, Beppu T, Horinouchi S (1995) Cloning, sequencing, and characterization of the gene encoding the smallest subunit of the three-component membrane-bound alcohol dehydrogenase from Acetobacter pasteurianus. J Bacteriol 177:5048–5055
6. Rojas-Sossa JP, Murillo-Roos M, Uribe L, Uribe-Lorio L, Marsh T, Larsen N, Chen R, Miranda A, Solis K, Rodriguez W, Kirk D, Liao W (2017) Effects of coffee processing residues on anaerobic digestion microorganisms and corresponding digestion performance. Bioresour Technol 245:714–723
7. Queirós D, Sousa R, Pereira S, Serafim LS (2017) Valorization of a pulp industry by-product through the production of short-chain organic acids. Fermentation 3:20
8. Dogan E, Dunaev T, Erguder TH, Demirer GN (2008) Performance of leaching bed reactor converting the organic fraction of municipal solid waste to organic acids and alcohols. Chemosphere 74:797–803
9. Erguder TH, Demirer GN (2016) Organic acid production from the organic fraction of municipal solid waste and cow manure in leaching bed reactors. Environ Eng Manag J 15:2487–2495
10. Ma H, Liu H, Zhang L, Yang M, Fu B, Liu H (2017) Novel insight into the relationship between organic substrate composition and volatile fatty acids distribution in acidogenic co-fermentation. Biotechnol Biofuels 10:137
11. Lee WS, Chua ASM, Yeoh HK, Ngoh GC (2014) A review of the production and applications of waste-derived volatile fatty acids. Chem Eng J 235:83–99
12. Cavinato C, Frison N, Herrero N, Gottardo M, Ros C, Strazzera G, Cherubin A, Fatone F, Pavan P, Bolzonella D (2017) Volatile fatty acids production from organic waste for biorefinery platforms. In: 5th international conference on sustainable solid waste management, Athens, 21–24 June 2017
13. Liu H, Wang J, Liu X, Fu B, Chen J, Yu H (2012) Acidogenic fermentation of proteinaceous sewage sludge: effect of pH. Water Res 46:799–807
14. Nielsen PH, Jahn A (1999) Extraction of EPS. In: Wingender J, Neu TR, Flemming HC (eds) Microbial extracellular polymeric substances: characterization, structure and function. Springer, Berlin
15. Zhang B, Zhang LL, Zhang SC, Shi HZ, Cai WM (2005) The influence of pH on hydrolysis and acidogenesis of kitchen wastes in two-phase anaerobic digestion. Environ Technol 26:329–340
16. Jankowska E, Chwiałkowska J, Stodolny M, Oleskowicz-Popiel P (2015) Effect of pH and retention time on volatile fatty acids production during mixed culture fermentation. Bioresour Technol 190:274–280
17. Jiang J, Zhang Y, Li K, Wanga Q, Gong C, Li M (2013) Volatile fatty acids production from food waste: Effects of pH, temperature, and organic loading rate. Bioresour Technol 143:525–530
18. Yu HQ, Fang HHP (2002) Acidogenesis of dairy wastewater at various pH levels. Water Sci Technol 45:201–206
19. Horiuchi JI, Shimizu T, Tada K, Kanno T, Kobayashi M (2002) Selective production of

organic acids in anaerobic acid reactor by pH control. Bioresour Technol 82:209–213

20. Feng L, Wang H, Chen Y, Wang Q (2009) Effect of solids retention time and temperature on waste activated sludge hydrolysis and short-chain fatty acids accumulation under alkaline conditions in continuous-flow reactors. Bioresour Technol 100:44–49
21. Xiong H, Chen JH, Wang H, Shi H (2012) Influences of volatile solid concentration, temperature and solid retention time for the hydrolysis of waste activated sludge to recover volatile fatty acids. Bioresour Technol 119:285–292
22. Demirer GN, Chen S (2005) Two-phase anaerobic digestion of unscreened dairy manure. Process Biochem 40:3542–3549
23. Speece RE (1996) Anaerobic biotechnology for industrial wastewaters. Arachae Press, Nashville, USA
24. APHA (2012) Standard methods for the examination of water and waste water, 22nd edn. American Public Health Association, Washington, DC
25. Dogan E, Demirer GN (2009) Volatile fatty acid production from organic fraction of municipal solid waste through anaerobic acidogenic digestion. Environ Eng Sci 26:1443–1450
26. Çecchi F, Battistoni P, Pavan P, Fava G, Mata-Álvarez J (1994) Anaerobic digestion of OFMSW and BNR processes: a possible integration–Preliminary results. Water Sci Technol 30:65–72
27. Maharaj I, Elefsiniotis P (2001) The role of HRT and low temperature on the acid-phase anaerobic digestion of municipal and industrial wastewaters. Bioresour Technol 76:191–197
28. Zhuo G, Yan Y, Tan X, Dai X, Zhou Q (2012) Ultrasonic-pretreated waste activated sludge hydrolysis and volatile fatty acid accumulation under alkaline conditions: effect of temperature. J Biotechnol 159:27–31
29. Longo S, Katsou E, Malamis S, Frison N, Renzi D, Fatone F (2015) Recovery of volatile fatty acids from fermentation of sewage sludge in municipal wastewater treatment plants. Bioresour Technol 175:436–444
30. Traverso P, Pavan P, Bolzonella D, Innocenti L, Cecchi F, Mata-Alvarez J (2000) Acidogenic fermentation of source separated mixtures of vegetables and fruits wasted from supermarkets. Biodegradation 11:407–414
31. Yu HQ, Fang HHP (2003) Acidogenesis of gelatin-rich wastewater in an upflow anaerobic reactor: influence of pH and temperature. Water Res 37:55–66
32. Demirer GN, Othman M (2008) Two phase thermophilic acidification and mesophilic methanogenesis anaerobic digestion of waste activated sludge. Environ Eng Sci 25:1291–1300
33. Guerrero L, Omil F, Mendez R, Lema JM (1999) Anaerobic hydrolysis and acidogenesis of wastewaters from food industries with high content of organic solids and protein. Water Res 33:3281–3290
34. Micolucci F, Gottardo M, Bolzonella D, Pavan P (2014) Automatic process control for stable bio-hythane production in two-phase thermophilic anaerobic digestion of food waste. Int J Hydrog Energy 39:17563–17572
35. Lu J, Gavala HN, Skiadas IV, Mladenovska Z, Ahring BK (2008) Improving anaerobic sewage sludge digestion by implementation of a hyperthermophilic prehydrolysis step. J Environ Manag 88:881–889
36. Cavinato C, Bolzonella D, Fatone F, Cecchi F, Pavan P (2011) Optimization of two-phase thermophilic anaerobic digestion of biowaste for hydrogen and methane production through reject water recirculation. Bioresour Technol 102:8605–8611
37. Yu J (2001) Production of PHA from starchy wastewater via organic acids. J Biotechnol 86:105–112
38. Rincón B, Sánchez E, Raposo F, Borja R, Travieso L, Martín MA, Martín A (2008) Effect of organic loading rate on the performance of anaerobic acidogenic fermentation of two-phase olive mill solid residue. Waste Manag 28:870–877
39. Sans C, Mata-Alvarez J, Cecchi F, Pavan P, Bassetti A (1995) Volatile fatty acids production by mesophilic fermentation of mechanically sorted urban organic wastes in a plug-flow reactor. Bioresour Technol 51:89–96
40. Angenent LT, Karim K, Al-Dahhan MH, Wrenn BA, Domiguez-Espinosa R (2004) Production of bioenergy and biochemicals from industrial and agricultural wastewater. Trends Biotechnol 22:477–485
41. Eggeman T, Verser D (2005) Recovery of organic acids from fermentation broths. Appl Biochem Biotechnol 122:605–618
42. Gluszcz P, Jamroz T, Sencio B, Ledakowicz S (2004) Equilibrium and dynamic investigations of organic acids adsorption onto ion-exchange resins. Bioprocess Biosyst Eng 26:185–190
43. Weier AJ, Glatz BA, Glatz CE (1992) Recovery of propionic and acetic acids from fermentation broth by electrodialysis. Biotechnol Prog 8:479–485

44. Saha B, Chopade SP, Mahajani SM (2000) Recovery of dilute acetic acid through esterification in a reactive distillation column. Catal Today 60:147–157
45. Huang C, Xu T, Zhang Y, Xue Y, Chen G (2007) Application of electrodialysis to the production of organic acids: State-of-the-art and recent developments. J Membr Sci 288:1–12
46. Joglekar HG, Rahman I, Babu S, Kulkarni BD, Joshi A (2006) Comparative assessment of downstream processing options for lactic acid. Sep Purif Technol 52:1–17
47. Matsumoto M, Otono T, Kondo K (2001) Synergistic extraction of organic acids with tri-n-octylamine and tri-n-butylphosphate. Sep Purif Technol 24:337–342
48. Eyal AM, Canari R (1995) pH dependence of carboxylic and mineral acid extraction by amine based extractants: effect of pKa amine basicity, and diluent properties. Ind Eng Chem Res 34:1789–1798
49. Tamada JA, Kertes AS, King CJ (1990) Extraction of carboxylic acids with amine extractants. 1. Equilibria and law of mass action modeling. Ind Eng Chem Res 29:1319–1326
50. Yang S, White SA, Hsu S (1991) Extraction of carboxylic acids with tertiary and quaternary amines: effect of pH. Ind Eng Chem Res 30:1335–1342
51. Sabolova E, Schlosser S, Martak J (2001) Liquid-liquid equilibria of butyric acid in water + solvent systems with trioctylamine as extractant. J Chem Eng Data 46:735–745
52. Alkaya E, Kaptan S, Ozkan L, Uludag-Demirer S, Demirer GN (2009) Recovery of Acids from anaerobic acidification broth by liquid-liquid extraction. Chemosphere 77:1137–1142
53. Gungor-Demirci G, Demirer GN (2004) Effect of initial COD concentration, nutrient addition, temperature and microbial acclimation on anaerobic treatability of broiler and cattle manure. Bioresour Technol 93:109–117
54. Bengtsson S, Hallquist J, Werker A, Welander T (2008) Acidogenic fermentation of industrial wastewaters: Effects of chemostat retention time and pH on volatile fatty acids production. Biochem Eng J 40:492–499
55. Hong C, Haiyun W (2010) Optimization of volatile fatty acid production with cosubstrate of food wastes and dewatered excess sludge using response surface methodology. Bioresour Technol 101:5487–5493
56. Gottardo M, Micolucci F, Mattioli A, Faggian S, Cavinato C, Pavan P (2015) Hydrogen and methane production from biowaste and sewage sludge by two phases anaerobic codigestion. Chem Eng Trans 43:379–384
57. Hong SK, Shirai Y, Aini ARN, Hassan MA (2009) Semi-continuous and continuous anaerobic treatment of palm oil mill effluent for the production of organic acids and polyhydroxyalkanoates. Res J Environ Sci 3:552–559
58. Beccari M, Bertin L, Dionisi D, Fava F, Lampis S, Majone M, Valentino F, Vallini G, Villano M (2009) Exploiting olive oil mill effluents as a renewable resource for production of biodegradable polymers through a combined anaerobic–aerobic process. J Chem Technol Biotechnol 84:901–908

Chapter 21

Producing Oleaginous Organisms Using Food Waste: Challenges and Outcomes

Singaram Jayanthi and Arun Kumar Thalla

Abstract

With organic or food waste being one of the main constituents of the total urban waste generated, it not only makes it essential to seek means for its safe disposal but at the same time reiterates the huge potential that lies with the proper utilization of such a widely available resource. Oleaginous microbes that are effective in producing or storing oil would use food waste rich in carbohydrates, lipids, and proteins, and this oil in turn could be an alternative feedstock for the production of biofuels. However, there are few challenges in the process. The various challenges in this process and methods to address them are discussed in the present chapter.

Key words Food waste, Oleaginous, Lipids, Biodiesel

1 Introduction

Food that is left uneaten or unused is termed as food waste. Food is squandered or lost through the supply chain from initial generation to final household utilization. Food waste in general includes food that has been thrown away on partial utilization or completely not used, covering for example fruits, vegetable, fruit scraps, roots and tubers, egg shell, cooked food, meat, bone, and fish, among others. In a survey reported by Food and Agricultural Organization (FAO) of the United Nations, it has been estimated that roughly 1.3 billion tons of food, which approximately measures one-third of the food produced in the world for human consumption, is being wasted. Such losses and waste roughly amounts to US$ 680 billion in industrialized countries and US$ 310 billion in developing countries.

The generated waste cannot be dumped in the yard, and it may lead to a heavy loss in economy, also contributing to immense quantities of solid waste. The solid waste in turn may result in leachate degrading the nature of soil and also contaminate groundwater. Additionally, great amounts of food waste are gathered from

Venkatesh Balan (ed.), *Microbial Lipid Production: Methods and Protocols*, Methods in Molecular Biology, vol. 1995, https://doi.org/10.1007/978-1-4939-9484-7_21, © Springer Science+Business Media, LLC, part of Springer Nature 2019

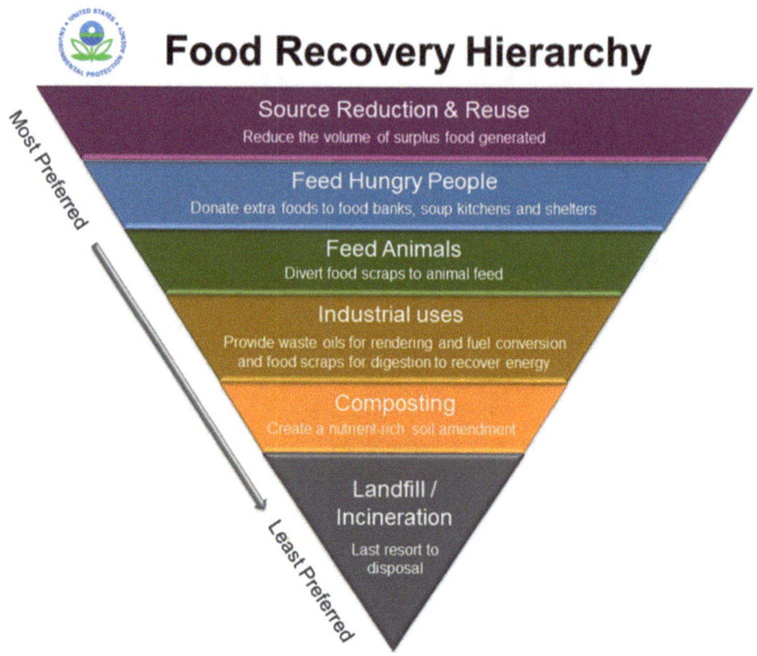

Fig. 1 Food recovery hierarchy [1]

the agricultural sector, food industries, hotels, hostels, and residential buildings. Alternatively, food waste disposal could adopt the strategy proposed by the Environmental Protection Agency as described in Fig. 1. However, the main drawback in such accomplishments is an improper knowledge on how to handle food scraps.

Additionally, the legislation governing the policies and regulations on the use of the excess food or food scraps as animal feed greatly vary from state to state. Some may even interdict feeding animals with food scraps, and some may have regulations on what is to be fed. Feeding hungry people with excess food produced may not be always possible, as it has its own difficulties such as conveyance and distribution in time. Hence, the current scenario calls for utilization of the excess food waste in an efficient way for production of bio-based fuels. Biodiesel is a nonpetroleum-based diesel fuel, which consists of the mono-alkyl esters of long chain fatty acids derived from renewable lipid sources. Biodiesel is produced by transesterification with a monohydric alcohol (usually methanol) which has many important technical advantages over petrodiesel, such as inherent lubricity, low toxicity, superior flash point, negligible sulfur content, and lower exhaust emissions. Oleaginous microorganisms can accumulate lipids more than 20% of their dry biomass, which is the main component for biodiesel production. Large quantities of food waste can be cultured with oleaginous yeast for biodiesel production.

2 Oleaginous Microbes

Oleaginous literally means rich in oil or capable of producing oil. Oleaginous microbes include bacteria, fungi, yeast, and microalgae that are capable of accumulating substantial amounts of oil or lipid, ranging from 20% to 80% or even more of its biomass when grown in controlled culture conditions. This concept of single-cell oil, the so-called lipid-accumulating microbes, opens some opportunities in the management of food waste. Food is generally rich in nutrients; thus, ultimately it would be used by microbes in the degradation of organic waste and thus accumulate oil, which in turn could be an alternative feedstock for the production of biofuels. Eukaryotic microbes can synthesize TAGs, which exhibit a similar composition as that of vegetable oil [2]. Generally, bacteria store polyhydroxyalkanoates, and only few oleaginous prokaryotic species such as *Mycobacterium*, *Streptomyces*, *Nocardia*, and *Rhodococcus* are capable of synthesizing small amounts of special lipids. Based on this potential, the following sections will focus on the challenges related to culturing oleaginous microbes in food waste.

2.1 General Steps in Lipid Production from Food Wastes

- Collection of food waste.
- Pretreatment of the waste material.
- Preparation of the hydrolysate.
- Detoxification of the hydrolysate.
- Fermentation.
- Cell wall disruption.
- Lipid extraction.

2.2 Lipid Production

Jintao Lin et al. [3] studied lipid production by *Lipomyces starkeyi* cells in glucose solution without additional nutrients. A two-stage fermentation process was used for lipid production by *Lipomyces starkeyi* AS 2.1560 in a glucose-only solution. In the first stage, cells were cultured in a nutrient-rich medium for propagation. In the second stage, cells were resuspended in the glucose solution to achieve high cellular lipid contents. The effects of the inoculum age, cell density, and initial glucose concentration on lipid production were briefly studied. Their results demonstrated that cell propagation and lipid accumulation processes can be spatially separated, allowing for further optimizations to improve both processes. The two-stage fermentation method is expected to have a great potential in the development of more efficient processes for converting renewable materials into biofuel and related products.

Wu et al. [4] studied lipid production by *Rhodosporidium toruloides* Y4 using different substrate feeding strategies in a 15-l bioreactor which were found to greatly impact lipid productivity, and they concluded that the repeated fed-batch process was the most appealing method to enhance microbial lipid production.

2.3 Pretreatment of Food Waste

Microbes without appropriate pretreatment procedures as such cannot use food waste to convert complex carbohydrates into simpler forms of monosaccharides. Cellulose and hemicelluloses are hardly hydrolyzed by yeast, but filamentous fungi strains like *Microsphaeropsis*, *Phomopsis*, and *Cephalosporium*, among others, can store 21–35% of its dry weight as lipids by degrading food and utilizing these substrates as carbon source.

2.4 Pretreatment Goals

- Produce highly digestible solids enhancing sugar yields during enzyme hydrolysis.
- Avoid sugar degradation.
- Minimize the formation of inhibitors for subsequent fermentation steps.
- Recover lignin for conversion into valuable coproducts.
- Minimize heat and power requirements, thus making the process cost-effective.
- Be scalable to commercial production.

The various pretreatment methodologies available for food waste are summarized below in the Table 1.

3 Factors Affecting Microbial Lipid Production

3.1 Carbon–Nitrogen Ratio

The lipid yield mainly depends upon the type of substrate used. The substrate should have the minimum carbon–nitrogen ratio of 70 for lipid production. The carbon–nitrogen ratio of the substrate is directly proportional to the lipid yield [5].

3.2 Nitrogen Limitation

Lipid accumulation in oleaginous yeasts occurs at a surplus of carbon source with one limiting element, usually nitrogen limitation [4]. Common nitrogen sources for yeast accumulating lipids include yeast extract, peptone, and ammonium sulfate. It has been reported that using yeast extract containing broth as a nitrogen source is the most common way to culture yeasts and that this can contribute to high amounts of cell biomass and lipid content, for example around 24% of lipid in *Rhodotorula glutinis* and 22% in *Lipomyces starkeyi* [6].

3.3 Sulfate Limitation

Sulfate limitation in the medium is another influencing factor that can affect the biosynthesis of some special amino acids and cellular cofactors such as cysteine, biotin, thiamine, and iron–sulfur clusters. Sulfate limitation influences cell growth and induces lipid accumulation [7]. It has been reported that sulfate limitation can directly influence the fatty acid composition of total lipids. Main fatty acids components like cocoa butter equivalent content was

Table 1
Various pretreatment methods

Pretreatment method	Processes	Possible changes in biomass	Notable remarks
Physical pretreatments	*Milling:* – Ball milling – Two-roll milling – Hammer milling – Colloid milling – Vibro energy milling *Irradiation:* – Gamma-ray irradiation – Electron-beam irradiation – Microwave irradiation *Others:* – Hydrothermal – High pressure steaming – Expansion – Extrusion – Pyrolysis	– Increase the accessible surface area and pore size – Decrease cellulose crystallinity – Decrease degrees of polymerization	– Most of the methods are highly energy demanding – Most methods cannot remove lignin – Inadequate for industrial uses – No chemicals are generally required for these methods
Chemical and physicochemical pretreatments	*Explosion:* – Steam explosion – Ammonia fiber explosion (AFEX) – CO2 explosion – SO2 explosion *Alkali:* – Sodium hydroxide – Ammonia – Ammonium sulfite *Acid:* – Sulfuric acid – Hydrochloric acid – Phosphoric acid	– Increase the accessible surface area – Partial or nearly complete delignification – Decrease cellulose crystallinity – Decrease degrees of polymerization – Partial or complete hydrolysis of hemicelluloses	– These methods are among the most effective and include the most promising processes for industrial uses – Rapid treatment rate – Typically need harsh conditions – There are chemical requirements

(continued)

Table 1
(continued)

Pretreatment method	Processes	Possible changes in biomass	Notable remarks
	Gas: – Chlorine dioxide – Nitrogen dioxide – Sulfur dioxide *Oxidizing agents:* – Hydrogen peroxide – Wet oxidation—Ozone		
Biological pretreatments	*Fungi and Actinomycetes*	– Delignification – Decreasethedegrees of polymerization of cellulose – Partial hydrolysis of hemicelluloses	– Low energy requirement – No chemical requirement – Mild environmental conditions – Very low treatment rate – Not considered for commercial application

increased from 47 % to 63 % when *Rhodosporidium toruloides* was cultured under sulphate limiting conditions [4].

3.4 Minerals and Other Growth Factors

Mineral elements such as magnesium, potassium, and calcium, among others, are essential for yeast growth and/or lipid accumulation at low concentrations [8]. Other compounds also affect lipid accumulation in yeast cells; for instance, inositol, pantothenic acid, and biotin omission results in a decrease of the total lipid amount [9]. It has been reported that the absence of vitamin B6 led to a decrease of the total lipid level with a great reduction on palmitoleic acid by *Hanseniaspora valbyensis* [10].

3.5 Inoculum Concentration

By increasing the inoculum concentration, the lipid yield also increases. The time of the fermentation is usually reduced when the inoculum concentration is increased. Generally, an inoculum concentration of 10% is used as seeding for higher lipid production.

3.6 Concentration of Inhibitors

The lipid yield is drastically reduced when the concentration of inhibitors like furfural compounds increases in the fermentation medium; therefore, detoxification is an essential step prior to fermentation.

3.7 Sugars

The sugars in food waste should be monomers so that the microorganisms can utilize them easily for their metabolic activity. Monomeric six-carbon sugar components are more metabolized than others in the case of oleaginous strains.

3.8 pH

The optimal pH for growing each yeast strain is different and varies from pH 3.0 to pH 7.0 [9]. However, it has been shown that the optimal pH for lipid accumulation is generally lower than that for growth. For example, when using sewage sludge to accumulate lipids by *Lipomyces starkeyi*, a highest lipid content was observed at pH 5.0 while the highest growth was found at pH 6.5 [10]. Another case is the highest lipid content found at pH 3.2 for *Trichoderma reesei* growing in glucose substrate while the highest growth was observed at pH 4.0 [11, 12].

3.9 Temperature

The optimal temperature for growth of different yeasts is variable but usually around 25 °C. A temperature too high or too low will affect cell growth and lipid accumulation. During yeast cell lipid accumulation, dropping temperature from optimal growth generally results in an increase of lipid content and influences lipid composition. Usually, the melting point of unsaturated fatty acids is lower than saturated fatty acids and for short chain fatty acids lower than long chain fatty acids. Thus, a temperature decrease results in increasing levels of unsaturated and short chain fatty acids, for example increasing the ratio of linoleic acid to oleic acid [9].

3.10 Aeration

Aeration is a very important factor for yeast cell growth, as well as for total lipid amounts. The amount of dissolved oxygen in the culture could highly influence the fatty acids composition of lipids. Under oxygen-limited conditions, the glyceride fraction greatly varies and the amounts of phospholipids and sterols decrease. These variations directly result in increasing amounts of saturated fatty acids, which become the main component of lipids. In an aerated growth media, cellular viability increased and free fatty acids were oxidized to unsaturated fatty acids since yeast cells need unsaturated fatty acids for continued growth. For example, the use of excess oxygen increased linolenic acid content in a *Candida utilis* culture [13] and palmitoleic acid content in *Saccharomyces cerevisiae* [14].

3.11 Culturing Time

Previous studies have shown that lipid accumulation in yeast cells increase in the exponential growth phase as lipids are used for membrane lipid synthesis to support cellular growth. When cells reach their maximum size, they switch to lipid accumulation mode as an oil bubble inside the cell. Cell lipid content reached highest value at stationary phase, and upon nutrient depletion fatty acid accumulation rate decreases and cells utilize the accumulated lipid for existence. Therefore, cells should be harvested at the early stationary phase to prevent lipid degradation [15].

3.12 Cell Wall Disruption and Lipid Extraction Methods

Cell disruption is a process used to recover biological molecules that are intracellular. A number of methods are currently available for cell disruption depending on the envisaged molecule and bioprocess. All cell disruption methods must ensure that labile materials are not denatured during the process or hydrolyzed by the enzymes present in the cell. These methods can be broadly classified into three types: physical methods; chemical and enzymatic methods; and mechanical methods.

4 Physical Methods

4.1 Thermolysis

This method is easy, economic and requires no additional specialized equipment. It can be used only if the product is thermostable. The organism is exposed to a heat shock treatment that will kill it and as a result will disturb the cellular integrity without affecting the product.

4.2 Osmotic Shock

Most of the mammalian cells have a plasma membrane with active transporters to maintain the osmotic balance. The maintenance of osmosis is an active process with expenditure of energy. A prolonged exposure of a cell to a hypotonic liquid, such as water, causes an osmotic imbalance and ultimately causes lysis of the cell. In this process, due to inflow of water, the cell swells and ultimately bursts to release the products. According to the Hoff's equation, osmotic pressure (Π) is directly proportional to the concentration of solute and temperature.

$$\Pi = (n/V)\ RT = [Ci]RT$$

where R is the gas constant, T the absolute temperature, and $[Ci]$ the molar concentration of solute.

4.3 Sonication

A sonicator generates the ultrasound waves of frequency more than 20 kHz to cause cell disruption by cavitation. The interaction of ultrasound with liquid causes compression and decompression very rapidly. The bubble formed in liquid compresses several thousand atmospheres and gives shock waves to the cell wall or plasma membrane to cause cell lysis. Generation of ultrasonic waves in liquid causes rapid change in temperature and may cause thermal denaturation. Hence, ultrasonication medium needs to be cooled, and long treatment durations should be avoided.

5 Chemical and Enzymatic Methods

5.1 Alkali Treatment

This is a harsh but effective chemical treatment to lyse cells. Alkali treatment causes lipid saponification, which disturbs the lipid packing and affects the cell wall integrity [16].

5.2 Use of Detergents

The addition of a detergent solution to the cell causes solubilization of the lipid to form micelles. The effect of detergents on the cell wall increases linearly with concentration. The detergent concentration that causes an abrupt change in lipid solubility and forms micelles is known as critical micelle concentration. Examples of detergents include SDS, CTAB, Triton X-100, saponin, and digitonin [17].

5.3 Permeabilization

A partial cell wall disruption or permeabilization is achieved by an organic solvent such as toluene. The organic solvent is absorbed by the cell wall, resulting in its swelling and ultimate rupture. However, at low concentrations, the organic solvent permeabilizes the cell walls without disturbing the cellular integrity. This process allows using the cell as a reaction vessel to catalyze a reaction and get the desired product [16].

5.4 Enzyme Digestion

Enzymatic methods are specific, gentle, and most effective but costly. Lysozyme is commercially available to treat bacteria to release intracellular products. In addition to lysozyme, there are three other types of bacteriolytic enzymes, glycosidases, acetylmuramyl-L-alanineamidase, and endopeptidase. Few proteases are also found to be bacteriolytic. Yeast cell lysis requires a mixture of different enzymes such as glucanase, protease, mannanase, or

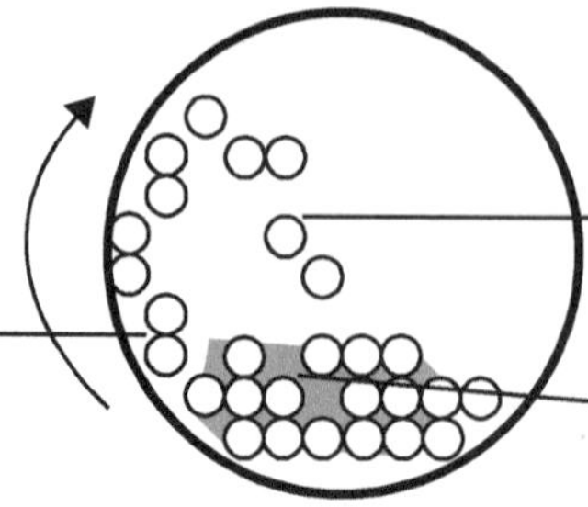

Fig. 2 Bead mill disruption [18]

chitinase. Plant cells can be lysed by cellulose and pectinase. In most of the enzyme mediated cell lysis method, the rupture of cell wall depends on the osmotic pressure of the external medium. In few cases, enzymatic digestion is performed to remove the external cell wall and then in second step the protoplast is disrupted by gentle agitation.

5.5 Mechanical Methods

The simplest mechanical cell disruption methods used at laboratory scale are the warring blender and pestle–mortar. Both are effective toward animal and plant tissues, as well as for filamentous organisms. At an industrial scale, bead mills and high-pressure homogenizers are used to carry out cell disruption.

5.6 Bead Mill Disruption

The bead mill (horizontal or vertical) as shown in Fig. 2 consists of a grinding cylinder with a central shaft fitted with a number of impellers, which can move in clockwise or counterclockwise direction with the help of a motor. The grinding cylinder is filled with beads made up of glass, alumina, titanium carbide, zirconium oxide, or zirconium silicate. There is an inlet to supply the cell suspension and an outlet to collect the sample after the process. When the bead mill runs, the cells experience shear forces between the moving beads and cells. The rate and degree of cell disruption depends on cell type, thickness of cell wall, localization of product, type and agitation speed of impeller, bead size, its density and loading, residence time, and temperature. Cell disruption in a bead mill and release of a product is a first order kinetics and it may be given by first order equation:

$$\ln\left[C_{\max}/(C_{\max}-C)\right]=-kt$$

where $C_{\max}$ is the concentration of product that can be released from a given amount of cell suspension, C is the concentration of product released at a given time "t," and k is the first order constant. This relationship holds true only for batch mode of operation. The value of k depends on type of impeller, bead size and loading, speed of agitation, and temperature.

5.7 High-Pressure Homogenizer

A high-pressure homogenizer consists of a high-pressure positive displacement pump connected to an adjustable discharge valve with a restricted opening. The cell suspension is sent into the homogenizer through a small homogenizing valve at a very high pressure (200–1000 atmospheric pressure). Cells passing through a small valve experience stress developed within the fluid and as a result get disrupted.

5.8 Lipid Extraction Methods

To extract the lipids from the sample resulting from disrupted cells, organic solvents such as chloroform, methanol, *n*-hexane, and isopropyl alcohol are generally used. The three most common lipid extraction methods are discussed in the following section.

5.8.1 Chloroform–Methanol Method

The chloroform–methanol lipid extraction method was proposed by Folch et al. [19]. The procedure is briefly explained below. Dissolve freeze-dried pellets of yeast biomass in 3.2 ml of 4 M HCl and incubate at 55 °C for 2 h. Then, add 8 ml of chloroform-–methanol mixture (1:1 as molarity), transfer it into glass tubes and shake at 20 °C for 3 h. Subsequently, centrifuge samples at 2000 × *g* for 15 min to obtain clear chloroform phase (lower phase). Next, carefully transfer the lower phase into new tubes by using rubber head dropper. Then, add 4 ml chloroform into the chloroform phase and shake vigorously. Following, centrifuge the new tubes once more. Afterward, collect the clear chloroform and transfer it into preweighted glass tubes. Next, gas the tubes with N_2 for 2 h to evaporate all chloroform. Finally, weigh the glass tubes after evaporation to calculate lipid weight.

5.8.2 Bligh and Dyer Method

Bligh and Dyer [20] proposed the following procedure for lipid extraction. The procedure is briefly described below. For each 1 ml of sample, add 3.75 ml 1:2 (v/v) $CHCl_3$–MeOH and vortex. Then, add 1.25 ml $CHCl_3$ and vortex. Finally, add 1.25 ml dH_2O and vortex. Afterward, centrifuge at 168 × *g* for 5 min at room temperature to form a two-phase system (aqueous top, organic bottom). Next, recover the bottom phase by inserting a Pasteur pipette through the upper phase with gentle positive pressure (i.e., gentle bubbling) so that the upper phase does not get into the pipette tip. When the pipette tip is at the bottom of the tube, carefully withdraw the bottom phase through the pipette, making sure to avoid the interface or the upper phase (recover~90% of bottom phase).

5.8.3 Lipid Extraction Using n-Hexane and Isopropyl Alcohol

The oil extraction from yeast cells can be carried out in 50 L SS agitator tank using the method described by Markham et al. [21] with some modifications. Homogenized yeast cells and n-hexane are used at a ratio of 1:2 and stirred for 1 h at 950 rpm. The mixture should be allowed to stabilize for 10–12 h at ambient temperature.

Then, remove the bottom aqueous layer. Next, add isopropyl alcohol (IPA) at the ratio of 0.5 to separate the organic layer, which should be subsequently collected and concentrated to recover lipid.

6 Conclusion

Food waste is nutritionally rich and offers a unique microbial feedstock for the production of numerous valuable bioproducts. It is estimated that the food industry accounts for 8% of all the energy used in the USA, and studies have found that over 27% of this food is thrown away. Not only do these actions neglect a valuable source of energy, but they also present an inefficient use of the energy invested in producing the food in the first place. In fact, food comprises one of the largest components of waste in the USA. However, less than 3% of the food waste generated in the USA is recycled. The rest of this waste is decomposed in landfills where it produces foul odors, can pollute groundwater, requires large amounts of space, and generates large amounts of methane gas, which has 20 times the global warming potential of carbon dioxide. Recycling food waste, therefore, is one of the most environmentally friendly processes for generating biofuels. In fact, it is a much better alternative to growing crops for the purpose of generating biofuel. Studies on lipid production from a variety of wastes such as municipal waste, vegetable waste, garden trimmings, and wastewater are in progress by using a combination of yeast strains or modifying the yeast strains for a given waste to enhance the maximum yield.

References

1. USEPA Sustainable Management of Food, https://www.epa.gov/sustainable-management-food/food-recovery-hierarchy. Accessed 24 Apr 2018
2. Thevenieau F, Nicaud JM (2013) Microorganisms as sources of oils. OCL 20(6):D603
3. Lin J, Shen H, Tan H, Zhao X, Wu S, Hu C, Zhao ZK (2011) Lipid production by Lipomyces starkeyi cells in glucose solution without auxiliary nutrients. J Biotechnol 152 (4):184–188
4. Wu S, Zhao X, Shen H, Wang Q, Zhao ZK (2011) Microbial lipid production by *Rhodosporidium toruloides* under sulfate-limited conditions. Bioresour Technol 102(2):1803–1807
5. Wiebe MG, Koivuranta K, Penttilä M, Ruohonen L (2012) Lipid production in batch and fed-batch cultures of *Rhodosporidium toruloides* from 5 and 6 carbon carbohydrates. BMC Biotechnol 12(1):26
6. Kumar SV, Kumutha K, Krishnan PS, Gopal H (2010) Influence of nitrogen on lipid and biomass production by oleaginous yeast cultures. Asian J Biol Sci 5(1):87–91
7. Scott C, Hilton ME, Coppin CW, Russell RJ, Oakeshott JG, Sutherland TD (2007) A global response to sulfur starvation in Pseudomonas putida and its relationship to the expression of low-sulfur-content proteins. FEMS Microbiol Lett 267(2):184–193
8. Spencer JFT, Spencer DM, de Figueroa LIC (1997) Yeasts as living objects: yeast nutrition. In: Yeasts in natural and artificial habitats. Springer, Berlin, Heidelberg, pp 68–79
9. Rattray JB, Schibeci A, Kidby DK (1975) Lipids of yeasts. Bacteriol Rev 39(3):197
10. Haskell BE, Snell EE (1965) Effect of vitamin B6 deficiency on the composition of yeast lipids. Arch Biochem Biophys 112(3):494–505

11. Angerbauer C, Siebenhofer M, Mittelbach M, Guebitz GM (2008) Conversion of sewage sludge into lipids by Lipomyces starkeyi for biodiesel production. Bioresour Technol 99 (8):3051–3056
12. Brown DE, Hasan M, Lepe-Casillas M, Thornton AJ (1990) Effect of temperature and pH on lipid accumulation by Trichoderma reesei. Appl Microbiol Biotechnol 34(3):335–339
13. Babij T, Moss FJ, Ralph BJ (1969) Effects of oxygen and glucose levels on lipid composition of yeast *Candida utilis* grown in continuous culture. Biotechnol Bioeng 11(4):593–603
14. Valero E, Millán C, Ortega J (2001) Influence of oxygen addition during growth phase on the biosynthesis of lipids in *Saccharomyces cerevisiae* (M330-9) in enological fermentations. J Biosci Bioeng 92(1):33–38
15. Beopoulos A, Mrozova Z, Thevenieau F, Le Dall MT, Hapala I, Papanikolaou S, Nicaud JM (2008) Control of lipid accumulation in the yeast Yarrowia lipolytica. Appl Environ Microbiol 74(24):7779–7789
16. Analysis of biomolecules. http://nptel.ac.in/courses/102103045/module5/lec28/4.html. Accessed Sep 2017
17. Lichtenberg D, Ahyayauch H, Goñi FM (2013) The mechanism of detergent solubilization of lipid bilayers. Biophys J 105 (2):289–299
18. Cell Disruption. http://upendrats.blogspot.in/2012/09/cell-disruption.html. Accessed 24 Apr 2018
19. Folch J, Lees M, Sloane Stanley GH (1957) A simple method for the isolation and purification of total lipids from animal tissues. J Biol Chem 226(1):497–509
20. Bligh EG, Dyer WJ (1959) A rapid method of total lipid extraction and purification. Can J Biochem Physiol 37(8):911–917
21. Markham JE, Li J, Cahoon EB, Jaworski JG (2006) Separation and identification of major plant sphingolipid classes from leaves. J Biol Chem 281(32):22684–22694

Chapter 22

Microbial Surfactants: Alternative to Vegetable Oil Surfactants

Eduardo J. Gudiña and Lígia R. Rodrigues

Abstract

Biosurfactants, surface active molecules synthesized by microorganisms, represent a promising alternative to the synthetic surfactants in many different applications. Among them, rhamnolipids have attracted considerable attention in the last years due to their extraordinary surface-active properties and biological activities. Rhamnolipids are usually synthesized by the gram-negative bacterium *Pseudomonas aeruginosa* as complex mixtures of different congeners. In this chapter, we describe the most common techniques that can be used for the production, recovery and purification of rhamnolipids, using two sequential chromatographic techniques to recover and separate the monorhamnolipid and dirhamnolipid congeners.

Key words Surfactant, Biosurfactant, Rhamnolipid, *Pseudomonas aeruginosa*, Adsorption chromatography, Orcinol, Silica gel

1 Introduction

Surfactants are an important class of chemical compounds widely used in many different fields. They are amphiphilic molecules that by accumulating at the interfaces between fluid phases with different polarities reduce the surface and interfacial tensions. As a result, these compounds facilitate the formation of emulsions, dispersions and foam, making them essential in applications that require emulsification, lubrication, foaming, solubilization of immiscible compounds or phase dispersion. Their uses include the formulation of cleaning products and detergents, applications in food processing, agriculture, environmental remediation, and tertiary oil recovery, among others. The total worldwide production of surfactants is estimated to be over 15 million tons per year, and it is expected to reach about 24 million tons and 42,120 million USD per year by 2020 [1].

Most conventional surfactants available nowadays are derived from petrochemical sources, which besides contributing to environmental pollution due to their low biodegradability consume

Venkatesh Balan (ed.), *Microbial Lipid Production: Methods and Protocols*, Methods in Molecular Biology, vol. 1995, https://doi.org/10.1007/978-1-4939-9484-7_22,

nonrenewable resources [2]. In the recent years, an increase in environmental awareness has led to much more interest in the use of renewable-based, biodegradable and more environmental friendly surfactants. The market for this "green" alternatives to synthetic surfactants was 344 kilo tons in 2013, and it is expected to reach 462 kilo tons and 2308 million USD by 2020 [3]. Among them, biosurfactants, surface active molecules produced by diverse groups of microorganisms, have attracted considerable attention as alternatives to their synthetic counterparts. Biosurfactants exhibit low toxicity, high biodegradability, high selectivity, low critical micelle concentrations (*cmc*), and effectiveness at extreme temperatures, pHs, and salinities, being also a better alternative than the bio-based surfactants derived from vegetable oils. Due to these properties, biosurfactants comprise a variety of potential applications in several industries [4–6].

Rhamnolipids, a class of glycolipid biosurfactants, are among the most effective biosurfactants due to their low minimum surface tension values (28–30 mN/m), high emulsifying activity (emulsifying indexes up to 60–70%) and low *cmcs* (10–200 mg/L). Rhamnolipids are usually synthesized by the gram-negative bacterium *Pseudomonas aeruginosa* and related species as complex mixtures, and more than 60 different congeners have been identified to date [1, 4, 7]. Their hydrophilic moiety consists of one (monorhamnolipids) or two (dirhamnolipids) rhamnose molecules, while the hydrophobic part is represented by two (or more rarely one) β-hydroxy fatty acids, saturated or unsaturated, of different chain lengths (C_8–C_{24}). The most common congeners are L-rhamnosyl-β-hydroxydecanoyl-β-hydroxydecanoate (Rha-C_{10}-C_{10}) and L-rhamnosyl-L-rhamnosyl-β-hydroxydecanoyl-β-hydroxydecanoate (Rha-Rha-C_{10}-C_{10}). Their physicochemical properties (e.g., solubility, surface activity) and applications depend on the type and proportion of congeners present in the mixture, which vary according to the bacterial strain, the composition of the culture medium and the cultivation conditions [1, 4, 8]. Besides their surface activity, rhamnolipids exhibit additional properties when compared with chemical surfactants, including antimicrobial and antitumor activities, which expands the spectrum of their potential applications [9–11].

The methods provided in this chapter represent the most common techniques that can be used for the production, recovery and purification of rhamnolipids. The method proposed for the rhamnolipids recovery is through adsorption chromatography using the polystyrene resin Amberlite® XAD®-2. Although other methodologies are described for this purpose (e.g., acidic precipitation, extraction with solvents), the one herein proposed allows for recovering most of the rhamnolipids present in the cell-free supernatant in one step, and also the product obtained is purer. Subsequently, the monorhamnolipid and dirhamnolipid congeners present in the crude rhamnolipid mixture will be separated through silica gel chromatography.

2 Materials

Prepare all solutions and culture media using distilled water and analytical grade reagents. Prepare and store all reagents at room temperature, unless indicated otherwise. Diligently follow all waste disposal regulations when disposing of waste materials.

2.1 Rhamnolipid Production

1. A rhamnolipid-producing *Pseudomonas aeruginosa* strain.
2. Luria–Bertani (LB) liquid medium: 10 g/L NaCl, 10 g/L tryptone, 5 g/L yeast extract, pH 7.0. Sterilize by autoclaving (121 °C for 15 min). To prepare LB solid medium add agar (20 g/L) before autoclaving.
3. 100 and 500 mL Erlenmeyer flasks.
4. Orbital incubator (37 °C, 200 rpm).
5. Centrifuge.
6. Tensiometer.
7. Hydrochloric acid (36%, w/w).

2.2 Rhamnolipid Recovery

Rhamnolipid recovery will be performed through adsorption chromatography.

1. A 125 mL glass chromatography column (e.g., 40 cm length, 2 cm inner diameter) with a porous support at the bottom.
2. Amberlite® XAD®-2 resin (particle size 20–60 mesh; mean pore size 90 Å) (Sigma-Aldrich Co., USA).
3. Equilibration buffer: 0.1 M potassium phosphate buffer, pH 6.1. Dissolve 2.78 g of K_2HPO_4 and 11.43 g of KH_2PO_4 in 900 mL of distilled water and adjust the solution to a final volume of 1 L. This solution can be stored at 4 °C for several days.
4. Peristaltic pump.
5. Rotary evaporator.
6. Freeze dryer.
7. Hydrochloric acid (36%, w/w).
8. Methanol.

2.3 Purification of Monorhamnolipids and Dirhamnolipids

1. A 125 mL glass chromatography column (e.g., 40 cm length, 2 cm inner diameter) with a porous support at the bottom.
2. Silica gel 60 (70–230 mesh, particle size 63–200 μm).
3. Peristaltic pump.
4. Rotary evaporator.

5. Freeze dryer.
6. Fume hood.
7. Thin layer chromatography (TLC) plates (silica gel 60 or equivalent) and TLC chamber.
8. Spray reagent bottle compatible with sulfuric acid.
9. Oven (105 °C).
10. Orcinol reagent: dissolve 38 mg of orcinol in 9.4 mL of distilled water, using heat if necessary. Cool in ice and add slowly 10.6 mL of sulfuric acid (96%, w/w). Allow to cool to room temperature. This solution can be stored at 4 °C in the dark for several weeks.
11. Chloroform.
12. Methanol.

3 Methods

3.1 Rhamnolipid Production

1. Aseptically pick a single *P. aeruginosa* colony from a freshly streaked LB agar plate and transfer it into a 100 mL Erlenmeyer flask containing 20 mL of LB liquid medium. Incubate overnight at 37 °C and 200 rpm (*see* **Note 1**).
2. The next day use 2 mL of this preculture to inoculate 500 mL Erlenmeyer flasks containing 200 mL of LB liquid medium (*see* **Note 2**). Incubate the flasks at 37 °C and 200 rpm (*see* **Note 3**).
3. Aseptically take 5 mL samples every 24 h in order to evaluate rhamnolipid production.
4. Centrifuge the samples at 9000 × *g* for 15 min and transfer the cell-free supernatants to clean tubes, making sure the pellet is undisturbed.
5. Dilute the cell-free supernatant samples 10 and 100 times using distilled water (*see* **Note 4**).
6. Measure the surface tension of the cell-free supernatant samples (ST) and the corresponding dilutions (ST^{-1}, ST^{-2}) at room temperature using a tensiometer, following the instructions provided by the manufacturer of your equipment (*see* **Note 5**). The highest rhamnolipid production is achieved when the ST^{-2} value attains its minimum value and remains stable along the time (*see* **Note 6**).
7. Once the highest rhamnolipid production is achieved centrifuge the cultures at 9000 × *g* for 15 min. Recover the cell-free supernatant making sure that the pellet is undisturbed (*see* **Note 7**).

3.2 Rhamnolipid Recovery

This protocol allows the recovery of the crude rhamnolipid mixture from the cell-free supernatant previously obtained. All the steps are performed at room temperature.

3.2.1 Cell-Free Supernatant Preparation

1. Transfer 250 mL of cell-free supernatant to a 500 mL glass beaker.
2. Adjust the pH to 6.1 by dropwise addition of hydrochloric acid (36%, w/w) with constant vigorous stirring (*see* **Note 8**).

3.2.2 Column Preparation

1. Transfer the appropriate amount of Amberlite® XAD®-2 resin (dry form) to a glass beaker (*see* **Note 9**).
2. Suspend the resin in methanol (the volume of methanol must be enough to cover the resin bed by 2.5–5 cm).
3. Stir the resin gently to ensure complete mixing and allow the mixture to stand for 15 min.
4. Carefully decant most of the methanol and replace it with distilled water. Stir the resin gently and allow it to stand for 15 min.
5. Add distilled water to the clean glass chromatography column up to approximately 2.5 cm of height.
6. Slowly pour the resin slurry into the column. As the column fills, drain excess of water through the bottom of the column, but do not allow the liquid level to fall below the top of the resin bed. Allow the resin to settle and continue the process until the column is filled up to two-thirds of its height. Leave about 5 cm of water above the top of the resin bed (*see* **Note 10**).

3.2.3 Rhamnolipid Recovery

1. Equilibrate the resin by passing 300 mL of equilibration buffer at a flow rate of 3 mL/min.
2. Pass 250 mL of cell-free supernatant (prepared as described above) through the column at a flow rate of 3 mL/min. Collect samples of the eluent at 50 mL intervals and measure the surface tension to get sure that the rhamnolipids remain attached to the resin (the surface tension value of the eluent should be equal or higher than 40 mN/m) (*see* **Note 11**).
3. Wash the column with 500 mL of distilled water at a flow rate of 3 mL/min to remove nonadsorbed compounds. Collect samples of the eluent at 100 mL intervals and measure the surface tension to ensure that rhamnolipids remain attached to the resin (the surface tension value of the eluent should be equal or higher than 50 mN/m).
4. Elute the rhamnolipids adsorbed to the resin with methanol (flow rate 3 mL/min). The volume of methanol necessary will depend on the amount of rhamnolipids present in the cell-free

supernatant (*see* **Note 12**). Collect samples of the eluent at 100 mL intervals, dilute the samples 10 times with distilled water, and then measure the surface tension. Surface tension values around 30 mN/m indicate the presence of rhamnolipids. Continue this process until the surface tension values increase up to 50 mN/m, which indicates the absence of rhamnolipids in the sample (*see* **Note 13**).

5. Wash the column with 300 mL of distilled water at a flow rate of 3 mL/min to remove the methanol and discard the eluent (*see* **Note 14**).
6. Combine all the methanol fractions containing rhamnolipids and remove the methanol using a rotary evaporator at room temperature.
7. Resuspend the honey-like crude rhamnolipid mixture in 25–50 mL of distilled water.
8. Freeze-dry the sample to obtain the crude rhamnolipid mixture in a powder form.
9. The freeze-dried rhamnolipids can be stored at room temperature in tightly closed tubes in a desiccator.

3.3 Purification of Monorhamnolipids and Dirhamnolipids

This method can be used to purify rhamnolipids (i.e., to remove fatty acids, nonpolar pigments and other impurities that are usually coextracted with rhamnolipids), as well as to separate monorhamnolipid and dirhamnolipid congeners. All the steps are performed at room temperature and must be performed in a fume hood.

3.3.1 Column Preparation

1. Transfer the appropriate amount of silica gel 60 (enough to fill the column up to two-thirds of its capacity) to a glass beaker (*see* **Note 15**).
2. Add sufficient chloroform to cover the resin bed by 2.5–5 cm.
3. Stir the resin gently and allow it to stand for 15 min.
4. Slowly pour the mixture of chloroform–silica gel into the column. As the column fills, drain excess of chloroform through the bottom of the column, but do not allow the liquid level to fall below the top of the resin bed. Allow the resin to settle and continue the process until the column is filled up to two-thirds of its capacity. Leave about 5 cm of chloroform above the top of the resin bed.

3.3.2 Purification of Monorhamnolipids and Dirhamnolipids

1. Dissolve 1 g of the freeze-dried crude rhamnolipid mixture in 10 mL of chloroform and pour it into the column.
2. Wash the column with chloroform at a flow rate of 3 mL/min until no color persists in the eluent (This will remove neutral lipids and nonpolar pigments present in the crude rhamnolipid sample.). Discard the eluent.

3. Wash the column with 500 mL of chloroform–methanol (180:5, v/v) at a flow rate of 3 mL/min. Analyze the eluent through TLC as described below. At this point the eluent must contain only monorhamnolipids. Continue the elution process with the same solvent mixture until it is not possible to detect the presence of monorhamnolipids on the TLC.
4. Wash the column with 500 mL of chloroform–methanol (180:7.5, v/v). Analyze the eluent through TLC. At this point, the eluent must contain a mixture of monorhamnolipids and dirhamnolipids. Continue washing the column with the same solvent mixture until it is not possible to detect the presence of rhamnolipids on the TLC.
5. Wash the column with 500 mL of chloroform–methanol (180:20, v/v). Analyze the eluent through TLC. At this point, the eluent must contain only dirhamnolipids. Continue washing the column with the same mixture until it is not possible to detect the presence of dirhamnolipids on the TLC (*see* **Note 16**).
6. Combine the different fractions recovered (based on their profiles) to obtain pure monorhamnolipid and dirhamnolipid fractions.
7. Remove the solvents of each sample using a rotary evaporator at room temperature.
8. Resuspend the monorhamnolipid and dirhamnolipid fractions in 25 mL of distilled water.
9. Freeze-dry the samples to obtain pure monorhamnolipids and dirhamnolipids in a powder form.
10. The freeze-dried monorhamnolipids and dirhamnolipids can be stored at room temperature in tightly closed tubes in a desiccator.

3.3.3 Thin Layer Chromatography Detection

1. Prepare a solvent system consisting of chloroform–methanol-–water (65:25:4, v/v/v) and transfer into the TLC chamber (to achieve a height of 1 cm). Allow the chamber to equilibrate for 1 h.
2. Apply 5–10 μL of rhamnolipid sample onto a TLC plate, 2 cm above the bottom of the plate. Apply 1 μL fractions with intermediate drying.
3. A sample of previously purified (or commercial) rhamnolipids should be used as a standard for comparison.
4. When the loaded samples are completely dry, place the TLC plate in the chromatographic chamber and develop it until the solvent front is about 2 cm from the top of the plate. With this solvent system, monorhamnolipids and dirhamnolipids are separated into two bands.

5. Remove the plate from the chamber and allow it to air-dry in a fume hood.
6. Spray the TLC plate uniformly with orcinol reagent and place it in an oven at 105 °C for about 5 min (until the yellow-red spots became visible). The spot nearer the point of origin corresponds to the dirhamnolipid congeners, whereas the spot further from the point of origin corresponds to the monorhamnolipid congeners.

4 Notes

1. Several *P. aeruginosa* strains are opportunistic pathogens for humans.
2. Alternative culture media have been reported to produce rhamnolipids using different *P. aeruginosa* strains. Some of these media contain different agro-industrial by-products and wastes. However, these culture media have been optimized to be used with specific *P. aeruginosa* strains, and their use with other strains may require a previous optimization. Some of these media contain water-immiscible substrates (vegetable oils, hydrocarbons) which act as inducers of rhamnolipid production. However, their use difficult the subsequent recovery and purification of rhamnolipids.
3. Rhamnolipid production by *P. aeruginosa* strains is inefficient under oxygen-limiting conditions, and it is favored by the use of high agitation rates.
4. Preparing the cell-free supernatant dilutions can lead to the formation of foam, which will interfere with the subsequent surface tension measurements. To avoid the formation of foam, the samples must be mixed gently. If foam is formed when preparing the dilutions, the samples must be centrifuged to remove it before measuring the surface tension.
5. Different types of tensiometers are commercially available. Perform the surface tension measurements following the instructions provided by the manufacturer of your equipment. Depending on the type of tensiometer used, the volume of sample required to perform the measurements can vary, but at least 5–10 mL are usually necessary.
6. The surface tension of LB medium is around 50 mN/m. Rhamnolipids reduce the surface tension of water to values around 30 mN/m. Usually the surface tension of the cell-free supernatants attains its minimum value after 24 h of growth, and after that it remains nearly constant until the end of the fermentation. This means that the amount of rhamnolipid present in the culture medium is equal or higher than the

cmc. In those conditions, it is necessary to measure the surface tension of the cell-free supernatants diluted 10 and 100 times with distilled water to study the rhamnolipid production along the fermentation course.

7. The cell-free supernatant can be sterilized by autoclaving (121 °C for 15 min) if the extraction is not performed immediately (in order to avoid its contamination or the degradation of the rhamnolipids), as rhamnolipids are not negatively affected by exposure to those conditions.
8. Under acidic conditions rhamnolipids are present in their protonated form, being less soluble in water. The addition of hydrochloric acid to the cell-free supernatant results in the precipitation of rhamnolipids. Before placing the cell-free supernatant into the column it is necessary to allow the complete dissolution of all the precipitates, which can take several hours. For that purpose, a vigorous stirring must be applied, which results in the formation of a considerable amount of foam. Therefore, the volume of the beaker used to perform this step must be at least twice the volume of cell-free supernatant.
9. The column must be filled only up to two-thirds of its capacity; this will allow room for expansion during the elution step (Methanol, which is used in the elution step, makes the resin to swell.).
10. Once hydrated, do not allow the resin to dry during preparation or subsequent use. A constant head of at least 2.5 cm of water or solvent above the resin bed will prevent resin dehydration.
11. Depending on the concentration of rhamnolipids present in the cell-free supernatant, the resin can get saturated before processing the entire sample. In this case, the volume of cell-free supernatant to be passed through the column must be reduced.
12. About 500 mL of methanol are necessary to elute 1 g of crude rhamnolipid mixture.
13. The surface tension of methanol is around 25 mN/m. Consequently, measuring the surface tension of the eluent is not useful to assess the presence of rhamnolipids. However, the surface tension of methanol diluted ten times with distilled water is around 50 mN/m. If rhamnolipids are preset in the sample, the surface tension value will be around 30 mN/m, whereas surface tension values close to 50 mN/m indicate the absence of rhamnolipids.
14. Once packed, the column can be reused several times. At the end of each utilization resuspend the resin by vigorously

shaking the column and wash the resin with 100 mL of distilled water to remove impurities. Repeat this process up to five times and allow the resin to settle. At this point the column is ready to be used again, starting with the equilibration step. The column can be stored at room temperature until the next use.

15. The amount of silica gel required depends on the amount of crude rhamnolipid to be purified. Usually the amount of silica gel should be 20 times the sample weight (e.g., for 1 g of crude rhamnolipid use at least 20 g of silica gel).

16. It is not possible to perform the separation of monorhamnolipid and dirhamnolipid congeners using only two solvent systems. Slight modifications in the composition of the solvent mixtures may be necessary to achieve an appropriate separation of monorhamnolipid and dirhamnolipid congeners, depending on the composition of the crude rhamnolipid mixture produced by the *P. aeruginosa* strain used.

Acknowledgments

The authors acknowledge the financial support from the Portuguese Foundation for Science and Technology (FCT) under the scope of the strategic funding of UID/BIO/04469/2019 unit and BioTecNorte operation (NORTE-01-0145-FEDER-000004), the project MultiBiorefinery (POCI-01-0145-FEDER-016403), and Gudiña E postdoctoral grant CEB-BPD/01/2015/07, all funded by the European Regional Development Fund under the scope of Norte2020—Programa Operacional Regional do Norte. Rodrigues LR also acknowledges her sabbatical leave fellowship (SFRH/BSAB/142991/2018) funded by FCT.

References

1. Gudiña EJ, Rodrigues AI, de Freitas V, Azevedo Z, Teixeira JA, Rodrigues LR (2016) Valorization of agro-industrial wastes towards the production of rhamnolipids. Bioresour Technol 212:144–150
2. Vaz DA, Gudiña EJ, Alameda EJ, Teixeira JA, Rodrigues LR (2012) Performance of a biosurfactant produced by a *Bacillus subtilis* strain isolated from crude oil samples as compared to commercial chemical surfactants. Colloids Surf B Biointerfaces 89:167–174
3. Gudiña EJ, Rodrigues AI, Alves E, Domingues MR, Teixeira JA, Rodrigues LR (2015) Bioconversion of agro-industrial by-products in rhamnolipids toward applications in enhanced oil recovery and bioremediation. Bioresour Technol 177:87–93
4. Abdel-Mawgoud AM, Lépine F, Déziel E (2010) Rhamnolipids: diversity of structures, microbial origins and roles. Appl Microbiol Biotechnol 86:1323–1336
5. Henkel M, Müller MM, Kügler JH, Lovaglio RB, Contiero J, Syldatk C, Hausmann R (2012) Rhamnolipids as biosurfactants from renewable resources: concepts for next-generation rhamnolipid production. Process Biochem 47:1207–1219
6. Pereira JFB, Gudiña EJ, Costa R, Vitorino R, Teixeira JA, Coutinho JAP, Rodrigues LR (2013) Optimization and characterization of biosurfactant production by *Bacillus subtilis* isolates towards microbial enhanced oil recovery applications. Fuel 111:259–268

7. Lovaglio RB, Silva VL, Ferreira H, Hausmann R, Contiero J (2015) Rhamnolipids know-how: looking for strategies for its industrial dissemination. Biotechnol Adv 33:1715–1726
8. Zhang L, Pemberton JE, Maier RM (2014) Effect of fatty acid substrate chain length on *Pseudomonas aeruginosa* ATCC 9027 mono-rhamnolipid yield and congener distribution. Process Biochem 49:989–995
9. Bharali P, Saikia JP, Ray A, Konwar BK (2013) Rhamnolipid (RL) from *Pseudomonas aeruginosa* OBP1: a novel chemotaxis and antibacterial agent. Colloids Surf B Biointerfaces 103:502–509
10. Christova N, Tuleva B, Kril A, Georgieva M, Konstantinov S, Terziyski I, Nikolova B, Stoineva I (2013) Chemical structure and in vitro antitumor activity of rhamnolipids from *Pseudomonas aeruginosa* BN10. Appl Biochem Biotechnol 170:676–689
11. Gudiña EJ, Rangarajan V, Sen R, Rodrigues LR (2013) Potential therapeutic applications of biosurfactants. Trends Pharmacol Sci 34:667–675

Chapter 23

Evaluation of Bacterial Lipid Production: Quantitative and Qualitative Measurements: Tips and Guidelines

Sima Modiri, Hossein Shahbani Zahiri, and Kambiz Akbari Noghabi

Abstract

Over the last decade, finding bacterial strains with ability to accumulate high concentrations of lipids has gained increasing interest, since these lipids may be used in different industries. Here we describe two methods for evaluation of lipid accumulation in cyanobacteria, following by our personal reflection on issues surrounding the use of these methods. First, we present the Bligh and Dyer protocol as a traditional extraction method using organic solvents for quantitative determination of lipids and next Nile red, a selective fluorescent stain, that has been used as a rapid approach for both qualitative and quantitative measurement of lipids.

Key words Bacterial lipid, Bligh and Dyer, Nile red staining, Qualitative and quantitative measurement

1 Introduction

Bacterial storage lipids, including polyhydroxyalkanate (PHA), triglycerol ($C_9H_{20}O_7$), and wax esters, are biodegradable compounds that are used in food production, household goods, cosmetics, packaging, and medicine [1]. Moreover, lipids, as valuable compounds, have potential to replace fossil fuels in various industries. The ability to accumulate significant amounts of triacylglycerides (TAGs) as intracellular lipid deposits can be found in both gram-positive and -negative bacteria, although this ability is more often associated with gram-positive bacteria. Cyanobacteria are a special case among gram-negative bacteria and have the ability to produce a lot of membrane lipids and diacylglycerols in the thylakoid membranes of their chloroplasts. These lipids can be modified in different ways and utilized in various sectors of biotechnology [2]. Figure 1 shows thylakoid-associated lipid droplets in the cyanobacterial strain *Synechococcus* sp. HS01 [3].

According to Fig. 2a, while *Pseudanabaena* sp. SK03 has the highest percentage of total lipids, *Synechococcus* sp. HS01 exhibited

Venkatesh Balan (ed.), *Microbial Lipid Production: Methods and Protocols*, Methods in Molecular Biology, vol. 1995, https://doi.org/10.1007/978-1-4939-9484-7_23,

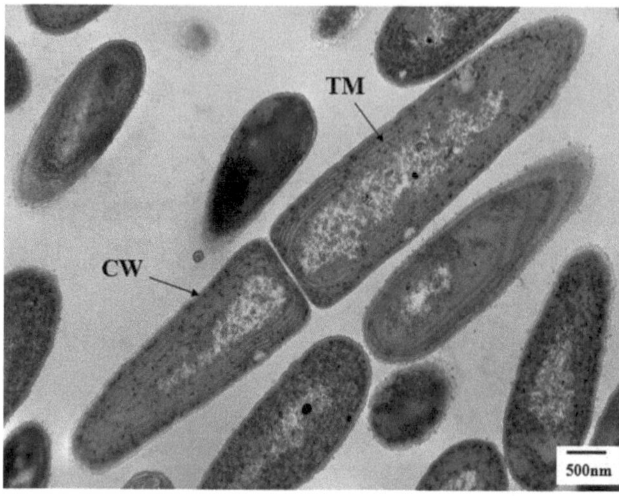

Fig. 1 Transmission electron microscopy (TEM) image of the cyanobacterial strain *Synechococcus* sp. HS01 grown in BG-11 medium. *CW* cell wall, *TM* thylakoid membrane

the highest lipid productivity, as a result of the difference in biomass production, making *Pseudanabaena* sp. SK01 a superior lipid-producing strain. Therefore, the use of appropriate methods to measure lipid levels and precise measurement of effective factors including growth rate, lipid yield, lipid content, and lipid productivity are essential in selection of a lipid-producing strain for commercial purposes.

The rate of lipid production is determined using different qualitative and quantitative methods. The traditional methods including extraction with organic solvents or using supercritical fluid extraction allow for quantitative measurement of lipids in cells [4]. However, these methods are time and energy-dependent, thus making them difficult to be used for screening of large number of strains [5]. One of the best methods for lipid extraction using organic solvents is Bligh and Dyer method [6], in which small amounts of solvent is used with recovery yields of ≥95% [7].

In recent years, fluorescent-based methods have been also been used as effective and rapid approaches for qualitative and quantitative measurement of lipids [8, 9]. Fluorescence microscopy and fluorescence spectroscopy are used to detect and count different fluorescent molecule types. Fluorescence microscopy techniques may provide fluorescent images and are used as qualitative methods for comparison of the lipid contents, while fluorescence spectroscopy methods may provide quantitative determination of lipid concentrations [9]. Nile red, a soluble-lipid fluorescent dye, is a common dye used in fluorescence-based methods to assess the lipid contents of mammalian and microbial cells [10–13]. Nile red staining method has several advantages such as simplicity, high

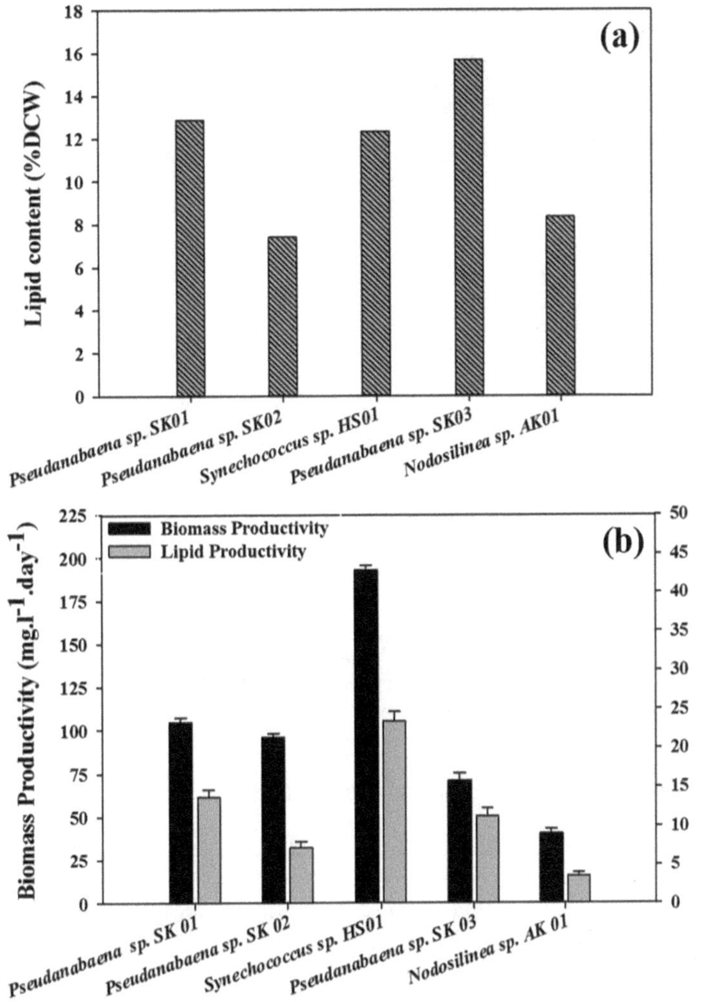

Fig. 2 Lipid content, biomass production, and lipid productivity of five strains of cyanobacteria grown in BG-11 medium (Modiri et al. 2015). (**a**) Lipid accumulation in percent dry cell weight (% dcw); (**b**) Biomass and lipid production rates in mb/L/day

sensitivity, small sample size, and low response time [14]. In this chapter, we present the results according to the modified procedure of Bligh and Dyer (Fig. 3) and Nile red staining methods (Fig. 4) for measurement of lipids in microbial cells (especially cyanobacteria) as considered in our experimental protocol.

2 Materials

1. Purified deionized water is used to prepare all the solutions, and the reagents are entirely of analytical grade.

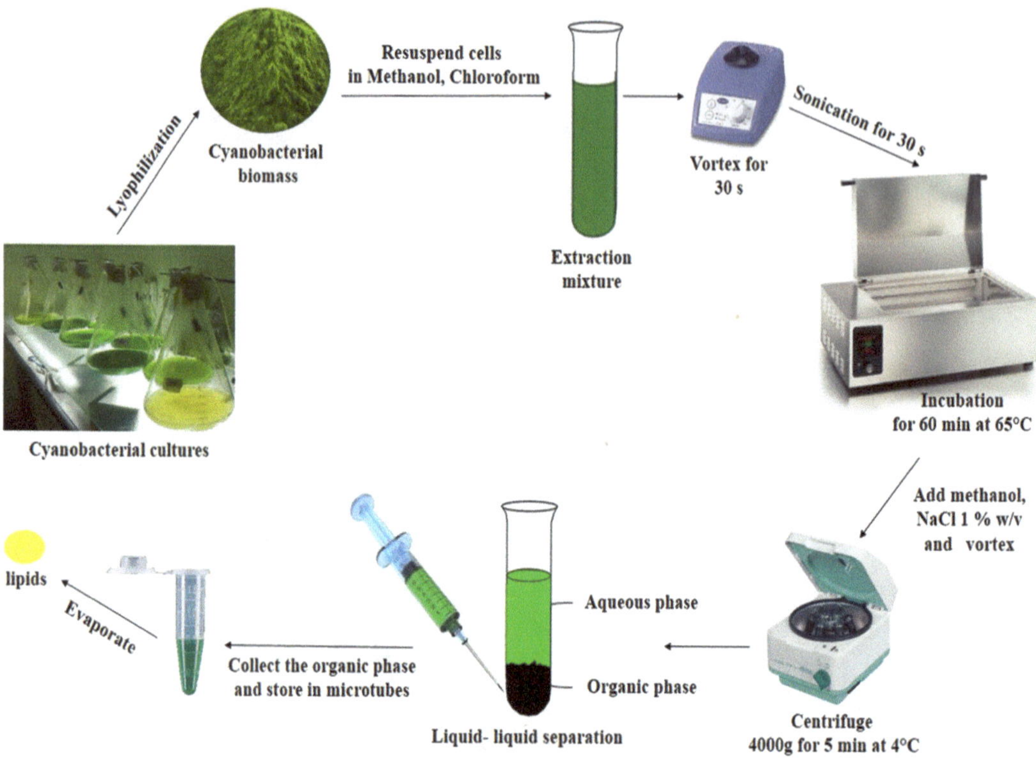

Fig. 3 Overview of modified Bligh and Dyer method

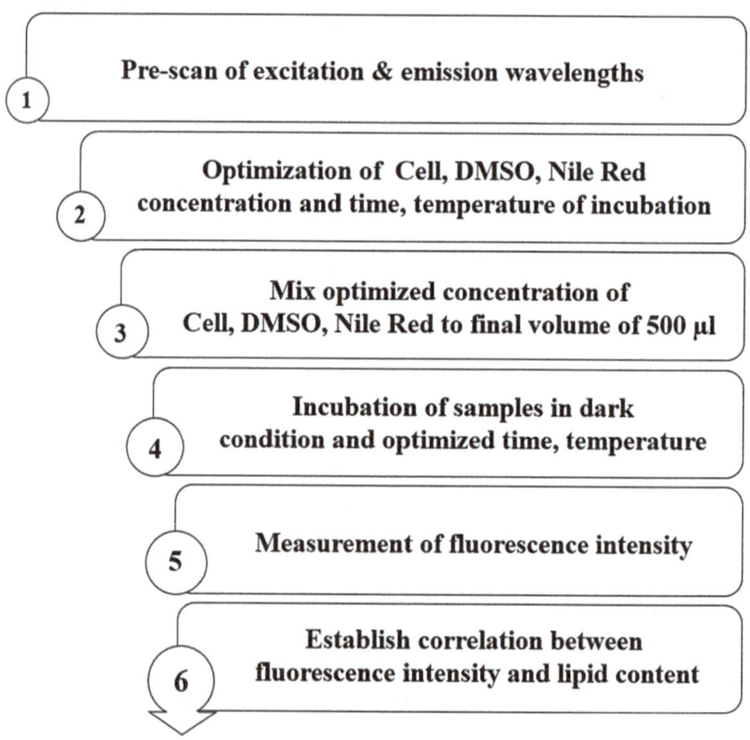

Fig. 4 Flowchart of an optimal procedure for Nile red staining

2. *PBS (1×)*: Add following components (8 g NaCl, 0.2 g KCl, 1.44 g Na_2HPO_4, and 0.24 g KH_2PO_4) to deionized water and bring volume to 1 L. After mixing it all up, adjust the pH of the solution to 7.4 (with HCl), sterilizing solution in autoclave (121 °C, 20 min). Put the buffer in a refrigerator (4 °C) before using.
3. *NaCl 1% (w/v)*: A 1% solution of NaCl is 100 mg NaCl in 10 ml distilled water.
4. *Nile red stock solution*: Nile red is a purple powder whose stock is prepared in acetone at a concentration of 1 mg/ml and stored at −20 °C in complete darkness.
5. Chloroform ($CHCl_3$) and methanol (CH_3OH) should be kept in the refrigerator before using.

3 Methods

3.1 Guidelines for Modified Bligh and Dyer Method

1. Two milliliters of bacterial cells is taken from growth medium in stationary phase cultures (*see* **Notes 1–3**) (Fig. 3).
2. To separate bacterial cells from culture, vials containing samples are centrifuged (4500 × *g* for 10 min) and the supernatant holding medium is discarded (*see* **Note 4**).
3. For waste separation, bacterial cells are washed twice with PBS 1× buffer and the buffer is separated by centrifugation (as in **step 2**) (*see* **Note 5**).
4. The samples should be freeze-dried (lyophilized) in appropriate conditions.
5. Freeze-dried cells (100 mg) are mixed with a total of 6 mL of solvent in the sequence: chloroform (4 mL), methanol (2 mL), thoroughly vortexed for 30 s immediately following the addition of each solvent (*see* **Notes 6–9**).
6. To lyse the cells, the resulting mixture is sonicated for 30 s (*see* **Note 10**).
7. The resulting mixture is incubated at 65 ° C for 1 h.
8. Cool the tube to reach ambient temperature.
9. 2 mL of methanol and 3.6 mL 1% NaCl (w/v) solution is added to the mixture and vortexed entirely (*see* **Note 11**).
10. The test tube is then centrifuged at 4000 × *g* for 5 min.
11. Weigh an empty test tube (m_1), and slowly separate the lower chloroform lipid–containing layer by using the syringe and add it into the test tube (*see* **Notes 12–14**).
12. Put the tube in a rotary evaporator (to remove the organic solvent) until a fine dried lipid sample is obtained.

13. Once again, weight the tube (m_2) and calculate the lipid contents C_L (% DCW) in accordance with the following formula (Fig. 2a) (*see* **Note 15**):

$$C_L = \frac{m_2 - m_1}{100} \times 100$$

14. Lipid productivity (P_L) (mg L^{-1} h^{-1}) is calculated for each strain according to the following formula [15] (*see* **Note 16**):

$$P_L = C_L \times \frac{\Delta DCW}{t}$$

- ΔDCW is the accumulated dry cell weight from inoculation to harvest, *t* is the culture time.

3.2 Guidelines for the Nile Red Staining Method

1. First, a prescan of excitation and emission based on Nile red, type of solvent, and spectrofluorometer device must be taken to make certain the wavelength that afford best combination is chosen (Fig. 4) (*see* **Notes 17** and **18**).
2. The Nile red concentration, time and temperature of incubation, concentration of stain carrier, and cell concentration should be optimized. As an example, the range of typical values for each of these cases with their optimized amounts to measure lipid contents in cyanobacterial strain *Synechococcus* sp. HS01 are shown in Table 1 (*see* **Notes 19–24**).
3. Optimal amounts of DMSO, Nile red, and cells determined in the previous step are mixed to final volume 500 mL in deionized water.
4. The resulting solution is optimized in accordance with the time specified in **step 2**, and incubated in the dark.
5. Immediately after the incubation time, the intensity of fluorescence is measured with a spectrofluorometer using the defined excitation & emission wavelengths (**step 1**),

Table 1
Recommended parameters for Nile red staining procedure

Parameter	Range	*Synechococcus* sp. HS01
DMSO concentration	5–40% (v/v)	40% (v/v)
Cell concentration	>5 × 10^4 cells/mL	~10^8 (OD_{750} = 0.1)
Nile red concentration	0.01–100 μg/mL	5 μg/mL
Incubation duration	2–40 min	4 min
Temperature of incubation	20–40 °C	25 °C

- It is imperative to use unstained cells as blank for each Nile red test, as in the case of autofluorescence, it would be possible to make a distinction between it and Nile red fluorescence (*see* **Notes 25** and **26**).

6. A triglyceride (e.g., Triolein) must be used to make a standard curve. To this end, after staining different concentrations of triglycerides with Nile red and measuring the fluorescence intensities, draw a standard curve, by plotting the triglycerides concentrations on the x-axis, and fluorescence intensities assay on the y-axis. Based on the fluorescence intensity of the samples and comparing it with the standard curve, the concentration of samples is determined (*see* **Note 27**).

4 Notes

1. When analyzing bacterial cultures, it is suggested to harvest the cells in stationary growth phase so that the cells are given enough time to accumulate lipids.
2. Two milliliters of culture in the stationary phase of growth is suitable for most bacteria to provide a sufficient number of cells.
3. Compare the lipid production values with cell growth, so as to calculate lipid productivity.
4. Cell density determination is required at this stage. Measure the OD at 600 nm for most bacteria and 750 nm for cyanobacteria.
5. Centrifuge at 4 °C to pellet the bacterial cells quickly and efficiently.
6. At this stage, deionized water can be used instead of Phosphate buffered saline (PBS) for the cell washing, if the bacterial cells are high durable in deionized water,
7. Chloroform is toxic and should be used only within a chemical (fume) hood.
8. Note that solvents such as chloroform are not compatible with all types of plastic tubes and tips.
9. At this stage, different volumes of chloroform and methanol can be used, but maintain exactly a ratio of 2:1 chloroform–methanol.
10. At this stage, the tubes should be closed completely in order to prevent evaporation of solution before the next stage.
11. Depending on the cell type and the resistance of the bacterial cell wall, the duration of sonication can be different. The appropriate total sonication time for cyanobacterium *Synechococcus* sp. HS01 is 30 s.

12. If in **step 5** you have used the additional amounts of chloroform and methanol, in this stage varying amounts of methanol and NaCl solution should be applied so that the ratio of methanol–chloroform–NaCl is 1:1:0.9.
13. At this point, the solid contents of the cell stick to the end of the tube, and the solution containing NaCl solution and methanol is added to the upper phase. Care must be taken during separation of chloroform–lipid fraction so that lower layer does not mix with the upper phase.
14. If you are not able to completely separate the lower layer, then add 4 mL of chloroform to the tube and repeat **steps 9** and **10**. If different amounts of chloroform are used in **step 5**, use similar values at this stage.
15. The solution should be clear. In case of existence of cell particles, the solution should be filtered using a syringe filter (0.2 μm).
16. If the aim of study is to compare the lipid contents, a strain which has the highest percentage of lipids at this stage should be selected.
17. If the objective is to choose a suitable strain with biotechnological and industrial significance, the selection of strain should be based on lipid productivity (Fig. 2).
18. Use an excitation wavelength range between 450 and 590 nm and emission between 570 and 670 nm.
19. Various solvents such as acetone, ethanol, dimethylformamide (DMF), isopropanol, ethylene glycol, hexane, or chloroform can be used with glycerol and EDTA. However, DMSO is used as the most common stain carrier [16, 17].
20. A concentration of bacterial cells should be selected so that a linear correlation is established between concentration and fluorescence intensity [18].
21. The intensity of fluorescence initially increases with increasing Nile red concentration and then decreases with excessive intensities. Therefore, the optimal concentration should be taken.
22. After period of time, quenching of the Nile red fluorescence occurs, and the fluorescence will be diminished.
23. Nile red fluorescence quenching is time-dependent as well as strain-specific. This means that fluorescence quenching is different for each strain and after a while quenching occurs due to the reaction with cellular components [19].
24. Incubation must be performed in complete darkness.
25. For qualitative comparison, it is possible to use fluorescence microscopy instead of spectrofluorometer.

26. If the sample size is large or only small amounts of sample is available, the fluorescence intensity can be measured with a microplate reader; however, spectrofluorometry is more sensitive (can detect lipids at lower concentrations) and more accurate.
27. If the objective is to compare the lipid production based on the fluorescence intensity, maximum production can be attained at this stage.

References

1. Thomson N, Summers D, Sivaniah E, Abe H, Doi Y, Taguchi S et al (2010) Synthesis, properties and uses of bacterial storage lipid granules as naturally occurring nanoparticles. Soft Matter 6:4045
2. Liu X, Sheng J, Curtiss R (2011) Fatty acid production in genetically modified cyanobacteria. Proc Natl Acad Sci 108:6899–6904
3. Modiri S, Hajfarajollah H, Zahiri HS, Sharafi H, Noghabi KA, Zamanzadeh Z et al (2015) Lipid production and mixotrophic growth features of cyanobacterial strains isolated from various aquatic sites. Microbiology 161:662–673
4. Halim R, Danquah MK, Webley PA (2012) Extraction of oil from microalgae for biodiesel production: a review. Biotechnol Adv 30:709–732
5. Satpati GG, Pal R (2014) Rapid detection of neutral lipid in green microalgae by flow cytometry in combination with Nile red staining-an improved technique. Ann Microbiol 65:937–949
6. Bligh EG, Dyer WJ (1959) A rapid method of total lipid extraction and purification. Can J Biochem Physiol 37:911–917
7. Iverson SJ, Lang SL, Cooper MH (2001) Comparison of the Bligh and Dyer and Folch methods for total lipid determination in a broad range of marine tissue. Lipids 36:1283–1287
8. Abdo SM, Ahmed E, El-enin SA, El Din RS, El Diwani G, Ali G (2014) Qualitative and quantitative determination of lipid content in microalgae for biofuel production. J Algal Biomass Util 5:23–28
9. Gikonyo B (2013) Advances in biofuel production: algae and aquatic plants. CRC Press (Taylor & Francis Group), Boca Raton FL
10. Chen W, Zhang C, Song L, Sommerfeld M, Hu Q (2009) A high throughput Nile red method for quantitative measurement of neutral lipids in microalgae. J Microbiol Methods 77:41–47
11. Genicot G, Leroy JLMR, Van Soom A, Donnay I (2005) The use of a fluorescent dye, Nile red, to evaluate the lipid content of single mammalian oocytes. Theriogenology 63:1181–1194
12. Evans CT, Ratledge C, Gilbert SC (1985) A rapid screening method for lipid-accumulating yeast using a replica-printing technique. J Microbiol Methods 4:203–210
13. McGinnis KM, Dempster TA, Sommerfeld MR (1997) Characterization of the growth and lipid content of the diatom *Chaetoceros muelleri*. J Appl Phycol 9:19–24
14. Castro GR, Larson BK, Panilaitis B, Kaplan DL (2005) Emulsan quantitation by Nile red quenching fluorescence assay. Appl Microbiol Biotechnol 67:767–770
15. Lv JM, Cheng LH, Xu XH, Zhang L, Chen HL (2010) Enhanced lipid production of *Chlorella vulgaris* by adjustment of cultivation conditions. Bioresour Technol 101:6797–6804
16. Alemán-Nava GS, Cuellar-Bermudez SP, Cuaresma M, Bosma R, Muylaert K, Ritmann BE et al (2016) How to us Nile red, a selective fluorescent stain for microalgal neutral lipids. J Microbiol Methods 128:74–79
17. Castell LL, Mann R (1994) Optimal staining of lipids in bivalve larvae with Nile red. Aquaculture 119:89–100
18. Rumin J, Bonnefond H, Saint-Jean B, Rouxel C, Sciandra A, Bernard O et al (2015) The use of fluorescent Nile red and BODIPY for lipid measurement in microalgae. Biotechnol Biofuels 8:42
19. Pick U, Rachutin-Zalogin T (2012) Kinetic anomalies in the interactions of Nile red with microalgae. J Microbiol Methods 88:189–196

Chapter 24

Production of Oleaginous Organisms or Lipids Using Sewage Water and Industrial Wastewater

Farha Deeba, Vikas Pruthi, and Yuvraj S. Negi

Abstract

Worldwide, wastewater produced from sewage and industry poses a serious risk to the surrounding environment. As a way to address this problem, an integrated approach for cultivation of oleaginous microorganisms on wastewater leading to effective removal of hazardous components and sustainable production of biodiesel is proposed. Oleaginous yeasts have the unique ability to utilize wastewater as feedstock and accumulate large amounts of triacylglycerols within their cellular compartments at stationary phase (144 h). The lipids stored in an oleaginous microbe can be visualized by fluorescence microscopy and converted into biodiesel through transesterification after extraction. Here, we describe the batch cultivation of oleaginous yeast on sewage and industrial wastewater at 25 °C. High lipid accumulation with efficient removal of toxic chemicals can be achieved by utilizing this integrated method.

Key words Oleaginous yeast, Wastewater, Integrated approach, Lipid accumulation, Nile red, Biodiesel

1 Introduction

An integrated method is an effective way for reduction of extremely toxic chemicals from industrial and sewage wastewater with simultaneous production of energy as a by-product via oleaginous microorganisms [1, 2]. The use of oleaginous microorganisms decreases the dependency of western societies on vegetable sources and fossil fuels [3]. These oil-rich microorganisms include several bacteria, fungi, yeasts, and microalgae. Lipids originating from oleaginous microbes are generally used as material in fine chemical production, personal care products, paints, coatings, and biodiesel, thus lessening the reliance on mineral and vegetable oil [3, 4]. Among various oleaginous microbes, production of oils by using oleaginous yeasts has attracted significant attention due to its high productivity. Moreover, oleaginous yeasts have the ability to grow on renewable carbon sources and do not require light for their cultivation unlike photoautotrophic microalgae. In addition, they are seasonally

Venkatesh Balan (ed.), *Microbial Lipid Production: Methods and Protocols*, Methods in Molecular Biology, vol. 1995, https://doi.org/10.1007/978-1-4939-9484-7_24,

independent, easy to maintain at their growth conditions, and have increased final cell density which both reduces input costs and enables 24 h per day production [5–7]. Oleaginous yeasts can store huge quantities of oil inside their lipid bodies in the form of triacylglycerols (TAGs) with 20–80% lipid content range depending upon culture conditions and species which can be converted to fatty acid methyl esters (FAMEs) easily through transesterification [5, 8]. Several oleaginous yeasts strains such as *Lipomyces lipofer*, *Cryptococcus curvatus*, *Rhodosporidium toruloides*, *Rhodotorula glutinis*, *Trichosporon pullulans*, *Candida*, and *Yarrowia lipolytica* have been studied for producing lipids on wastewater. Furthermore, screening is performed in an attempt to isolate new oleaginous yeast species with high lipid titer [5, 7, 9, 10]. But, the cost of feedstock is a major obstacle and is responsible for up to 50% of the biodiesel production costs [11]. Essential nutrients and elements present in wastewater can be utilized by oleaginous yeasts as a substrate which reduces production costs [1, 2].

In this chapter methods that support the screening of oleaginous yeasts are detailed. TAG accumulation is visualized by intracellular labeling of the yeast strain with fluorescent Nile red (493/505 nm) stain using fluorescent microscope at different time intervals. The TAG accumulation is directly correlated to the lipid body size and can be monitored without altering the cellular integrity [12]. Growth conditions and nutritional requirements of oleaginous yeasts to produce maximum TAGs in wastewater are optimized and compared to GSM (glucose synthetic media) used in industries. Lipid obtained is characterized by TLC (thin layer chromatography) and FTIR (Fourier transform infrared). The transesterification is carried out using BF_3–methanol, and the FAME obtained is verified by GC-MS (gas chromatography–mass spectroscopy) analysis. Recently, studies on the role of oleaginous yeasts grown on industrial wastewater for integrated hazardous content removal and sustainable production of biodiesel have been carried out. For example, cultivation of the oleaginous yeast *Rhodosporidium kratochvilovae* HIMPA1 on pulp and paper mill effluent gives 8.56 g/L of lipid yield with effective decrease in hazardous components from effluent that is, chemical oxygen demand (COD), biological oxygen demand (BOD), total dissolved solids (TDS), lignin, phenol, and color [1]. In another study, the oleaginous yeast *Debaryomyces etchellsii* strain BM1 reported a lipid yield of 1.2 g/L with significant removal of COD from olive mill wastewater and expired soft drinks [13]. In a study performed by Karakaya et al., *Rhodotorula glutinis* showed significant phenol reduction from raw olive mill wastewater [14].

Using oleaginous yeasts for biodiesel production could prove cost-effective when low-cost substrates such as sewage and industrial wastewater are utilized as media for their growth compared to synthetic media. The basic steps involved in lipid production by

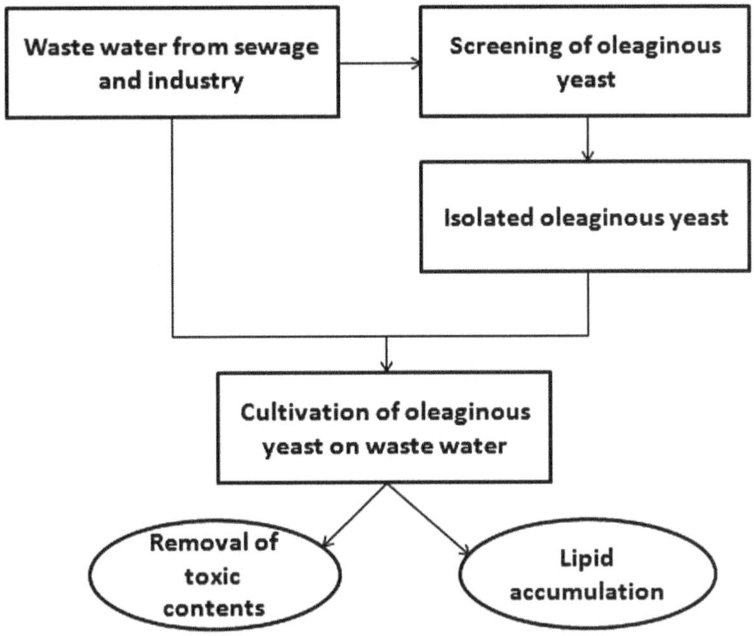

Fig. 1 Integrated method for lipid production and toxic content removal from wastewater by oleaginous yeasts

oleaginous yeasts on wastewater are outlined in Fig. 1. Firstly, the oleaginous yeasts are isolated from wastewater followed by their cultivation on wastewater to investigate their lipid agglomeration and toxic removal capability. The aim of this chapter is to reveal the potential of the utilization of sewage and industrial wastewater and oleaginous yeasts that meet the aforementioned criteria for suitable production of lipids for biodiesel application and wastewater treatment.

2 Material

All solutions are prepared using ultrapure water produced by purifying deionized water and analytical grade reagents. Prepare and store all reagents at 4 °C before use. The wastewater used in this study is obtained from paper mill industry.

2.1 Screening of Oleaginous Yeasts

1. Yeast extract–peptone–dextrose (YEPD) agar plates: agar 18 g/L, yeast extract 3 g/L, glucose 10 g/L, peptone 5 g/L; malt extract 3 g/L (*see* **Note 1**).
2. Agar plates of nitrogen limiting medium (NLM) is prepared by using glucose 20 g/L; vitamins: PABA 0.2 g/L, niacin 0.4 g/L, pyridoxine–HCl 0.4 g/L, riboflavin 0.2 g/L, inositol,2 g/L, thiamine 0.4 g/L, folic acid 0.4 g/L, calcium pantothenate 0.002 g/L, D-biotin, 0.002 g/L; trace elements (mg/L):

$MnSO_4$ 0.4 mg/L, $FeCl_3$, 0.2 mg/L, $NaMO_3$ 0.2 mg/L, $ZnSO_4$ 0.4 mg/L, KI, 0.1 mg/L, $CuSO_4$ 0.04 mg/L, boric acid 0.5 mg/L; salt solutions $CaCl_2$ 0.1 mg/L, $(NH_4)_2SO_4$ 0.1 g/L, $MgSO_4$ 0.5 g/L, and KH_2PO_4 1 g/L (pH 7) [5, 7].

2.2 Nile Red Staining

1. Materials: centrifuge, fluorescent microscope, eppendorf, saline water, *phosphate-buffered* saline (PBS), Nile red stain (9-diethylamino-5*H*-benzo[a]-phenoxazine-5-one), dimethyl sulfoxide (DMSO), glass slides, glass coverslip.
2. PBS is prepared by dissolving 0.24 g of KH_2PO_4, 1.44 g of Na_2HPO_4, 0.2 g of KCl, and 8 g of NaCl in 800 mL ultrapure water. The pH is adjusted to 7.4 with HCl and volume to 1 L with ultrapure H_2O. Sterilize by autoclaving.
3. Nile red solution is prepared by dissolving 0.1 mg Nile red stain per mL of DMSO (*see* **Note 2**).

2.3 Nutrient Removal

2.3.1 Inductively Coupled Plasma (ICP) and Ion Chromatography (IC)

1. 10% v/v HNO_3 is prepared using 10 mL of HNO_3 in 90 mL of ultrapure water (*see* **Note 3**).
2. pH meter.
3. ICP equipment.

2.3.2 BOD

1. Wastewater sample, specialized BOD bottle.

2.3.3 COD

1. 0.1 N sodium sulfate prepared by dissolving 15.811 g of sodium sulfate in 1000 mL of ultrapure water, wastewater sample (*see* **Note 4**).
2. Starch solution (1%), potassium iodide solution, 0.1 N potassium dichromate solution, conical flask (100 mL).
3. Burette and ice bath.

2.3.4 Protein Measurement

1. Reagent A: 2% Na_2CO_3 in 0.1 N NaOH.
2. Reagent B: 0.5% $CuSO_4 \cdot 5H_2O$ (1 mL) + 1.0% sodium potassium tartrate (1 mL).
3. Reagent C: 48 mL of reagent A + 2 mL of reagent B (*see* **Note 4**).
4. Water bath (30 °C), test tubes (30 mL), volumetric flask (50 mL).
5. UV spectrophotometer.

2.4 Batch Cultivation

1. The 2% seed culture is prepared by growing yeast strains on YEPD media for 24 h.
2. The GSM is prepared using yeast nitrogen base (YNB) (w/o amino acids) 6.7 g/L; amino acid mixture: L-tyrosine 0.1 g/L L-aspartic acids 0.1 g/L, L-glutamic acid 0.1 g/L, L-

threonine 0.2 g/L, L-serine 0.4 g/L, L-valine 0.15 g/L, L-phenylalanine 0.05 g/L, L-isoleucine 0.03 g/L, L-leucine 0.03 g/L, L-methionine 0.02 g/L, L-arginine 0.02 g/L, L-histidine 0.02 g/L, L-tryptophan 0.02 g/L; salt solutions: $(NH4)_2SO_4$ 5.0 g/L and KH_2PO_4 1 g/L; carbon source: glucose, 20 g/L.

3. Incubator shaker (*see* **Note 5**), wastewater, autoclave, Erlenmeyer flask (500 mL).

2.5 Lipid Extraction

1. Teflon-capped glass vials (50 mL), centrifuge, distilled water, probe sonicator (700 W, 20 kHz).
2. 2:1 v/v chloroform–methanol solution is prepared by mixing 20 mL of chloroform and 10 mL of methanol.
3. 0.034% $MgCl_2$ is prepared by dissolving 0.034 g of $MgCl_2$ in 100 mL of ultrapure water.
4. 2 N KCl is prepared by dissolving 14.912 g of KCl in 100 mL of ultrapure water. Add 20 mL of 2 N KCl to 5 mL of methanol to prepare 4:1 (v/v) 2 N KCl–methanol solution.
5. The chloroform–methanol–water (3:48:47, v/v/v) is prepared by adding 3 mL of chloroform, 48 mL of methanol and 47 mL of ultrapure water (*see* **Note 4**).

2.6 TLC

1. The hexane–diethyl ether–acetic acid (85:15:1, v/v/v) solution is prepared by dissolving 85 mL of hexane, 15 mL of diethyl ether, and 1 mL of acetic acid.
2. $MnCl_2$ methanolic solution is prepared by combining 0.63 g of $MnCl_2{\cdot}4\ H_2O$, 60 mL of methanol, 60 mL of water and 4 mL of sulfuric acid (*see* **Note 6**).

2.7 FTIR Spectrometry

1. Potassium bromide (*see* **Note 7**), capillary tube.
2. Triolein, extracted lipids.

2.8 Transesterification of Lipids

1. 14% boron trifluoride–methanol reagent, extracted lipid.
2. Boiling water bath, centrifuge.
3. Hexane.
4. Ultrapure water.
5. Anhydrous sodium sulfate (*see* **Note 8**).

3 Method

3.1 Screening of Oleaginous Yeasts

1. Wastewater collected is serially diluted using 0.9% sterilized saline water and cultured on agar plates of YEPD at 25 °C for 48 h.

2. The colonies appearing on agar plates are selected and reinoculated on NLM agar plates [5].
3. The plates are incubated at 25 °C. The selected oleaginous yeast colonies showing increased amount of TAG accumulation are then grown on NLM broth to evaluate lipid droplet and cell size by Nile red stain at stationary phase (144 h).

3.2 Nile Red Staining

1. 100 μL of the yeast cell culture is centrifuged at 9000 × *g* for 8 min. The pellet is resuspended in 1 mL of PBS and vortexed for 1 min at room temperature (25 °C).
2. Add 6 μL of Nile red (1 mg/mL in dimethyl sulfoxide) to the above cell suspension and incubated in the dark for 8 min at room temperature (25 °C). Wash pellet thrice with 0.9% saline water (unsterilized) before visualization (*see* **Notes 9** and **10**).
3. Cell diameter (major axis recorded at 50 μm scale bar) of about 100 cells is estimated using imaging software (*see* **Note 11**).

3.3 Nutrient Removal Analysis

1. Wastewater is collected in bulk from the industry. Remove solid particles through sedimentation and filtration using 0.2 μm filters. Wastewater should be stored at 4 °C in the dark after filtration until needed for experiments. The pH of the wastewater is monitored using a pH meter and adjusted to pH 7 with hydrochloric acid (35%).
2. The nutrient removal capacity is analyzed at the beginning and end of the experiment by measuring physicochemical parameters such as nitrogen, phosphate, COD, BOD and various other trace elements (potassium, sodium, iron, manganese, zinc, and calcium).
3. At the end of the experiment, yeast biomass is obtained by centrifugation at 3000 × *g* for 10 min at 4 °C and the wastewater supernatant obtained is filtered using 0.2 μm filters. Filtrates are subjected to physicochemical analysis. The removal percentage of nutrients is calculated by using the following equation:

$$\%\text{Nutrient removal efficiency} = \left[{}^{(S_o - S_f)}/_{S_o}\right] \times 100$$

where S_o and S_f are initial concentration of wastewater before consumption and final concentration of wastewater after consumption, respectively [2].

3.3.1 ICP Analysis

Quality testing of the wastewater collected is done to evaluate PO_4^- for total phosphorous (TP), NO_3^-, NO_2^-, and NH_4^+ for total nitrogen (TN), and trace elements such as potassium, sodium, iron, manganese, zinc, and calcium using ion chromatography.

1. Water samples are filtered and stabilized to a pH below 2 using HNO_3 prior to analysis for trace element detection. Samples are prepared using ultrapure water (*see* **Note 12**).
2. About 10 mL of acidified sample is transferred to a test tube for quality estimation using argon as carrier gas. Sample flow rate is 2.0 mL/min and total analysis time per sample is approximately 2.5 min (*see* **Note 13**).
3. Calibration curves for NO_3^-, NO_2^-, NH_4^+, TN, PO_4^-, and TP are prepared using ultrapure water with concentrations ranging from 0 to 10 mg/L.
4. Calibration curves for multielement standards are prepared in 10% v/v HNO_3 and contained 6 mg/L of each element. The blank solution used is 10% v/v HNO_3 solution.

3.3.2 BOD Measurement

1. The specialized 300 mL BOD bottles are filled with water sample to be tested and an airtight seal is used with no air space (*see* **Note 14**). A control or blank bottle is filled with dilution water.
2. A dissolved oxygen (DO) meter is used to measure the initial dissolved oxygen concentration (mg/L). The bottle is then placed into a dark incubator at 20 °C for 5 days.
3. After 5 days of incubation, the DO meter is used again to measure a final dissolved oxygen concentration (mg/L).
4. The final DO value is subtracted from the initial DO value and the resulting value is the BOD concentration (mg/L) (*see* **Note 15**).

3.3.3 COD Measurement

1. Take 10 mL of water sample in three 100 mL conical flasks marked as T1, T2, and T3. Simultaneously prepare blanks 1, 2, 3 in 100 mL conical flasks using distilled water.
2. Add 5 mL of potassium dichromate solution to each conical flask and place them in water bath at 100 °C for 1 h.
3. After cooling the sample to room temperature, add 5 mL potassium iodide and 10 mL sulfuric acid to each flask.
4. Titration is done using 0.1 N sodium thiosulfate until the blue color disappears completely.

   Calculation:

$$\text{COD as O2 mg/L} = \frac{(A - B) \times N \times 8000}{\text{mL sample}}$$

   where A = mL sodium thiosulfate used for blank, B = mL sodium thiosulfate used for sample, N = normality of sodium thiosulfate, and 8000 = milliequivalent weight of oxygen × 1000 mL/L.

3.3.4 Protein Measurement

1. Folin–Ciocalteu reagent was diluted with deionized water (1:1).
2. Add 5.5 mL of the reagent C and 0.5 mL of the Folin–Ciocalteu reagent–water (1:1) to 1.0 mL wastewater sample.
3. Vortex the mixture well and incubate in the dark for 30 min.
4. Absorbance is measured at 660 nm by UV spectrophotometer [15, 16]. Calibration curves are prepared using bovine serum albumin (BSA) at concentrations of 0.05–1 mg/mL (*see* **Note 16**). Plot the absorbance against protein concentration to get a standard calibration curve.
5. Check the absorbance of the sample collected and find out its concentration using the standard curve plotted.

3.4 Batch Cultivation

1. Take 250 mL prefiltered, autoclaved wastewater and add 7% glucose to utilize as a growth medium for cultivating oleaginous microorganism (*see* **Notes 17** and **18**).
2. Batch cultivation is performed by inoculating wastewater (250 mL) with 2% of the seed culture in Erlenmeyer flask (500 mL) at 25 °C for 144 h with 200 revolutions per minute (rpm) in an incubator shaker at a pH 7 (*see* **Note 19**).
3. The yeast growth is measured by taking absorbance at 600 nm. A calibration curve was plotted with the absorbance versus cell dry weight. Use glucose synthetic medium (GSM) supplemented glucose as control.
4. The cell dry weight is measured by using the following process: [1] 6 mL of culture broth is centrifuged at 5000 × *g* for 8 min [2]. The cells are washed thrice with 6 mL of deionized water [3]. The pellet is dried at 55–75 °C to constant weight (*see* **Note 20**).

3.5 Lipid Extraction

1. The culture broth is transferred to Teflon-capped glass vials (50 mL), centrifuged at 4000 × *g* for 6 min, and the supernatant discarded. The pellet is washed thrice with distilled water.
2. The yeast cells are lysed by ultrasonication at 20 kHz for 4 min.
3. Add 10 mL of chloroform–methanol (2:1; v/v) and stir for 1 h.
4. Filter the extract and add 5 mL of 0.034% $MgCl_2$ followed by centrifugation at 4000 × *g* for 6 min.
5. Add 1 mL of 2 N KCl–methanol (4:1, v/v) to the lower organic phase and vortex for 2 min. Add 5 mL of artificial upper phase (chloroform–methanol–water; 3:48:47, v/v/v) to the above mixture. Continue adding the artificial layer (5 mL) to the mixture until the phase boundary becomes clear.

6. The lower layer of chloroform is transferred to a round-bottom flask (25 mL), and the lipid yield is determined gravimetrically [17] (*see* **Note 21**).
7. After harvesting, the lipid yield (g/L) is measured gravimetrically and averaged results were obtained. The lipid content (%), biomass productivity (mg/L/h), and lipid productivity (mg/L/h) of the oleaginous yeasts are measured by the following equations:

$$\text{Lipid content (\%)} = \frac{\text{Total lipid (g/L)}}{\text{Cell dry weight (g/L)}} \times 100$$

$$\text{Biomass productivity(mg/L/h)} = \frac{\text{Cell dry weight (g/L)}}{\text{Cultivation time (h)}} \times 100$$

$$\text{Lipid productivity (mg/L/h)} = \frac{\text{Total Lipid (g/L)}}{\text{Cultivation time (h)}} \times 100$$

3.6 TLC

1. A TLC plate is spotted with approximately 1–2 μL of lipid sample and triolein using capillary tubes (*see* **Note 22**).
2. The solvent system of hexane–diethyl ether–acetic acid (85:15:1, v/v/v) was used as mobile phase and compared with triolein as standard.
3. The plate is dipped into $MnCl_2$ methanolic solution to develop the chromatograms.
4. The plate is dried and heated for 15 min at 120 °C [5, 6, 18].

3.7 FTIR Spectrometry

1. The extracted lipid is confirmed by FTIR spectrometer (*see* **Note 23**).
2. Thin transparent pellets of potassium bromide are prepared by grinding and mixing the powdered potassium bromide followed by pressing with hydraulic press (125 kg/cm^3) under vacuum.
3. A small quantity of the sample is deposited on a transparent pellet of potassium bromide with the help of a capillary to form a thin film.
4. The spectra are recorded from 4000–400 cm^{-1} with a spectral resolution of 4 cm^{-1} at 16 scans (*see* **Notes 24** and **25**).
5. Triolein is used as standard for comparison with the lipid extracted from oleaginous yeasts.

3.8 Transesterification of Lipids

1. Transesterification of the lipid sample is carried out using the method of Morrison and Smith [19].
2. Add 1 mL of boron trifluoride–methanol reagent per 4–16 mg of lipid sample (*see* **Note 26**).

3. The reaction mixture is heated in a boiling water bath at 80 °C for 20 min.
4. Add 2 mL of hexane–water (2:1 v/v) preceded by centrifugation at 3000 × *g* until both layers are clearly seen.
5. The upper hexane layer is collected for FAME analysis.

3.9 GC-MS

1. FAME quantification is carried out using a GC-MS equipped with DB-5MS capillary column (30 m × 0.25 mm × 1 μm).
2. The operating conditions are as follows: injection volume 1 μL, splitless injection mode, helium as a carrier gas at 1 mL/min, initial column temperature set at 50 °C (1.5 min), injector temperature of 250 °C, detector temperature of 280 °C, temperature gradient: oven temperature started at 180 °C for 1 min and increased at the rate of 25 °C/min to 220 °C for 1 min and finally ramped to 250 °C at 15 °C/min, hold for 3 min. The temperatures are maintained at 250 °C and 200 °C for the mass transfer line and ion source, respectively. The FAME detection is done using a 50–600 m/z scan mode and 70 eV electron ionization [5].
3. Methyl heptadecanoate is used as the internal standard. The quantification and identification of FAMEs are accomplished by using (NIST11.L database) library search reports.

3.10 Estimation of Biodiesel Fuel Properties

Predictive equations based on fatty acid composition were used for the calculation of critical biodiesel properties. The physical properties of biodiesel such as high heating value (HHV), cetane number (CN), saponification value (SV), iodine value (IV), density (D), kinematic viscosity (kV), cold filter plugging property (CFPP), degree of unsaturation (DU), long chain saturation factor (LCSF), and oxidative stability (OS) are calculated using the following series of empirical correlations: [12].

$$\mathrm{SV} = \sum 560 \times \frac{F}{M}$$

$$\mathrm{IV} = \sum 254\ N \times \frac{F}{M}$$

$$\mathrm{CN} = 46.3 + \frac{5458}{\mathrm{SV}} - (0.255 \times \mathrm{IV})$$

$$\mathrm{LCSF} = (0.1 \times \mathrm{C16}) + (0.5 \times \mathrm{C18})$$

$$\mathrm{HHV} = 49.43 - 0.041\ (\mathrm{SV}) - 0.015\ (\mathrm{IV})$$

$$\mathrm{Density} = 0.8463 + 4.9/\sum M + 0.0118 \times \sum N$$

$$\mathrm{DU}\ (\%) = \mathrm{MUFA} + (2 \times \mathrm{PUFA})$$

$$\mathrm{OS} = 117.9295/(\%\mathrm{C18}:2 + \%\mathrm{C18}:3) + 2.5905$$

$$\ln\ (\mathrm{KV}) = -12.503 + 2.496 \times \ln\left(\sum M\right) - 0.178 \times \sum N$$

$$\mathrm{CFPP} = (3.417 \times \mathrm{LCSF}) - 16.477$$

Here, *M* represents molecular mass, *N* represents the number of double bonds and *F* represents the percentage of each fatty acid constituent, PUFA indicates the weight percentage of the polyunsaturated fatty acids, MUFA indicates the weight percentage of the monounsaturated fatty acids, and C16 and C18 represent weight percentages of saturated fatty acids (SFA) of individual chain length.

4 Notes

1. The YEPD plates are supplemented with chloramphenicol 1 μg/mL to avoid bacterial growth.
2. The Nile red solution preparation as well as staining of cells is carried out in dark color vials because light can affect the Nile red solution which in turn affects the staining of cells.
3. The standard solutions are prepared from the stock solution in 10% HNO_3.
4. All solutions should be prepared fresh each time before usage.
5. The temperature of the incubator is maintained at 25 °C, and the incubator is fumigated to avoid any contamination.
6. $MnCl_2$ solution is used as the staining solution. The TLC plates are submerged into the staining solution to visualize the bands.
7. Potassium bromide should be kept in a vacuum oven at 40 °C for 24 h before use to remove moisture.
8. The impurities and moisture are removed when the FAME mixture is passed through a filter containing anhydrous sodium sulfate.
9. Washing with saline water removes unbound Nile red from the solution mixture.
10. Stained cells are visualized under a fluorescent microscope (EVOS-FL, Advance Microscopy Group, AMG, USA). Intracellular lipid bodies are visualized by Nile red staining method to know the highest lipid-accumulating phase of oleaginous microorganisms.
11. Imaging software (e.g., Image J 1.49 a) can be used for cell and lipid body size estimation.
12. Trace metals are ubiquitous in the environment; therefore, it is necessary to protect from external contamination such as dust particles. All areas/materials involved in analysis and

preparation of sample must be dust-free to minimize the chance of contamination. Blend the samples uniformly prior to analysis.

13. Optimization will be necessary depending on the instruments being used, and instrument parameters may vary.
14. During a BOD test for proper biological activity, the wastewater sample should be free from chlorine. The wastewater sample containing chlorine should be supplemented with sodium sulfite (a dechlorination chemical) before testing. Centrifuging the wastewater samples prior to analysis helps remove insoluble debris which may produce uncertainty in data.
15. If the wastewater sample required dilution, the BOD concentration value is multiplied by the dilution factor.
16. The corresponding multiplication is carried out for the diluted sample.
17. The samples should be filtered through a 0.45 μm pore diameter membrane filter before analysis.
18. Autoclave the wastewater before use at 121 °C for 30 min.
19. Batch cultivation is performed to investigate the effect of wastewater on the lipid and growth accumulation of the selected oleaginous yeasts.
20. All experiments are carried out in triplicate. The repeat measurements are taken to assess the precision of the experimental analysis and average values with standard deviations that are depicted. The duplicate samples are prepared independently of each other and analyzed randomly along with all other samples. The one-way analysis of variance (ANOVA) is also used to determine the significant differences among independent groups, if comparison is to be made between two or more oleaginous yeast species.
21. Extraction of lipids is done following the modified protocol of Bligh and Dyer. For lipid measurements, the organic phase is transferred before solvent removal to a round bottom-flask and solvent is evaporated at 40 °C under vacuum in a rotary evaporator.
22. A small volume of sample (1–2 μL) should be spotted on TLC plate to avoid merging of different samples.
23. FTIR spectrometer of PerkinElmer equipped with a deuterated triglycerine sulfate (DTGS) detector is used to analyze different functional groups present in the lipid sample. The parameters of the instrument should be optimized to ensure system suitability.

24. Insert the sample in the instrument, make sure the sample is covering the window, and ensure that the sample is aligned with the IR source.
25. If a spectrum has too much baseline noise, increase the number of scans or reduce the resolution.
26. The reaction conditions are optimized first to obtain the high yield of FAMEs. The yield of FAME obtained may vary with the change in reaction conditions such as catalyst-to-lipid ratio, temperature, time, and the type of microorganism.

Acknowledgments

This work was supported by the Ministry of Human Resource Development (MHRD), Government of India.

References

1. Patel A, Arora N, Pruthi V, Pruthi PA (2016) Biological treatment of pulp and paper industry effluent by oleaginous yeast integrated with production of biodiesel as sustainable transportation fuel. J Clean Prod 142:2858–2864
2. Arora N, Patel A, Sartaj K, Pruthi PA, Pruthi V (2016) Bioremediation of domestic and industrial wastewaters integrated with enhanced biodiesel production using novel oleaginous microalgae. Environ Sci Pollut Res 23:20997–21007
3. Lamers D, Biezen NV, Martens D, Peters L, Zilver EVD, Dreumel NJ-N, Wijffels RH, Lokman C (2016) Selection of oleaginous yeasts for fatty acid production. BMC Biotechnol 16:45
4. Beopoulos A, Verbeke J, Bordes F, Guicherd M, Bressy M, Marty A, Nicaud JM (2014) Metabolic engineering for ricinoleic acid production in the oleaginous yeast *Yarrowia lipolytica*. Appl Microbiol Biotechnol 98:251–262
5. Deeba F, Pruthi V, Negi YS (2017) Fostering TAG accumulation in novel oleaginous yeast *Cryptococcus psychrotolerans* IITRFD utilizing groundnut shell for improved biodiesel production. Bioresour Technol. https://doi.org/10.1016/j.biortech.2017.04.001
6. Deeba F, Pruthi V, Negi YS (2016) Converting paper mill sludge into neutral lipids by oleaginous yeast *Cryptococcus vishniaccii* for biodiesel production. Bioresour Technol 213:96–102
7. Patel A, Pravez M, Deeba F, Pruthi V, Singh RP, Pruthi PPA (2014) Boosting accumulation of neutral lipids in *Rhodosporidium kratochvilovae* HIMPA1 grown on hemp (*Cannabis sativa Linn*) seed aqueous extract as feedstock for biodiesel production. Bioresour Technol 165:214–222
8. Vicente G, Bautista LF, Rodriguez R, Gutierrez FJ, Sadaba I, Ruiz-Vazquez RM, Torres-Martinez S, Garre V (2009) Biodiesel production from biomass of an oleaginous fungus. Biochem Eng J 48:22–27
9. Patel A, Sindhu DK, Arora N, Singh RP, Pruthi V, Pruthi PA (2015) Biodiesel production from non-edible lignocellulosic biomass of *Cassia fistula* L. fruit pulp using oleaginous yeast *Rhodosporidium kratochvilovae* HIMPA1. Bioresour Technol 197:91–98
10. Patel A, Pruthi V, Singh RP, Pruthi PA (2015) Synergistic effect of fermentable and non-fermentable carbon sources enhances TAG accumulation in oleaginous yeast *Rhodosporidium kratochvilovae* HIMPA1. Bioresour Technol 188:136–144
11. Li XF, Xu H, Wu QY (2007) Large-scale biodiesel production from microalga *Chlorella protothecoides* through heterotrophic cultivation in bioreactors. Biotechnol Bioeng 98:764–771
12. Arora N, Patel A, Pruthi PA, Pruthi V (2016) Boosting TAG accumulation with improved biodiesel production from novel oleaginous microalgae *Scenedesmus* sp. IITRIND2 utilizing waste sugarcane bagasse aqueous extract (SBAE). Appl Biochem Biotechnol 180:109–121
13. Arous F, Frikha F, Triantaphyllidou I-E, Aggelis G, Nasri M, Mechichi T (2016) Potential utilization of agro-industrial wastewaters

for lipid production by the oleaginous yeast *Debaryomyces etchellsii*. J Clean Prod 133:899–909

14. Karakaya A, Laleli Y, Takaç S (2012) Development of process conditions for degradation of raw olive mill wastewater by *Rhodotorula glutinis*. Int Biodeter Biodegr 75:75–82
15. Lowry OH, Rosebrough NJ, Farr AL, Randall RJ (1951) Protein measurement with the Folin-phenol reagent. J Biol Chem 193:265–275
16. Tazeen R, Mular SM, Khan ZH, Khan ND (2017) Comparative analysis of sugar and protein in three different fresh and decaying fruits pomegranate, orange and sweet lime (musambi). IJAR 3:562–564
17. Bligh EG, Dyer WJ (1959) A rapid method for total lipid extraction and purification. Can J Biochem Physiol 37:911–917
18. Fei W, Wang H, Fu X, Bielby C, Yang H (2009) Conditions of endoplasmic reticulum stress stimulate lipid droplet formation in *Saccharomyces cerevisiae*. Biochem J 424:61–67
19. Morrison WR, Smith LM (1964) Preparation of fatty acid methyl esters and dimethylacetals from lipids with boron fluoride–methanol. J Lipid Res 5:600–608

Index

Venkatesh Balan (ed.), *Microbial Lipid Production: Methods and Protocols*, Methods in Molecular Biology, vol. 1995, https://doi.org/10.1007/978-1-4939-9484-7,

Kolibri 360 GmbH
Ferdinand-Jühlke-Straße 7
99095 Erfurt, Deutschland
produktsicherheit@kolibri360.de